PHOTOCONDUCTIVITY
AND RELATED PHENOMENA

PHOTOCONDUCTIVITY AND RELATED PHENOMENA

Edited by

J. MORT

and

D.M. PAI

Joseph C. Wilson Center for Technology
Xerox Corporation, Webster, N.Y., U.S.A.

ELSEVIER SCIENTIFIC PUBLISHING COMPANY
AMSTERDAM — OXFORD — NEW YORK   1976

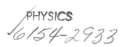

ELSEVIER SCIENTIFIC PUBLISHING COMPANY
335 Jan van Galenstraat
P.O. Box 211, Amsterdam, The Netherlands

Distributors for the United States and Canada:

ELSEVIER/NORTH-HOLLAND INC.
52, Vanderbilt Avenue
New York, N.Y. 10017

Library of Congress Cataloging in Publication Data
Main entry under title:

Photoconductivity and related phenomena.

  Includes bibliographies and indexes.
  1. Photoconductivity. I. Mort, J. II. Pai,
Damodar Mangalore, 1934-
QC612.P5P48   537.6'2   76-16160
ISBN 0-444-41463-0

With 197 illustrations and 1270 references

Copyright © 1976 by Elsevier Scientific Publishing Company, Amsterdam

All rights reserved. No part of this publication may be reproduced, stored in a retrieval system, or transmitted in any form or by any means, electronic, mechanical, photocopying, recording, or otherwise, without the prior written permission of the publisher,
Elsevier Scientific Publishing Company, Jan van Galenstraat 335, Amsterdam

Printed in The Netherlands

PREFACE

Photoconductivity is the enhancement of electrical conductivity of matter produced by the motion of carriers created by absorbed radiation. In the dark, under conditions of thermal equilibrium, the thermally generated carriers are distributed among the available energy states according to Fermi statistics. These electrons and holes occupying conduction states determine the dark electrical conductivity of the material. Under constant illumination a steady-state photocurrent is reached in which the rate of photogeneration is balanced by the various recombination processes through which the carriers tend to relax to the normal equilibrium distribution. The phenomenon thus involves absorption, photogeneration, recombination and transport processes in an intimate relationship. This is one reason for the critical role that photoconductivity has played in the development and understanding of the physics of the solid state. Concurrent with the basic studies of photoconductivity have been the emergence and successful exploitation of a wide range of technologies and devices utilizing this phenomenon. Indeed, as is often the case, the basic and applied science have developed in synergy. Both disciplines have been driven by the need for and the development of better characterized materials with controlled properties. This has required substantial efforts and contributions on the part of synthetic and physical chemists and materials scientists.

The initial focus of interest was on the crystalline covalent solids and the study of these materials dominated the scientific literature in the 1940's and 1950's, although some notable exploratory studies of the phenomena in molecular solids were carried out. Nonetheless, much of the detailed quantitative study and the concepts of photoconductivity which emerged in this period were based on covalent crystalline solids and this has been the subject of numerous review articles and the books "Photoconductivity in Solids" by R.H. Bube (Wiley, New York 1960), "Photoconductivity and Related Processes" by A. Rose (Interscience Publishers, New York 1963) and "Photoelectric Effects in Semiconductors" by S.M. Ryvkin (Consultants Bureau, New York 1964). Generally speaking, the study of photoconductivity in covalent solids has progressed from the detailed studies aimed at obtaining a fundamental understanding of the phenomenon to a detailed study of the specific identity and microscopic description of imperfections, both defects and impurities, which control carrier lifetimes and hence photoconductor characteristics. The immensely complex issues of surface states, including chemical and physical adsorption, have been important areas of investigation which have received renewed impetus through advances in theoretical and experimental techniques such as X-ray and UV photoelectron spectroscopy and auger electron spectroscopy for characterizing the surfaces of materials in great detail. Very sophisticated advances in device technology have also taken place, one notable example being the development of charge-coupled devices.

In contrast, recent detailed studies of photoconductive phenomena in molecular solids, the amorphous state and liquids have revealed unusual features. A full understanding of the phenomena in these materials therefore requires judicious application of the traditional concepts of photoconductivity. This state of affairs occurs because of the substantial differences which exist between these classes of materials and covalent crystalline photoconductors. In molecular solids and liquids, for example, the interactions between the molecules which constitute the condensed states are relatively weak. Thus even in molecular crystals, the concept of allowed energy bands is of limited validity and excitations and interactions localized on individual molecules play a predominant role. As a result, intramolecular relaxation and electron-phonon interaction processes become particularly important. The diversity and reactivity of molecular solids and liquids also pose formidable problems in achieving the stringent purity control necessary for determining meaningful photoelectronic properties.

Increasingly in the last few years, attention has turned to understanding the properties of the disordered or amorphous state. Again, in addition to the intrinsically interesting scientific questions, a strong motivating force was the perceived and, in the case of electrophotography, the demonstrated technological applications of such materials. In this respect, the prime characteristic was the ability to produce with relative ease and low cost, large-area, defect-free films. In the case of chalcogenides, for example, this was achieved by vacuum deposition. The analogous materials in the case of organic solids are polymers where films can be cast from solution. As a result, there is considerable activity in studying the photoelectronic properties of amorphous materials including both covalently bonded and molecular solids.

Even for covalently bonded amorphous solids rather basic questions must be asked because of the profound effect disorder has on photoelectronic properties. Carrier mobilities are typically much smaller than in the crystalline state, reflecting significantly reduced mean free paths. The translational and, in the case of compounds, compositional disorder introduce large densities of states within the band gap. These can drastically curtail carrier lifetimes and thus photosensitivity. In some materials considerable fluctuation in observed properties can occur, depending on preparation conditions, although in the case of a-Se and a-Ge significant progress has been made in understanding and controlling such variations.

In the case of amorphous molecular solids, amorphous organic polymers being members of this generic class, the effects of disorder are expected to be even more apparent. Since the band picture is of limited applicability even in the crystalline state, the introduction of disorder is believed to result in the complete breakdown of a band model. As a result, the photoelectronic, including photogeneration and charge transport, properties are best described as inter-molecular charge transfer processes involving neutral molecules and molecular ions. In this situation, the photoelectronic properties can be viewed as a particular manifestation of solid state reduction and oxidation reactions quite familiar to chemists.

In the case of liquids, particularly hydrocarbon liquids, there has been considerable progress recently in making definitive measurements of transport of electronic carriers produced by photoionization. Electrical studies in liquids are inevitably plagued by the presence of ions and only the most stringent purification and controlled experimental conditions reveal the underlying electronic processes. Here again, the specific molecular identity is apparently important in determining the electronic parameters observed.

The progress in understanding the photoelectronic properties of materials has always required parallel and co-ordinated efforts in materials science including crystal growth, purification and characterization. As attention has turned to molecular materials and particularly the disordered state, it is increasingly apparent that materials characterization can be no longer viewed as the application of relatively standard and routine techniques. The characterization of these types of materials stands as a formidable research challenge to the materials scientist and in the case of molecular solids, where molecular design and engineering of photoelectronic properties are potentially possible, to the synthetic chemist. Future advances in the understanding and applications of photoelectronic properties of materials will involve a closer interdisciplinary collaboration between chemists and physicists than ever before.

In this book we have invited authors who have made major contributions to the various aspects of photoconductivity and its related phenomena to discuss the most recent developments in various classes of matter including technological applications. In view of the scope of this book, the authors were obviously limited to discussing only a few selectively chosen topics in any detail. Each chapter, however, includes a detailed bibliography and, viewed as a supplement to existing texts, it is our hope that it will be of value to newcomers and active workers in the field alike.

Webster, N.Y. J. Mort
November, 1975 D.M. Pai

LIST OF CONTRIBUTORS

D. Bonnet	Battelle-Institut e.V., Frankfurt, W. Germany
R.H. Bube	Department of Materials Science and Engineering, Stanford University, Stanford, Calif. 94305, U.S.A.
F. Dolezalek	Battelle-Institut e.V., Frankfurt, W. Germany
R.C. Enck	Xerographic Technology Department, Joseph C. Wilson Center for Technology, Xerox Corporation, Webster, N.Y. 14580, U.S.A.
W.D. Gill	IBM Research Laboratory, San Jose, Calif. 95193, U.S.A.
H. Inokuchi	Institute for Molecular Science, Okazaki 444, Japan
P.G. LeComber	Carnegie Laboratory of Physics, University of Dundee, Dundee, Scotland
Y. Maruyama	Department of Chemistry, Faculty of Science, Ochanomizu University, Tokyo 112, Japan
G. Pfister	Webster Research Center, Joseph C. Wilson Center for Technology, Xerox Corporation, Webster, N.Y. 14580, U.S.A.
H. Scher	Webster Research Center, Joseph C. Wilson Center for Technology, Xerox Corporation, Webster, N.Y. 14580, U.S.A.
F.W. Schmidlin	Webster Research Center, Joseph C. Wilson Center for Technology, Xerox Corporation, Webster, N.Y. 14580, U.S.A.
W.F. Schmidt	Hahn-Meitner-Institut für Kernforschung Berlin GmbH, Ber. Strahlenchemie, 1 Berlin 39, W. Germany, and Fachbereich Physik, Freie Universität, Berlin, W. Germany
M.P. Shaw	Department of Electrical Engineering, Wayne State University, Detroit, Mich., U.S.A.
M. Silver	Physics Department, University of North Carolina, Chapel Hill, N.C., U.S.A.
W.E. Spear	Carnegie Laboratory of Physics, University of Dundee, Dundee, Scotland

CONTENTS

Preface v

Chapter 1. Contacts and Photoinjection Currents
by M. Silver and M.P. Shaw 1
1.1. Introduction 1
1.2. The metal—bulk material interface 2
1.3. Current-voltage characteristics in the dark 6
1.4. Photoinjection over a step-function potential barrier ... 10
1.5. Photoinjection in the presence of an image barrier 15
1.6. Space-charge-limited currents with no trapping 20
References 24

Chapter 2. Experimental Techniques
by F.K. Dolezalek 27
2.1. Introduction 27
2.2. Photoconductivity relaxation 29
2.3. Transit-time techniques 33
2.4. Magnetic field effects 57
References 63

Chapter 3. Theory of Time-Dependent Photoconductivity in Disordered Systems
by H. Scher 71
3.1. Introduction 71
3.2. Electrical transport via localized states as a random walk ... 74
3.3. The continuous-time random walk (CTRW) 76
3.4. Hopping time distribution function 78
3.5. AC conductivity and time-dependent pair luminescence ... 81
3.6. Transient photoconductivity 90
3.7. Comparison with experiment on hopping systems 103
3.8. Onsager theory and relation to hopping transport 107
3.9. Langevin recombination 111
3.10. Conclusion and Summary 112
References 114

Chapter 4. Covalent Semiconductors
by R.H. Bube 117
4.1. Introduction 117
4.2. Imperfection control of lifetime 118
4.3. Achieving maximum photoconductivity performance ... 126
4.4. Photo-Hall and photothermoelectric effects in crystals ... 128
4.5. Analysis of trapping 133
4.6. Photoadsorption effects 135
4.7. Photochemical changes 137

4.8. Photoconductivity in polycrystalline films 140
4.9. Heterojunctions 144
References 147

Chapter 5. Molecular Crystals
by H. Inokuchi and Y. Maruyama 155
5.1. Introduction 155
5.2. Photogeneration processes 156
5.3. Trapping and recombination processes 165
5.4. Photo-carrier transport processes 168
5.5. Related phenomena 176
References 182

Chapter 6. Amorphous Tetrahedrally Bonded Solids
by W.E. Spear and P.G. LeComber 185
6.1. Introduction 185
6.2. Photoconductivity in glow discharge silicon 186
6.3. Photoconductivity in evaporated a-Si 206
6.4. Photoconductivity in a-Ge 207
6.5. Photoconductivity in other tetrahedrally bonded amorphous materials 211
6.6. Concluding remarks 212
References 213

Chapter 7. Amorphous Chalcogenides
by R.G. Enck and G. Pfister 215
7.1. Introduction 215
7.2. Photogeneration efficiency 216
7.3. Steady-state photoconductivity 238
7.4. Electronic transport properties 254
References 297

Chapter 8. Polymeric Photoconductors
by W.D. Gill 303
8.1. Introduction 303
8.2. Charge transport 305
8.3. Charge generation and recombination 323
8.4. Injection experiments 330
8.5. Conclusions 331
References 332

Chapter 9. Non-Polar Liquids
by W.F. Schmidt 335
9.1. Introduction 335
9.2. Radiation-induced conductivity 337
9.3. Photoionization and photoeffect 361
9.4. Transport properties of charge carriers 368
9.5. High energy radiation detectors with non-polar liquids 377
9.6. Purification techniques and measurement cells 380
References 384

Chapter 10. Photoelectronic Semiconductor Devices
by D. Bonnet 389
10.1. Introduction 389

10.2. Light sensors and detectors 390
10.3. Imaging systems 400
10.4. Optical memories and image storage 412
10.5. Image intensifiers and converters 415
10.6. Solar cells 415
References. 418

Chapter 11. Electrophotography
by F.W. Schmidlin. 421
11.1. Introduction. 421
11.2. Physical basis of charged pigment electrophotography 426
11.3. Statistical physics of single toner development 431
11.4. The electric field in xerography 436
11.5. Macroscopic image transformations in charge pigment xerography 440
11.6. The fundamental processes of latent image formation. 449
11.7. Photoreceptor material requirements. 458
11.8. Discussion of practical xerographic photoreceptors 469
11.9. Conclusions 475
References. 476

Author Index 479

Subject Index 497

CHAPTER 1

CONTACTS AND PHOTOINJECTION CURRENTS

M. SILVER and M.P. SHAW

1.1. Introduction
1.2. The metal—bulk material interface
1.3. Current—voltage characteristics in the dark
1.4. Photoinjection over a step-function potential barrier
1.5. Photoinjection in the presence of an image barrier
1.6. Space-charge-limited currents with no trapping

1.1 INTRODUCTION

In this chapter, we will review the role that contacts play in determining the manifestation of photoinjection-conductive phenomena and show how the contacts can determine the current and response time associated with these processes. To understand the photoinjection-conductive behavior of bulk insulators and semiconductors we must first electrically define the bulk material and its local environment, which consists of the contacts, leads and support components. Were the bulk material a metal, then the attachment (touching) of metallic leads would produce small interfacial fields [1, 2]; the contacts would (a) be of extremely low resistance and (b) affect the bulk behavior in only a minor way. If the bulk is a semiconductor or insulator, however, then the contact region can have a substantial effect on the behavior of the bulk. The contact region will impose boundary electric fields and may produce a nonlinear "contact" resistance, R_c. (We are assuming here that it is possible to conveniently separate R_c out of the total resistance of the system.) When dealing with good insulators or with alloyed contacts to semiconductors, R_c is often small compared to the bulk resistance, R_b. In this case, one major effect of the contact is simply the imposition of a boundary condition on the bulk electric field profile. Since we also want to package, mount, or support the sample, there will be a package capacitance C_p and package inductance L_p present. The leads themselves will of course have some inductance L_l and some resistance R_l. There must be a battery V_B, from which the bulk electric field E is derived, and, in general, a load resistor R_L may be included that represents the internal resistance of the battery and/or an actual load in the circuit. All the above parameters must

References pp. 24—25

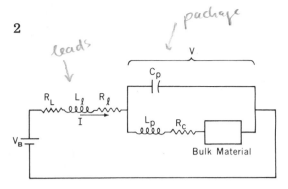

Fig. 1.1. Lumped element approximation of the local circuit environment of bulk material.

be included in a simple lumped element approximation of the local environment of the bulk material. Figure 1.1 shows this configuration. Our prime concern in this chapter is the role that the boundary conditions play in determining the observed current—voltage, $I(V)$, characteristics. (The influence of the reactive components, although often significant, will not be discussed further.) Assuming that the complete transport properties of the bulk can be predicted and/or measured as a function of frequency, how do we determine the boundary conditions? To answer this question, we must investigate one of the most complex problems in all of solid state physics: the metal—semiconductor (or insulator) interface problem. In what follows we will often refer to the semiconductor or insulator interchangeably as "bulk material", "bulk" or "material".

In Section 1.2 we discuss the status of our present knowledge concerning the metal—bulk material interface and in 1.3 we relate various interfacial electric field distributions to the current—voltage characteristics observed in the dark. Sections 1.4 and 1.5 are concerned with photoinjection from the metal into the bulk over a step function potential barrier and when an image barrier is present. Section 1.6 is a discussion of space-charge-limited currents in the absence of trapping effects. The status of existing theories and comparisons between theory and experiment will be discussed throughout the chapter.

1.2 THE METAL—BULK MATERIAL INTERFACE

To begin with we can ask: what are the most important phenomenological features of the metal—bulk material interface? How do they depend upon the characteristics of the material and the preparation of its surface? Clearly, considerable care must be given to preparing the surface before the metal is sputtered or evaporated onto it. If not, reproducible interface properties will not be observed. We therefore assume that the surfaces are free from avoidable contaminants. From studies of interfacial potential barriers, Φ_B, for a large number of clean semiconductors and insulators [3], two empirical rules have emerged. The first rule involves the variation of Φ_B with a change in metal for a given material. The second rule involves the magnitude of Φ_B for metals on covalently bonded materials.

With regard to the first rule, it has been suggested that Φ_B be written as the difference between two quantities [4]. One quantity is representative of the metal—vacuum interface (the work function, Φ_M), the other is representative of the material—vacuum interface (the electron affinity, χ_s). Φ_M is the energy necessary to remove an electron from the top of the Fermi sea and put it into the vacuum. χ_s is the energy gained by taking an electron in vacuum and moving it to the bottom of the conduction band of the bulk material. Then $\Phi_B = \Phi_M - \chi_s$. Since values of Φ_M are not available for all the metals of interest, we will use instead the electronegativity [5], χ_M, of the metal, which is readily available and closely related to Φ_M (except for a constant K). We expect from this argument that, for a given material, Φ_B would increase linearly with unity slope as χ_M is varied. This, in fact, occurs if the material has a high degree of ionic bonding. If, however, the material has predominantly covalent bonding, it is found that Φ_B is substantially independent of χ_M. For intermediate ionicities the slope falls between 0 and 1.

Assuming that the behavior of the more ionic materials can be understood via the above simple model, it remains to explain the behavior of the more covalently bonded materials. Since, in this book, we are also interested in the more covalently bonded materials, this problem is of fundamental importance to us. Much intense theoretical effort has gone into this endeavor [6—11], but to date it appears that a complete theory explaining the second empirical rule does not exist. This rule is that, with few exceptions, for barriers on covalent material, $\Phi_B \simeq (2/3)\mathcal{E}_g$ [12], where $\mathcal{E}_g$ is the band gap. (A discussion of the experimental determination of barrier heights can be found in the text by Milnes and Feucht [13].) Mead's explanation [14] of the result is that covalently bonded materials have a large density of surface states centered about $(2/3)\mathcal{E}_g$ below the bottom of the conduction band. This pins the Fermi level of the metal (any metal) about one third of the way up the band gap from the valence band.

Since in the photoinjection problem we will be dealing primarily with good insulators, we should expect that, in general, the resistance of the contact will often be much less than the resistance of the bulk material ($R_c \ll R_b$). Under these conditions, if the bulk material has a linear $I(V)$ characteristic in the dark, then any nonlinearity produced by the contacts will often be masked by the properties of the material [15]. The total dark $I(V)$ of the bulk material-plus-contact characteristic may be linear over a wide range of voltage; the system may be "ohmic". In practice, the word "ohmic" has apparently come to mean in some areas that $R_c \ll R_b$ in materials having linear $I(V)$ curves; it says nothing about the nature of the interface fields. It is unfortunate that this has occurred, since other disciplines reserve the word "ohmic" to mean that accumulation layers of charge exist at the boundaries of the bulk material. Since a prediction of the photoinjection-conductive behavior of bulk material requires a knowledge of the electric fields at the boundaries (these can be, and often are, substantial even though $R_c \ll R_b$), in this chapter we will minimize the use of the word "ohmic" and refer directly to the specific field distribution under consideration. We suggest that in the future the word "ohmic" should either be reserved for a specific field distribution (accumulation layers of charge at the contacts) or abandoned altogether since it tends to cause unnecessary confusion.

What types of electric field distributions can occur near the interface when a metal is placed in contact with a bulk material in the dark? Let us for the moment assume that the disturbance in the equilibrium carrier density distribution produced by the presence of the metal is negligible deep inside the bulk. If this were not the case, the depletion layers produced could easily extend over the entire length of the specimen. An identical contact on the opposite side of the specimen would produce the same effect except that it would be "inverted"; field cancellation occurs and the resulting electric field, $E(x)$, becomes similar to the case shown in Fig. 1.2(b). In either case, Fig. 1.2 depicts schematically the most important of the variety of possible equilibrium $E(x)$ profiles at the contact which will be the cathode when a bias voltage is applied to an n-type semiconductor (these profiles can be obtained by solving the Poisson equation including both electrons and holes). Different types of contacts and contacting procedures could produce any of these field distributions at the cathode side of a particular n-type semiconductor. Note in Fig. 1.2 that when a bulk field E_b is applied, $E(x)$ increases positively downward. Thus, Fig. 1.2(a) signifies an accumulation layer at the cathode and Fig. 1.2(b) signifies a depletion layer (from a consideration of the Poisson equation). In the remainder of the chapter, we will invert the fields so that an increase in E_b is represented by a positive upwards increase in $E(x)$.

How do we represent and treat these relatively low resistance contacts analytically? (We may be able to ignore R_c, but not the boundary conditions imposed upon the bulk.) In practice, the bulk material will have a mobile carrier concentration below about 10^{15} carriers/cm^3 and the interface to the circuit to which it is connected is a metal with approximately 10^{21} carriers/cm^3. The boundary is therefore a very narrow region where the carrier density changes very sharply. Even if these barriers could be completely characterized for a given system in equilibrium [13], we would eventually still have to consider hot electron effects within the junction in the presence of a current. In general, calculations of hot electron effects are made for regions of uniform field and carrier concentration, and these are not applicable to the boundary region.

Since there are usually severe difficulties in accurately controlling,

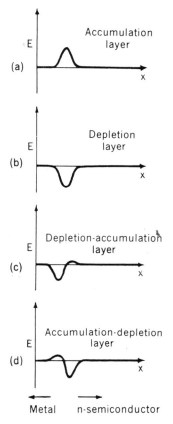

Fig. 1.2. Example of electric field distributions for various metal—n-type semiconductor contacts.

experimentally determining, and theoretically calculating the conditions present at a metallic contact, it is important to seek a simple model for the boundary. One of the simplest approaches is to assume that we know the electric field at the boundary of the bulk material. Even with this knowledge, the problem is complex. For any given system, we need to know the ratio R_c/R_b, the type of charge layer at the interface and the bias dependence of the contact field. Furthermore, the separability of R_c and R_b is often not a viable procedure. All of these problems arise in the dark. If we now illuminate the metal and cause photoinjection to occur, how are these parameters altered? We clearly have a problem that is by no means simple. What we would like to do with the limited space available in this chapter is to offer some physical insight, present simple examples and discuss a specific case in some detail. We hope to categorize some of the more important situations that often arise in practice.

References pp. 24—25

1.3 CURRENT–VOLTAGE CHARACTERISTICS IN THE DARK

It is important first to characterize contacts (a) and (b) shown in Fig. 1.2 via the form of the dark $I(V)$ characteristics that are associated with each contact. A discussion of the $I(V)$ characteristics in the presence of photons incident on the metal contact then forms the major theme of this chapter. To determine the $I(V)$ characteristics, we assume that (a) both the cathode and anode contacts to the bulk material are identical and (b) the influence of each contact on the properties of the bulk occurs sufficiently close to each contact so that in equilibrium the contact effects do not extend appreciably into the bulk. For the moment, we neglect tunneling and image charge effects.

Contact (a) in Fig. 1.2, an accumulation layer of charge, is often referred to as an "ohmic" or "majority carrier injecting" contact. The word "injecting" here should not be confused with either "minority carrier injection" or the injection involved in photoinjection. In photoinjection, carriers in the metal are energetically raised by photons to energies sufficient for them to overcome whatever barrier exists at the interface and to enter into the insulator. Photoinjection can, and most often does, occur over a contact that is "non-injecting" or "blocking" in the dark. Contacts that are majority carrier injecting in the dark when a bias is applied are only rarely formed when a metal contacts an insulator. An injecting contact usually results at $p - p^+$, $n - n^+$, $n^+ - i$ or $p^+ - i$ interfaces, where i denotes intrinsic material. Bias-induced injection phenomena are best studied in configurations such as the PIN diode.

A contact that is perfectly injecting can supply all the carriers that the bulk demands for any applied field. Here, for a linear bulk $I(V)$ curve, the current varies linearly with voltage at low electric fields but then varies as the square of the voltage when the space-charge-limited current (SCLC) regime is reached [see Fig. 1.3(a)]. To understand why a nonlinear (superlinear) region of conduction results, the following argument is useful [16]. Injecting contacts act as a reservoir of carriers at an interface that continuously keeps adjusting itself to meet the current levels demanded by the bulk. What keeps the space charge from spreading throughout the bulk? Charge, in fact, is being continuously drawn into the bulk where it is neutralized by dielectric relaxation and recombination. Assume that (a) dielectric relaxation occurs much faster than recombination (we set the recombination time $\tau_r = \infty$) and (b) the processes of being drawn into the bulk (drift) and then neutralizing via dielectric relaxation occur separately rather than simultaneously and (c) that electrons are injected from the cathode. We now consider a volume element near the anode. An electric field is now applied to produce carrier drift for a period equal to the cathode-to-anode transit time t_T. Charge will flow from the cathode reservoir and the volume element will charge up to some charge density ρ_0. Now remove the drift mechanism and turn on the

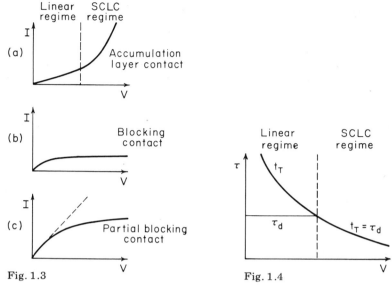

Fig. 1.3. I(V) characteristics for various metal—semiconductor contacts.

Fig. 1.4. Dielectric relaxation time and transit-time *versus* voltage.

dielectric relaxation mechanism for an equal period t_T. Since $\rho = \rho_0 \exp(-t/\tau_d)$, where τ_d is the dielectric relaxation time, then ρ_0 is diminished by the fraction $\exp(-t_T/\tau_d)$. For $t_T \gg \tau_d$ transport occurs essentially without space charge; we are in the low field linear regime. At $t_T = \tau_d$ we enter into the space-charge-limited current (SCLC) regime.

Suppose that we next increase the applied bias even further, making t_T shorter. Here the carriers are drawn more rapidly into the interior and their density increases. This raises the conductivity, which in turn shortens τ_d, exactly in proportion to the decreased transit time. Carriers are neutralized as fast as they arrive and arrive as fast as they are neutralized. The condition $t_T = \tau_d$ therefore not only marks the onset of the SCLC regime, it also defines the regime itself. Figure 1.4 shows how the linear and SCLC regime are separated according to the above argument. Equating t_T and τ_d yields a current that is proportional to the square of the electric field, thus demonstrating the non-linearity. To show the precise relationship [17] between the current density j and V the equation of total injected current is solved simultaneously with the Poisson equation to yield $j = (9/8)\kappa\kappa_0\mu V^2/L^3$, where $\kappa\kappa_0$ is the permittivity, μ the mobility of the injected carriers and L the length of the specimen.

In the above argument, the condition $t_T > \tau_d$ defines the region where linear conduction can exist for linear bulk $I(V)$ characteristics (field-independent conductivity). However, transport in this regime may also be

References pp. 24—25

accompanied by space-charge effects if τ_r becomes sufficiently small. We examined above only a single carrier model in which the sole mechanism for removing the drifting carriers from the stream was through dielectric relaxation. The consequences of introducing two types of carriers with a characteristic τ_r, and making τ_r less than τ_d, has been examined in great detail by van Roosbroeck and coworkers [18]. The condition $\tau_r < \tau_d$ defines the "relaxation" regime (as opposed to the more usual "lifetime" regime applicable to most semiconductor devices). Many insulators fall into the relaxation category.

Contact (b) in Fig. 1.2 has a depletion layer of charge associated with it. Most metal—insulator and metal—semiconductor contacts are of this form. The contact is "blocking" and carriers cannot be readily injected through the contact region from the metal into the bulk; for $R_c \gg R_b$ it is simply the well-known Schottky barrier [13, 19, 20, 21] and the current—voltage characteristics are basically those expected from emission over a barrier [13]. The first-order approach to the solution of this problem is to neglect all tunnelling effects and image charge barrier lowering. The result is similar to that obtained for the emission of electrons over a barrier into vacuum. The characteristics of the junction are given by

$$j = j_A [\exp(q V/kT) - 1], \qquad (1.1)$$

where $j_A = A T^2 \exp(-\Phi_B/kT)$. A is the Richardson constant for a free electron effective mass. The emission model assumes that electrons that are emitted into the insulator have no difficulty moving into the bulk. This is not generally true, since the depletion region is often rather wide. Furthermore, inclusion of the image force causes the barrier to be lowered as the applied reverse bias is increased. We will discuss this effect in more detail in a later section.

Application of a bias to two Schottky contacts on either side of the bulk will forward bias one contact and reverse bias the other. The reverse-biased Schottky blocks most of the carriers from reaching the bulk. The carriers that are predominant in the transport process under reverse bias are those emitted into the bulk, generated in the space-charge (depletion) region and generated in the bulk within about a diffusion length away from the space-charge region. The associated currents are small, exceedingly so in insulators. Figure 1.3(b) shows the reverse bias $I(V)$ characteristic that might be expected from a blocking contact dominated by diffusion currents and injection over the barrier. Note the sublinear, saturated form of the curve. Figure 1.3(b) can also be associated with the $E(x)$ curve shown in Fig. 1.2(b) and is of the form expected for what has been called a "partially" blocking contact. Partially blocking contacts on semiconductors most often occur when Schottky barriers are modified and reduced by annealing and alloying so that $R_c \ll R_b$ [22]. However, this condition can also be achieved quite easily on good insulators without alloying [15]. The form of the $I(V)$ characteristic will still have features similar to that of the blocking Schottky barrier. However, for the partially blocking contact the onset of the sublinearity and tendency towards saturation will generally occur at higher values of applied bias. Thus, partially blocking contacts to bulk material having linear $I(V)$ characteristics can produce total $I(V)$ characteristics that will also remain linear to quite high fields. Eventually, however, a sublinearity will appear [15], often of the form $V^{1/2}$. Linearity of an $I(V)$ curve over a range of fields certainly does not mean that the contacts are injecting. On the contrary, depletion regions probably are situated at the contacts. The sublinearity produced by partially blocking contacts ($R_c \ll R_b$) is due to the fact that as the bias is increased, more and more of the voltage drop appears across the contact region as the width of the depletion region spreads and the field inside it increases. The voltage drop does not divide itself in a way such that the drop across the bulk region increases uniformly with bias. The field in the bulk and hence the current does not rise as fast as it did for low voltages [15].

It is worth mentioning at this point that it is also possible for an accumulation layer to become a depletion layer (and vice versa) as a function of applied bias. A specific contact is not necessarily fixed in its behavior. To see this, simply imagine a linear bulk in contact with a nonlinear R_c whose current-density—field, $j(E)$, characteristics are shown in Fig. 1.5. For any value of current below the "crossover" current j_c the bulk field is below the contact field. Above j_c, the opposite holds. A transition in the charge layer at the interface takes place at the point (j_c, E_c).

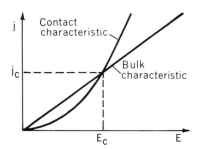

Fig. 1.5. Example of a contact and bulk current-density—field characteristic. The curves cross at (j_c, E_c).

Contacts (c) and (d) in Fig. 1.2 are variations of contacts (a) and (b). Their $I(V)$ characteristics are expected to be modifications of those discussed above and to depend upon the specific details of the preparation of the interface.

It is clear from the above discussion that we have used the term "blocking" contact to mean a Schottky barrier where $R_c \gg R_b$. We have also used the term "partially blocking" contact to mean a Schottky barrier where

References pp. 24—25

$R_c \ll R_b$. Again, we believe that the term "ohmic" causes confusion when it is used. We suggest that the best description is simply to use the word "Schottky" (which implies charge depletion at the interface) and indicate how R_c compares with R_b.

1.4 PHOTOINJECTION OVER A STEP-FUNCTION POTENTIAL BARRIER

Now that we have briefly outlined some of the general properties of interfaces, we shall next discuss a few specific cases appropriate to the photoinjection of carriers from metals into insulators. These considerations will culminate in a discussion of the space-charge-limited and electrode-limited currents that can flow in insulators under photoinjection conditions.

For ease of discussion, we first (a) ignore the image potential and (b) consider a fixed step-function potential barrier. This type of contact represents the *non-physical case* where the fields under zero bias are zero everywhere except for a δ function spike *at* the interface. It is important to study, however, since it provides insight into the solutions when realistic nonuniform fields are present. When there is "internal" photoemission (photons incident on the metal cause carriers to be injected into the insulator) a space-charge layer will be formed in the insulator. The magnitude of the layer depends upon the magnitude of the internal photoemission current, which in turn depends upon the rate of electron excitation in the electrode, the kinetic energy of the electrons and the conduction band minimum of the insulator.

A simplified view of the situation is shown in Fig. 1.6. A photon of energy

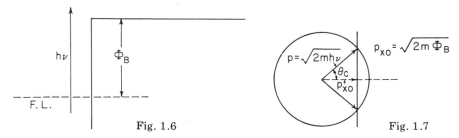

Fig. 1.6. Schematic of photoinjection over a barrier from a metal into an insulator.

Fig. 1.7. Schematic of the escape cone for an electron of energy $h\nu$ over a barrier Φ_B.

$h\nu > \Phi_B$ may excite an electron over the interfacial barrier. In the simplest case, the momentum distribution of the photoexcited electron is isotropic; thus only a fraction of the excited electrons enter the insulator. (This well-known problem was discussed in detail by Fowler [23]. A very useful discussion has also been presented by Williams [24].) The major point is that only those electrons having a sufficiently large component of momentum directed into the insulator can overcome the barrier. This leads to a "cone of momentum" which the electrons must possess in order to enter the insulator. A schematic representation of this is shown in Fig. 1.7. The escape cone is defined by the critical angle θ_c. In order to escape, θ must be less than θ_c, which is given by:

$$\theta_c = \cos^{-1}\left(\frac{2m^*\Phi_B}{p^2}\right)^{1/2} \tag{1.2}$$

and the probability for escape is then:

$$F_e(p) = \tfrac{1}{2}[1 - (2m^*\Phi_B/p^2)^{1/2}], \tag{1.3}$$

where $p^2/2m^* = h\nu$. The more familiar form of this expression is in terms of the photon energy:

$$F_e(h\nu) = \tfrac{1}{2}[1 - (\Phi_B/h\nu)^{1/2}]. \tag{1.4}$$

Equation (1.4) is valid if the excitation occurs from a narrow band in the metal. However, if the excitation occurs from a wide band, eqn. (1.4) must be integrated over all possible states:

$$F_T(h\nu) = \int_{h\nu_0}^{h\nu} F(h\nu)\,\mathrm{d}(h\nu) \simeq K(h\nu - h\nu_0)^2, \tag{1.5}$$

where $h\nu_0 = \Phi_B$. It is only the wide-band case which yields the result that the square root of the photoinjected current is proportional to $h\nu$ [23].

In the above discussion we have assumed that m^* is the same in both the metal and bulk material. When the effective mass changes across the junction the problem becomes more complex, resulting in substantially different conclusions. An example of this is seen by examining the case where the transverse component of momentum is conserved. Here:

$$p_1 \sin \theta_1 = p_2 \sin \theta_2, \tag{1.6}$$

where 1 and 2 refer to the metal and insulator respectively. Consider a possible case where there is no change in direction of the injected electron in traversing the barrier, so that $\sin \theta_1 = \sin \theta_2$. Squaring each side of eqn. (1.6) and writing $\epsilon = p^2/2m^*$ yields

$$\epsilon_1 m_1^* = \epsilon_2 m_2^*. \tag{1.7}$$

Thus, if $m^* \propto 1/\epsilon$, the escape cone is 2π for all ν and the yield is 50%. This is much greater than that given by eqn. (1.5) for $m^* =$ constant. Caution must therefore be exercised when evaluating the yield due to photoinjection of carriers into insulators where m^* changes.

We have so far considered the yield of photoinjected carriers assuming that those that entered the insulator at the interface remained in the insulator.

References pp. 24—25

This would be the case if scattering and energy loss processes were absent in the insulator. As it turns out, in most cases the yield will be determined more by the scattering processes than by the escape cone. In what follows, we assume that the injected current, after taking the escape cone into consideration, is j_0. The simplest treatment of the net yield due to scattering is by J.J. Thomson [25]. He assumed that (a) the field was uniform throughout the sample, (b) the amount of charge was too small to significantly modify the field and (c) the injected charge had the same energy distribution as the charge moving in the insulator. From current balance:

$$j_0 = n_0 e v_0/(6\pi)^{1/2} + j, \tag{1.8}$$

where v_0 is the average speed of the carriers and $j = e n_0 \mu E$.
Thus,

$$j = j_0/[1 + (v_0/(6\pi)^{1/2} \mu E)]. \tag{1.9}$$

This case applies only to very low fields, where (a) $eE\lambda \ll kT$ and (b) the energy of the electrons is comparable to kT. (λ is the mean free path for momentum exchange scattering.) When the energy of the injected electrons is large compared to kT, Theobald [26] and Loeb [27] proposed that v_0 should have the value appropriate to this energy rather than $(3/2)kT$. Bekiarian [28] extended their ideas by including the reflection of electrons at the cathode and by deriving an equation similar to eqn. (1.9) by solving a Boltzmann equation.

The next important case for uniform fields occurs for $eE\lambda \gg kT$ (first discussed by Young and Bradbury [29] in 1933). Here, we cannot solve a Boltzmann equation; the problem is treated ballistically. This problem involves the probability that an electron will be back-scattered into the electrode, a probability that is a function of the potential at the position of scattering and the kinetic energy of the electron. It is therefore an "escape cone problem" but differs from the case previously discussed. The problem now involves integrating over all scattering possibilities. It is useful to examine this problem in detail since it has important implications when nonuniform fields are examined.

A schematic representation of the problem is shown in Fig. 1.8. Photoexcited charged particles of density n_0 enter the insulator from the metal with kinetic energy ϵ_0. The initial velocity of these particles makes an angle ϕ with the normal to the electrode. The scattering is assumed to be isotropic and the scattering probability is $\exp[-x/\lambda\cos\phi]$ where $x/\cos\phi$ is the path length in the direction of the particle's motion. If, at every collision, the probability of a particle being back-scattered into the electrode is $\omega(x)$, then the number of particles left moving into the insulator is

$$n = n_0 \exp[-\omega(x) x/\lambda \cos\phi]. \tag{1.10}$$

To simplify the problem we (a) ignore the curvature of the path of the

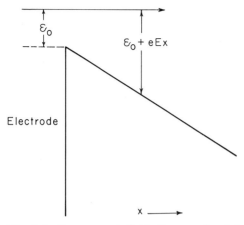

Fig. 1.8. Energy and field diagram for injection into an insulator under a high applied field.

entering particles, (b) assume that only the first collision is important and (c) assume that all the particles enter with their mean normal component of velocity. The last approximation means that eqn. (1.10) may be written as

$$n = n_0 \exp[-\omega(x) x/\overline{\lambda\cos\phi}]. \tag{1.11}$$

When a particle is scattered at x it can return to the electrode provided that

$$\cos^2\theta > \left(\frac{eEx}{\epsilon_0 + eEx}\right) \tag{1.12}$$

This limits the fraction of particles that can return. The fraction of particles for isotropic scattering is obviously proportional to the solid angle between $\theta = 0$ and $\theta = \theta_c$, where

$$\theta_c = \cos^{-1}\left(\frac{eEx}{\epsilon_0 + eEx}\right)^{1/2}.$$

Therefore:

$$\omega(x) = \frac{1}{2} \int_0^{\theta_c} \sin\theta \, d\theta = \frac{1}{2}\left[1 - \left(\frac{U(x)}{\epsilon_0 + U(x)}\right)^{1/2}\right], \tag{1.13}$$

where $U(x) = e\int E dx$ ($= Eex$ for a uniform field).

The total return probability is the integrated return probability for scattering at all x and

$$R = n_0 \int_0^L \exp\left[-\frac{\int \omega(x)dx}{\overline{\lambda\cos\phi}}\right] \frac{1}{\overline{\lambda\cos\phi}} \omega(x)dx. \tag{1.14}$$

References pp. 24—25

Since $\overline{\cos\phi} = 1/2$ and for small x, $\omega(x) \simeq 1/2$, then

$$R \simeq \int_0^x \exp(-x/\lambda) \left[1 - \left(\frac{U(x)}{\varepsilon_0 + U(x)} \right)^{1/2} \right] \frac{dx}{\lambda}. \tag{1.15}$$

The net current j is then

$$j = j_0(1 - R). \tag{1.16}$$

For small fields, this reduces to a very simple form:

$$j \simeq 0.9 \left[\frac{U(\lambda)}{\epsilon_0} \right]^{1/2}, \tag{1.17}$$

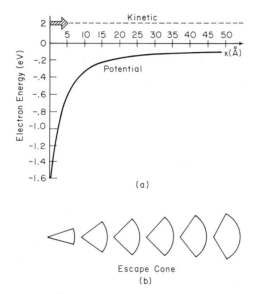

Fig. 1.9. Energy and escape cone diagram for injection over an image potential barrier.

a result that is different from eqn. (1.9) because we are in a different regime; i.e., for the Young and Bradbury case [eqn. (1.17)] $U(\lambda) \gg kT$, while for the Thomson case [eqn. (1.9)] $U(\lambda) \ll kT$.

The Young and Bradbury model is easily adapted to the case where one has a nonuniform field such as an image plus an applied field. The case for just an image field is depicted in Fig. 1.9. Figure 1.10 shows the potential profile for both an image and an applied field. (Also shown in Fig. 1.10 are current components which will be discussed later.) Since the return cone is greater than 2π for $x < x_m$, the easiest way to treat the problem is to integrate from 0 to x_m using:

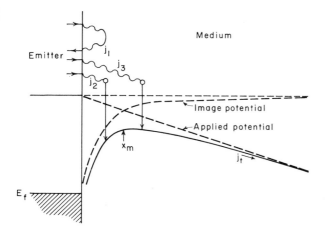

Fig. 1.10. Schematic diagram for injection into an insulator including back-scattering. The applied and image fields are shown.

$$\omega = \frac{1}{2}\left[1 + \left(\frac{U(x)}{\epsilon_0 + U(x)}\right)^{1/2}\right] \quad (1.18)$$

and then integrate from x_m to L using eqn. (1.13), where $x_m = (e/4\kappa_0 E)^{1/2}$. Surprisingly, this procedure produces essentially a constant current for $\lambda = 100$Å and 70Å $< x_m < 700$Å. Thus, momentum exchange scattering alone cannot account for very rapidly varying currents.

The Young and Bradbury procedure as indicated above involves only the first scattering event. Inclusion of multiple scattering does not change the picture significantly. Subsequent scattering events almost cancel themselves out because the fraction of particles in a small return cone is proportional to the return cone but have a large probability of being scattered into the large escape cone. This loss of returned particles is just balanced by the large fraction of particles in the large escape cone which have a small probability of being scattered into the small return cone. Thus, only the first scattering event is important.

1.5 PHOTOINJECTION IN THE PRESENCE OF AN IMAGE BARRIER

There are many instances in which the injected current is a rapidly varying function of applied voltage [30, 31]. The explanation of this behavior involves inelastic scattering of the injected particles in the region of a potential barrier. The case most often treated is a barrier due solely to an image field, although any realistic barrier, such as that produced by both an image field and a depletion layer of charge at the contact (Schottky), can be dealt with in a similar manner. Since inelastic processes are important and the

References pp. 24—25

range of the potential small, the logical approximation to use is that $eE\lambda < kT$. In this case, one can solve a Boltzmann equation or equivalently solve the continuity of current equation with

$$j_c = ne\mu E - eD \frac{\partial n}{\partial x}. \tag{1.19}$$

Figure 1.10 shows schematically the possible events involving inelastic scattering and their effect on the net current. The scalloped lines indicate momentum exchange scattering; the sharp drop depicts an inelastic scattering event. In the figure, we show three possible fates of an injected carrier. In process (1) momentum exchange scattering occurs which returns the particles to the electrode without appreciable loss in energy. The current associated with these back-scattering events will be designated j_1 for a total injected current of j_0. Process (2) also involves momentum exchange scattering, but the carrier loses all of its kinetic energy through some inelastic process taking place at $x < x_m$. This current is designated j_2 and does not contribute to the net current j. The final process, (3), involves an inelastic loss of kinetic energy at $x > x_m$. It is j_3 which makes the major contribution to the net current j.

The net yield of current Y is schematically:

$$Y = [j_0 - (j_1 + j_2)]/j_0 = j_3/j_0, \tag{1.20}$$

since $j_1 + j_2 + j_3 = j_0$. The magnitude of j_3 depends upon the nature of the inelastic processes and the applied field through $x_m^2 = e/4\kappa\kappa_0 E$. We will formulate the problem through general energy relaxation processes and internal potentials $U(x)/e$. However, to illustrate the physics, we will apply the general formulae to the specific case of:

$$U(x)/e = -(V/L)x - e/4\kappa\kappa_0 x, \tag{1.21}$$

where the energy relaxation process is monomolecular and can be described by a single lifetime τ. We assume that the nonthermalized carriers can be described by the general current equation

$$j_h = -D_h(\partial \rho_h(x)/\partial x) - (\mu_h/e)(\partial U(x)/\partial x)\rho_h(x). \tag{1.22}$$

To further simplify the problem, we assume only one nonthermalized state $\epsilon_h - U(x) \gg kT$ and one thermalized state $\epsilon_t - U(x) \simeq kT$, where ϵ_h and ϵ_t are the average energy of the hot and thermalized carriers respectively. Since we are distinguishing between two states of the carriers, in the steady state:

$$\nabla \cdot j = 0 = \nabla \cdot (j_h + j_t) \tag{1.23}$$

and:

$$\frac{\partial \rho_h}{\partial t} = -\nabla \cdot j_h + \dot{\rho}_h = 0$$

$$\frac{\partial \rho_t}{\partial t} = -\nabla \cdot j_t - \dot{\rho}_h = 0,$$
(1.24)

where $\dot{\rho}_h$ is a source term.

Equation (1.24) states that the only source of thermal carriers is the hot carriers which disappear by thermalizing. Integrating eqn. (1.23) yields:

$$j - j_h = j_t = -D_t \partial \rho_t / \partial x - (\mu/e)(\partial U(x)/\partial x)\rho_t. \tag{1.25}$$

Integrating eqn. (1.25) results in

$$\rho_t = \exp[-U(x)/kT] \int_0^{x'=x} [(j_h - j)/D_t] \exp(U(x')/kT) dx' + \rho_t(0). \tag{1.26}$$

For a very long sample of thickness L and a high applied voltage, $-U(L)/kT$ is very large and in order that $\rho_t(L)$ is not enormously large, the integral must be very small. Taking the limit that the integral must be equal to zero at $x = L$ and taking $L = \infty$ yields

$$j = \frac{\int_0^\infty j_h(x) \exp[U(x)/kT] \, dx}{\int_0^\infty \exp[U(x)/kT] \, dx}. \tag{1.27}$$

Equation (1.27) can be integrated by parts:

$$j = \frac{\int_0^\infty \dot{\rho}_h \int_0^{x'=x} \exp[U(x)/kT] dx' \, dx}{\int_0^\infty \exp[U(x)/kT] \, dx}. \tag{1.28}$$

Equation (1.28) is the one-dimensional analogue of the well-known Onsager formula [32] for escape from a three-dimensional potential well.

For photoinjection, eqn. (1.27) is a little easier to deal with because we can determine $j_h(\lambda)$ easily from the boundary condition

$$j_h(\lambda) = j_0 - j_1. \tag{1.29}$$

To obtain $j_h(x)$, eqn. (1.24) must be solved. To do this, we must know the form of $\dot{\rho}_h$ and obtain $\rho_h(x)$. The important point is that under the conditions that the motion of the hot electrons is diffusive, it is $j - j_1$ which enters into eqn. (1.27) at $x = \lambda$ and not just j_0. This means that for very large applied fields eqn. (1.27) saturates to $j = j_h(\lambda) = j_0 - j_1$. When the field

References pp. 24—25

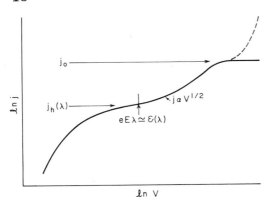

Fig. 1.11. Schematic of the current—voltage characteristics for injection into an insulator including the image field.

is increased further so that $(V/L)e\lambda \simeq \epsilon(\lambda)$, the current will increase again because j_1 will decrease. The decrease is due to the reduction in the return cone, as illustrated by the Young and Bradbury method. Consequently, the current—voltage characteristics shown in Fig. 1.11 are anticipated. We have also shown schematically (by the dotted line) a further increase in current at higher fields due to Schottky emission and other possible high field effects. (The details of these effects are beyond the scope of what we intend to treat here.)

The simple analysis presented here is also applicable to the case of thermionic emission into insulators. Emtage and O'Dwyer [33] discussed this problem but ignored j_h. They arrived at an intractable solution where $\rho_t(x)$ exhibits a singularity at $x = \infty$ because they chose $\rho_t(0) = 2e(2\pi mkT/h^2)^{3/2}$. Had they included the fact that the injected carriers were not thermalized with regard to the potential in the insulator, they would have arrived at the same solution as presented above. (The thermionic emission case is considered here because of the interest in using electrolytic electrodes as contacts to insulators [34].)

For the simple case where $\dot{\rho}_h = \rho_h/\tau$, where τ is very small so that $(D_h\tau)^{1/2} \ll x_m$ and where the image field is the only distortion of the total field from V/L, eqn. (1.27) reduces to a simple form;

$$j = j_h \exp[-x_m/(D_h\tau)^{1/2}]. \tag{1.30}$$

These conditions apply, for example, at low temperatures for injection into liquid helium [35] and injection into the polymer poly(N-vinyl carbazole), PVK, from semiconductor electrodes [31]. The results for PVK are shown in Fig. 1.12.

The case of $\dot{\rho}_h = -\rho_h/\tau$, with image fields only, applies to injection into very dense gases such as H_2 and N_2 [36]. Figure 1.13 shows some results for H_2. The $\dot{\rho}_h = -\rho_h/\tau$ case along with the image field leads to a relatively simple solution. In this case [36]:

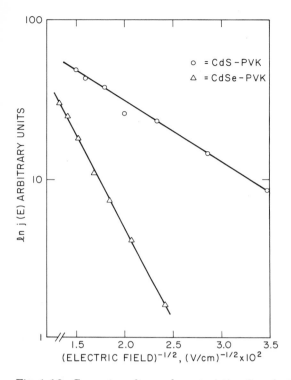

Fig. 1.12. Current—voltage characteristics for photoinjection into PVK from a semiconducting contact. (From Salaneck [31])

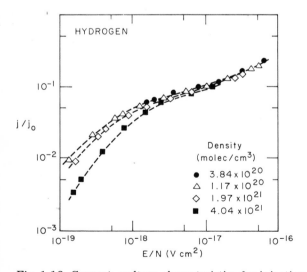

Fig. 1.13. Current—voltage characteristics for injection into hydrogen.

References pp. 24—25

$$j/j_0 = \frac{1}{1+(1-r)C(\lambda)\bar{v}_0\tau/x_0} \cdot \frac{\int_0^\infty \exp(-x/x_0)\exp[U(x)/kT]\,dx}{\int_0^\infty \exp(U(x)/kT)\,dx}, \qquad (1.31)$$

where r is a reflection coefficient at the electrode, $C(\lambda)\bar{v}_0$ is the average X component of the velocity of $\rho_h(\lambda)$ and $x_0 = (D_h\tau)^{1/2}$. (The derivation of eqn. (1.31) is given in detail in ref. 36.) In Fig. 1.13 the solid lines show the fit to the experimental data using eqn. (1.31).

1.6 SPACE-CHARGE-LIMITED CURRENTS WITH NO TRAPPING

We have determined above the j versus V characteristics for photoinjection currents for the case where the current is electrode limited, where there is no trapping, but where there is severe nonuniformity of the field at the injecting contact. The final topic which we will consider is the case of the space-charge-limited currents with no trapping. Diffusion will be included and the field at the injecting contact will be treated as not negligible. The transient case for times less than the arrival time of the first wave of carriers will be specifically analyzed, but inferences will be made regarding the steady state. The transient case to be discussed is one in which the injection process has produced its steady-state reservoir of thermalized charge and then a step-function voltage is applied at time $t = 0$. (The response time of the circuit is assumed to be zero.) We will only treat the case where (a) L is very large compared with the range of the nonuniformity of the electric field at the injecting cathode contact due to the reservoir of charge and (b) where the anode is noninjecting. The image charge will not be included although this problem should also be treated because of the interest in the transition from space-charge-limited to electrode-limited currents with increasing applied voltage. The injection current producing the reservoir is $j_r = j_0 - j_1$.

The large L case lends itself to an approximate analytic solution and the details are easily interpretable [37]. The small L and low applied voltage case was numerically treated by Rosental and Lember [38]. Another interesting case was treated by Baru and Temnitskii [39] who considered nonuniform photogeneration with diffusion neglected.

The basic equations are well known and are given in c.g.s. units

$$j(x,t) = e\mu n(x,t)E(x,t) - eD\,\partial n(x,t)/\partial x \qquad (1.32)$$

$$\partial E(x,t)/\partial x = (4\pi e/\kappa\kappa_0)n(x,t) \qquad (1.33)$$

$$\partial j(x,t)/\partial x = -e\partial n(x,t)/\partial t \qquad (1.34)$$

Integrating eqn. (1.34) twice over all space for t less than the arrival time of charge at the anode yields:

$$j(t) = (e\mu/8\pi L)\{E_L^2(t) - E_0^2(t)\} + \mu n_0 kT/L, \tag{1.35}$$

where L and 0 refer to the anode and cathode region and n_0 is the density of charge in the reservoir near the injecting contact. The magnitude of n_0 is found from:

$$j_r = en_0 c|\bar{v}_0|(1-r), \tag{1.36}$$

but here the $\bar{v}_0$ is that associated with the thermalized carriers. In general, c should be very close to the Thomson value of $(6\pi)^{-1/2}$. When there is no applied voltage both E_L and j vanish. Therefore,

$$E_0(0) = -(8\pi n_0 kT/\kappa\kappa_0)^{1/2}. \tag{1.37}$$

Remembering that the Debye length $x_0 = (\kappa\kappa_0 kT/2\pi n_0 e^2)^{1/2}$ and the total charge Q_0 in the Debye cloud is $en_0 x_0$, eqn. (1.35) can now be written as

$$j(t) = e\mu/8\pi L\{[E_L(t) - E_0(t)][E_L(t) + E_0(t)]\} - \mu Q_0 E_0(0)/2L. \tag{1.38}$$

Since the total charge in the volume is $\kappa\kappa_0[E_L(t) - E_0(t)]/4\pi$, we let this be $Q(t)$ and eqn. (1.38) is simplified to

$$J(t) = (\mu/2L)[Q(t)\{E_L(t) + E_0(t)\} - Q_0 E_0(0)]. \tag{1.39}$$

Following Rosental and Lember [38], we assume that $n_0(t) = n_0(0)$.
The problem can be done self-consistently by solving for $J(t)$ for $n_0(t) = n_0(0)$ and then changing the boundary condition [eqn. (1.36)] to

$$j_r - j(t) = n_0(t) c|\bar{v}_0|(1-r). \tag{1.40}$$

$j(t)$ is normally very small compared with j_r. The added difficulty of including the correction is probably not worth the effort and is therefore ignored.

The problem now is to calculate $Q(t)$. We make the assumption that since n_0 is a constant, x_0 is also constant and therefore it is apparent that $Q(t)$ increases as long as $E_0(t) \neq E_0(0)$. $E_0(0)$ is the magnitude of the field when the system is in equilibrium and the net injection current is zero. Consequently:

$$dQ(t)/dt = en_0\mu[E_0(t) - E_0(0)]. \tag{1.41}$$

or since

$$E_0(t) = E_L(t) - 4\pi Q(t)/\kappa\kappa_0,$$

$$Q(t) = Q_0 + (\mu Q_0/x_0)\int_0^t \{E_L(t) - E_0(0) - 4\pi Q(t)/\kappa\kappa_0\}dt. \tag{1.42}$$

Remembering that $j(t) = (\kappa\kappa_0 \partial E_L/\partial t)/4\pi$, to simplify the problem by using reduced units, we can write eqns. (1.39) and (1.42) as:

$$\partial F_l/\partial T = [\eta(T)/2]\{\partial F_l(T) - \eta(T)\} + \alpha^2/2 \tag{1.43}$$

and:

References pp. 24—25

$$\eta(t) = \alpha\left[1 + \gamma \int_0^T \{F_l(T) + \alpha - \eta(T)\}\, dT\right], \tag{1.44}$$

where $\eta(T) = 4\pi Q(t)L/V$, $F_l = (E_L L)/V$, $\gamma = L/x_0$, $\alpha = 4\pi Q_0 L/\kappa\kappa_0 V$ and $T = (\mu V t)/L^2$.

When γ is very large, it is easy to show that $F_l(T)$ is a slowly-varying function of time while $\eta(t)$ may change substantially. While eqns. (1.43) and (1.44) can be solved numerically, treating F_l as constant in eqn. (1.44), leaving it time dependent in eqn. (1.43) produces a very simple equation

$$\partial F_l/\partial T = \alpha F_l \exp(-\alpha\gamma T) + (F_l^2/2)[1 - \exp(-2\alpha\gamma T)]. \tag{1.45}$$

From eqn. (1.45) one can calculate $F_l(T)$ and therefore $j(t)$, since $j(t)$ is proportional to $\partial F_l/\partial T$. (When $\gamma \gg 1$, solutions of eqn. (1.45) agree with numerical solutions to eqns. (1.43) and (1.44) to within a few percent.)

It is interesting to compare eqn. (1.45) with the results of Many and Rakavy [40], who ignored diffusion and made $E_0(t) = 0$. Under their approximations,

$$\partial F_l/\partial T = F_l^2/2. \tag{1.46}$$

It is apparent that for $\alpha\gamma \gg 1$, eqn. (1.45) reduces to eqn. (1.46) even for $T \ll 1$. A plot of $j(t)$ derived from eqn. (1.45) is shown in Fig. 1.14 for $\gamma = 10^2$ and various values of α. For $\alpha\gamma > 1$ the difference between these results and those of Many and Rakavy is apparent only for times less than about 10% of a transit time for $\alpha > 0.2$. This means that, even with $Q_0 = CV/5$, space-charge-limited conditions are achieved within $t = t_T/10$. Consequently, over a wide range of α the results presented here coincide with the Many–Rakavy [40] theory over much of the time domain. Their results (particularly for large α) are therefore useful for times greater than a transit time.

The condition $\alpha\gamma \ll 1$ applies for high voltages where the current is electrode, rather than space-charge, limited. The relationship $\alpha\gamma \ll 1$ is equivalent to

$$V \gg 2\frac{kT}{e}\left(\frac{L}{X_0}\right)^2. \tag{1.47}$$

If we designate V_c as the critical voltage for the transition from space-charge to electrode limitation, then a reasonable value to choose for V_c is

$$V_c = 10(kT/e)(L/X_0)^2. \tag{1.48}$$

This is exactly the condition derived by Adirovich [41]. Equation (1.48) is equivalent to:

$$V_c = 5n_0 eL/C, \tag{1.49}$$

and therefore a measurement of V_c gives n_0 directly. Since $n_0 \simeq j_r \langle |v_0| \rangle^{-1}$,

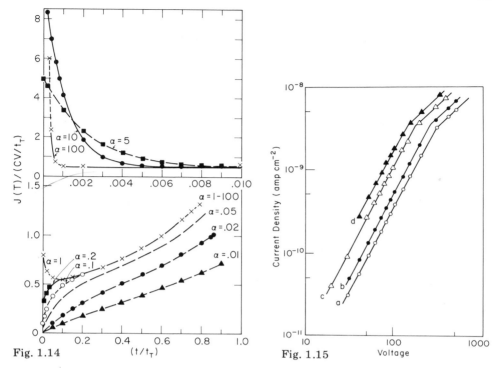

Fig. 1.14. Space-charge-limited transients for various injection levels.
Fig. 1.15. Space-charge-limited currents in fluid helium.

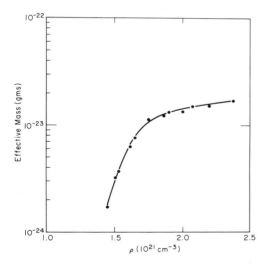

Fig. 1.16. Derived variation of the effective mass of electrons in fluid helium.

References pp. 24—25

we can also determine the average magnitude of the random velocity. Further, since:

$$\langle m^* \rangle = (3/kT)/\langle |v_0| \rangle^2, \tag{1.50}$$

the effective mass of the carrier can also be determined. This was done for dense fluid helium by Howell and Silver [42] using a tunnel cathode as a limited source for j_r. The current—voltage characteristics for various helium fluid densities are shown in Fig. 1.15. It is interesting how sharp the transition is from space-charge to electrode limitation. Figure 1.16 shows the variation of the effective mass, derived from Fig. 1.15, with helium density. (This technique should also be applicable to many other systems.)

We cannot, in this short chapter, do justice to the role trapping plays in determining injection currents. The extension to include trapping should follow directly from the nontrapping case. There are, however, two effects of trapping which are worth mentioning. The first involves the well-known effect that traps decrease the Debye length. In this case, x_0 becomes very small compared with L. Consequently, even though $\alpha \ll 1$, because γ is so large, $\alpha\gamma$ can be comparable to 1. This means that since $j_0^+ \propto \alpha$, the current may start from a very low value but in a time greater than $t_T/\alpha\gamma$ reach its space-charge-limited value. The curve for $\alpha = 0.2$ in Fig. 1.14 is an illustration of this point. This effect was also noted by Rosental [43]. The consistency between Rosental's results and those predicted here suggest that the simplified model presented here can be adapted to the case of trapping as well.

The second case of interest is one in which less than band gap light is used to photoinject carriers from an electrode into an insulator. Levinson, Burshtein and Many [44] considered this problem for the case of metal contacts to anthracene. They point out that two processes can partly or completely mask the photoemission from the metal contacts: (a) carrier excitation out of surface traps which are occupied under thermal equilibrium and (b) photoexcitation out of bulk traps which are continuously being filled by thermal carrier injection from the contact. Unless care is taken to eliminate these processes, misleading conclusions may be drawn from the experimental results. These complications further illustrate the difficulty in determining energy levels, etc. from injection currents.

REFERENCES

1 A.J. Bennett and C.B. Duke, Phys. Rev., 160 (1967) 541.
2 A.J. Bennett and C.B. Duke, Phys. Rev., 162 (1967) 578.
3 C.A. Mead, Solid State Electron., 9 (1966) 1023.
4 A. Many, Y. Goldstein and N.B. Grover, Semiconductor Surfaces, North-Holland, Amsterdam, 1965, p. 131.

5   L. Pauling, The Nature of the Chemical Bond, 3rd edn., Cornell Univ. Press, Ithaca, New York, 1960, p. 93.
6   J. Bardeen, Phys. Rev., 71 (1947) 717.
7   V. Heine, Phys. Rev., 138A (1965) 1689.
8   J.C. Phillips, Solid State Commun., 12 (1973) 861.
9   J.C. Inkson, J. Phys. C6 (1973) 1350.
10 B. Pellegrini, Phys. Rev., B7 (1973) 5299.
11 E. Louis and F. Yndurain, Phys. Status Solidi, 57b (1973) 175.
12 C.A. Mead and W.G. Spitzer, Phys. Rev., 134A (1964) 713.
13 A.G. Milnes and D.L. Feucht, Heterojunctions and Metal Semiconductor Junctions, Academic Press, New York, 1972.
14 C.A. Mead, Proc. IEEE, 54 (1966) 307. See also D.E. Eastman and J.L. Freeouf, Phys. Rev. Lett., 34 (1975) 1624.
15 F.W. Schmidlin, G.G. Roberts and A.I. Lakatos, Appl. Phys. Lett., 13 (1968) 353.
16 E.A. Fagen, Personal Communication.
17 N.F. Mott and R.W. Gurney, Electronic Processes in Ionic Crystals, Oxford Press, London, 1948, p. 172.
18 W. van Roosbroeck and H.C. Casey, Jr., Phys. Rev., B5 (1972) 2154.
19 H.K. Henisch, Semiconductor Contacts, Oxford University Press, London, 1965.
20 C.R. Crowell and S.M. Sze, Solid State Electron., 9 (1966) 1035.
21 J.D. Levine, J. Appl. Phys., 42 (1971) 3991.
22 M.P. Shaw, P.R. Solomon and H.L. Grubin, IBM J. Res. Dev., 13 (1969) 587; and P.R. Solomon, M.P. Shaw, H.L. Grubin and R. Kaul, IEEE Trans. Electron Devices, ED—22 (1975) 127.
23 R.H. Fowler, Statistical Mechanics, Cambridge Univ. Press, London, 1955, pp. 358 ff.
24 R. Williams, Semiconductors and Semimetals, Vol. 6, Academic Press, New York, 1970, pp. 112 ff.
25 J.J. Thomson and G.P. Thomson, Conduction of Electricity Through Gases, Cambridge Univ. Press, Cambridge, England, 1928, p. 466.
26 J.K. Theobald, J. Appl. Phys., 24, (1953) 123.
27 L.B. Loeb, Basic Processes of Gaseous Electronics, Univ. of Calif. Press, Berkeley, California, 1955, p. 604.
28 P.A. Bekiarian, J. Phys. (Paris), 29 (1963) 434, P.A. Bekiarian, J.L. Delcroix and P. Ricateau, C.R. Acad. Sci. (Paris), 265 (1967) 238.
29 L.A. Young and M.E. Bradbury, Phys. Rev., 43 (1934) 34.
30 D.G. Onn and M. Silver, Phys. Rev., A3 (1971) 1773.
31 W. Salaneck, Appl. Phys. Lett., 22 (1973) 11.
32 L. Onsager, Phys. Rev., 54 (1938) 554.
33 P.R. Emtage and J.J. O'Dwyer, Phys. Rev. Lett., 16 (1966) 356.
34 M. Pope and William Weston, in K. Masuda and M. Silver (Eds.), Energy and Charge Transfer in Organic Semiconductors, Plenum Press, New York, 1974, p. 10.
35 D.G. Onn and M. Silver, Phys. Rev., 183 (1969) 295.
36 P. Smejtek, M. Silver, K.S. Dy and D.G. Onn, J. Chem. Phys., 59 (1973) 1374.
37 M. Silver, Solid State Commun., 15 (1974) 1785.
38 A. Rosental and L. Lember, Phys. Status Solidi, 39 (1970) 19.
39 V.G. Baru and Y.N. Temnitskii, Sov. Phys. Semicond., 5 (1972) 1862.
40 A. Many and G. Rakavy, Phys. Rev., 126 (1962) 1980.
41 E.K. Adirovich, Solid State Phys., 2 (1960) 1282.
42 J. Howell and M. Silver, Phys. Rev. Lett., 34 (1975) 921.
43 A. Rosental, Phys. Lett., 46A (1973) 270.
44 L. Levinson, Z. Burshtein and A. Many, J. Mol. Cryst., 26 (1974) 329.

CHAPTER 2

EXPERIMENTAL TECHNIQUES

F.K. DOLEZALEK

2.1. Introduction
2.2. Photoconductivity relaxation
 2.2.1. Pulsed excitation
 2.2.2. Modulated excitation
2.3. Transit-time techniques
 2.3.1. Ambipolar transits
 2.3.2. Unipolar transits
  (a) Fixed space charge
  (b) High injection levels
  (c) Trapping
  (d) Recombination
  (e) Diffusion
  (f) Special transit-time methods for liquids
  (g) Excitation pulse sources
2.4. Magnetic field effects
 2.4.1. Hall and photo-Hall effect
 2.4.2. Photoelectromagnetic effect

2.1 INTRODUCTION

In the fifteen years which have elapsed since Bube's monograph [1] appeared, the field of photoconductivity has been considerably enriched by the extension of the investigations to the less well-ordered forms of condensed matter and to a wider range of compounds, particularly the organic and other molecularly bonded materials, and narrow band gap semiconductors. The availability of high intensity light sources has permitted the observation of new phenomena such as electron—hole drop formation [2], multiphoton excitation processes [3], and the photon drag effect [4]. It has also been possible to extend the useful wavelength range far into the infrared [5].

Since most experimental techniques [1, 6, 7] employed in these investigations were already well described in Bube's book there seems to be no point in repeating them here in an updated form. There has been, however, a substantial development of the methods which employ time-varying excitation for the elucidation of trapping parameters and mobilities. In the first part of this chapter, we will therefore make an attempt to present a comprehensive survey of techniques which are based on transient or periodically modulated excitation. This will be supplemented by a description of measurements of the Hall and the photoelectromagnetic effects which are also employed in mobility and life-time determinations.

The phenomenon of photoconductivity has been employed as an efficient tool in the study of the basic electrical transport parameters of semiconductors and insulators. The increase in conductivity which is usually observed under illumination depends on the product of the density of the mobile charge carriers generated and their velocity in an applied electric field. It is often observed that the conduction is mainly due to one carrier, say, the electrons, which are then called the majority carriers. The conductivity increase due to illumination is then given by:

$$\Delta \sigma = e \Delta n \mu \qquad (2.1)$$

where $\Delta \sigma$ is the conductivity increase due to excess electrons of total density Δn which move with an overall drift mobility μ. The dependence of the charge carrier density on the wavelength of the exciting radiation can give information on the nature of the transitions involved and the pertinent energy differences. It may thus be possible to find out whether the conductivity change observed is due to excitation from impurity levels [5], intra- or interband transitions or internal photoemission [8—11], from an illuminated contact. Such measurements can be carried out by using illumination which is either of fixed intensity or chopped at low frequencies to permit more sensitive detection. Alternatively, wavelength modulation [12] can be used to increase the energy resolution of the photoconductivity spectrum.

In photoconduction experiments which use steady excitation it is normally necessary to refer to another independent experiment in order to disentangle the factors Δn and μ whose product appears in eqn. (2.1). In some cases, it may be possible to determine one of these quantities directly using, for example, electron spin resonance [13] and acoustoelectric interaction [14] measurements for the density and the drift mobility, respectively. The usual procedure, however, is to determine the microscopic mobility, μ_0, which is the mobility of the carriers between trapping events from the Hall effect (see Section 2.4.1) and to calculate the density of untrapped excess carriers from $\Delta \sigma/(e\mu_0)$. In the trap-free case, the microscopic mobility, μ_0, is, of course, identical with the "macroscopic" drift mobility, μ, used in eqn. (2.1) which includes trapping events.

Methods using time-varying excitation have to be used, however, where Hall effect measurements are too involved or not meaningful as in the short-mean-free-path materials where the Hall effect reflects the structure rather than the mobility of the medium [15]. As will be shown in Section 2.2 of this chapter, the response of a photoconductor to pulsed or modulated excitation can be used to evaluate the lifetime and the mobility of the majority carrier. If trapping is involved, an investigation of the temperature dependence will permit the pertinent parameters to be evaluated. Section 2.3 deals with yet another kind of pulsed photoconductivity measurements which enable the observation of the transit of a charge carrier bunch across the sample. In experiments of this type, the sign, drift mobility, and density of the charge carriers can be unambiguously evaluated from a single measurement. The transient waveforms observed can be influenced, for example, by the level and the form of the excitation, by nonuniform fields in the sample, by trapping, and by diffusion. In cases where only one of these perturbations is present in a mild form, transit-time measurements can be used to determine the required parameters. The final two sections are concerned with the effects of a magnetic field on the motion of charge carriers. In Section 2.4.1, Hall and photo-Hall effect measurements are reviewed with particular reference to the problems encountered in the more insulating materials. Section 2.4.2 deals with the photoelectromagnetic effect which is useful in investigations of very small minority carrier diffusion lengths and lifetimes.

2.2 PHOTOCONDUCTIVITY RELAXATION

The response of a photoconductor to time-varying excitation may be employed as an efficient tool to unravel the kinetics of the generation of carriers, their interaction with traps, and, finally, their recombination. Both pulsed or periodic excitation waveforms should give essentially the same results [16]. Pulse techniques, described in Section 2.2.1, are usually preferred, probably because they are less elaborate and time-consuming. Recent advances in electronic instrumentation, however, would appear to make modulation methods a little more attractive, especially in applications where their great sensitivity is of advantage.

The speed at which a photoconductor reacts to a short pulse of radiation will, in general, be determined not only by the material of the sample itself but also by other factors, *e.g.* carrier injection from, or loss to, the contacts. For a detailed discussion of these effects, and the analysis of transient photocurrents in general, the reader is referred to Chapters 3 and 8 of Bube's book [1] and, with specific reference to low resistivity semiconductors, to Bray and Many's article [17]. A detailed review of imperfection center spectroscopy of CdS-type materials by transient photoconduction methods was given by Böer and Niekisch [18].

In Sections 2.2.1 and 2.2.2 on pulsed and modulated excitation only two typical cases will be considered: firstly, a high resistivity sample in which only one carrier is mobile, and, secondly, a low resistivity extrinsic semiconductor sample. Both samples possess injecting contacts.

Photoconduction in insulating samples fitted with noninjecting electrodes is often hampered by the gradual build-up of polarization. This decreases the current to a low level and generally tends to produce spurious results. In this

References pp. 63—69

case, it is necessary to restrict measurements to periods of time which are short enough to prevent appreciable polarization. If the sample is shorted and thoroughly discharged between measurements by flooding it with light [see also Section 2.3.2.(a)] one can take readings under virtually polarization-free conditions. In cases where electronic pulse averaging must be used to improve the accuracy of the results, fully automated systems such as described by Borders and Hodby [19] and Smith [20] are necessary. The sensitivity achieved by Borders and Hodby was such that photocurrent pulses down to 5 electrons and current level changes of one part in 1000 could be detected. In this way, it was possible to measure the effects of cyclotron resonance and spin-polarized scattering on the photoconductivity of polar compounds [21].

2.2.1 Pulsed excitation

In typical high sensitivity photoconductors such as sensitized CdS (see Chapter 4) or other II—VI compounds, the minority carriers — holes — are immediately immobilized at deep recombination centers. If injecting contacts are used, the excess current produced by photo-excitation is not terminated by the early capture of the holes but continues to flow until all majority carriers have recombined. If the latter are extracted through a contact before this happens they are replenished from the opposite contact so that charge neutrality is maintained. Suppose the sample is first excited at a constant rate producing g_0 mobile carriers per second and cm^3. The excitation is then cut off after the steady-state excess conductivity $\Delta\sigma$ has been reached. If the density of electrons in traps, n_t, is much smaller than the free electron density, n, the current will decay exponentially with a decay time given by the electron lifetime, τ_0. This allows calculation of the microscopic mobility, μ_0, of the majority carriers from

$$\Delta\sigma = eg_0\tau_0\mu_0. \tag{2.2}$$

Trap saturation, i.e. $n \gg n_t$, may be reached at high excitation levels or temperatures. If, however, $n \ll n_t$, the decay is lengthened by the comparatively slow release of electrons from traps. If trapping occurs by a discrete level, its depth $\Delta\epsilon$ below the conduction band may be found by measuring the dependence of the photoconductive decay time τ on the absolute temperature T [22]. A plot of τ versus $1/T$ then allows $\Delta\epsilon$ to be evaluated from the slope of the straight line obtained.

If, on the other hand, a distribution of trap levels has to be assumed, at least part of the distribution can be determined from the photoconduction decay using DeVore's procedure [23]: the value of the photocurrent, I, under the steady excitation g_0, and its initial decay, dI/dt, are measured just after excitation cut-off. The trap density, $N(\epsilon_{nf})$, at the position of the electron quasi-Fermi level, ϵ_{nf}, (measured downwards from the conduction band edge) is then calculated from:

$$N(\epsilon_{nf})kT = -g_0 I/(dI/dt), \qquad (2.3)$$

where kT is the thermal energy. The energy ϵ_{nf} can be calculated from the ratio of the effective density of states at the conduction band edge, N_c, and the density of free carriers, n, according to [1, 18]

$$\epsilon_{nf} = kT \ln(N_c/n) \qquad (2.4)$$

Since n enters only logarithmically in this expression, an educated guess of its magnitude will usually suffice [18]. By varying the generation rate, g_0, or the temperature, T, the quasi-Fermi level can be scanned over a limited region below the conduction band, thus permitting $N(\epsilon_{nf})$ to be determined in this range.

The second typical case which we will consider is a low resistivity extrinsic semiconductor, say n-Si, whose dielectric relaxation time is very much shorter than the majority carrier transit time. The illumination should be restricted to part of the sample length to avoid contact effects or minority carrier sweep-out [17, 24]. If the influence of surface recombination can be excluded and if interfering traps are saturated by a high level of d.c. background radiation, the measured decay time after excitation cut-off will be the minority carrier lifetime, τ_m, which in this case coincides with the majority carrier lifetime. From the results, the majority carrier mobility can be calculated using eqn. (2.2), if the excess conductivity, $\Delta\sigma$, and the generation rate, g_0, have been measured under uniform excitation in a separate experiment. A variety of other determinations of the minority carrier lifetime have evolved, many of which were reviewed by Bemski [25]. They include measurements by the Haynes—Shockley experiment (see Section 2.3.1), diffusion length determinations [26], the photoelectromagnetic effect (see Section 2.4.2), and others [27—29].

2.2.2 Modulated excitation

The use of frequency-swept modulated excitation, first introduced by Fassbender and Lehmann [30], has subsequently been extensively applied by Niekisch [18] and Ryvkin [7] to II—VI photoconductors, and by Schultz [31] and van der Pauw [32] to Ge and Si.

In the experiment, a d.c. voltage is applied to the photoconductor which is excited with (mostly sinusoidally) modulated irradiation producing a generation rate, g:

$$g = g_0(1 + m \cos \omega t), \qquad (2.5)$$

where g_0 is the average carrier generation rate, $m \ll 1$ is the modulation depth, ω the frequency, and t the time. The amplitude (and possibly also the phase) of the resulting a.c. photocurrent is measured as a function of the excitation frequency.

For a typical sensitized photoconductor where only the majority carrier is mobile, the a.c. photocurrent is constant at low frequencies but drops as $1/\omega$ beyond a frequency ω_0:

$$\omega_0 = \frac{1}{\tau}, \qquad (2.6)$$

where τ is the photoconductive decay time, as before. The latter quantity will depend on the excitation intensity if traps interact with the carriers. In this case, the photoconductivity decay time, τ, can be much longer than the recombination lifetime of the majority carriers in the trap-free case, τ_0. The decay time τ observed approaches τ_0 only at high excitation intensities where the traps are saturated. If the average generation rate, g_0, and the corresponding average excess conductivity, $\Delta\sigma$, can also be determined under these conditions, the majority carrier mobility, μ, may be calculated [33] from eqn. (2.2).

For low intensities and frequencies, i.e. $\tau \gg \tau_0$ and $\omega\tau \ll 1$, the density of trapped carriers, n_t, is in equilibrium with the density of free carriers, n. If a distribution of trap levels is assumed, the phase angle, ϕ, between photocurrent and excitation may then be used to determine the energetic distribution of the trap density, $N(\epsilon_{nf})$, from [18]

$$g_0 \tan\phi = \omega k T N(\epsilon_{nf}) \qquad (2.7)$$

As before, variation of the generation rate [33—35] or, better still, the temperature [34], permits $N(\epsilon_{nf})$ to be evaluated over a limited range of energies below the conduction band. If, on the other hand, a discrete set of traps predominates, the measurement of τ as a function of the temperature will, as before, permit the trap depth, $\Delta\epsilon$, to be derived from a plot of τ versus $1/T$ [22].

The above methods can only be applied in a straightforward manner to cases where the photocurrent is due to a single carrier, the decay is truly exponential ($m \ll 1$) and the intensity dependence of the current is indeed linear. In other situations, the evaluation is more complex, as pointed out by Tokarsky et al. [34] and Cheroff et al. [35]. The equipment required for the application of intensity-modulated photoconductivity consists of a suitably intense excitation source which can be modulated over a broad frequency range and the complementary set of sensitive a.c. voltmeters. A variety of light sources and modulation methods has been described in Chapter 3 of Ryvkin's monograph [7], but only a few of these will give constant modulation up to the MHz range. However, continuous-wave tunable lasers, light-emitting diodes [36], or electron beam irradiation-sources [33, 37, 38] can easily be modulated up to much higher frequencies [see also Section 2.3.2.(g)]. For the detection of the a.c.-photocurrent, tuned or phase-sensitive amplifiers may be used to increase the signal-to-noise ratio, permitting detection of a.c. currents below 10^{-11} A [33].

Detectors which allow measurements of both the amplitude and the phase are superior because the latter's variation at the critical frequency ω_0 is much more pronounced than the more gradual decrease of the amplitude.

While the aforementioned methods have mainly been applied to the typical II—VI photoconductors [18], similar equipment can also be used to measure the photoconductive decay times and surface recombination velocities of semiconductors such as Ge [31] and Si [32, 39, 40]. Recently, microwave modulated electron beam excitation has been employed to determine the high field electron velocities in epitaxial GaAs by measuring the photocurrent phase delay due to transit time effects [37]. An application of intensity-modulated excitation to the measurement of minority carrier mobility and diffusion will be described in Section 2.3.1.

2.3 TRANSIT-TIME TECHNIQUES

If the time resolution of pulsed photoconduction experiments is high enough, it may, in certain circumstances, be possible to observe the effects of carrier transits across the sample. In this case, it is possible to measure directly the transit time and hence velocity and mobility of the carriers responsible for the excess current observed. The idea underlying such experiments is to observe the progress of a strongly nonuniform excess charge carrier distribution, *e.g.* a charge packet or front, as it is swept through the sample by a drift field. The resulting transient current as measured in the external circuit, or by chemical methods [41] may then hopefully bear a prominent time mark which can be analyzed in terms of the carrier transit time. Nonuniform spatial distributions of moving carriers can be produced by generating excess carriers with step function, pulsed, or modulated excitation or, alternatively, by injecting excess carriers through the rapid application of a field to a suitable contact. A necessary condition for the observation of a discernible time mark on the photocurrent waveform is, of course, that the carrier-density gradient is not eroded during transit by diffusion, dielectric relaxation (exception: minority carriers in low resistivity materials), deep trapping, or recombination.

Ideally, unipolar conduction would be preferable for such experiments as this greatly simplifies the analysis of the transient current waveforms. However, single-carrier transits are impossible in low resistivity semiconductors where the dielectric relaxation time is given by:

$$\tau_{\rm rel} = \kappa \kappa_0 \rho, \tag{2.8}$$

where $\kappa \kappa_0$ is the permittivity and ρ is the specific resistance, because $\tau_{\rm rel}$ is much shorter than the transit times so that any excess charge is immediately neutralized. In this case, [42—44], which will be discussed in more detail in Section 2.3.1, only the minority carrier mobility of extrinsic material can be

References pp. 63—69

measured, *e.g.* by the Haynes–Shockley method [45], or related transit-time methods [46].

According to the theory [42, 43], a neutral pulse of excess electrons and holes in an extrinsic material (say, n-type) cannot be separated by an applied field but travels as holes would [44] with an ambipolar mobility of:

$$\mu_a = \mu_p \mu_n (n_0 - p_0)/(n\mu_n + p\mu_p), \tag{2.9}$$

where μ_p, μ_n are the mobilities, $n = n_0 + \delta$ and $p = p_0 + \delta$ are the densities, and n_0, p_0 the equilibrium densities of the carriers. Note that $\mu_a = 0$ for intrinsic material, *i.e.* $n_0 = p_0$. For truly extrinsic material, say n-type, $n_0 \gg p_0$, and for small excess charge densities, *i.e.* $\delta \ll n_0$, it follows from eqn. (2.9) that the ambipolar mobility is equal to the minority carrier mobility, μ_m.

The situation is less complicated in the better insulating media where the equilibrium dielectric relaxation time is usually much longer than both carrier transit times. Here, an initially neutral pulse composed of both carrier polarities can be separated by a sufficiently strong applied field so that transits of either carrier polarity may be observed in the same sample. These experiments will be described in Section 2.3.2, while ambipolar transits are discussed in the following section.

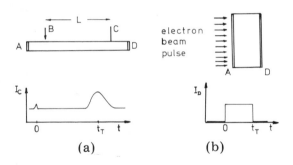

Fig. 2.1. Ambipolar transits. (a) Haynes–Shockley experiment; (b) modern version.

2.3.1 Ambipolar transits

The geometry of the well-known Haynes–Shockley [45] experiment is shown in Fig. 2.1(a). A long semiconductor rod of strongly extrinsic, say p-type, material is fitted with ohmic p^+-contacts A and D at both ends and an electric field is applied between them so that a strong hole current is generated. At point B, a few excess electrons are produced either by a forward-biased point contact or a pulse of ionizing radiation. These electrons, and an equal number of holes, will then drift toward the anode, D, with an ambipolar mobility according to eqn. (2.9). During the packet's transit through the sample a slightly higher current flows owing to the increased conductivity, but this is difficult to detect [47] on top of the much stronger majority carrier current injected at A. In the Haynes—Shockley experiment the detection problem was elegantly solved by the use of a small reverse biased contact, denoted by C in Fig. 2.1(a), which collects minority carriers while discriminating their opposite numbers. In the collector circuit, we observe the waveform also shown in the figure; the time t_T elapsed between the pulses due to injection at B and arrival at C yields the minority carrier mobility according to:

$$\mu_m = L^2/t_T V, \tag{2.10}$$

where V/L is the field. The "arrival pulse" broadens and shrinks in height with increasing drift time owing to diffusion [48, 49] and recombination [45]. The former effect may be used to evaluate the diffusivity; the dependence of the area under the arrival pulse on the drift time yields the minority carrier lifetime [50]. Carrier production by radiation [49, 51] is particularly suitable for this type of experiment as it enables the drift length, L, to be readily varied by scanning the excitation across the sample. If the excitation can be periodically modulated, the dependence of the phase lag of the a.c. current detected at C on L may be used to determine both the mobility and the diffusivity of the minority carriers [40, 52].

Recently, Neukermans et al. [46] observed minority carrier transits in a sandwich electrode configuration which is shown in Fig. 2.1(b). It differs from the previously described experiment in two points: excitation is now by a 12 keV electron pulse at the field contact A, and the current due to the electron—hole packet must now be sensitively detected on top of the much stronger majority carrier current. This was accomplished with the aid of a sampling oscilloscope whose fast rise and recovery times, large dynamic range, and signal averaging capabilities are ideally suited to this type of experiment.

A different approach was adopted in "extraction" experiments where a fast rise time drift field is used to extract the equilibrium minority carrier density faster than it can be replenished. The transit time and the lifetime of the minority carriers can then be obtained by analysing the resulting triangular transient current waveform [53, 54].

2.3.2 Unipolar transits

This type of experiment, though anticipated in the work of the Göttingen photoconductivity school [55, 56] and in the time-of-flight methods employed in gases [57], has been developed to its full potential only in the last two decades [58, 59]. The method has now found widespread application in low-conductivity crystalline [59, 60] and amorphous [61—63] solids and liquids [63—67] of inorganic and organic [68, 69] composition. Transit

References pp. 63—69

times have been measured in the range from nanoseconds to seconds [64], at temperatures down to 0.3 K [66], pressures up to 7 kbar [70], and in materials whose mobilities were as diverse as 10^4 cm^2/V.s in semiconductors [71] and 10^{-6} cm^2/V.s in some polymers [72, 73].

The principle of typical transit-time measurements may be described on the basis of Fig. 2.2 (a—c) as follows: the sample is sandwiched between electrodes, and through one of these a short pulse of strongly absorbed radiation excites a thin layer of carrier pairs. Depending on the polarity of an applied electric field, either positive or negative carriers are drawn into the bulk. If carrier lifetime permits, they will reach the opposite electrode in a time t_T. During the transit of the pulse, of charge Q, a constant current:

$$I = Q/t_T \tag{2.11}$$

flows in the external circuit. This permits Q to be measured and the carrier drift mobility, μ, to be determined from t_T and the applied field, V/L, using eqn. (2.10). Depending on the time constant of the external circuit, RC, either a rectangular pulse, Fig. 2.2(a), or its integral, shown in Fig. 2.2(b), can be observed. A rectangular current pulse is only expected for systems in which a unique mobility or mean velocity per unit field can be defined or expected for the drifting charge sheet. As discussed in Chapter 3 this is not generally true since where a stochastic transport mechanism is operative the transient current pulse shape is more complex. The analysis of transport in these types of systems is discussed in Chapter 7. The time constant is the product of the resistance, R, and the capacitance, C', of the lower sample electrode to earth; it determines the bandwidth, $i.e.$, the resolution, and also the voltage level of the waveform. "Current pulses" which are observable with $RC' \ll t_T$ are usually easier to analyze intuitively than the integrated waveform ($RC' \gg t_T$). The latter is preferred at low signal levels where waveform resolution may be traded against an order of magnitude or more in signal height.

This type of experiment where $RC' \ll t_T$ is known as the "current" or "short-circuit" mode to distinguish it from the "voltage" or "open-circuit" mode where $RC' \gg t_T$. Another experimental procedure consists of corona charging a photoconductor and subjecting it to appropriate flash excitation. The latter arrangement closely resembles the experimental conditions in typical applications of high resistivity photoconductors such as in electrophotography or vidicon tubes. Furthermore, it is less vulnerable to electric breakdown and thus is particularly useful at high fields and/or with thin films (see chapters by Bonnet and Schmidlin). A typical arrangement is as follows: the electrodeless top surface of the sample is corona-charged to a voltage, V, which can be monitored by an electrostatic sensor, S. An excitation pulse absorbed in a narrow region below the surface will then result in the transfer of a charge Q to the bottom contact and this produces a linear drop in the surface potential $V(t)$, as depicted in Fig. 2.2(d). If the

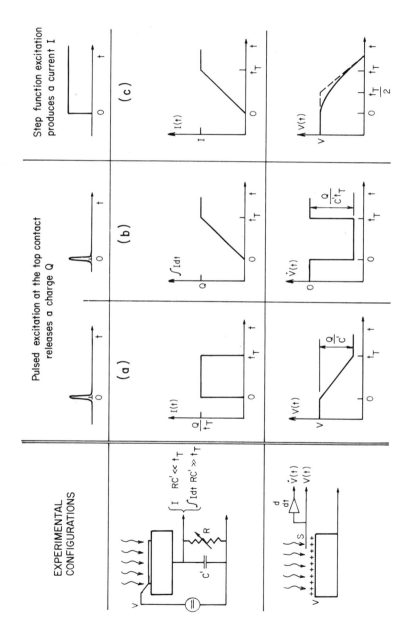

Fig. 2.2. Unipolar transits. Specimen geometries and idealized waveforms for pulse excitation, (a), (b), (d), (e) and step-function excitation, (c), (f).

References pp. 63—69

signal-to-noise ratio permits, the voltage signal may be differentiated electronically to yield the discharge rate waveform, $\dot{V}(t)$, which resembles the current mode waveform. It should be noted that in the small signal limit, *i.e.* $Q \ll CV$, where C is the geometrical capacitance of the excited area of the sample, the voltage-mode waveform, $V(t)$, is proportional to the integral of the current-mode signal $I(t)$ [74]. The deviations appearing at high injection levels [74] will be discussed in Section 2.3.2.(b).

While pulse excitation is usually easier to implement in practice, step-function excitation [75] as shown in Fig. 2.2(c) also allows the transit time t_T and the generated current, I_0, to be measured. In some materials, continuous injection may also be achieved by a suitable contact to which a step-function field voltage has been applied [75—78].

The transient waveforms shown in Fig. 2.2 will only be observed under idealized conditions, the most stringent of which being that the deep trapping lifetime of the carrier is long compared to the transit time, t_T. Furthermore, the electric field in the sample must be uniform throughout the transit period and strong enough so that diffusion effects are negligible. When radiation is used to generate injecting conditions at the upper surface, the duration or rise time of the excitation must be much smaller than t_T. If possible, carrier generation should be confined to a narrow region below the surface to facilitate interpretation of the transient waveforms [79]. If one of these conditions is not satisfied, the transient shape will be altered in a characteristic manner and this may be used, as shown in the following, to determine additional photoconductivity parameters apart from the drift mobility. Sections 2.3.2.(a) and 2.3.2.(b) deal with distortions of the internal field due to fixed space charge and high injection levels, respectively. The evaluation of trapping and recombination parameters from the transient waveforms then follows in Sections 2.3.2.(c) and 2.3.2.(d), while diffusion effects are treated in Section 2.3.2.(e). The special techniques employed in liquids and sources of short excitation pulses are reviewed in Sections 2.3.2.(f) and 2.3.2.(g), respectively.

(a) Fixed space charge

As frequently observed in pulsed photoconductivity measurements in insulating solids, the shape and magnitude of the transients are gradually obliterated after a few passages of charge packets through the specimen. This is due to the accumulation of trapped space charge in the sample which distorts the internal field. The effects will become severe as soon as the trapped charge approaches the magnitude of the charge on the contacts, CV. As Spear described in his review [58], polarization build-up can be minimized by keeping the charge in transit as small as possible and by carefully discharging the sample after every reading. This may be accomplished by applying a number of excitation pulses or penetrating d.c. illumination under zero field conditions. An automated version of the neutralization procedure which employs pulsed fields and multiple discharge pulses has also been developed [58, 65].

Another cause of sample polarization may be dark-current injection from the contacts in the time interval between the application of the drift field and the excitation. In this case, one can either try to shorten the time interval by employing pulsed fields or use less injecting, or even "blocking" contacts (see Chapter 1). The latter are usually made by separating the contacts from the sample using very thin plastic films.

Space-charge problems of a different kind are encountered in many of the semiconductor structures investigated by transit-time methods where a net density of ionized dopant centers, $|N_D - N_A|$, produces a built-in field which is noticeable at low applied fields. Here, the applied voltage is not large compared with V_1, the voltage necessary to fully deplete the specimen [59]:

$$V_1 = L^2 e |N_D - N_A|/\kappa\kappa_0, \qquad (2.12)$$

where L is the width of the depletion region of the sample.

The influence of a uniform space charge on current [80] or charge [81] transients in the short-circuit mode has been calculated by several workers for the small signal case. The current waveform up to $t = t_T$ is given by:

$$I(t) = (Q/t_T)[1 \mp 1/(2\alpha)] \exp[\pm t/(\alpha t_T)] \qquad (2.13)$$

where Q is the moving charge, t_T the transit time in the space-charge free case, and $\alpha = |CV/q|$ is the ratio of the charge on the contacts to the total space charge, q, resident in the sample. In eqn. (2.13) the upper signs refer to the case where the injected charge is of the same sign as the fixed space charge. The exponentially rising and decaying waveforms described by the expression can, in the experiment, be mistaken for the manifestations of high injection levels [Section 2.3.2.(b)] and trapping [Section 2.3.2.(c)] respectively. This emphasises again the point that efficient space-charge neutralization is necessary if meaningful results are to be extracted from the detailed shape of a transient.

It can be shown [59] that in fully-depleted semiconductor devices αt_T is identical with the dielectric relaxation time, defined in eqn. (2.8):

$$\tau_{\text{rel}} = \kappa\kappa_0/(e|N_D - N_A|\mu_1), \qquad (2.14)$$

of the specimen at zero field. Here μ_1 is the majority carrier mobility. Thus, in depleted semiconductor structures with a suitably long carrier lifetime, the low voltage transient shape may be analyzed in terms of the defect center density $|N_D - N_A|$ and, to a certain degree, its uniformity. This procedure was used by Sigmon et al. [81] for the assessment of radiation damage in Si.

The distribution of fixed space charge is also of great interest in the insulating materials used in electrophotography. Scharfe and Tabak [82] showed that, under certain conditions, the drift of an injected sheet of carriers can be used as a probe of the distribution of fixed space charge. These workers employed the equivalent of the following parametric expressions for the electric field $E(x)$ at a depth x below the surface:

$$E(x) = [L/(\mu Q)] I(t) \tag{2.15}$$

$$x = (L/Q) \int_0^t I(t) \, dt \tag{2.16}$$

and analyzed the waveforms obtained in amorphous Se using a probe-charge Q of mobility μ. This procedure yielded a linear function for $E(x)$ which indicated that the space charge was uniform [82].

The above method will not, however, give meaningful results for probing carriers of sign opposite to that of the fixed space charge if Langevin-type bimolecular recombination cannot be ruled out. As will be shown in Section 2.3.2.(d) the condition for the observation of appreciable space-charge-field effects, $\alpha \leqslant 1$, is tantamount to efficient Langevin-type recombination. In this special case, eqns. (2.13, 2.15 and 2.16) require revision.

(b) High injection levels
  (i) Step-function injection

If the contact permits the injection of more than a small fraction of a CV per transit time, the magnitude of the current flowing in an applied field will be influenced by the perturbations of the internal field created by the moving charge. In the limit of ample carrier supply at the contact, the current level will be determined entirely by the charge flowing in the bulk; it is then space-charge-limited [83] (SCL). The transient SCL case in the current mode was treated by Many and Rakavy [84] and Mark and Helfrich [85]. The current-mode waveform is shown in Fig. 2.3(a). It displays an instantaneous rise to a level $I(0)$:

$$I(0) = CV/2t_T = 0.5\kappa\kappa_0 A\mu V^2/L^3, \tag{2.17}$$

where A is the contact area and $\kappa\kappa_0$ the permittivity, and then rises rapidly to a cusp of height $2.7I(0)$ at an apparent transit time, $t_1 = 0.8t_T$, at which the first carriers arrive at the collecting contact. At this point, the current drops and underswings its final value of $9CV/(8t_T)$ which it attains within roughly another transit time. During the advance of the carrier front, injection at the contact continues and the voltage source in the external circuit has to supply more charge to make up for the increased capacitance due to the extra charge moving in the bulk. In the voltage mode, the injection of charge CV is tantamount to the eventual total discharge of the sample because there is no external voltage source which restores the internal field to its original value. The calculated discharge rate [86—88] for this process is shown in Fig. 2.3(b). The rate remains constant at $\dot{V}(t) = -V/(2t_T)$ up to the time $t = t_T$, where the leading carrier front reaches the collecting

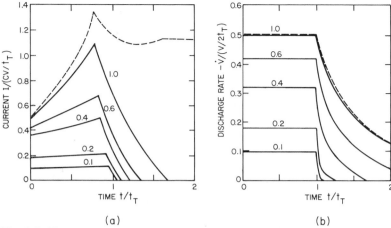

Fig. 2.3. Transient waveforms at high injection levels; (a) current mode; (b) voltage mode. Broken lines denote step function (SCL) and full lines denote pulse (SCP) injection of a charge βCV; the parameter is β.

electrode, and then drops to zero as

$$\dot{V}(t) = -[L^2/(2\mu)](1/t^2) \tag{2.18}$$

A particular feature of the voltage-mode SCL discharge is that after the injection of the charge CV at $t = 0$ no further injection can take place. In contrast to the current mode, it is thus irrelevant whether flash or step-function excitation is used to produce the injecting conditions at the contact. Both types of transient SCL measurements are especially valuable because the initial height of the transient yields values for the time $t_T = L^2/\mu V$, and thus of the mobility μ, even in the face of appreciable trapping [see Section 2.3.2.(c)] where the transients do not exhibit a discernible break at the arrival time of the carrier front. In this case, however, it is important to ascertain that one is dealing with truly injecting contacts which show a square voltage dependence and no intensity dependence of the initial current. Furthermore, the thickness dependence in the current and voltage mode should be L^{-3} and L^{-2}, respectively. The voltage mode has the unique advantage that the sample area need not be known and that, in the case of negligible trapping, the slope $L^2/2\mu$ of the t^{-2}-tail, eqn. (2.18), can be used as a further check on the mobility.

(ii) Pulse injection

Although the theoretical groundwork was first laid for conditions corresponding to step-function excitation [84, 85], most experiments were performed under the experimentally easier flash excitation. Schwartz and Hornig [89] first pointed to this fact and calculated the space-charge-perturbed (SCP) current caused by the pulse injection of a charge in the

References pp. 63—69

short-circuit mode. The analysis has been extended to include the effects of deep trapping [90, 91]. As can be seen from Fig. 2.3(a), the current waveform for $Q/(CV) \equiv \beta = 1$ is very similar to the SCL case, which is also shown. However, because no further charge is injected after $t = 0$, the SCP current rises less steeply to a cusp of only $2.2 I(0)$ at $t = 0.8 t_T$ and then drops rapidly to zero at $1.7 t_T$. The pulse shapes for $\beta < 1$ are also shown in Fig. 2.3(a), and they conform to [89, 90]:

$$I(t) = (Q/t_T)(1 - 0.5\beta) \exp[(\beta/t_T)t] \qquad (2.19)$$

up to the cusp time which lies between $0.8 t_T$ and t_T. Qualitatively similar waveforms were obtained for field-dependent mobilities [92], and for the special conditions in semiconductor detectors, where built-in internal fields, finite excitation depths and short relaxation times have to be taken into account. As indicated before, the voltage mode waveform for $Q = CV$ is identical with that for step function illumination [86–88]. Weaker injection with $Q < CV$, however, will give the smaller discharge rates also depicted in Fig. 2.3(b). According to an analysis made by Batra et al. [94], the discharge rate is given by:

$$\dot{V}(t) = -\beta(2 - \beta)(V/2 t_T) \qquad (2.20)$$

up to the time $t = t_T$, where it drops rapidly as:

$$\dot{V}(t) = -(V/2 t_T)[(t_T/t)^2 - (1 - \beta)^2] \qquad (2.21)$$

to zero at $t = t_T/(1 - \beta)$. Qualitatively similar waveforms will also be observed in the case of short lifetimes [95] with respect to deep trapping, and short dielectric relaxation times [96], provided that the corresponding time constants are not shorter than a few transit times. Under conditions of pulsed injection, the initial height of the transient can only be used as an independent check on the mobility value if the charge injected at $t = 0$ is equal to CV. If it is smaller, the current is still injection-limited and a mobility determined from the initial height may be very much in error. At the same time, the current may display typical SCL features, it may be proportional to the square of the voltage and show only a weak intensity dependence (over a limited range) [74].

Fully SCL transients in the current mode are difficult to observe because a contact is required which is capable of supplying one CV's worth of charge in a fraction of a transit time at the beginning of the transient and more than one CV per transit time afterwards. Strong injection can be achieved by applying a very rapidly rising voltage step [75, 77, 97, 98] to a sample with an injecting ("ohmic") contact, or, alternatively, by using a fixed field and making a normally blocking contact injecting by a strong step-function excitation [74, 98]. The only experimental verification of the ideal unipolar trap-free SCL transient in solids appears to have been given by Many et al. for iodine [75] and by Lemke and Müller for Si [77]. Most other claimed

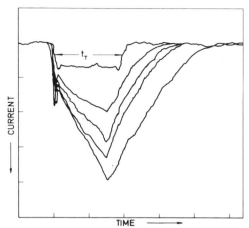

Fig. 2.4. Space-charge-perturbed (SCP) waveforms observed in a fully depleted Si surface barrier structure. (From Canali et al. [100])

observations of transient SCL currents show waveforms which decay more or less rapidly to zero after the cusp, an effect which may be due to the dwindling with time of the carrier reservoir at the contact and/or deep trapping. In the former case, the currents would be better described as SCP sheet currents [89] due to pulsed injection. Such currents are readily observed in all samples which are suitable for transit-time measurements by using strong pulse excitation. Fig. 2.4 shows an example of SCP currents in a Si surface barrier, taken from the work of Canali et al. [100]. Gibbons and Papadakis [101] analyzed SCP current waveforms obtained in sulphur and found that the total charge injected into the bulk as determined from the ratio of the cusp to the initial current agreed very well with the value obtained by integration. Voltage-mode SCL transients are relatively easy to observe in insulating materials such as amorphous Se and orthorhombic S which discharge only slowly in the dark [74, 86, 102]. Batra et al. [86], who measured the discharge rates in amorphous Se under perfect SCL conditions, found that the mobility values determined from the initial discharge rate, $-\mu V^2/2L^2$, the transit time, $L^2/\mu V$, or the time dependence of the tail of the transient, agreed very well. Fig. 2.5 shows their results for the tail, proving that the inverse square dependence on time is satisfied over more than four orders of magnitude [86]. There are benefits in raising the injection level to the space-charge-perturbed (SCP) or space-charge-limited (SCL) regime. In cases where too short deep-trapping and/or relaxation times result in unresolvable transits, the distinct cusp produced by SCP or even SCL conditions may still provide a recognizable break at approximately the transit time, t_T. This value can then be checked against the transit time derived from the initial SCL current $CV/2t_T$. The relative insensitivity of the current height to the excitation intensity under SCP conditions will also be an advantage if flash light

References pp. 63—69

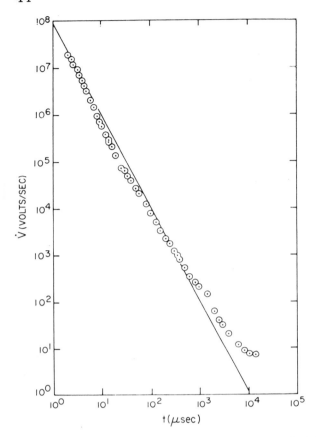

Fig. 2.5. Voltage-mode discharge of an amorphous Se film. The t^{-2}-dependence of the tail of the waveform is shown. (From Batra et al. [86])

sources, which are notorious for their fluctuations, are used. If, however, one is interested in measuring the carrier generation efficiency or carrier lifetimes, space-charge perturbations and the accompanying more rapid polarization of the sample should be avoided by keeping the injection level well below $0.1 CV$. The resulting loss of signal height can, if necessary, be compensated for — at least partly — by using with an integrating input as discussed in Section 2.3.2. For insulators, the measurement of discharge rate transients under SCL conditions is an extremely useful method which provides multiple access to the carrier mobility by a contactless method.

(c) *Trapping*

In transient photoconductivity studies, it proved convenient to distinguish between shallow and deep trapping events. The latter expression refers to those trapping centers from which a carrier has only a negligible chance of re-emission during the time interval of observation. It should be noted that this practical definition is somewhat arbitrary (depending even on the magnitude of the sample thickness), it should not be confused with Rose's [103] fundamental distinction between trapping and recombination centers. The interaction rate of free carriers with a species of traps is given by $1/\tau_t$, where τ_t is the "lifetime" or trapping time with respect to that species [1]

$$1/\tau_t = N_t S v_{\text{th}}. \qquad (2.22)$$

The rate is proportional to both the density, N_t, and the cross section, S, of the traps and is only weakly dependent on the carrier temperature through their thermal velocity, v_{th}. In low mobility media, where the carrier mean free path may be small compared to the lattice spacing, eqn. (2.22) has to be replaced by an equivalent expression [15]. In contrast to the trapping rate $1/\tau_t$, the release rate, $1/\tau_r$, depends strongly on the temperature [1]:

$$1/\tau_r = \nu \exp(-\Delta\epsilon/kT), \qquad (2.23)$$

where τ_r is the release time, $\nu \simeq 10^{11}$ to 10^{13} Hz is the attempt-to-escape frequency, and $\Delta\epsilon$ is the trap depth. If $\Delta\epsilon$ is small, the release rate may be increased by high electric fields owing to the Poole—Frenkel effect or tunnelling [59]. Shallow trapping by a particular set of traps will appreciably affect the observed photoconductive transient if a large fraction of the carriers interacts with these traps, i.e. $\tau_t < t_T$, and also spends an appreciable part of its transit through the sample at these sites, i.e. $\tau_r > \tau_t$. The effective transit time of the fraction which has suffered trapping will then be longer than t_T by a period of time proportional to the mean time lost in the traps, τ_r, so that [59]

$$t_{\text{eff}}/t_T = (\tau_t + \tau_r)/\tau_t. \qquad (2.24)$$

In view of the strong temperature dependence of the release time according to eqn. (2.23) t_{eff} will become appreciably longer than t_T as the temperature is lowered beyond the point at which $\tau_t = \tau_r$. If the trapping time τ_t is very short compared to t_T, a condition which can be fulfilled by using a suitable sample thickness and/or small fields, both τ_t and τ_r will be very small compared to t_{eff}. This is the case of multiple trapping where thermal equilibrium between trapped and free carriers can be assumed. Here, the principle of detailed balance [1, 104] applies:

$$\tau_r/\tau_t = (N_t/N_c) \exp(\Delta\epsilon/kT), \qquad (2.25)$$

where N_c is the effective density of states at the band edge or its equivalent in the non-crystalline media. Using eqn. (2.24) an effective mobility for multiple trapping is then defined as [1]:

$$\mu_{\text{eff}} = \mu_0 [1 + (N_t/N_c) \exp(\Delta\epsilon/kT)]^{-1}, \qquad (2.26)$$

where μ_0 is the mobility of the carriers between trapping events. As the

References pp. 63—69

temperature decreases, a mobility turnover according to eqn. (2.26) is to be expected, which permits the density of traps, N_t, and their depth, $\Delta\epsilon$, to be evaluated from a plot of μ_{eff} versus $1/T$ [1, 105]. If, however, τ_t is not small compared with t_T, many carriers will avoid trapping even at low temperatures, and the effective transit time of the remaining carriers will not be long compared to the release time, so that the thermal equilibrium, eqn. (2.25), may not be reached. This case was analyzed by a number of authors [59, 106—8], but the complexity of the expressions derived appears to discourage their application. In most cases, it is still possible to determine the temperature dependence of the effective transit time of the re-emitted carriers, and this yields the trap depth, $\Delta\epsilon$. In contrast to the multiple trapping case, however, the density of traps cannot be extracted from the experiment with any certainty.

So far we have only considered carrier interaction with a single shallow trap level but, in practice, this represents a fair approximation. Owing to the strong temperature dependence of carrier detrapping, shallow trapping effects are dominated by release from a very narrow energy interval of the gap. Thus, trap-controlled transit-time experiments usually yield only two shallow trapping parameters, i.e. $\Delta\epsilon$ and N_t in the multiple-trapping case, and one, τ_r/τ_t or $\Delta\epsilon$, in all other cases. Under favourable conditions τ_t may be determined from the detailed shape of the current transient [109, 110], and, if N_t and the thermal velocity are known, the cross section, S, can be calculated from eqn. (2.22). Note that under conditions of multiple trapping, the effective lifetime, τ_{eff}, with respect to another, deeper, set of traps is larger than the corresponding value, say τ_d, calculated from eqn. (2.22), because the carriers are protected from deep trapping during part of the effective transit time. It follows that $\mu_{\text{eff}}\tau_{\text{eff}} = \mu_0\tau_0 \equiv (\mu\tau)$, i.e. the $(\mu\tau)$-product is not affected by shallow trapping [103]. As noted previously, trapping rates depend only weakly on the temperature. The deep-trapping rate is proportional to the sum of all products N_tS of traps whose release time is much longer than t_{eff}. As the temperature decreases, the number of terms and hence the deep trapping rate may increase up to a point where it becomes so large compared to $1/t_{\text{eff}}$ that transits can no longer be resolved. This effect generally limits the observation of transits to relatively pure materials and short transit times. A turnover of the effective mobility has been observed in many materials, e.g. monoclinic Se [111], CdS [105, 112], orthorhombic S [113], doped anthracene [114], KCl [115], KBr [116], AgCl [117], and benzene [63, 118].

Hoesterey and Letson [114] measured the hole mobility in naphthacene-doped anthracene, their result is shown in Fig. 2.6. Analysis of the curves in terms of multiple trapping according to eqn. (2.26) yielded an activation energy of $\Delta\epsilon = 0.43$ eV. The ratios N_t/N_c evaluated from the graph agree well with the dopant concentrations employed in the preparation of the crystals.

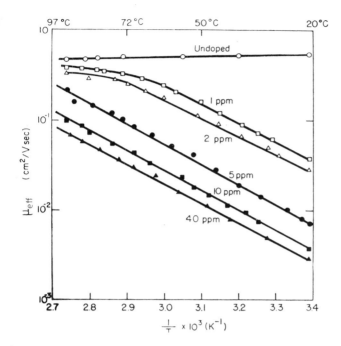

Fig. 2.6. Effective drift mobility of holes in pure and naphthacene-doped anthracene. (From Hoesterey and Letson [114])

In semiconductor detectors, it is usually impractical to increase carrier-trap interaction by making the free-carrier transit time, t_T, much longer than the trapping time, so that multiple trapping conditions were not always applicable. Instead, an approximative procedure was developed [59, 107—8] to determine the effective transit time, t_{eff}, from the trapped-carrier component of the transients. The trap depths determined by this method [110, 119] in Ge and Si were in good agreement with those measured by other means. In Ge(Li)–detectors, where particularly small trap depths of $\simeq 10$ meV were measured for both electrons and holes, it was found that the application of electric fields permitted the release time to be decreased by more than one order of magnitude [120]. This was attributed to Poole—Frenkel lowering of the potential barrier and, at very high fields and low temperatures, possibly also to tunnelling [59]. A pulse technique superficially similar to Spitzer's determination of minority carrier lifetimes [121] was employed by Lemke [122] for the measurement of release times of carrier traps in Fe-doped Si. Lemke used a long injection current period to saturate the crystal with trapped space charge. The field was then removed and trap emptying was probed by applying a comb-shaped burst of needle pulses. The height of the probing pulses was taken as a measure of the escaped carrier fraction and from this the release time has been determined directly.

References pp. 63—69

The deep trapping lifetime is important for two reasons: firstly, it generally limits the applicability of transit-time measurements and, secondly, it permits a calculation of the $(\mu\tau)$-product, also known as the carrier range. This is the carrier drift distance per field unit irrespective of shallow trapping and thus is a fundamental performance parameter in applications such as electrophotography, television pick-up tubes (vidicons), and others (see Chapter 11). Experimentally, one determines a deep trapping lifetime, τ_{eff}, from the decay of a long current-mode transient, produced by a low level excitation pulse, using the expression

$$I(t) = I(0) \exp(-t/\tau_{\text{eff}}). \tag{2.27}$$

In the integrated mode, a simple extrapolation technique described by Spear and Mort [105] may be used to evaluate τ_{eff} from the low voltage deviations of a plot of $1/t_T$ versus V. If the signal magnitude achievable under weak excitation is insufficient, one may be tempted to work at higher injection levels, where space-charge effects have to be considered. The transient will then display the partially antagonistic effects of both trapping and space-charge perturbation which may be difficult — if not impossible — to disentangle. Calculations of space-charge-limited or space-charge-perturbed transient shapes for both current and voltage modes, and for times shorter [84, 95] and longer [84] than the transit time of the first carriers, are available. However, it seems that only in the limit of SCL injection, *i.e.* step-function injection, have reliable trapping times been extracted from a transient SCL current experiment [75]. Another technique, developed by Gibbons and Spear [124], used interrupted transits to measure trapping lifetimes in orthorhombic sulphur. Shortly after a pulse-excited thin sheet of carriers has been injected into the bulk of the sample, the external voltage is removed for a predetermined period during which a proportion of the carriers are permanently trapped. If the field is then applied again, the surviving fraction of the carriers can be determined by measuring the height of the resulting transient. By varying the time position of the zero-field period, it is possible to measure the variation of the lifetime over the specimen thickness. The method can be expected to yield reliable results only if the charge in transit is very small and if the interruption point is not too close to either contact so that carrier loss to the electrodes due to image forces and diffusion is insignificant.

If the $(\mu\tau)$-product is to be determined in a medium where the quantum efficiency is known to be independent of the field, this can be achieved by observing the saturation of the integrated current transients at high fields and low excitation intensities. The $(\mu\tau)$-product can then be evaluated by fitting the results to the Hecht relationship [56]:

$$\int I \, dt = QG[1 - \exp(-1/G)], \tag{2.28}$$

where Q is the generated charge, I the current and $G \equiv (\mu\tau) V/L^2$ the photoconductive gain [1, 103]. Tabak and Scharfe [125] described a very simple experiment to determine the $(\mu\tau)$-product in cases where the injection efficiency is field-dependent. If weak, but uniform, step-function excitation is used at one contact, a steady electrode-limited current I is drawn across the sample in an applied field V/L. A small fraction of the current, $i = I t_T / \tau_{\text{eff}}$, will be trapped in the bulk leading to the eventual build-up of a space charge CV which will reduce the field at the excited contact. As a result, the injection efficiency and, hence, the current will drop at a time

$$t = CV/i = (\mu\tau) CV^2/(L^2 I). \tag{2.29}$$

The value of $(\mu\tau)$ determined with this method in amorphous Se was close to that derived from individual measurements of μ and τ using pulse excitation [125].

(d) Recombination

Under electrode-limited conditions, where the current level is determined by (insufficient) carrier generation at the excited electrode rather than by bulk space charge, transient currents permit the carrier generation to be studied directly. Using pulsed excitation, the quantity of charge Q drawn out of the excitation region may be determined from the initial current $I(0)$ and the discharge rate $\dot{V}(0)$ in the current and voltage mode, respectively, as

$$Q = I(0) t_T \quad \text{and} \quad Q = C\dot{V}(0) t_T. \tag{2.30}$$

In semiconductors, where the carrier generation yield does not depend on the applied field, the generated charge can also be determined from the integrated current response, even in the presence of appreciable trapping, by fitting the measured curve to the Hecht relationship, eqn. (2.28). In semiconductor counter and radiolysis work, the energy required to produce a charge carrier pair is of great interest, since this quantity enters into the calibration of radiation detectors, which employ these materials. In semiconductors, it is usually found that with high energy radiation, an energy equivalent to about two to three times the band gap is required for the production of a carrier pair, independent of the field. In low mobility media, however, carrier generation is usually much less efficient since the short mean free paths in these materials prevent many carrier pairs from escaping their Coulomb attraction, resulting in strong initial (geminate) recombination. Under these conditions, the carrier generation yield becomes field dependent, and generally conforms to a theory by Onsager [126] which was originally developed for ion production in liquids. Measurements by Batt et al. [127] and Chance and Braun [128] fully confirmed the Onsager predictions also for a low mobility solid such as anthracene, and there is recent evidence for their applicability to other solids, e.g. amorphous Se [129], orthorhombic S [130], PVK:TNF [131], and As_2S_3 [132]. A detailed discussion of this topic can be found in Chapters 3 and 7. Further information on the carrier generation process can

References pp. 63—69

be obtained from the temperature dependence and, where applicable, the dependence on the wavelength of the excitation light pulses [133–135]. Analysis of the details of the current waveform can reveal whether the carrier generation occurs at the surface of the sample or in the bulk [133, 136]. In anthracene, where carrier generation at long wavelengths proceeds via excitons, the rapidly, and the slowly, rising portions of the observed transient shapes were analyzed and showed reasonable values for the diffusivity and the lifetimes of the excitons involved [137, 138].

In low mobility media, transient current methods may also be used to study the recombination kinetics of the "free" carriers which have escaped initial recombination. This requires a relatively large recombination coefficient γ, where γ is defined by

$$d\rho_1/dt = -\gamma \rho_1 \rho_2 /e. \qquad (2.31)$$

ρ_1, ρ_2 being the densities of the recombining charges. Efficient recombination is to be expected if γ is given by an expression calculated by Langevin [139] for the case of diffusion-limited recombination of ions:

$$\gamma = e(\mu_1 + \mu_2)/(\kappa \kappa_0), \qquad (2.32)$$

where μ_1, μ_2 are the drift mobilities of the recombining charges. The Langevin expression applies to recombination in a large number of low mobility liquids and solids. It can also serve as an analytically convenient upper limit to recombination in materials of higher mobility, where the coefficients γ are usually several orders of magnitude smaller.

It will now be shown for two typical experimental conditions that even Langevin-type recombination can become appreciable only if the carrier density is high enough to cause strong space-charge distortion of the field in the sample. Consider a thin sheet of charge moving through a sample of thickness L and area A which possesses a uniform fixed space-charge density, ρ, of opposite sign. During transit, the sheet, of charge Q, $Q \ll CV$, and mobility μ, will see a constant charge density and recombine according to eqns. (2.31) and (2.32)

$$dQ/dt = -\mu \rho Q/(\kappa \kappa_0). \qquad (2.33)$$

The mobile charge will thus decay exponentially with a time constant, τ_{rec}, given by

$$\tau_{rec} = \kappa \kappa_0 /(\mu \rho). \qquad (2.34)$$

Recombination will be strong if τ_{rec} equals the transit time, and hence $CV = LA\rho = q$, the condition for appreciable distortion of the transient waveform caused by a fixed space charge as discussed in Section 2.3.2.(a).

Similar conclusions can also be drawn for the situation where both carriers are mobile. Consider a sample which has been uniformly excited by a short pulse of penetrating radiation. As Boag [140] and others [141–2] showed, the surviving fraction n/n_0 of the carrier density is given by:

$$n/n_0 = (t_s n_0 \gamma)^{-1} \ln(1 + t_s n_0 \gamma), \qquad (2.35)$$

where $t_s = L^2/[V(\mu_1 + \mu_2)]$. For appreciable recombination, say $n = n_0/2$, we get $(t_s n_0 \gamma) \simeq 2.5$, and with eqn. (2.32) this reduces to $Q \simeq 2.5 CV$, i.e. appreciable space-charge screening of the applied field [see Section 2.3.2.(b)].

We may thus conclude that "in-flight" recombination of charge packets is highly unlikely in semiconductors and only significant in low mobility media under SCP or SCL conditions. It would thus appear that evaluations of in-flight recombination experiments [141—144, 146] which do not explicitly take account of the screening effect of the space charge on the applied field will overestimate recombination and hence yield coefficients γ which may be estimated to be too high by a factor of 2 to 5. Better results can be expected from recombination experiments at zero applied field which will be described later in this section. In-flight recombination of free carriers in the volume of an anthracene crystal and the accompanying fluorescence emission were studied by Helfrich and Schneider [98] in an elegant experiment. A step-function field, applied to a crystal cemented between liquid injecting electrodes, resulted in current and fluorescence transients which were analyzed to yield the recombination coefficient and exciton parameters. The corresponding experiment using short flash illumination of opposite faces of a platelet sample is much more difficult for reasons described in the foregoing. In some materials such as S or amorphous Se, however, where one carrier polarity (holes) is much more mobile than the other, the effects of recombination upon the transient waveform of the less mobile carrier (electrons) can be observed in a simple experiment: a sheet of electrons is injected into the bulk by weak flash illumination of the sample under an applied field. Well before the electron charge reaches the collecting contact, the latter is excited by another, stronger, light source for a duration of time long compared to the hole transit time so that a total hole charge $Q_h > CV$ may be passed through the sample under essentially space-charge-free conditions. During the hole excitation, the electron sheet will experience a constant charge density and in analogy with eqn. (2.33) an exponential decay may be expected. The remaining fraction, Q_r/Q, of the initial electron charge, Q, is given by [130]

$$Q_r/Q = \exp[-Q_h/(CV)]. \qquad (2.36)$$

Figure 2.7, trace b, shows the electron transient observed in orthorhombic sulphur in such a double injection experiment. The step in the current waveform produced by the encounter was in close accord with eqn. (2.36) which proved that the recombination was well described by the Langevin expression, eqn. (2.32). Furthermore, SCP conditions are also not required in recombination experiments where the field is not applied until after the charge carriers have had ample time to recombine. In this type of experiment, the

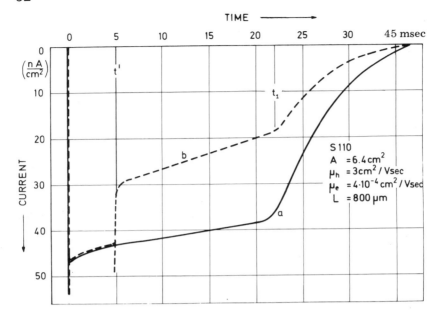

Fig. 2.7. "In-flight" recombination of electrons and holes in sulphur as observed in a double flash experiment. (a) electron transient without, and (b) with, excitation of holes at the opposite electrode at $t = t'$. The step in the current waveform is due to recombination which reduces the number of electrons in transit.

carriers are generated at essentially zero-field conditions and after a specified but varied time t_d an externally applied voltage quickly sweeps out the remaining carriers in a time much shorter than t_d [130, 136, 144, 147–9]. If the field delay time t_d is smaller than the lifetimes of both carriers with respect to deep trapping and if the initial carrier densities n_0 are equal, the remaining density will be

$$n = n_0/(1 + \gamma n_0 t_d). \tag{2.37}$$

This can be measured by integrating the observed current transient using high sweep-out fields. The coefficient γ determined from a plot of $1/n$ *versus* t_d for anthracene [136, 144, 147, 148], sulphur [130], and organic liquids [149] gave values close to the Langevin prediction. The experiment can also be used to determine the initial charge density n_0 and from this the zero-field carrier generation efficiency [149].

(e) Diffusion

The effects of diffusion on the gross transit shapes can be expected [150] to become noticeable if the field due to the thermal voltage (kT/e) is comparable to the applied field. Two effects may be anticipated. Firstly, there may be a reduction of the injection efficiency [135, 151] due to back-diffusion of excited carriers to the contact where they may be quenched. Secondly, a sharply defined carrier front produced by, say, pulse excitation, may be somewhat broadened during transit due to diffusion. The first effect is expected to become appreciable at small applied fields, V/L, or very thin excitation layer depths, d, where the "thermal field" is stronger than the applied electric field [135, 151]

$$kT/(de) \gg V/L \tag{2.38}$$

At room temperature and the usual range of fields above 10^3 V/cm, back-diffusion requires $d < 0.25\,\mu$m and thus applies only to cases of very high absorption. The second diffusion effect can be observed under carefully controlled experimental conditions by comparing the rise and fall times of transients in the current mode [71, 152, 153]. A prerequisite for such measurements is of course that other causes of carrier dispersion are minimized, e.g. space-charge effects [see Sections 2.3.2.(a) and 2.3.2.(b)], shallow trapping with a release time which is not small compared to the effective transit time, or with a distribution of release times, variations of the thickness or the excitation depth across the specimen area.

Ruch and Kino [71] showed that the field-dependent diffusivity D in GaAs can be calculated from the experimentally determined 5—95% rise and fall times, t_r, and, t_f, of the current:

$$D = (t_f^2 - t_r^2)\,21.6L^2/t_T^3 \tag{2.39}$$

Similar measurements were undertaken on Si [152] and anthracene [153]. From the diffusivity and the mobility as measured in the same experiment, the carrier temperature can be deduced using the Einstein relation. Note that a finite circuit response time and/or a finite excitation pulse length will only add an approximately constant term to both t_r^2 and t_f^2 in eqn. (2.39) and thus tend to cancel out [71]. This may be accompanied, however, by a substantial loss in accuracy when determining the diffusivity.

(f) Special transit-time methods for liquids

In view of the relatively long carrier lifetimes and low mobilities observed in liquids and gases, a great variety of more or less sophisticated transit-time techniques can be applied to these media apart from the well-known type which has been discussed in the previous sections [64, 154—6]. In some liquids, such as the rare gas liquids or some hydrocarbons which require high excitation energies, it may be very difficult to find a suitable pulsed light source. Here alpha particle [157], or electron beam [58, 65] excitation may be the only solution. The problem of introducing electrons into a specimen compartment which is under a gas pressure of several hundred torr was solved by Miller et al. [65] as shown in Fig. 2.8(a). They used a window made of a 6 μm thin PET film which was supported by a microscope mounting grid and a perforated disk through which the electron beam entered from

References pp. 63—69

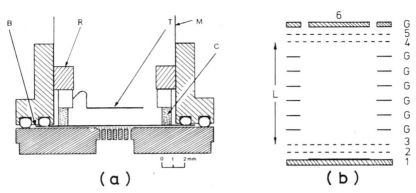

Fig. 2.8. Drift cells for liquids. (a) Cell for rare gas liquids and electron beam excitation. (From Miller et al. [65]). B: gold-coated PET film, R: PTFE clamping screw, T: disk electrode, M: PET tube, C: ceramic spacer; (b) double-shutter drift tube, 1: radioactive emitter electrode, 2—3 and 4—5: shutter grids, G: guard rings, 6: collector electrode.

below. This arrangement permitted drift velocities to be followed through the fully reversible liquid—solid transition in rare gas [65] and other liquids [156] at field strengths between 10 and 10^5 V/cm.

In radiolysis experiments, it is sometimes more advantageous to irradiate the full volume between the electrodes. According to an analysis by Hudson [158] of the case where one carrier is much more mobile than the other, the transient current response due to a fast step-function excitation saturates at the transit time of the fast carrier and thus allows the mobility to be determined [159]. If, however, the exciting radiation can be confined, by a slit, to the vicinity of one electrode, no restrictions apply to the carrier mobilities, and switching the excitation on or off will result in linearly rising or falling currents, respectively, with a clear break at the transit time [159], as shown in Fig. 2.2(c). If the carrier velocities are very low, it is necessary to avoid bulk liquid flow from being induced by the transits, as this may result in systematic errors. On the other hand, one may choose methods where the liquid is made to flow in a direction parallel to the contacts. If the flow velocity is sufficient, some carriers will not be able to reach the collecting electrode and be carried away by the stream. The resultant current anomalies may be used to determine the carrier mobility [160]. Probably the most accurate transit-time mobility measurements in liquids were performed by the "double shutter" technique [155, 161] originally developed for gases by Van de Graaff and Tyndall [57], which is analogous to Fizeau's light velocity measurement. In this method, whose geometry is shown in Fig. 2.8(b), charge carriers are continuously produced by a radioactive emitter 1 while an electronic grid shutter 2—3 breaks up the flow into charge packets which then travel in a constant drift field, V/L, of length L until they encounter a second periodic shutter, 4—5, and, when passed, are collected at electrode 6. The resulting current pulses can be sensitively measured by an electrometer. The time-averaged current will be maximal if the interval between the "on" periods of shutters 2—3 and 4—5 are adjusted such that they are exactly equal to the carrier transit time, $t_T = L^2/\mu V$. The accuracy is improved by using extra long drift spaces, which requires a series of guard rings [66, 162], denoted G in Fig. 2.8(b). These assure homogeneous field conditions. The great potential of the double shutter method was demonstrated by Schwarz [66] who, using a modified arrangement originally due to Meyer and Reif [161], made an exhaustive and accurate study of both positive and negative charge carrier mobilities in liquid helium. Mobilities varying over 6 orders of magnitude were measured, and it was possible to observe transits in fields as low as 30 mV/cm. For a detailed discussion of experimental results for liquids see Chapter 9.

(g) Excitation pulse sources

In the usual form of drift mobility experiments, a thin layer of charge carriers is generated at one of the electrodes in a time which is short compared to the transit time of the carrier polarity to be investigated. In most applications it is immaterial how the carriers are produced, and the experimenter is free to choose either light sources such as spark gaps, flash lamps, lasers, light-emitting diodes, or highly ionizing radiation of X-ray, nuclear, or electron-beam sources. Many of these excitation methods were reviewed by Spear [58] and, with special reference to their applicability to semiconductor detectors, by Martini *et al.* [59]. The number of photons required in a drift mobility measurement can easily be estimated, remembering that CV is the maximum charge [see Section 2.3.2.(b)] which can be injected at any one time. An upper limit to the number of photons N required per sample area A is then, for fields below 10^4 V/cm:

$$N/A = \kappa f \times 5 \times 10^9 \text{ photons/cm}^2, \qquad (2.40)$$

where κ is the relative permittivity and f the number of photons required to produce a fully dissociated carrier pair. The coefficient f is about unity in semiconductors but it is often found to be as large as 10^3 to 10^4 in low mobility media. The number of photons required per square centimeter thus lies between 10^{10} and 10^{14} photons for the space-charge-perturbed case, and lower by one or two orders of magnitude for space-charge-free measurements. Light sources of a suitable spectral range can be used to excite the specimen through a semi-transparent contact. In most cases, light of energy just above the band gap energy will give the highest carrier generation yield, and the excitation, then, is also confined to a thin absorption region near the contact. At the expense of a somewhat lower efficiency, however, lower light energies may also be used to produce carriers by photoemission from the contact [8—11, 64] or by injection from a thin sensitizing layer between the illuminated contact and the sample [163]. In the former case, it is important to ensure that the low energy light does not excite carriers at both electrodes.

References pp. 63—69

Open air [164], or pressurized spark gaps [165] supplying about 10^{10} to 10^{12} photons per nanosecond at pulse durations between a few nanoseconds and 100 nanoseconds are not too difficult to build [166]. Commercial models are also available [167—8] and one of them can even be triggered [167].

In recent years, the rapid development of coherent optics and flash photolysis have made available a great number of excellent flash lamps of high intensity and short duration. Xenon flash and stroboscope lamps are mostly built for extra high intensities, but short duration models exhibiting half widths of a few microseconds are also available. Other lamps, which were specifically tailored for fluorescence decay time measurements [169], supply about 10^{10} to 10^{11} photons within a flash time of a few nanoseconds duration and an energy up to 6 eV. Where a pulsed laser of suitable wavelength is available, it will form a very useful excitation source giving ample intensity at very short pulse widths [170]. Their high photon flux is a particular advantage in experiments where the carriers are to be generated by the very inefficient extrinsic processes using penetrating light [129, 137, 138]. Light-emitting-diode (LED) lamps, which are currently available in the wavelength range above 550 nm are convenient to operate, but give only about 10^8 photons per pulse length of typically 100 ns. In cases where suitably intense, or energetic, light sources are not available, or cannot be used for other reasons, high energy radiation like X-rays, or γ-rays, ionizing particles, or pulsed electron beams may be employed.

Pulsed X-ray sources of duration down to a few nanoseconds are commercially available. They can be used in mobility studies if it is known that one carrier is much more mobile than the other or in studies of carrier recombination in the bulk of a specimen [141, 171]. In cases where the magnitudes of the carrier mobilities are comparable, however, mobility measurements using X-ray [171] or γ-ray excitation [172] can only be interpreted if the radiation is collimated and confined to the vicinity of the electrode. Beta-particle sources have also been employed in drift mobility measurements [60, 174]; however, they suffer from high penetration depths, strongly fluctuating intensity, and a low carrier yield of only about 10^5 carrier pairs per particle. Alpha particles, on the other hand, produce a much higher and constant number of carriers in absorption depths of a few tens of a micron [59, 175]. Ionizing particles generally have the disadvantage of a comparatively low intensity and the lack of advance trigger signals which would permit the use of pulsed fields and modern averaging techniques. However, the simplicity of α-particle excitation recommends its use in applications where space is at a premium, *e.g.* in the crowded quarters of high pressure cells or very low temperature apparatus [66, 157]. When working with very short transit times, especially at low fields, the finite lifetime of the α-particle generated carrier reservoir may be a problem [59]. The locally very high carrier pair concentration created along the track of an α-particle will effectively shield the interior of the plasma so formed from the applied electric field. As a result, carriers may be released into transit for a long time after the excitation has occurred, thus producing a tail on the observed current pulse. If the mobility is measured in the integrated current mode, the rise time will be increased considerably resulting in the "plasma time" effects described by Canali et al. [59, 110].

Excitation by an electron beam pulse is not subject to plasma effects, because the excitation is usually uniformly distributed over the irradiated area. The use of electron beam excitation for drift mobility work has been pioneered by Spear [58, 61] who has demonstrated its versatility in many applications for both high [65] and low [113] mobility measurements in solids and liquids. Simple electron guns are not too difficult to construct [58] from cathode-ray oscilloscopes [176], television tubes or microwave triodes [177]. Special features such as ultra-short pulses [178], applicability to liquids [65], microwave frequency [37] or extremely linear [33] modulation, and extra high energy [179] also belong to the state of the art. A compact electron gun capable of supplying 6×10^3 electrons of 40 keV energy in a pulse duration of 70 ps was described by Alberigi Quaranta et al. [178] who used it in drift mobility measurements on semiconductor detectors. An experimental arrangement comprising an electron gun and the pulse equipment required for the prevention of polarization [see Section 2.3.2.(a)] in the more insulating media were described in detail by Spear [58]. The special advantages of electron beam excitation over other ionizing radiation may be summarized as follows:

(1) The penetration depth can be adjusted by changing the beam energy [58].

(2) The excitation is uniform over the contact area so that plasma time effects [59, 110] are avoided.

(3) There is usually an advance trigger signal available so that pulsed bias fields can be used to reduce sample heating in semiconductors [178] or polarization [58].

Since the time position of the electron pulse and hence of the signal is accurately known, it is even possible to recover signals from noise by averaging techniques [58, 180]. A disadvantage inherent in most electron guns is the fact that the specimen has to be mounted in a vacuum chamber. However, windows consisting of plastic foil [65] or mica can overcome this problem, so that even experiments at air pressure appear feasible.

2.4 MAGNETIC FIELD EFFECTS

2.4.1 Hall and photo-Hall effect

The transit-time methods previously described are most useful for an unambiguous evaluation of the charge transport parameters in insulators.

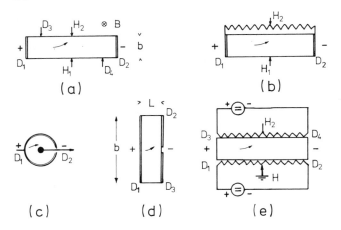

Fig. 2.9. Hall effect specimen geometries. (a) Conventional six-electrodes arrangement; (b) electronic misalignment compensation; (c) Corbino disk; (d) split electrode arrangement for Hall current measurements; (e) Redfield method. The curved arrows show the initial deflection of a positive carrier with the field directions as indicated.

However, these techniques are restricted to long-lifetime materials and cannot measure the majority carrier mobility in low resistivity samples. The latter quantity can be determined from measurements of the Hall effect. Consider a rectangular plate made of, say, p-type semiconductor material, whose geometry is shown in Fig. 2.9(a). A voltage is applied between the contacts D_1, D_2, and this establishes a drift field, E_D, in which a hole current of density j flows towards D_2. If a magnetic induction, B, is applied in a direction perpendicular to the plane of the paper, as shown, the carriers are deflected towards H_2 due to Lorentz forces. In the steady state, a Hall field E_H will be established which counteracts deflection:

$$E_H = RjB, \tag{2.41}$$

where R is the Hall coefficient. In the one-carrier case considered here, R is given by [1]

$$R = +K/(ep), \tag{2.42}$$

where p is the density of free (untrapped) holes, and K is a constant which depends mainly on the scattering mechanism in the material [181]. In the absence of a calculation, K is usually put equal to unity. At small magnetic fields, the conductivity σ of the sample will not deviate appreciably from its value at $B = 0$ where $\sigma = en\mu_0$ holds. Using eqns. (2.41) and (2.42) and $j = \sigma E_D$ we find that

$$E_H/E_D = +(R\sigma)B = \mu_0 B. \tag{2.43}$$

The product $(R\sigma)$, called the Hall mobility, denoted by μ^H in the following, can in our case be interpreted as the mobility μ_0 of the "free" majority carriers. Note that the direction of carrier drift and the applied drift field E_D make an angle θ, the so-called Hall angle, which is given by

$$\tan \theta = E_H/E_D = \mu^H B \qquad (2.44)$$

The simple expressions (2.42) and (2.44) apply only to one-carrier conduction and small Hall angles, i.e. $\theta \ll 1$; the full calculations for the general case of mixed conduction and higher fields can be found in several monographs [181]. In the typical Hall effect measurement, the sample geometry is as shown in Fig. 2.9(a). The mobility μ^H is determined from the measured Hall voltage V_H between H_1 and H_2:

$$V_H = \mu^H B E_D b, \qquad (2.45)$$

where b is the specimen width. The drift field is measured between D_1, D_2, or between two additional contacts D_3, D_4, in cases where the voltage drop at the current contacts D_1 and D_2 is too high. Typical photoconductors usually show very small Hall voltages and every effort must be made to prevent interference from spurious signals due to photovoltaic, thermoelectric, and galvanomagnetic effects [181], or improper alignment of the Hall electrodes. Since only the Hall effect and the usually negligible Ettinghausen effect are odd functions of both the current and the magnetic induction, a Hall voltage average, $\bar{V}_H$, defined by

$$4\bar{V}_H = V_H(B, E_D) + V_H(-B, -E_D) - V_H(B, -E_D) - V_H(-B, E_D) \qquad (2.46)$$

will be virtually free from interference other than noise. A variety of similar but less time-consuming procedures have been developed to strip off spurious effects: the three-contact arrangement shown in Fig. 2.9(b) provides an electronic alignment of the Hall contacts; the voltage developed between H_1 and H_2 will then be only $V_H/2$. A well-known procedure due to Van der Pauw [182] permits the Hall mobility and the conductivity to be determined from simple voltage measurements between four point contacts on a disk-shaped sample. Alternating drift fields may be used to overcome electrode polarization effects in ionic crystals [183] and also to reduce $1/f$-contact noise [184]. The Corbino-disk geometry shown in Fig. 2.9(c) has been employed in a.c. Hall effect measurements in amorphous semiconductors [185]. An a.c. field is applied between the contacts D_1 and D_2 at the circumference and the center of the disk, respectively, causing a radial current flow. If a magnetic field is applied there will also be a small circular current due to the Hall effect, which may be detected by a pick-up coil mounted coaxially with the sample. Even higher sensitivities can be achieved using the so-called double a.c. method introduced by Russell and Wahlig [186], where the sample current and the magnetic field alternate at different frequencies. The Hall signal is then detected at the sum or the difference frequency using a narrow bandwidth amplifier, which rejects the error signals mentioned earlier.

References pp. 63—69

Alternating magnetic fields can be produced by either rotating the sample in a steady magnetic field [187, 188] or by feeding a low frequency a.c. current to an electromagnet [186, 189, 190]. In the latter case, hysteresis and eddy-current losses tend to restrict the amplitude of the magnetic field to somewhat lower values. Alternating currents may be generated in two ways: by an a.c. drift field [186, 189, 190], in good photoconductors, by a d.c. field and modulated radiation [188]. The latter method suppresses the dark current and is thus not plagued by capacitive pick-up problems. While the sensitivity of single and double a.c. methods is excellent, permitting mobilities of 10^{-2} cm^2/V.s and below to be determined [190, 191], they remain restricted to the less resistive materials. In insulating solids, d.c. measurements [192, 193] or pulse measurements using electrometers are to be preferred because of the better impedance match to the sample.

Most of the techniques reviewed so far were based on measurements of the open-circuit Hall voltage. In other electrode configurations which are particularly suited to high resistivity or thin film [194] samples, one of which is shown in Fig. 2.9(d), the Hall voltage is short-circuited and one observes, instead, an asymmetry between the currents going to the electrodes D_2 and D_3. According to an analysis made by Dobrovolskii and Gritsenko [195] the Hall mobility μ^H is then calculated from the balance current, ΔI, between D_1 and D_2 and the total current, I, at D_1 according to

$$\mu^H B = \Delta I b/(IL). \qquad (2.47)$$

This split-electrode method has been applied to Se [196], SiO$_2$ [197], and to anthracene [198] in particular. However, due to the formidable experimental difficulties involved in Hall effect measurements in insulators such as anthracene, the results obtained with this material [193, 198, 199] are still rather inconsistent. Interest has been focused on anthracene because it was suggested [200] that a comparison of the experimentally determined Hall-to-drift-mobility ratio with existing theoretical predictions for band [200–201] or hopping [202] conduction might decide which of the two actually applies. A method capable of measuring the Hall effect in insulators under conditions of transient photoexcitation has been described by Redfield [203]. The experimental arrangement shown in Fig. 2.9(e) is as follows: a steady drift field is established in the sample by the voltage drop along two transparent resistance strips, D_1D_2 and D_3D_4, which are in close contact to the top and bottom faces of the sample. If a hole pulse is excited by flashing the positive end of the sample, the former will drift in a direction determined by the electric field and the Lorentz forces. However, only the lateral motion of the carriers toward one of the field plates D_3D_4 or D_1D_2 leads to the appearance of a displacement-current signal between the plates. As in steady state, the deflection and, hence, the signal will be zero if the potential difference between corresponding points on the two plates equals the Hall voltage. In the experiment, this is achieved by adjusting the ground point H_1 until the Hall signal at H_2 vanishes. In its modern form due to Smith [20] the method can measure mobilities down to about $1 cm^2/V.s$ under virtually space-charge-free conditions. The results obtained by the original method and the later modifications in diamond [204], alkali halides [205], and sulphur [204], show satisfactory agreement with other measurements. A unique though not yet fully exploited feature of the Redfield method is the possibility to separately measure the Hall mobility of either carrier polarity in the same sample.

2.4.2 Photoelectromagnetic effect

The photoelectromagnetic (PEM) effect [206, 207] can be described as the Hall effect associated with an optically generated diffusion current. The specimen geometry employed for the observation of the linear PEM effect is shown in Fig. 2.10. If a carrier density gradient is created by strongly absorbed illumination, a diffusion current is set up in the y-direction. An applied magnetic field deflects the diffusing electrons and holes in opposite directions, making the upper and lower sample faces positive and negative, respectively. A general theory of the PEM effect in weak magnetic fields has been given by van Roosbroeck [208]. Its predictions have been verified [209] in many semiconductors. Most of the early theoretical and experimental work on the effect was reviewed by Bube [1], Tauc [210], and others [7, 207, 211]. The magnitude of the short-circuit current observed in a sufficiently thick sample at weak magnetic fields can be derived using the following simplified arguments [211]. Suppose $\dot{N}$ carrier pairs per second escape recombination at the illuminated surface and diffuse into the bulk. The mean distance covered by the pairs before recombination occurs is given by L_a, the ambipolar diffusion length. While diffusing, electrons and holes are deflected parallel to the z-axis by amounts $\theta_n L_a$ and $\theta_p L_a$, respectively, where $\theta_n = \mu_n^H B$ and $\theta_p = \mu_p^H B$ are the Hall angles. If the distance between the electrodes is w, the total short-circuit current will be

$$I_{PEM} = e\dot{N}(\theta_n + \theta_p)L_a/w \tag{2.48}$$

The PEM effect can thus be used to measure the diffusion length of a carrier pair in a relatively simple experiment. Only the Hall mobilities of the carriers and their release rate, $\dot{N}$, must be known. The latter can be determined from a measurement of the photoconductivity [1, 207, 210, 212] produced at zero magnetic field or, if surface recombination is negligible, from the incident photon flux [213]. In contrast to the Hall effect, the steady-state PEM effect requires mobile carriers of both polarities, because otherwise the carrier-pair diffusion length L_a vanishes. This feature of the PEM effect which is also shared by the photovoltaic effect has been exploited to measure the onset of interband absorption without interference from exciton and impurity absorption. It was thus possible to measure the undisturbed

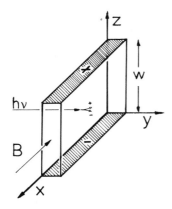

Fig. 2.10. Linear photoelectromagnetic effect (PEM).

Landau-level structure of Ge at high magnetic fields [214]. A PEM effect due to only one mobile carrier can only be observed under transient conditions. It is particularly strong in materials such as insulators where the dielectric relaxation time greatly exceeds the carrier lifetime [7]. In a low resistivity extrinsic semiconductor, L_a reduces to the diffusion length of the minority carriers, which depends on the lifetime, τ_m, and diffusivity, D_m, according to

$$L_m = (D_m \tau_m)^{1/2}. \tag{2.49}$$

If τ_m also governs the photoconductivity (no trapping) and if the difference between Hall and drift mobilities can be neglected, a particularly simple expression holds for the ratio of the PEM and photocurrents measured at zero electric field, E, and magnetic field, B, respectively [215],

$$I_{PEM}/I_{PC} = (B/E)(D_m/\tau_m)^{1/2}. \tag{2.50}$$

The determination of τ_m is less clear-cut, however, if trapping is involved [216] or in insulating materials [213, 217] where the full ambipolar expressions have to be used. So far we have only considered the PEM short-circuit current in a sufficiently thick sample under highly absorbed radiation. If one of these conditions is relaxed, e.g. in thin samples [208—9] or with penetrating illumination [218], the PEM effect can be employed to measure the surface recombination velocity. Under open-circuit conditions the PEM effect produces a circular current pattern in the sample. Using modulated excitation, the circulating currents can be detected by means of a pick-up coil, thus permitting contactless measurement of the effect [219].

ACKNOWLEDGEMENT

Financial support by Battelle-Institute funds is gratefully acknowledged.

REFERENCES

1. R.H. Bube, Photoconductivity of Solids, Wiley, New York, 1960.
2. Y. Pokrovskii, Phys. Status Solidi A, 11 (1972) 385;
   C. Benoît à la Guillaume, M. Voos and F. Salvan, Phys. Rev. B, 5 (1972) 3079;
   O. Christensen and J.C. McGroddy, Solid State Commun., 15 (1974) 811.
3. S. Rafi Ahmad and D. Walsh, J. Phys. D, 4 (1971) 1820;
   A. Bergmann and J. Jortner, Phys. Rev. B, 9 (1974) 4560;
   S. Jayaraman and C.H. Lee, J. Appl. Phys., 44 (1973) 5480;
   K. Yoshino, Y. Watanabe and Y. Inuishi, Jpn. J. Appl. Phys., 4 (1965) 312.
4. A.M. Danishevskii, A.A. Kastalskii, S.M. Ryvkin and I.D. Yaroshetskii, Sov. Phys. JETP, 31 (1970) 292;
   A.F. Gibson, M.F. Kimmit and A.C. Walker, Appl. Phys. Lett., 17 (1970) 75;
   A.F. Gibson and A.C. Walker, J. Phys. C, 4 (1971) 2209;
   J. Auth, D. Genzow, K.H. Herrmann and M. Wendt, Phys. Status Solidi B, 65 (1974) 293.
5. T.M. Lifschits and F. Nadya, Sov. Phys. Dokl., 10 (1965) 532;
   C.T. Sah, T.H. Ning, L.L. Rosier and L. Forbes, Solid State Commun., 9 (1971) 917;
   S.D. Seccombe and D.M. Korn, Solid State Commun., 11 (1972) 1539;
   E.M. Bykova, T.M. Lifschits and V.I. Sidorov, Sov. Phys. Semicond., 7 (1973) 671.
6. K. Lark-Horovitz and V.A. Johnson (Eds.), Methods of Experimental Physics, Vol. 6, Part B, Academic Press, New York, 1959.
7. S.M. Ryvkin, Photoelectric Effects in Semiconductors, Consultants Bureau, New York, 1964.
8. R. Williams, in R.K. Willardson and A.C. Beer (Eds.), Semiconductors and Semimetals, Vol. 6, Academic Press, New York, 1970, pp. 97—138.
9. J. Mort, in P.G. LeComber and J. Mort (Eds.), Electronic and Structural Properties of Amorphous Semiconductors, Academic Press, London, 1973, pp. 493—526.
10. J.M. Caywood, Mol. Cryst. Liq. Cryst., 12 (1970) 1—26.
11. A.M. Goodman, in E.M. Pell (Ed.), Proc. 3rd Int. Conf. on Photoconductors, Pergamon Press, Oxford, 1971, p. 69.
12. A.V. Yukhnevich, Sov. Phys. Semicond., 5 (1971) 326;
    T. Nishino and Y. Hamakawa, Phys. Status Solidi B, 50 (1972) 345;
    See also: M. Cardona, in Solid State Physics, Suppl. 11, Academic Press, New York, 1969, p. 101.
13. D. Geist and J. Meyer, in Proc. 10th Int. Conf. Phys. Semicond., Cambridge, Mass., 17—21 Aug. 1970, US Atomic Energy Commission, Wash., D.C., 1970, p. 597.
14. R.W. Smith, Phys. Rev. Lett., 9 (1962) 87;
    A.R. Hutson, ibid., p. 296;
    A.R. Moore and R.W. Smith, Phys. Rev. A, 138 (1965) 1250;
    J. Mort, Phys. Rev. Lett., 18 (1967) 540;
    A. Bers, J.H. Cafarella and B.E. Burke, Appl. Phys. Lett., 22 (1973) 399.
15. N.F. Mott and E.A. Davis, Electronic Processes in Non-Crystalline Materials, Clarendon Press, Oxford, 1971.
16. F. Stöckmann, Phys. Status Solidi, 2 (1962) 517.
17. R. Bray and A. Many, in K. Lark-Horovitz and V.A. Johnson (Eds.), Methods of Experimental Physics, Vol. 6, Part B, Academic Press, New York, 1959, pp. 78—109.
18. K.W. Böer and E.A. Niekisch, Phys. Status Solidi, 1 (1961) 191, 275.
19. J.A. Borders and J.W. Hodby, Rev. Sci. Instrum., 39 (1968) 722;
    J.W. Hodby, J.A. Borders and F.C. Brown, J. Phys. C, 3 (1970) 335;
    J.W. Hodby, J. Phys. E, 3 (1970) 229.
20. G.C. Smith, Rev. Sci. Instrum., 40 (1969) 1454.

21 J.W. Hodby, J.A. Borders, F.C. Brown and S. Foner, Phys. Rev. Lett., 19 (1967) 952.
22 R.H. Bube, Photoconductivity of Solids, Wiley, New York, 1960, p. 281.
23 H.B. DeVore, RCA Rev., 20 (1959) 79.
24 A.F. Gibson, Physica, 20 (1954) 1058.
25 G. Bemski, Proc. IRE, 46 (1958) 990.
26 U. Langmann, Solid State Electron., 16 (1973) 1011.
27 C.J. Hwang, J. Appl. Phys., 40 (1969) 3731.
28 M. Ettenberg, H. Kressel and S.L. Gilbert, J. Appl. Phys., 44 (1973) 827.
29 I.W. Pence and P.T. Greiling, Proc. IEEE, 62 (1974) 1030.
30 J. Fassbender and H. Lehmann, Ann. Phys., 6 (1949) 215.
31 B.H. Schultz, Philips Res. Rep., 10 (1955) 337.
32 L.J. Van der Pauw, Philips Res. Rep., 12 (1957) 364.
33 W. Stössel and W. Zimmermann, Phys. Status Solidi, 30 (1968) 311.
34 R.W. Tokarsky and D.E. Brodie, Can. J. Phys., 51 (1973) 1944.
35 G. Cheroff, J. Heer and S. Triebwasser, J. Phys. Chem. Solids, 22 (1961) 51.
36 R.W. Tokarsky and D.E. Brodie, Can. J. Phys., 50 (1972) 1685.
37 A.G.R. Evans and P.N. Robson, Solid-State Electron., 17 (1974) 805.
38 F. Lappe, Z. Phys., 154 (1959) 267.
39 E. Harnik, A. Many and N.B. Grover, Rev. Sci. Instrum., 29 (1958) 889.
40 J. Boucher, C.R. Acad. Sci. B, 275 (1972) 259.
41 J.R. Haynes and W. Shockley, Phys. Rev., 82 (1951) 935;
    U. Heukeroth and P. Süptitz, Phys. Status Solidi, 13 (1966) 285.
42 W. Shockley, Electrons and Holes in Semiconductors, Van Nostrand, New York, 1950, pp. 328—333.
43 W. van Roosbroeck, Phys. Rev., 91 (1953) 282; ibid., 123 (1961) 474.
44 J.L. Scales and A.L. Ward, J. Appl. Phys., 39 (1968) 1692.
45 J.R. Haynes and W. Shockley, Phys. Rev., 81 (1951) 835;
    J.R. Haynes and W.C. Westphal, Phys. Rev., 85 (1952) 680.
46 A. Neukermans and G.S. Kino, Appl. Phys. Lett., 17 (1970) 102; Solid State Commun., 8 (1970) 987;
    J.L. Su, Y. Nishi, J.L. Moll. and A. Neukermans, Solid-State Electron., 13 (1970) 1115.
47 A. Many, Proc. Phys. Soc., 67B (1954) 9.
48 Transistor Teachers' Summer School, Phys. Rev., 88 (1952) 1368.
49 R. Lawrance and A.F. Gibson, Proc. Phys. Soc., 65B (1952) 994.
50 J.A. Hornbeck and J.R. Haynes, Phys. Rev., 97 (1955) 311;
    R.L. Watters and G.W. Ludwig, J. Appl. Phys., 27 (1956) 489.
51 G.W. Ludwig and R.L. Watters, Phys. Rev., 101 (1956) 1699.
52 D.G. Avery and J.B. Gunn, Proc. Phys. Soc., B68 (1955) 918;
    M. Daurie and J. Boucher, C.R. Acad. Sci. B, 277 (1973) 727.
53 R. Bray, Phys. Rev., 100 (1955) 1047.
54 J.B. Arthur, W. Bardsley, M.A.C.S. Brown and A.F. Gibson, Proc. Phys. Soc., 68B (1955) 43.
55 B. Gudden and R.W. Pohl, Z. Phys., 6 (1921) 248; ibid., 16 (1923) 170.
56 K.H. Hecht, Z. Phys., 77 (1932) 235.
57 R.J. Van de Graaff, Philos. Mag., 6 (1928) 210;
    A.M. Tyndall, L.H. Starr and C.F. Powell, Proc. Roy. Soc. (London), 121 (1928) 172;
    A. Hornbeck, Phys. Rev., 83 (1951) 374.
58 W.E. Spear, J. Non-Cryst. Solids, 1 (1969) 197.
59 M. Martini, J.W. Mayer and K.R. Zanio, in R. Wolfe (Ed.), Applied Solid State Science, Vol. 3, Academic Press, New York, 1972, pp. 181—261.

60   F.C. Brown, Phys. Rev., 97 (1955) 355;
     R. Van Heyningen, Phys. Rev., 128 (1962) 2112.
61   W.E. Spear, Proc. Phys. Soc (London), 70B (1957) 669.
62   N.F. Mott and E.A. Davis, Electronic Processes in Non-Crystalline Materials, Clarendon Press, Oxford, 1971, Ch. 7.
63   H. Seki, in J. Stuke and W. Brenig (Eds.), Amorphous and Liquid Semiconductors, Taylor and Francis, London, 1974, p. 1015.
64   O.H. LeBlanc, Jr., J. Chem. Phys., 30 (1959) 1443.
65   L.S. Miller, S. Howe and W.E. Spear, Phys. Rev., 166 (1968) 871.
66   K.W. Schwarz, Phys. Rev. A, 6 (1972) 837.
67   R.M. Minday, L.D. Schmidt and H.T. Davis, J. Chem. Phys., 54 (1971) 3112.
68   R.G. Kepler, Phys. Rev., 119 (1960) 1226;
     O.H. LeBlanc, Jr., J. Chem. Phys., 33 (1960) 626.
69   O.H. LeBlanc, Jr., in D. Fox, M.M. Labes and A. Weissberger (Eds.), Physics and Chemistry of the Organic Solid State, Vol. III, Interscience, New York, 1967, pp. 133—191.
70   R.G. Kepler, in J.J. Brophy and J.W. Buttrey (Eds.), Organic Semiconductors, McMillan, New York, 1962, p. 1;
     T. Kajiwara, H. Inokuchi and S. Minomura, Bull. Chem. Soc. Jpn., 40 (1967) 1055;
     F.K. Dolezalek and W.E. Spear, J. Non.-Cryst. Solids, 4 (1970) 97.
71   J.G. Ruch and G.S. Kino, Phys. Rev., 174 (1968) 921.
72   J. Mort and A.I. Lakatos, J. Non.-Cryst. Solids, 4 (1970) 117.
73   E.H. Martin and J. Hirsch, J. Appl. Phys., 43 (1972) 1001.
74   J. Mort and H. Scher, J. Appl. Phys., 42 (1971) 3939.
75   A. Many, S.Z. Weisz and M. Simhony, Phys. Rev., 126 (1962) 1989.
76   W. Helfrich, in D. Fox, M.M. Labes and A. Weissberger (Eds.), Physics and Chemistry of the Organic Solid State, Vol. III, Interscience, New York, 1967, p. 1.
77   H. Lemke and G.O. Müller, Phys. Status Solidi, 24 (1967) 127.
78   M.A. Lampert and P. Mark, Current Injection in Solids, Academic Press, New York, 1970.
79   J. Hirsch, Phys. Status Solidi A, 25 (1974) 575.
80   G. Bertolini, in G. Bertolini and A. Coche (Eds.), Semiconductor Detectors, North-Holland, Amsterdam, 1968, Section 3.2;
     A.M. Hermann, J. Appl. Phys., 44 (1973) 926.
81   T.W. Sigmon, J.F. Gibbons and C.B. Norris, Jr., Appl. Phys. Lett., 14 (1969) 90.
82   M.E. Scharfe and M.D. Tabak, J. Appl. Phys., 40 (1969) 3230.
83   N.F. Mott and R.W. Gurney, Electronic Processes in Ionic Crystals, Clarendon Press, Oxford, 1948.
84   A. Many and G. Rakavy, Phys. Rev., 126 (1962) 1980.
85   P. Mark and W. Helfrich, J. Appl. Phys., 33 (1962) 205.
86   I.P. Batra, K.K. Kanazawa and H. Seki, J. Appl. Phys., 41 (1970) 3416.
87   I.P. Batra, B.H. Schechtmann and H. Seki, Phys. Rev. B, 2 (1970) 1592.
88   I. Chen, J. Appl. Phys., 43 (1972) 1137.
89   L.M. Schwartz and J.F. Hornig, J. Phys. Chem. Solids, 26 (1965) 1821.
90   A.C. Papadakis, J. Phys. Chem. Solids, 28 (1967) 641.
91   I.P. Batra and H. Seki, J. Appl. Phys., 41 (1970) 3409.
92   W.D. Gill and K.K. Kanazawa, J. Appl. Phys., 43 (1972) 529.
93   A. Taroni and G. Zanarini, J. Phys. Chem. Solids, 30 (1969) 1861;
     W. Seibt, Solid-State Electron., 16 (1973) 1017.
94   I.P. Batra, K.K. Kanazawa, B.H. Schechtmann and H. Seki, J. Appl. Phys., 42 (1971) 1124.
95   H. Seki and I.P. Batra, J. Appl. Phys., 42 (1971) 2407.
96   K.K. Kanazawa, I.P. Batra and H.J. Wintle, J. Appl. Phys., 43 (1972) 719.

97 S.R. Hofstein, Appl. Phys. Lett., 10 (1967) 291.
98 W. Helfrich and W.G. Schneider, J. Chem. Phys., 44 (1966) 2902.
99 M.E. Michel-Beyerle, W. Harengel and J. Kinder, Phys. Status Solidi A, 20 (1973) 563.
100 C. Canali, G. Ottaviani, A. Taroni and G. Zanarini, Solid-State Electron., 14 (1971) 661.
101 D.J. Gibbons and A.C. Papadakis, J. Phys. Chem. Solids, 29 (1968) 115.
102 J. Mort, I. Chen, R.L. Emerald and J.H. Sharp, J. Appl. Phys., 43 (1972) 2285.
103 A. Rose, Concepts in Photoconductivity and Allied Problems, Interscience, New York, 1963.
104 A. Rose, RCA Rev., 12 (1951) 362.
105 W.E. Spear and J. Mort, Proc. Phys. Soc., 81 (1963) 130.
106 W.E. Tefft, J. Appl. Phys., 38 (1967) 5265.
107 K.R. Zanio, W.M. Akutagawa and R. Kikuchi, J. Appl. Phys., 39 (1968) 2818.
108 J.W. Mayer, in G. Bertolini and A. Coche (Eds.), Semiconductor Detectors, North-Holland, Amsterdam, 1968, Chap. 5.
109 R.M. Blakney and H.P. Grunwald, Phys. Rev., 159 (1967) 658, 664.
110 C. Canali, M. Martini, G. Ottaviani, A. Alberigi Quaranta and K.R. Zanio, Nucl. Instrum. Methods, 96 (1971) 561.
111 W.E. Spear, J. Phys. Chem. Solids, 21 (1961) 110.
112 A.R. Moore and R.W. Smith, Phys. Rev. A, 138 (1965) 1250.
113 A.R. Adams and W.E. Spear, J. Phys. Chem. Solids, 25 (1964) 1113.
114 D.C. Hoesterey and G.M. Letson, J. Phys. Chem. Solids, 24 (1963) 1609.
115 H. Hirth and U. Tödheide-Haupt, Phys. Status Solidi, 31 (1969) 425.
116 K. Möstl, Phys. Status Solidi A, 21 (1974) 123.
117 R. van Heyningen, Phys. Rev., 128 (1962) 2112.
118 H. Hirth and F. Stöckmann, Phys. Status Solidi B, 51 (1972) 691.
119 M. Martini and T.A. McMath, Nucl. Instrum. Methods, 79 (1970) 259.
120 M. Martini and T.A. McMath, Appl. Phys. Lett., 17 (1970) 362.
121 W.G. Spitzer, T.E. Firle, M. Cutler, R.G. Shulman and M. Becker, J. Appl. Phys., 26 (1955) 414.
122 H. Lemke, Phys. Status Solidi, 16 (1966) 413, 427; Phys. Status Solidi A, 1 (1970) 283.
123 H. Lemke and G.O. Müller, Phys. Status Solidi A, 1 (1970) 287.
124 D.J. Gibbons and W.E. Spear, J. Phys. Chem. Solids, 27 (1966) 1917.
125 M.D. Tabak and M.E. Scharfe, J. Appl. Phys., 41 (1970) 2114.
126 L. Onsager, Phys. Rev., 54 (1938) 554.
127 R.H. Batt, C.L. Braun and J.F. Hornig, J. Chem. Phys., 49 (1968) 1967; Appl. Opt. Suppl., 3 (1969) 20.
128 R.R. Chance and C.L. Braun, J. Chem. Phys., 59 (1973) 2269.
129 R.C. Enck, Phys. Rev. Lett., 31 (1973) 220;
D.M. Pai and R.C. Enck, Phys. Rev. B, 11 (1975) 5163.
130 F.K. Dolezalek and W.E. Spear, J. Phys. Chem. Solids, 36 (1975) 819.
131 P.J. Melz, J. Chem. Phys., 57 (1972) 694.
132 D. Blossey and R. Zallen, Phys. Rev. B, 9 (1974) 4306.
133 W.E. Spear and A.R. Adams, J. Phys. Chem. Solids, 27 (1966) 281.
134 D.M. Pai and S. Ing, Phys. Rev., 173 (1968) 729.
135 J.M. Caywood and C.A. Mead, J. Phys. Chem. Solids, 31 (1970) 983.
136 M. Silver and R. Sharma, J. Chem. Phys., 46 (1967) 692.
137 H. Killesreiter and R. Braun, Phys. Status Solidi B, 48 (1971) 201.
138 R.R. Chance and A. Prock, Phys. Status Solidi B, 57 (1973) 597.
139 P. Langevin, Ann. Chim. Phys., 28 (1903) 289, 433; see also M. Lax, Phys. Rev., 119 (1960) 1502.
140 J.W. Boag, in G.J. Hine and G.L. Brownell (Eds.), Ionization Chambers in Radiation Dosimetry, Academic Press, New York, 1956, Chap. 4.

141 R.G. Kepler and F.N. Coppage, Phys. Rev., 151 (1966) 610.
142 N. Karl and G. Sommer, Phys. Status Solidi A, 6 (1971) 231.
143 H. Seki, J. Appl. Phys., 43 (1972) 1144.
144 C. Bogus, Z. Phys., 207 (1967) 281.
145 G. Delacote, C.R. Acad. Sci. B, 262 (1966) 958.
146 R.C. Hughes, Appl. Phys. Lett., 21 (1972) 196.
147 C. Bogus, Z. Naturforsch. Teil A, 21 (1966) 667.
148 R. Morris and M. Silver, J. Chem. Phys., 50 (1969) 2969.
149 W.F. Schmidt, Z. Naturforsch. Teil B, 23 (1968) 126.
150 R.B. Schilling and H. Schachter, Solid-State Electron., 10 (1967) 689; see also M. Silver, Solid State Commun., 15 (1974) 1785.
151 J.M. Caywood, C.A. Mead and J.W. Mayer, Nucl. Instrum. Methods, 79 (1970) 329.
152 T.W. Sigmon and J.F. Gibbons, Appl. Phys. Lett., 15 (1969) 320.
153 N. Karl, E. Schmid and M. Seeger, Z. Naturforsch. Teil A, 25 (1970) 382.
154 P.K. Ghosh and W.E. Spear, J. Phys. C, 1 (1968) 1347.
155 R.M. Minday, L.D. Schmidt and H.T. Davis, J. Chem. Phys., 50 (1969) 1473.
156 R.J. Loveland, P.G. LeComber and W.E. Spear, Phys. Rev. B, 6 (1972) 3121.
157 H.D. Pruett and H.P. Broida, Phys. Rev., 164 (1967) 1138.
158 D.E. Hudson, US AEC—Report, MDDC—524, 1946.
159 W.F. Schmidt and A.O. Allen, J. Chem. Phys., 50 (1969) 5037 and 52 (1970) 4788.
160 G. Kleinheins, Phys. Lett. A, 28 (1969) 498;
R. Coelho, in N. Klein, D.S. Tannhauser and M. Pollak (Eds.), Conduction in Low Mobility Materials, Taylor and Francis, London, 1971, p. 439.
161 L. Meyer and F. Reif, Phys. Rev., 110 (1958) 279;
F. Reif and L. Meyer, Phys. Rev., 119 (1960) 1164.
162 P.H. Tewari and G.R. Freemann, J. Chem. Phys., 49 (1968) 4394.
163 S.N. Ing, Jr. and Y.S. Chiang, J. Chem. Phys., 46 (1967) 478, 487;
H. Meier, Die Photochemie der organischen Farbstoffe, Springer, Berlin, 1963, Chap. 12; Y. Maruyama and K. Funabashi, J. Chem. Phys., 56 (1972) 2342.
164 H. Fischer, J. Opt. Soc. Am., 51 (1961) 543;
W.D. Gill, R.E. MacDonald and D.B. Greene, Rev. Sci. Instrum., 39 (1968) 1114.
165 J. Yguerabide, Rev. Sci. Instrum., 36 (1965) 1734;
H.G. Franke and H.S. Schmeing, Nucl. Instrum. Methods, 52 (1967) 171.
166 J.T. D'Alessio, P.K. Ludwig and M. Burton, Rev. Sci. Instrum., 35 (1964) 1015.
167 Nanolite, Impulsphysik, D—2 Hamburg-Rissen, West Germany.
168 Nanosecond lamp, Applied Photophysics Ltd, 20 Albemarle Street, London W1X 3HA.
169 R.C. Mackey, S.A. Pollack and R.S. Witte, Rev. Sci. Instrum., 36 (1965) 1715 (available from TRW Instrum., El Segundo, Calif.);
I.B. Berlman, O.J. Steingraber and M.J. Benson, Rev. Sci. Instrum., 39 (1968) 54 (available from Yissum R+D Co., Hebrew University, Jerusalem, Israel).
170 D. Von der Linde, Appl. Phys., 2 (1973) 281;
P.A. Tove, G. Anderson, G. Ericson and R. Lidholt, IEEE Trans. Electron Devices, ED17 (1970) 407.
171 R.C. Hughes, Phys. Rev. Lett., 30 (1973) 1333; J. Chem. Phys., 55 (1971) 5442.
172 P.S. Winokur, J.M. McGarrity, M.L. Roush and J. Silverman, 1973 Ann. Rep. Natl. Acad. Sci. on Conf. on Electrical Insulation and Dielectric Phenomena, Wash. DC, 1974, pp. 120—129.
173 P.P. Webb, H.L. Malm, M.G. Chartrand, R.M. Green, E. Sakai and I.L. Fowler, Nucl. Instrum. Methods, 63 (1968) 125.
174 G. Schubert and E. Schnürer, Phys. Status Solidi, 32 (1969) 679.
175 A. Alberigi Quaranta, M. Martini, G. Ottaviani, G. Redaelli and G. Zanarini, Solid-State Electron., 11 (1968) 685.

176 J.G. Ruch and G.S. Kino, Appl. Phys. Lett., 10 (1967) 40.
177 C.B. Norris and J.F. Gibbons, IEEE Trans. Electron Devices, ED14 (1967) 38.
178 A. Alberigi Quaranta, C. Canali and G. Ottaviani, Rev. Sci. Instrum., 41 (1970) 1205.
179 B.M. Kramer and R. Stille, Rev. Sci. Instrum., 41 (1970) 230.
180 H.R. Zullinger, C.B. Norris, T.W. Sigmon and R.H. Pehl, Nucl. Instrum. Methods, 70 (1969) 125.
181 A.C. Beer, in F. Seitz and D. Turnbull (Eds.), Galvanomagnetic Effects in Semiconductors, Solid State Physics, Suppl. 4, Academic Press, New York, 1963;
E.H. Putley. The Hall Effect and Related Phenomena, Butterworths, London, 1960.
182 L.J. Van der Pauw, Philips. Res. Rep., 13 (1958) 1;
A.H.M. Kipperman and G.A. Van der Leeden, Solid State Commun., 6 (1968) 657;
P.M. Hemenger, Rev. Sci. Instrum., 44 (1973) 698.
183 J.R. MacDonald and J.E. Robinson, Phys. Rev., 95 (1954) 44.
184 E.M. Pell and R.L. Sproull, Rev. Sci. Instrum., 23 (1952) 548;
J.M. Lavine, *ibid.*, 29 (1958) 970.
185 G.P. Carver, Rev. Sci. Instrum., 43 (1972) 1257;
G.P. Carver and R.S. Allgaier, J. Non-Cryst. Solids, 8—10 (1972) 347.
186 B.R. Russell and C. Wahlig, Rev. Sci. Instrum., 21 (1950) 1028.
187 F.M. Ryan, Rev. Sci. Instrum., 33 (1962) 76;
A.M. Hermann and J.S. Ham, *ibid.*, 36 (1965) 1553;
A.A. Andreev, M. Manadaliev and A.I. Shelykh, Instrum. Exp. Tech. (USSR), 15 (1972) 568.
188 I. Eisele and L. Kevan, Rev. Sci. Instrum., 43 (1972) 189.
189 J.L. Levy, Phys. Rev., 92 (1953) 215.
190 T. Kaneda, S. Kobayashi and K. Shimoda, Jpn. J. Appl. Phys., 12 (1973) 1335.
191 N.Z. Lupu, N.M. Tallan and D.S. Tannhauser, Rev. Sci. Instrum., 38 (1967) 1658.
192 H.E. MacDonald and R.H. Bube, Rev. Sci. Instrum., 33 (1962) 721;
G.H. Heilmeier and S.E. Harrison, Phys. Rev., 132 (1963) 2010;
J. Dresner, J. Phys. Chem. Solids, 25 (1964) 505.
193 A.I. Korn, R.A. Arndt and A.C. Damask, Phys. Rev., 186 (1969) 938.
194 H.F. van Heck, Solid-State Electron., 10 (1967) 268;
D. Bednarczyk and J. Bednarczyk, Acta Phys. Pol. A, 39 (1971) 295.
195 V.N. Dobrovolskii and Yu.I. Gritsenko, Sov. Phys. Solid State, 4 (1963) 2025.
196 H. Gobrecht, A. Tausend and G. Clauss, Z. Phys., 176 (1963) 155.
197 A.M. Goodman, Phys. Rev., 164 (1967) 1145.
198 G. Delacote and M. Schott, Solid'State Commun., 4 (1966) 177;
R. Pethig and K. Morgan, Nature, 214 (1967) 266;
M. Schadt and D.F. Williams, Phys. Status Solidi, 39 (1970) 223.
199 G.C. Smith, Phys. Rev., 185 (1969) 1133.
200 O.H. LeBlanc, Jr., J. Chem. Phys., 39 (1963) 2395.
201 L. Friedman, Phys. Rev. A, 133 (1964) 1668.
202 L. Friedman and T. Holstein, Ann. Phys. (USA), 21 (1963) 494;
L. Friedman, J. Non-Cryst. Solids, 6 (1971) 329.
203 A.G. Redfield, Phys. Rev., 91 (1953) 753; *ibid.*, 94 (1954) 526.
204 G.C. Smith, Phys. Rev., 185 (1969) 1133.
205 F.C. Brown, Phys. Rev., 92 (1953) 502;
A.G. Redfield, Phys. Rev., 94 (1954) 537;
R.K. Ahrenkiel and F.C. Brown, Phys. Rev. A, 136 (1964) 223;
C.H. Seager and D. Emin, Phys. Rev. B, 2 (1970) 3421.
206 I.K. Kikoin and M.M. Noskov, Phys. Z. Sowjetunion, 5 (1934) 586; *ibid.*, 6 (1934) 478.

207 O. Garreta and J. Grosvalet, in A.F. Gibson, R.E. Burgess and P. Aigrain (Eds.), Progress in Semiconductors, Vol. 1, Heywood, London, 1956, p. 165.
208 W. Van Roosbroeck, Phys. Rev., 101 (1956) 1713.
209 T.M. Buck and F.S. McKim, Phys. Rev., 106 (1957) 904.
210 J. Tauc, Photo and Thermoelectric Effects in Semiconductors, Pergamon, Oxford, 1962, pp. 193—212.
211 H.S. Sommers, Jr., in K. Lark-Horovitz and V.A. Johnson (Eds.), Methods of Experimental Physics, Vol. 6, Academic Press, New York, 1959, Part. B, Section 12.2.
212 F. Aducci, C. Cingolani, M. Ferrara, A. Minafra and P. Tantalo, J. Appl. Phys., 45 (1974) 5000.
213 H.S. Sommers, Jr., R.E. Berry and I. Sochard, Phys. Rev., 101 (1956) 987.
214 A. Barbarie and E. Fortin, Can. J. Phys., 50 (1972) 1593.
215 T.S. Moss, Proc. Phys. Soc. (London) B, 66 (1953) 993.
216 R.N. Zitter, Phys. Rev., 112 (1958) 852;
A. Amith, Phys. Rev., 116 (1959) 793;
W. Van Roosbroeck, Phys. Rev., 119 (1960) 636;
S.S. Li and C.I. Huang, J. Appl. Phys., 43 (1972) 1757.
217 J. Auth, Z. Phys. Chem. (Leipzig), 217 (1961) 159;
G. Diemer and W. Hoogenstraaten, J. Phys. Chem. Solids, 2 (1957) 119.
218 D.L. Lile, Phys. Rev. B, 8 (1973) 4708.
219 J. Hlavka and I. Sirucek, Rev. Sci. Instrum., 44 (1973) 1410.

CHAPTER 3

THEORY OF TIME-DEPENDENT PHOTOCONDUCTIVITY IN DISORDERED SYSTEMS

HARVEY SCHER

3.1. Introduction
3.2. Electrical transport via localized states as a random walk
3.3. The continuous-time random walk (CTRW)
3.4. Hopping time distribution function
3.5. AC conductivity and time-dependent pair luminescence
3.6. Transient photoconductivity
    3.6.1. The current I(t) for hopping carriers
    3.6.2. Absorbing boundary in CTRW
    3.6.3. Qualitative discussion of dispersed transport and question of mobility
3.7. Comparison with experiment on hopping systems
3.8. Onsager theory and relation to hopping transport
3.9. Langevin recombination
3.10. Conclusion and Summary

3.1 INTRODUCTION

The review of photoconductivity theory in this chapter will be largely restricted to recent developments in the area of highly insulating molecular solids. The emphasis will be on disordered molecular solids that are either organic or inorganic. The restriction to developments in this area of materials is largely in consonance with a major thrust of the present book. In addition, in the opinion of this author, research in disordered materials, including liquids, represents one of the most exciting topics of photoconductivity studies in recent years. In particular, transient photoconductivity has become one of the few reliable probes of the electrical properties of these insulating materials.

The phenomenon of photoconductivity consists of a few generic processes (1) generation, by external excitation, of mobile carriers, (2) transport of those mobile carriers, (3) recombination. Most descriptions of these processes in solids have rested largely on the "scaffold" of electronic energy bands. The bands refer to disjoint ranges of electron energy where the density of states is finite and quasi-continuous. The bands are separated by "forbidden gaps" which can contain allowable energy levels introduced by impurities or crystal defects.

References pp. 114—115

In the band framework, therefore, in simplest terms, generation consists of exciting electrons across the forbidden gap, *i.e.*, promoting charge from a nonconducting state (valence band) to a conducting state (conduction band). Both the electron in the conduction band and the hole in the valence band are then, free to contribute to the electrical conductivity. The residual coulomb attraction (exciton effects) between the electron and hole is usually a small perturbation on the carrier motion in the familiar covalently bonded semiconductors (*e.g.* Ge, Si). In molecular solids, in the optical excitation range, exciton effects are dominant. We shall return to this point.

Transport occurs strictly in the extended band states. The free-like motion of the carriers are interrupted sporadically by scattering, among the band states, due to phonons or impurities. Direct recombination of electrons and holes (at moderate free carrier densities) is an improbable process. The recombination usually occurs through the intermediary of the impurity or defect levels in the gap. In general, these levels correspond to localized electronic states and act as "traps" for the electrons or holes in the extended band states. When a carrier makes a transition and is trapped at these localized levels, it no longer contributes to the conductivity. However, the carrier can be released and return to the band. If the probability, for recombination, at the level, with a carrier of the opposite sign, is greater than the probability of release to the band, the level is called a recombination center. Conversely, if the probability for release is greater than recombination, the level is simply called a trap. The distinction is mainly based on the proximity of the level to the band edge.

The theory of photoconductivity in solids can be grouped into two areas. One area is phenomenological — it deals with the kinetics of generation of free carriers, the capture and release of free carriers from trapping levels and the ultimate recombination of these carriers. A small sampling of the parameters that are inputs into the kinetic equations are g, the generation rate, τ_n, the trapping time, τ_r, the release time, τ_0 the microscopic lifetime, and the number and types of trapping levels, recombination centers, etc. In the computation of the conductivity, one introduces the various transport coefficients, for each sign of carrier, μ_0, the microscopic mobility, D, the diffusion constant. This phenomenological area is fully and completely described in a number of textbooks such as Bube [1], Rose [2], Jonscher [3], Ryvkin [4] and in a series of review articles [5]. The book by Ryvkin carries the reader through the complications of the phenomenology of photoconductivity with a series of "case studies" while the book by Bube has more emphasis on experimental photoconductivity.

The second area of photoconductivity theory is concerned with first-principles calculations of transitions, using the appropriate quantum mechanical wave functions. The transitions can be grouped according to the nature of the initial and final state wave function: (1) Band-to-band transitions with the absorption of a photon [6], the elementary process of light absorption in a solid. (2) The intra-band transitions [7] due to interactions with phonons and impurities. These transitions determine the microscopic mobility. (3) The band-to-localized-level transition describing the capture process. (4) The inverse transition of localized level to band that determines the release process. (5) The localized-level-to-localized-level transition. This transition has recently been included in the discussion of the recombination kinetics in amorphous materials [8] (see also Chapters 6 and 7).

The material of this second area is seen to cover a large subset of solid state physics, in particular, the optical and electrical-transport properties of semiconductors. There is extensive literature available for the interested reader [6, 9]. Our purpose in the present article will not be to review either of these important areas, which have been extensively discussed in the references already cited. Our point of departure is the particular challenge that a proper understanding of photoconductivity in insulating molecular solids (especially disordered molecular solids) presents to the type of theoretical framework we have just outlined.

We will concentrate on a group of ideas which center about the *important role of localized states in both generation and transport*. In sharp distinction to the band model, we will consider carrier motion proceeding directly from one localized state to another. We will develop a model that has as its *basic building block, the quantum mechanical transition rate between localized sites*. We will apply this approach in Sections 3.5, 3.6 and Section 3.8 to transport and generation, respectively.

In molecular solids, the excitation, due to a photon absorption, tends to remain localized. Therefore, the generation of free carriers is a process which involves the interaction of the carriers with both the external electric field and strong internal attractive Coulomb fields. As stated above for the covalently-bonded semiconductors, the Coulomb effects are less important. One can view a covalently bonded semiconductor as one large "extended" molecule. A molecular solid represents almost the opposite limit: a small unit molecular species in a weakly interacting aggregate. The bands in a molecular crystal are usually narrow and very susceptible to internal potential fluctuations.

We will discuss photogeneration under the conditions of localized excitation when we deal with the Onsager mechanism in Section 3.8. In that section, it will be shown that the Onsager process represents a continuum limit of a carrier hopping from one (localized) state to another in a Coulomb well modified by an external electric field. We will set the stage for this demonstration by a thorough discussion of electrical transport in disordered molecular solids in Sections 3.2—3.7. In particular, we will introduce the mathematical formalism that has been developed to describe hopping transport among randomly distributed localized states. With some modifications, the theory can be applied to carrier motion with extensive trapping and detrapping.

References pp. 114—115

3.2 ELECTRICAL TRANSPORT VIA LOCALIZED STATES AS A RANDOM WALK

The simplest way to see a connection between electrical conductivity and a random walk is through the use of linear response theory [10]. In general, the computation of electrical conductivity $\sigma(\omega)$ is a quantum mechanical one, involving the probability *amplitudes* for transitions, between electron states, caused by interaction with an applied field (oscillating at angular frequency ω) and a scattering mechanism (*e.g.*, phonons, impurities, etc.). A random walk (RW) is a classical process involving a sequence of transitions, between states, where each transition is governed by a probability function. The connection between $\sigma(\omega)$ and a RW is therefore an approximate one and we shall exhibit the nature of these approximations below.

In linear response theory [11], the conductivity is expressed in terms of a velocity correlation function

$$\sigma(\omega) = \frac{ne^2}{kT} \int_0^\infty \exp(i\omega t) \langle v(t) v(0) \rangle \, dt \tag{3.1}$$

The ensemble average, indicated by the brackets in eqn. (3.1), is defined by:

$$\langle v(t) v(0) \rangle = \text{Trace}\,[v(t) v(0) \tilde{\rho}] \tag{3.2}$$

where $\tilde{\rho}$ is the density matrix of the field-free ensemble. The expression for $\sigma(\omega)$ in eqn. (3.1) is quite general, however, to facilitate our description of carrier transport in a *system of localized states*, we transform the expression for $\sigma(\omega)$ in eqn. (3.1) to one using the position operator $r(t)$ instead of the velocity $v(t)$ ($\equiv dr/dt$). Employing a simple integration by parts [10] we have

$$\sigma(\omega) = -\frac{ne^2}{kT} \omega^2 \int_0^\infty \exp(i\omega t) \langle [r(t) - r(0)]^2 \rangle \, dt. \tag{3.3}$$

In evaluating the ensemble average of the squared displacement, we first specialize to a basis set of localized wave functions $\phi_s(r)$, where s denotes the site position. We proceed by making a basic assumption that r, and $\tilde{\rho}$ are diagonal in this representation:

$$\langle s | r | s' \rangle \simeq \delta_{s,s'} s. \tag{3.4}$$

$$\langle s | \tilde{\rho} | s' \rangle \simeq \delta_{s,s'} f(s). \tag{3.5}$$

where $f(s)$ is the initial distribution. The approximation in eqn. (3.5) is good to lowest order in the overlap integral (between sites), and at moderate to low frequencies. It is this basic assumption, for a system of localized states, that converts the calculation of $\sigma(\omega)$ into a classical one. It is easy to show [10] with eqn. (3.5) that eqn. (3.3) becomes:

$$\sigma(\omega) = -\frac{ne^2}{kT} \omega^2 \sum_{s, s_0} (s - s_0)^2 \, \tilde{P}(s, \omega | s_0) f(s_0), \tag{3.6}$$

$$\tilde{P}(s,\omega|s_0) = \int_0^\infty \exp(i\omega t) P(s,t|s_0) \, dt \tag{3.7}$$

where $P(s,t|s_0)$ is the probability that a carrier starting at site s_0 at $t = 0$ is found at site s at time t. It follows from $r(t) = \exp(iHt) r(0) \exp(-iHt)$ and eqn. (3.4) that:

$$P(s,t|s_0) = |\langle s | \exp(iHt) | s_0 \rangle|^2, \tag{3.8}$$

in terms of the system Hamiltonian, H. However, it is clear that $P(s,t|s_0)$ is the same kind of object which plays the central role in the theory of random walks.

One can now pursue the calculation of $\sigma(\omega)$ in eqn. (3.6) in a number of ways: (1) determine $P(s,t|s_0)$ directly from eqn. (3.8) with a suitable Hamiltonian, (2) assume a transport equation obeyed by $P(s,t|s_0)$ and obtain the solution, or (3) calculate $P(s,t|s_0)$ from a suitable random walk that models the carrier motion.

In general, it is not feasible to calculate $P(s,t|s_0)$ directly from the quantum mechanical definition $|\langle s | \exp(iHt) | s_0 \rangle|^2$, because for large time t one must expand the time development operator $\exp(iHt)$ out to high orders in H. In other words, for large t the carrier propagates from s_0 to s via many paths, including ones with a large number of multiple hops among the random distribution of localized sites $\{s_i\}$. The computation and summing of all these contributions is a formidable task. The definition in eqn. (3.8) is, however, useful for small time Δt where one can define [12]:

$$w_{s',s} = \frac{1}{\Delta t} |\langle s' | \exp(iH\Delta t) | s \rangle|^2 \tag{3.9}$$

as the transition probability per unit time from s to s'. To lowest order in Δt only single hops are important. For a suitable H the rates $w_{s',s}$ were calculated by Miller and Abrahams [13] and we shall use their expression later. From small Δt one can build up to large t to calculate $P(s,t)$ with a transport equation. We can assume $P(s,t|s_0)$ obeys a Master equation:

$$\frac{\partial P(s,t|s_0)}{\partial t} = \sum w_{s,s'} P(s',t|s_0) - \sum w_{s',s} P(s,t|s_0) \tag{3.10}$$

where the $w_{s,s'}$ are defined in eqn. (3.9) and s_0 appears only in the initial condition $P(s,0|s_0) = \delta(s,s_0)$. One can easily show that the solution of eqn. (3.10) is equivalent to solving a random walk on a random network of sites [*cf.* Appendix B, ref. 10 and Section 3.8]. So far, the solution of this problem has evaded us, but the form of the problem exhibited in eqn. (3.10) will be useful in our discussion of the Onsager theory in Section 3.8.

We now continue with a calculation of $P(s,t|s_0)$ for a continuous-time random walk (CTRW) and show how the CTRW can incorporate a process similar to hopping transport among a random distribution of localized sites.

References pp. 114—115

3.3 THE CONTINUOUS-TIME RANDOM WALK (CTRW)

We wish to consider a stochastic process that contains the essential physical features of hopping transport in a disordered molecular solid and yet is tractable enough to calculate $P(s, t|s_0)$. The process should also contain hopping in a (ordered) molecular crystal as a special case.

In the absence of band transport, an amorphous insulator might be considered to be a network of localized sites for electrons or holes. We can divide the material into a regular lattice of equivalent cells, with each cell containing many randomly distributed localized sites available for hopping carriers.

Carrier transport is a succession of carrier hops from one localized site to another and finally from one cell to another. The hopping time is defined to be the time interval between a carrier arrival on successive sites.

If one follows the motion of a single carrier, one notes a wide fluctuation (or dispersion) of the hopping times as the carrier progresses from site to site. This sequence is a stochastic process with the probability to remain at a single site as the significant random variable. In any stochastic process, one generates the distribution function of the random variable that controls the individual events in the sequence. We denote this distribution function as $\psi(s, t)$. The probability that the hopping time (from one cell to another) occurs in the interval $(t, t + \Delta t)$ and the hop results in a displacement s is equal to $\psi(s, t)\Delta t$.

Therefore, we want a CTRW on a discrete lattice with a distribution function $\psi(s, t)$ governing the intercell events. This process is a generalization of one developed by Montroll and Weiss [14] as part of a study of random walks on a lattice.

We will outline the calculation of $P(s, t|s_0)$ for the CTRW and apply the results to a.c. conductivity and transient photoconductivity in Sections 3.5 and 3.6, respectively.

In a CTRW, with time as a continuous variable, one must proceed carefully. $P(s, t|s_0)$ is the probability of the carrier being found at s at time t if it started from s_0 at $t = 0$. We must allow for the possibility that the carrier could arrive at s at an earlier time $\tau < t$ and remain at s for at least the time interval $t - \tau$. We, therefore, introduce an auxiliary function; let $R_n(s, t)\Delta t$ be the probability for a carrier to just arrive at s between t and $t + \Delta t$ in n hops, if it started at $t = 0^+$ and s_0, (we will suppress the s_0 dependence for brevity) where:

$$s = s_1 \hat{a}_1 + s_2 \hat{a}_2 + s_3 \hat{a}_3 \qquad (3.11)$$

and s_i is equal to an integer, $\hat{a}_i$, the unit primitive translation vectors of the lattice. The RW are restricted to be on infinite lattices or on finite lattices (N^3 distinct points) with periodic boundary conditions.

The central aspect of any RW is the step-by-step generation of the probability to arrive at a given site. The probability to reach the site s in $n + 1$ steps is simply related to the previous one in n steps at some other site

$$R_{n+1}(s, t) = \sum_{s'} \int_0^t d\tau \psi(s - s', t - \tau) R_n(s', \tau). \tag{3.12}$$

The function of immediate interest is:

$$R(s, t) \equiv \sum_{n=0}^{\infty} R_n(s, t) \tag{3.13}$$

the probability per unit time to reach s in time t, independent of the number of steps to get to s. Thus, summing eqn. (3.12) over n and inserting the initial condition:

$$R_0(s, t) = \delta_{s,0} \delta(t - 0^+) \tag{3.14}$$

one obtains

$$R(s, t) - \sum_{s'} \int_0^t d\tau \psi(s - s', t - \tau) R(s', \tau) = \delta_{s,0} \delta(t - 0^+). \tag{3.15}$$

The form of eqn. (3.15) lends itself to solution by transform techniques, which reduces eqn. (3.15) to an algebraic one.

One takes the Laplace transform of eqn. (3.15) to obtain:

$$\tilde{R}(s, u) - \sum_{s'} \tilde{\psi}(s - s', u) \tilde{R}(s', u) = \delta_{s,0}, \tag{3.16}$$

where

$$\tilde{\psi}(s, u) = \int_0^{\infty} dt \exp(-ut) \psi(s, t). \tag{3.17}$$

The solution of eqn. (3.16) is accomplished with the use of Fourier transforms ($k_i = 2\pi m_i/a_i N$, m_i integer):

$$U(k, u) = \sum_s \tilde{R}(s, u) \exp(-ik \cdot s), \tag{3.18}$$

with the result:

$$\tilde{R}(s, u) = N^{-3} \sum_k \frac{\exp(ik \cdot s)}{1 - \Lambda(k, u)}, \tag{3.19}$$

where:

$$\Lambda(k, u) = \sum_s \tilde{\psi}(s, u) \exp(-ik \cdot s), \tag{3.20}$$

which can be called the generalized structure function of the CTRW.

References pp. 114–115

The final part of the solution involves the relation between $P(s, t)$ and $R(s, \tau), \tau \leqslant t$:

$$P(s, t) = \int_0^t R(s, \tau) \Phi(t - \tau) d\tau, \qquad (3.21)$$

where $\Phi(t)$ is the probability that the walker remains fixed in the time interval $[0, t]$:

$$\Phi(t) = 1 - \int_0^t \psi(\tau) d\tau, \qquad (3.22)$$

with

$$\psi(t) \equiv \sum_s \psi(s, t). \qquad (3.23)$$

Taking the Laplace transform of eqn. (3.21) we obtain a simple final expression:

$$\tilde{P}(s, u) = \tilde{R}(s, u)[1 - \tilde{\psi}(u)]/u, \qquad (3.24)$$

where $\tilde{R}(s, u)$ is equal to the right-hand side of eqn. (3.19).

Hence, one has obtained the Laplace transform of $P(s, t)$, our basic propagator, as a function of $\Lambda(k, u)$, the transform of $\psi(s, t)$, the single hop distribution function ($\tilde{\psi}(u) \equiv \Lambda(0, u)$). To determine $\sigma(\omega)$ as in eqn. (3.6) one can use the transform, $\tilde{P}(s, u)$, directly and we shall demonstrate this in Section 3.5. To calculate the current $I(t)$ in a transient photoconductivity measurement, one must take the inverse Laplace transform of eqn. (3.24) and obtain $P(s, t)$. In this latter case, one cannot evaluate $\mathcal{L}^{-1} P(s, u)$ without specifying a definite $\psi(s, t)$. We will do this first in the next section.

3.4 HOPPING TIME DISTRIBUTION FUNCTION [15]

Up to this point in our discussion, we have solved the transport part of the hopping conduction problem by obtaining $\tilde{P}(s, u)$. In the following two sections, we shall determine $\sigma(\omega)$ and $I(t)$, respectively, from $\tilde{P}(s, u)$. However, the generalized hopping time distribution $\psi(s, t)$ has not yet been determined for a situation intrinsic to hopping in a disordered solid. All the dynamics of the motion are incorporated into $\psi(s, t)$. The simplification, inherent in the structure of the CTRW model, allows one to focus on the basic fluctuating quantity in the hopping motion: the transition rate between the sites, $W(r)$, or more specifically, the basic fluctuating quantity is the probability of remaining on a site ($\simeq \exp(-W(r)t)$). In a system of *localized electron states*, the $W(r)$ is a sensitive function of the intersite separation and the intersite energy level fluctuation. Thus, moderate spatial fluctuations, in the distribution of site positions, can produce enormous hopping time fluctuations (or more accurately, dispersion). We shall show that these wide hopping time dispersions are characterized by a $\psi(t)$ [cf. eqn. (3.23)] that has slowly varying dependence on t. We will further show subsequently that this slow time variation in $\psi(t)$ when incorporated into a calculation of $\tilde{P}(s, u)$ [or $P(s, t)$] can account for all the novel features of time dependent hopping transport in disordered solids.

We show this connection between the time dependence of $\psi(s, t)$ and the spatial disorder of the hopping sites by explicit calculation. The function $\psi(s, t)$ is a weighted measure, at any one cell, of the different configurations a carrier will encounter as it hops through a random network of sites, from one cell to another. We have already remarked that the dispersion in actual displacement at each hop is mild compared to the dispersion in the *hopping times* (produced by the spatial variations). Therefore, for an explicit calculation of $\psi(s, t)$ it will suffice to have one site per cell.

We consider the probability $Q(t|\{s_i\})$ to leave a site relative to a specific configuration of other sites $\{s_i\}$ and then average over all possible random configurations to determine $\psi(s, t)$. This probability can decrease in time via all the parallel decay channels for the carrier to transfer to surrounding sites:

$$\frac{dQ}{dt} = -Q \sum_j W(r_j), \qquad (3.25)$$

thus,

$$Q(t) = \exp\left(-t \sum_j W(r_j)\right). \qquad (3.26)$$

In computing the configuration average of eqn. (3.26), one makes use of the well-known technique [16] for calculating the characteristic function of a sum of random variables:

$$\Phi(t) \equiv \langle Q(t|\{r_i\})\rangle = \left\langle \exp\left\{-t \sum_j W(r_j)\right\}\right\rangle$$

$$= \exp\left(-\int d^3r\, p(r)\{1 - \exp[-W(r)t]\}\right), \qquad (3.27)$$

where $p(r)d^3r$ is the probability a site is located in a volume d^3r centered about r. To effect an evaluation of eqn. (3.27) we choose:

$$W(r) = W_0 \exp(-\mathcal{E}/kT) \exp(-r/R_0) \equiv W_M \exp(-r/R_0) \qquad (3.28)$$

which is a prototype transition rate for electron transfer between localized sites. An excellent approximation to the integrand in eqn. (3.27) can be made in the limit of large $W_M t\, (\equiv \tau)$,

References pp. 114—115

$$1 - \exp[-W(r)t] = \begin{cases} 1 & W(r)t > 1 \\ 0 & W(r)t < 1. \end{cases} \qquad (3.29)$$

Due to the exponential dependence of $W(r)$ on r, the transition from 1 to 0 in eqn. (3.29) is very rapid as a function of r. One obtains:

$$\ln \Phi(t) = -4\pi \int_0^{r_\tau} p(r) r^2 \, dr, \qquad (3.30)$$

with

$$r_\tau = R_0 \ln \tau. \qquad (3.31)$$

For a totally random distribution:

$$p(r) = n, \qquad (3.32)$$

where n is the density of sites. Inserting eqn. (3.32) and eqn. (3.31) into eqn. (3.30) one has:

$$\Phi(t) = \exp(-(\eta/3)(\ln \tau)^3) \qquad (3.33)$$

hence:

$$\frac{\psi(t)}{W_M} = \frac{-d\Phi(\tau)}{d\tau} = \eta (\ln \tau)^2 / \tau^{1 + (\eta/3)(\ln \tau)^2}, \qquad (3.34)$$

with

$$\eta = 4\pi n R_0^3. \qquad (3.35)$$

The significance of eqn. (3.29) is clear: for a given t one sums up all the transition rates $\geqslant 1/t$. The long tail in $\psi(t)$ in eqn. (3.34) is related to the fact that one can change $W(r)$ significantly by a small variation in r, then for any t one can always "find" a site separation such that $W(r)t \simeq 1$ is satisfied. For a perfectly ordered crystal, with fixed intersite separation a, it follows from eqn. (3.26) that

$$\psi(t) = W(a) \exp(-W(a)t). \qquad (3.36)$$

The contrast between eqn. (3.36) and eqn. (3.34), exhibits the basic difference between an ordered and disordered system. In the ordered case, the probability to leave a site at a time $t > [W(a)]^{-1}$ is very small. While in a disordered system, the probability remains reasonably finite that for any arbitrary time (within a certain range) one can encounter a site separation r such that $t \simeq W(r)^{-1}$. Hence, $\psi(t)$ for a disordered system decays slowly in time and eqn. (3.34) points to a decay somewhere between exponential and purely algebraic.

We will calculate $\sigma(\omega)$ for the general distribution [15]:

$$\psi(s, t) = nvW(s) \exp\left(-W(s)t\right) \left\langle \exp\left(-t \sum_{r_j \neq s} W(r_j)\right)\right\rangle$$
$$\simeq nvW(s) \exp(-W(s)t) \Phi(t) \tag{3.37}$$

where v is the lattice cell volume. However, we will show that the use of eqn. (3.37) instead of eqn. (3.34) simply replaces a constant mean squared displacement σ_{rms}^2 by $\sigma_{rms}^2(\omega)$, the frequency dependent mean squared displacement [not to be confused with $\sigma(\omega)$, the conductivity]. For a range of ω on the order of a few decades one can approximate $\sigma_{rms}^2(\omega) \approx \sigma_{rms}^2$. The *net effect* of this replacement is to use a distribution

$$\psi(s, t) \simeq \tilde{p}(s) \psi(t). \tag{3.38}$$

The replacement shown in eqn. (3.38) is particularly appropriate and convenient in the calculation of $I(t)$, where typically one is interested in a time range of two or three decades.

3.5 A.C. CONDUCTIVITY AND TIME-DEPENDENT PAIR LUMINESCENCE

It is now a simple task to calculate $\sigma(\omega)$ from the linear response relation in eqn. (3.6). We first make the connection:

$$\tilde{P}(s, \omega | s_0)_{s_0 = 0} = \tilde{P}(s, u)|_{u = i\omega + \epsilon} \equiv \tilde{P}(s, i\omega), \tag{3.39}$$

i.e., we evaluate the Laplace transform of $P(s, t)$, calculated in Section 3.3 at $u = i\omega + \epsilon$, where ϵ can be understood as a small damping frequency to the rest of the universe [11]. The initial condition $f(s_0) = \delta_{s_0, 0}$ is assumed from the actual physical system, *e.g.*, impurity conduction in semiconductors. The site s_0 can be taken as the one having the lowest energy state in the vicinity of a local minimum in the random potential. Hopping on a lattice does not mean we start out with a uniform initial distribution $f(s_0) = 1/N^3$. We are modelling a hopping transport in a spatially and energetically disordered system. The restriction to a lattice only has meaning in the context discussed in the preceding section (Section 3.3), namely, that the intersite displacements have a very mild dispersion compared to the enormous dispersion in the hopping times. Hence, a proper calculation of the hopping time distribution inserted into a RW on a lattice will reproduce the major features of time dependent carrier hopping, as we shall see below.

$$\sigma(\omega) = \frac{ne^2}{kT} \frac{(i\omega)^2}{6} \frac{1 - \psi(i\omega)}{i\omega} \sum_s s^2 \tilde{R}(s, i\omega). \tag{3.40}$$

The sum over s in eqn. (3.40) can be accomplished easily by recognizing:

$$\sum_s s^2 \tilde{R}(s, i\omega) = -\frac{\partial^2 U}{\partial k^2}(k, i\omega)|_{k=0}, \tag{3.41}$$

where $U(k, u)$ is defined in eqn. (3.18) and $k_i = 2\pi r_i/N$.
From eqn. (3.18),

$$-\frac{\partial^2 U}{\partial k^2}(k, i\omega)|_{k=0} = \frac{-\partial^2}{\partial k^2}[1 - \Lambda(k, i\omega)]^{-1}|_{k=0}$$

$$= -\frac{\partial^2 \Lambda}{\partial k^2}(k, i\omega)|_{k=0}[1 - \Lambda(0, i\omega)]^{-2} - 2\left(\frac{\partial \Lambda(k, i\omega)}{\partial k}\right)^2\bigg|_{k=0}$$

$$- [1 - \Lambda(0, i\omega)]^{-3}. \tag{3.42}$$

Using the definition in eqn. (3.20),

$$\Lambda(0, i\omega) = \tilde{\psi}(i\omega), \tag{3.43}$$

$$i\frac{\partial \Lambda(k, i\omega)}{\partial k}\bigg|_{k=0} = \sum_s s\, \tilde{\psi}(s, i\omega), \tag{3.44}$$

$$-\frac{\partial^2 \Lambda}{\partial k^2}(k, i\omega)|_{k=0} = \sum_s s^2\, \tilde{\psi}(s, i\omega). \tag{3.45}$$

As the average fluctuations in eqn. (3.3) are those of a field-free equilibrium ensemble and the system under consideration is assumed to lack any intrinsic asymmetry, the first moment of the generalized structure factor $\Lambda(k, i\omega)$ vanishes. This condition will be relaxed in the next section. Thus, substituting eqn. (3.43), eqn. (3.45), and eqn. (3.42) into eqn. (3.41), one finally has

$$\sigma(\omega) = \frac{ne^2}{kT}\frac{1}{6}\sigma^2_{\text{rms}}(\omega)\, i\omega \tilde{\psi}(i\omega)/[1 - \tilde{\psi}(i\omega)], \tag{3.46}$$

$$\sigma^2_{\text{rms}}(\omega) \equiv \sum_s s^2\, \tilde{\psi}(s, i\omega)/\tilde{\psi}(i\omega). \tag{3.47}$$

The conductivity for the CTRW in eqn. (3.46) is a simple function of the zeroth and second spatial moment of the Fourier Transform of the basic hopping distribution function $\psi(s, t)$. The reason $\tilde{\psi}(i\omega)$ occurs in the denominator of eqn. (3.46) is a reflection of the fact that we are considering a sum over many different paths to each point and we also sum over all the points of the lattice.

The total effect of using a general $\psi(s, t)$ instead of the expression in eqn. (3.38), is to produce a frequency dependence in the mean square displacement $\sigma^2_{\text{rms}}(\omega)$. If one restricts the application of $\sigma(\omega)$ in eqn. (3.46) to a frequency range (or a time range) of a few decades, it is a good approximation to set $\sigma^2_{\text{rms}}(\omega) \approx \sigma^2_{\text{rms}}$, a constant ($\simeq$ the square of the mean nearest neighbor distance), for a hopping system characterized by transition rates $W(r)$ of form shown in eqn. (3.28). It is in this sense that we pursue explicit calculation of $\sigma(\omega)$ and $I(t)$ in the next section with $\psi(s, t) = \tilde{p}(s)\psi(t)$.

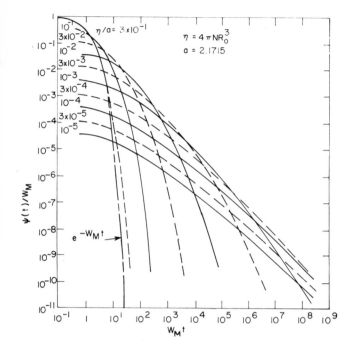

Fig. 3.1. Hopping-time distribution function $\psi(t)$ calculated for a random medium. The parameter η is a dimensionless measure of the localized site density. The broken line curve (—·—·—·—) is $\psi(t)/W_M = \exp(-W_M t)$. (From Scher and Lax [15].)

One now notes that $\sigma(\omega)$ for hopping transport is related to a spectral analysis of the hopping time distribution. In Fig. 3.1, we show a family of $\psi(t)/W_M$ curves as a function of $\tau(\equiv W_M t)$ with the parameter $\eta(\equiv 4\pi n R_0^3)$ designating each curve. For contrast, we also show $\psi(t)/W_M = \exp(-W_M t)$, an exponential decay curve, which is characteristic of an ordered system. One can observe that with increasing η, $\psi(t)$ tends toward the exponential decay curve. Increasing η corresponds to decreasing the spatial fluctuations of the system. Conversely, with decreasing η (increasing spatial fluctuation) the fall-off in time of $\psi(t)$ becomes slower. Hence, the larger the fluctuations affecting the intersite transition rate the wider the hopping distribution. The curves in Fig. 3.1 (other than $\exp(-W_M t)$) are a plot of eqn. (3.34) for larger τ ($\geqslant 10$). The small τ behavior of:

$$\psi(t)/W_M = 2\eta - (\tfrac{1}{4}\eta + 4\eta^2)\tau + \ldots \tag{3.48}$$

can be obtained from a power series expansion of eqn. (3.27) in τ.

There is a direct physical connection between $\psi(t)$ and the total light intensity $F(t)$ emitted in time dependent pair fluorescence in semiconductors [17—19]. If $F(t)$ is normalized to unity then $F(t) = \psi(t)$. For these types of

References pp. 114—115

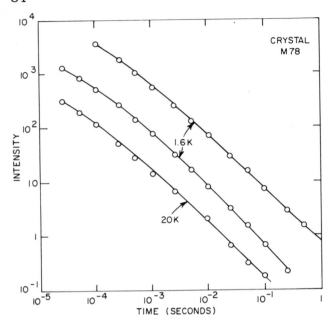

Fig. 3.2. The total light decay curves of GaP crystal M78 at 20 K and 1.6 K. Two flash-light intensities were used at 1.6 K. The displacement of the 20 K and 1.6 K curves is of no significance. (From Thomas et al. [18].)

measurements the semiconductor is prepared at $t = 0$ with the incidence of strongly absorbed light ($h\nu > \mathcal{E}_G$, the band gap energy). The electrons are trapped on donors and the holes are trapped on acceptors. Fluorescence occurs with the photon emitted from radiative recombination between a pair consisting of trapped electron and hole. The energy of the emitted photon $h\nu_e$ depends on the separation r between the donor and acceptor

$$h\nu_e = \mathcal{E}_G - (\mathcal{E}_D + \mathcal{E}_A) + \frac{e^2}{4\pi\kappa\kappa_0 r} \tag{3.49}$$

where $\mathcal{E}_G$ is the band gap energy and $\mathcal{E}_D$ ($\mathcal{E}_A$) is the ionization energy of the donor (acceptor). At short times after the semiconductor is prepared, the close pairs will mainly contribute to the fluorescence and if the emitted light is sampled at longer time, the distant pairs are mainly recombining. Thus, in this fluorescence process the light given off in recombination is simply a signal that a hop has occurred and the time dependence of the total number of photons emitted, $F(t)$, is a measure of the spectrum of hopping times between pairs of sites with random intersite separations. Measurements on GaP of $F(t)$ are shown in Fig. 3.2. Moreover, if one measures the energy spectrum of the fluorescence at a fixed time $F(\mathcal{E}, t)$ one can correlate with eqn. (3.49) the energy of the peak in the spectrum with the mean separation

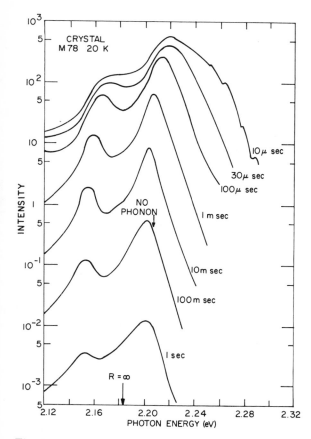

Fig. 3.3. The photoluminescence spectra observed from GaP crystal M78 after a pulse from the Van de Graaff machine, which saturates the pair system. The point $R = \infty$ marks the energy at which Si and S pairs at infinite separation would radiate. (From Thomas et al. [18].)

$\bar{r}$ of the pairs making the dominant contribution to the fluorescence at the time t. Measured time dependent spectra in GaP are shown in Fig. 3.3. It is obvious that $F(\mathcal{E}, t)$ corresponds to $\psi(s, t)$. Colbow [17] measured $\bar{r}(t)$ in GaP and found $\bar{r}(t)$ increased with increasing time and for a span of five orders of magnitude in emission time, $\bar{r}(t)$ changed by a factor $\simeq 2.5$, as shown in Fig. 3.4(a). Enck and Honig [19] found similar features in their fluorescence studies of P-doped n-Si and correlated it with their theoretical spectra shown in Fig. 3.4(b). This is direct experimental confirmation of the wide dispersion of hopping times and the small variation in intersite displacement at each carrier transfer event.

Luminescence can be considered to be the inverse of photoconductivity. In the present formulation, $\sigma(\omega)$ for a hopping system has been shown to be

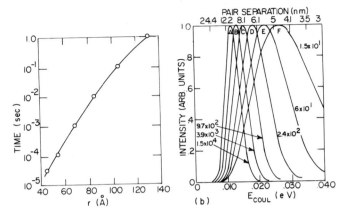

Fig. 3.4. (a) Time of light emission after photoexcitation *versus* calculated mean electron–hole separation responsible for this light. (From Colbow [17].)
(b) Theoretical spectral line shapes for the no-phonon line, using anisotropic donor radii $a_1 = 2.11$ nm and $a_2 = 0.93$ nm. Values of the parameter $W_0 t$ are marked on the curves. (From Enck and Honig [19].)

intimately connected to the Fourier analysis of the time dependence of pair luminescence. Thus, steady-state a.c. conductivity $\sigma(\omega)$ for hopping carriers is related to a time dependent photoconductivity process. In the next section, we will deal directly with transient photoconductivity.

We first wish to calculate $\sigma(\omega)$ directly from $\psi(i\omega)$ and compare the results to low temperature a.c. conductivity measurements, by Pollak and Geballe, on impurity-doped n-Si.

Qualitatively, one can derive the frequency dependence of $\sigma(\omega)$ from an inspection of Fig. 3.1. For the exponential $\psi(t)$, which obtains in a system with *one* transition rate:

$$\tilde{\psi}(i\omega) = \int_0^\infty \frac{\psi(t)}{W_M} \exp(-i\Omega\tau)\,d\tau = \int_0^\infty \exp[-(i\Omega+1)\tau]\,d\tau \qquad (3.50)$$

$$= (1+i\Omega)^{-1}, \quad \Omega \equiv \omega/W_M$$

which leads to a slow frequency variation for $\Omega < 1$. In fact, for an exponential $\psi(t)$ one obtains a frequency independent $\sigma(\omega)$ [insert eqn. (3.50) into eqn. (3.46)]

$$\sigma(\omega) = \frac{ne^2}{kT} \frac{\sigma_{rms}^2}{6} W_M. \qquad (3.51)$$

Hence, the frequency dependent behavior of $\sigma(\omega)$ that one observes in impurity conduction [20] and in amorphous semiconductors [21] is related to a *spectrum of transition rates*.

A convenient alternate expression for $\sigma(\omega)$ can be written as:

$$\bar{\sigma}(\Omega) = \tilde{\Phi}(\Omega)^{-1} - i\Omega \tag{3.52}$$

where:

$$\bar{\sigma}(\Omega) \equiv \sigma(\omega) \bigg/ \left(\frac{ne^2}{kT} \frac{\sigma_{rms}^2}{6} W_M \right) \tag{3.53}$$

and $\tilde{\Phi}(\Omega)$ is the Fourier transform of $\Phi(\tau)$. In the region of slowly varying $\Phi(\tau)$, i.e.:

$$\Phi(t) \simeq \tau^{-\alpha}, \quad 0 < \alpha < 1 \tag{3.54}$$

we have

$$\text{Re } \bar{\sigma}(\Omega) \simeq \tilde{\Phi}(\Omega)^{-1} \simeq \Omega^{1-\alpha}. \tag{3.55}$$

Thus, the larger the α ($\alpha \to 1$) the slower the frequency dependence of Re $\bar{\sigma}(\Omega)$. Exact calculation [15] has shown the same behavior for Im $\bar{\sigma}(\Omega)$. We would expect from this qualitative discussion that $\sigma(\omega)$ should vary as $\omega^{1-\alpha}$ for the higher values of ω, and $\sigma(\omega)$ should tend to a constant value as ω is lowered. One notes from eqn. (3.33) that $\alpha \propto \eta$, so that for large values of η the frequency variation of $\sigma(\omega)$ should be compressed.

Figures 3.5 and 3.6, show $\bar{\sigma}(\Omega)$ for two different values of η. These plots were determined from a careful numerical evaluation of $\tilde{\psi}(i\omega)$ in ref. 15. All the qualitative features of our discussion of $\bar{\sigma}(\Omega)$ are confirmed. It should be remarked that this calculation of $\sigma(\omega)$ for hopping transport is

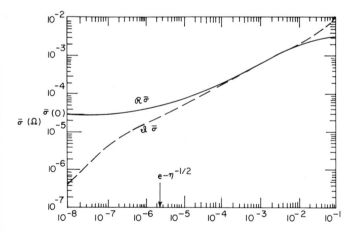

Fig. 3.5. Real $\mathcal{R}\bar{\sigma}$(solid line) and imaginary $\mathcal{I}\bar{\sigma}$(dashed line) parts of the dimensionless conductivity $\bar{\sigma}(\Omega) \equiv \sigma(\omega)/(\frac{1}{6}\sigma_{rms}^2 W_M\, ne^2/kT)$, $\Omega \equiv \omega/W_M$, with W_M in eqn. (3.28). The plot is for $\eta \equiv 4\pi N_D R_0^3 = 6 \times 10^{-3}$, where $R_0 = \frac{1}{2}a$; a is the effective Bohr radius. In the above, $r_{maj} \equiv (\frac{4}{3}\pi N_D)^{-1/3}$. Indicated in the plot are the values for $\bar{\sigma}(0) \simeq \eta^{1/4} \exp(-\frac{2}{3}\eta^{-1/2})/\sqrt{\pi}$ and $\Omega_c \simeq \exp(-\eta^{-1/2})$ the frequency characterizing the transition to d.c. (From Scher and Lax [15].)

References pp. 114—115

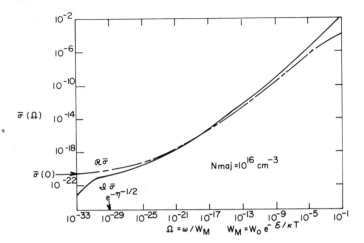

Fig. 3.6. Real $\Re\bar{\sigma}$(dot—dashed line) and imaginary $\Im\bar{\sigma}$(solid line) parts of the dimensionless conductivity $\bar{\sigma}(\Omega) \equiv \sigma(\omega)/(\frac{1}{6}\sigma_{rms}^2 W_M ne^2/kT)$, $\Omega \equiv \omega/W_M$, with W_M defined in eqn. (3.28). The plot is for $\eta \equiv 4\pi N_D R_0^3 = 2.22 \times 10^{-4}$, where $R_0 = \frac{1}{2}a$; a is the effective Bohr radius. The value $N_{maj} = N_D = 10^{16}$ cm^{-3} corresponds to $R_0 = 12$ Å. Indicated in the plot are the values for $\bar{\sigma}(0) \simeq \eta^{1/4} \exp(-\frac{2}{3}\eta^{-1/2})/\sqrt{\pi}$ and $\Omega_c \simeq \exp(-\eta^{-1/2})$, the frequency characterizing the transition to d.c. (From Scher and Lax [15].)

the first and only one to obtain within the same theoretical framework the familiar $\sigma(\omega) \propto \omega^s$ at higher ω and a *continuous transition* to the d.c. limit as $\omega \to 0$.

The d.c. limit $\sigma(0)$ can be obtained analytically:

$$\bar{\sigma}(0) = \eta^{1/4} \exp(-\tfrac{2}{3}\eta^{-1/2})/\pi^{1/2} \qquad (3.56)$$

which yields the same dominant majority concentration dependence as Miller and Abrahams [13]. However, the d.c. limit $\bar{\sigma}(0)$ is sensitive to the spectrum of transition rates at small values of $W(r)$, *i.e.* at large separations. A recent modification [22] to include a nearest-neighbor distribution function for $p(r)$ in eqn. (3.30) has resulted in a:

$$\ln \bar{\sigma}(0) \propto -\eta^{-1/3} \qquad (3.57)$$

dependence for small values of η with a continuous transition to the η-dependence in eqn. (3.56) with increasing η. The η-dependence of $\bar{\sigma}(0)$ in eqn. (3.57) is similar to one resulting from an application of percolation theory [23] to the d.c. hopping conductivity problem.

The theoretical curve Re $\bar{\sigma}(\Omega)$ of Fig. 3.5 is reproduced in Fig. 3.7 along with experimental values of Re $\bar{\sigma}(\Omega)$ measured by Pollak and Geballe [20] on a high density sample ($N_D = 2.7 \times 10^{17}$ cm^{-3}). There are no adjustable parameters used in the comparison in Fig. 3.7. Moreover, since Ω is a function of temperature through the factor $W_M (\equiv W_0 \exp(-\Delta/kT))$, the

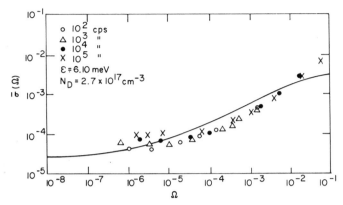

Fig. 3.7. Comparison of the theoretical (solid line) and experimental values of $\text{Re}\bar{\sigma}(\Omega)$. The dimensionless conductivity $\bar{\sigma}(\Omega) \equiv \sigma(\omega)/[(N_A e^2/kT)(\frac{1}{6}\sigma_{rms}^2 W_M)]$, $\Omega \equiv \omega/W_M$, with W_M obtained from the transition rate $W(r) = W_M \exp(-2r/a)$; a is the effective Bohr radius and $W_M \propto \exp(-\mathcal{E}/kT)$. Data for $\text{Re}\sigma(\omega)$ taken from Pollak and Geballe, ref. 20, Fig. 5. $N_A = 0.8 \times 10^{15}\,\text{cm}^{-3}$, $N_D = 2.7 \times 10^{17}\,\text{cm}^{-3}$, $\mathcal{E} = 6.1\,\text{meV}$, $\sigma_{rms} = r_D \equiv (\frac{4}{3}\pi N_D)^{-1/3}$. (From Scher and Lax [15].)

data points are seen to lie on a single curve which is both a simultaneous display of the frequency and temperature dependence of $\bar{\sigma}(\Omega)$. Thus the scaling law in T, ω exhibited by the data is in good agreement with the theoretical plot. Comparisons with $\sigma(\omega)$ measured from lower density samples ($N_D \simeq 10^{16}\,\text{cm}^{-3}$) and the Ω^s part of the theoretical curve in Fig. 3.6 has been made in ref. 15. The comparison again has been very satisfactory and the analysis has uncovered some unusual behavior in $\sigma(\omega)$ for these samples: the possibility for two-channel hopping in P-doped n-Si.

In amorphous semiconductors [21], even at room temperature, one observes $\sigma(\omega)$ in the dark with a similar frequency dependence as impurity conduction in crystalline semiconductors. In fact, the $\sigma(\omega) \propto \omega^s$ where $s \lesssim 1$ is almost universally observed in a wide variety of organic and inorganic amorphous solids [24]. It is still ambiguous to assign this frequency dependence to carrier hopping. Time variations in the local dipole moment in these materials could also contribute to $\sigma(\omega)$. However, some progress in identifying the dark $\sigma(\omega)$ with carrier hopping at the Fermi level has been achieved [25] in amorphous Si.

In those amorphous semiconductors that exhibit a $\sigma(\omega) \propto \omega^s$ in the dark, there is *no evidence*, to date, of an a.c. photoconductivity that shows:

$$\Delta\sigma(\omega) \propto \omega^s \tag{3.58}$$

where

$$\Delta\sigma(\omega) \equiv \sigma_{\text{photo}}(\omega) - \sigma_{\text{dark}}(\omega). \tag{3.59}$$

References pp. 114—115

Even a system like TNF—PVK, which clearly shows hopping conduction in an $I(t)$ measurement, exhibits no frequency dependent $\Delta\sigma(\omega)$ [21].

3.6 TRANSIENT PHOTOCONDUCTIVITY [26, 27, 28]

The measurement of the transient photocurrent $I(t)$ (initiated by a short pulse of strongly absorbed light) under constant voltage conditions, is a direct monitor of the time dependence of the space average of the conduction current of the transiting packet of carriers. These measurements on a wide class of organic and inorganic amorphous insulators have exhibited unusual or anomalous transport behavior. Instead of a fairly well defined square pulse shape of current, which is characteristic of a narrow sheet of charge sweeping across the sample at a constant mean velocity (as shown in Fig. 3.8), one finds that $I(t)$ has a long tail, i.e., the current slowly decays

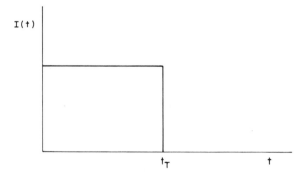

Fig. 3.8. Idealized transient-current trace measured with the technique described in Chapters 2 and 7 of this book. The trace depicts a sheet of charge moving with constant velocity until it leaves the sample at time t_T. (From Scher and Montroll [28].)

over a decade in time. This long tail indicates a wide dispersion of carrier transit times which can be understood in terms of a statistical spread in the number and types of events which limit the carrier motion in the disordered material. However, the striking feature (or anomaly) of all the measurements is the shape invariance of $I(t)$ with respect to changes in a number of variables that affect the transit time, e.g., electric field E, sample thickness L, temperature T and pressure p. The invariance of the current shape for each sample over a few decades change in the transit time has been designated as universality. The novel conceptual problem is that universality is incompatible with the usual ideas of statistical spreading. If one assumes that one is dealing with a large number of events, the spreading group of moving carriers must behave as a propagating Gaussian packet. However, it can be shown

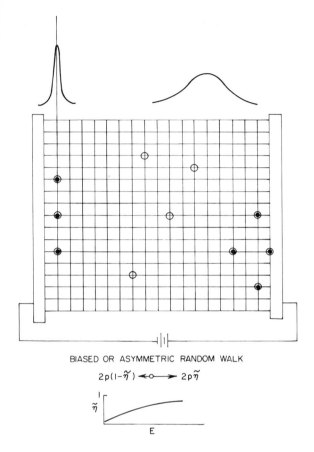

Fig. 3.9. Schematic diagram for the random-walk model of transient photoconductivity. The diagram is a composite in time. The carriers (●) are injected as a narrow distribution at $t = 0$ in the localized sites (○). The packet spreads and propagates to the right, $t > 0$, with a spatial bias $\bar{\eta}$, towards an absorbing barrier. (From Scher and Montroll [28].)

that a Gaussian cannot possibly account for the universality property over the transit time range considered in many of the measurements.

A review of the experimental technique of the time-of-flight method and the measurement results on the As—Se glasses are presented in this book in Chapters 2 and 7.

We will show that a model describing the dynamics of the carrier packet as executing a continuous time random walk (CTRW) with a $\psi(t)$, described in Section 3.4, characteristic of hopping in a disordered system [eqn. (3.34)] can account quantitatively, in a natural way, for the unusual results of the $I(t)$ measurements described above. In particular, *the wide dispersion of hopping times*, which accounts for the frequency dependence of $\sigma(\omega)$ in

References pp. 114—115

impurity conduction and the time dependence of $F(t)$ in radiative recombination between donor—acceptor pairs, will be shown to lead to the non-Gaussian properties of $I(t)$, i.e., the scaling laws for the current shape and the thickness dependence of the transit time. Moreover, the nature of the transit time and its relation to the concept of a drift mobility for these very dispersed packets will be clearly exhibited by the theoretical calculation.

The CTRW that we discussed in Section 3.3 must now be generalized to include an asymmetric spatial bias which can be an arbitrary function of the field E, i.e. $\tilde{p}(s)$ is a function of E and

$$\bar{l}(E) = \sum_s s\tilde{p}(s) \neq 0. \tag{3.60}$$

The carriers move in the sample, as shown schematically in Fig. 3.9 until they encounter the opposite electrode where they recombine with the other sign charge on the capacitor plate. So, we must include in our CTRW a completely absorbing boundary. The calculation of the effects of the boundary condition must be treated very carefully. In the ideal current shape shown in Fig. 3.8 the sharp drop in $I(t)$ is due to the total absorption of the sheet of charge as it intercepts the opposite electrode. The feature of the $I(t)$ curve used to determine the transit time is due to the effect of the boundary. Hence, with the dispersed packets in the CTRW one must calculate the effect of the absorbing boundary, exactly, so as not to produce any spurious mark (or lack of one) on the theoretical $I(t)$.

3.6.1 The current $I(t)$ for hopping carriers

We start first with a definition of $I(t)$ in terms of the drifting packet. In general[*]:

$$I(t) = \frac{1}{L} \int_0^L j_c(x, t)\,dx, \tag{3.61}$$

the current in the external circuit for constant voltage conditions, is equal to the spatial mean of the conduction current in the sample. This mean conduction current for an assembly of hopping charges is proportional to the time rate of change of the spatial mean of the carrier packet:

$$I(t) = en\frac{d\langle l \rangle}{dt} \tag{3.62}$$

---

[*] The total current $I(t) = j_c(x, t) + \epsilon \partial E(x, t)/\partial t$, the conduction plus displacement current. One integrates $I(t)$ and uses $d\left[\int_0^L dx E(x, t)\right]/dt = 0$, the constant voltage condition, to arrive at eqn. (3.61).

where

$$en = F\xi/L, \tag{3.63}$$

F is the incident flux of strongly absorbed light and ξ is the quantum efficiency. The position of the sites do not change in time, thus for hopping carriers:

$$\frac{d\langle l \rangle}{dt} = \sum_l l \frac{\partial P}{\partial t}(l, t), \tag{3.64}$$

where $P(l, t)$ is the same function discussed in Sections 3.2—3.5.

To show the connection between $I(t)$ defined in eqn. (3.62) and $\sigma(\omega)$ in Section 3.5, we consider the response to a step-function (in time) electric force eE,

$$eE(t) = \begin{cases} eE & t > 0 \\ 0 & t < 0 \end{cases} \tag{3.65}$$

This step time dependence could arise from switching on a field or suddenly creating charged carriers in a constant field. The Fourier transform of the current in response to eqn. (3.65) is:

$$\tilde{I}(i\omega) = \sigma(\omega) E/i\omega \tag{3.66}$$

where $E/i\omega$ is the Fourier transform of eqn. (3.65). Now, using eqn. (3.62) for $I(t)$

$$\tilde{I}(i\omega) = \int_0^\infty \exp(i\omega t) I(t) \, dt = en \sum_l l \int_0^\infty \exp(i\omega t) \frac{\partial P}{\partial t}(l, t) \, dt$$

$$= en \sum_l l [i\omega \tilde{P}(l, i\omega) - \delta_{l,0}], \tag{3.67}$$

one has, after inserting eqn. (3.67) into eqn. (3.66)

$$\sigma(\omega) = ne \sum_l l(i\omega)^2 \tilde{P}(l, i\omega)/E. \tag{3.68}$$

The sum over l can be evaluated in the same way it was done in Section 3.5

$$\sum_l l \tilde{P}(l, i\omega) = \frac{[1 - \tilde{\psi}(i\omega)]}{i\omega} i \frac{\partial U}{\partial k}(k, i\omega)\bigg|_{k=0} = \frac{\sum_l l \tilde{\psi}(l, i\omega)}{i\omega[1 - \tilde{\psi}(i\omega)]}. \tag{3.69}$$

Thus,

$$\sigma(\omega) = ne \left[\frac{\sum_l l \tilde{\psi}(l, i\omega)/\tilde{\psi}(i\omega)}{E} \right] \frac{i\omega \tilde{\psi}(i\omega)}{1 - \tilde{\psi}(i\omega)}. \tag{3.70}$$

If we now use $\tilde{\psi}(l, i\omega) = p(l) \tilde{\psi}(i\omega)$:

$$\sum_l \frac{l\tilde{\psi}(l, i\omega)/\tilde{\psi}(i\omega)}{E} = \frac{\bar{l}(E)}{E}, \qquad (3.71)$$

and for the *special case*, where $\bar{l}(E) \propto E$, one can show, from the Einstein relation, that

$$\frac{\bar{l}}{E} = \frac{e}{kT} \frac{\sigma_{rms}^2}{6}. \qquad (3.72)$$

Hence, using eqn. (3.72) and eqn. (3.71) in eqn. (3.70), one can observe that the $I(t)$ defined in eqn. (3.62), for the asymmetric CTRW, leads to the expression for $\sigma(\omega)$ derived in Section 3.5, from linear response theory, when $\bar{l}(E) \propto E$; in general, however, $\bar{l}(E)$ can be an arbitrary function of field (intrinsic field dependent mobility).

Before incorporating the absorbing boundary into the CTRW, we will illustrate and discuss the basic contrast in behavior of the propagator $P(l, t)$ and the $I(t)$ in an ordered and disordered system of local sites.

Taking the inverse Laplace Transform of eqn. (3.24) one has

$$P(l, t) = \frac{1}{2\pi i} \int_{c-i\infty}^{c+i\infty} (du/u) \exp(ut) [1 - \tilde{\psi}(u)] \tilde{R}(l, u); \qquad (3.73)$$

$\tilde{R}(l, u)$ can be calculated analytically, from its Fourier transform in eqn. (3.19), with:

$$\Lambda(k, u) = \lambda(k) \tilde{\psi}(u), \qquad (3.74)$$

$$\lambda(k) = \sum_l \tilde{p}(l) \exp(-ik \cdot l), \qquad (3.75)$$

and when the structure factor $\lambda(k)$ is specified. k and l in the above expression are henceforth normalized to the lattice spacing. The result is a complicated analytic form for the integrand in eqn. (3.73). The evaluation of the contour integral in eqn. (3.73) for a suitable $\tilde{\psi}(u)$ becomes tedious, if not impossible. We find it expedient to substitute the simple form of the Fourier Transform of $\tilde{R}(l, u)$ [eqn. (3.19)] into eqn. (3.73), with the result:

$$\gamma(k, t) = \frac{1}{2\pi i} \int_{c-i\infty}^{c+i\infty} (du/u) \exp(ut)[1 - \tilde{\psi}(u)][1 - \lambda(k) \tilde{\psi}(u)]^{-1} \qquad (3.76)$$

with

$$\gamma(k, t) \equiv \sum_l P(l, t) \exp(-il \cdot k). \qquad (3.77)$$

Our procedure is to evaluate $\gamma(k, t)$ in closed form and numerically invert the spatial Fourier Transform in eqn. (3.77) to obtain $P(l, t)$. The philosophy behind this approach is quite simple: As we are dealing with a problem that is discrete in the space variable l and continuous in time t, the finite spatial

Fourier Transform lends itself naturally to computer evaluation by the fast Fourier Transform algorithm [29], while the calculation of the inverse Laplace Transform in eqns. (3.73) and (3.76) often involves the subtle consideration of the nature of the singularities of $\tilde{P}(l, u)$ or $\tilde{\gamma}(k, u)$, for small u in the complex u-plane (as we shall show below in our calculation of $\mathcal{L}^{-1}\{\tilde{I}(u)\}$).

In general, we [26] believe it is an arduous task to work with both space and time continuous variables (e.g., in diffusion problems). The evaluation of the propagator $P(l, t)$ for the domain of greatest interest, large l and t, involves two simultaneous limits in, for example, eqn. (3.76): the limit $\lambda(k) \to 1$ (for small k) and the limit $\tilde{\psi}(u) \to 1$ (for small u). It is these simultaneous limits and their interaction that complicate transport problems and we believe can be facilitated by the use of discrete space.

We now choose two characteristically different $\psi(t)$ and evaluate $\gamma(k, t)$ for k in the same direction as E, i.e., $k = \{k_x, 0, 0\}$ for E in the x-direction. The restriction of $k = k_x \hat{i}$ has the effect of replacing the propagator $P(l, t)$ by the "effective one-dimensional" propagator:

$$P(l, t) \equiv \sum_{l_y, l_z} P(l, l_y, l_z, t) \tag{3.78}$$

where we drop the subscript x for brevity. The two $\psi(t)$ represent the ordered case:

$$\psi_1(t) = W \exp(-Wt) \tag{3.79}$$

and the disordered case:

$$\psi_2(t) = 4W_M \exp(\tau) i^2 \operatorname{erfc}(\tau^{1/2}). \tag{3.80}$$

where $i^2 \operatorname{erfc}(z)$ is the second repeated integral of the complementary error function [30]. The function $\psi_2(t)$ is to be considered a special case of the general class of $\psi(t)$ that exhibit a long tail:

$$\psi(t)/W_M \simeq [A \tau^{1+\alpha} \Gamma(1-\alpha)]^{-1}, \quad 0 < \alpha < 1 \tag{3.81}$$

for large τ, where $\Gamma(x)$ is the gamma function. The tail of $\psi(t)$ simulates the behavior of the hopping time distribution function [eqn. (3.34)] calculated in Section 3.4, for a finite region of τ (of the order of a few decades). The large τ limit of $\psi_2(t)$ is:

$$\psi_2(t)/W_M \simeq \pi^{-1/2} \tau^{-3/2}, \quad \tau \gg 1, \tag{3.82}$$

and, thus, is an example of eqn. (3.81) with $\alpha = 1/2$.

In Fig. 3.10, we show the propagator $P(l, t)$ corresponding to $\psi_1(t)$. In the limit of large τ one can observe Gaussian packet behavior. The peak of the packet and the mean position $\langle l \rangle$ are located at the same position and move with some *constant* velocity. The dispersion or spread of the packet about the mean, σ, varies as $\sigma \propto t^{1/2}$ so that

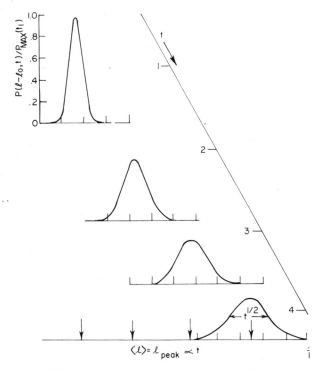

Fig. 3.10. Propagator for a carrier packet $P(l-l_0, t)$ versus l/N for a range of time $t/t_T = (2n-1)/10$, $n = 1, 2, 3, 4$. The "transit time" is defined by $\langle l(t_T)\rangle/N \simeq 1$. The random walk is based on $\psi(t)/W = \exp(-Wt)$. The plot is scaled by the peak value of $P(l-l_0, t)$ at the earliest time. The spatial bias factor $\bar{\eta} = 0.9$. (From Scher and Montroll [28].)

$$\frac{\sigma}{\langle l \rangle} \simeq t^{-1/2}. \qquad (3.83)$$

The $P(l, t)$ curves in Fig. 3.11 corresponding to $\psi_2(t)$, clearly show the effects of disorder. The peak of the packet remains fixed at the point of origin of the carriers, while the position of the mean $\langle l \rangle$ separates from the peak, in conformance to the packet spreading, and moves away with a velocity which decreases in time. It is evident that for these $P(l, t)$,

$$\sigma/\langle l \rangle \simeq \text{constant}. \qquad (3.84)$$

It is this feature of the motion that accounts for the universality of the current shape.

The most salient features of the packet propagation in a disordered system can be obtained directly from the mean $\langle l \rangle$, as a function of time, without recourse to the numerical evaluation of $P(l, t)$. The pictures of $P(l, t)$ in Fig. 3.11 add to the physical discussion of the propagation and most

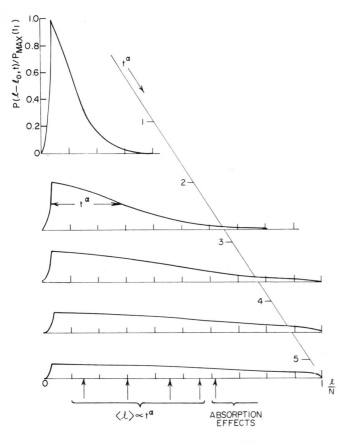

Fig. 3.11. Propagator for a carrier packet $P(l, t)$ in the presence of absorbing barriers at $l/N = 0, 1$, for a range of time $\tau_n^{1/2} = 0.2(2n-1)^{1/2}/\sqrt{\pi}$, $n = 1, 5$. The transit time t_T is defined in the text. The random walk is based on $\psi_2(t) \simeq \pi^{1/2}\tau^{-3/2}$, $\alpha = 0.5$. The spatial bias factor $\bar{\eta} = 0.6$. (From Scher and Montroll [28].)

dramatically show the non-Gaussian behavior for the CTRW with a $\psi(t)$ of the form of eqn. (3.81).

The Laplace Transform of $\langle l(t) \rangle$ is obtained from eqn. (3.69) [with $\tilde{\psi}(l, u) = \tilde{p}(l)\tilde{\psi}(u)$]:

$$\sum_l l\tilde{P}(l, u) = \frac{\bar{l}(E)\tilde{\psi}(u)}{u[1 - \tilde{\psi}(u)]} \simeq \frac{\bar{l}(E)}{cu^{1+\alpha}}, \quad \frac{u}{W_M} \ll 1 \tag{3.85}$$

where the small u behavior of $\tilde{\psi}(u)$:

$$\tilde{\psi}(u) \simeq 1 - cu^\alpha \tag{3.86}$$

has been inserted. A Tauberian theorem, concerning the asymptotic

References pp. 114—115

properties of Laplace Transforms, has been used to derive eqn. (3.86) from the t-dependence of $\psi(t)$ in eqn. (3.81). The use of Tauberian theorems was first introduced into the present transport study by Shlesinger [31]. Conversely, the asymptotic theorems can also derive the large τ behavior of $\langle l(t) \rangle$ from the small u properties of $\mathcal{L}\{\langle l(t) \rangle\}$ in eqn. (3.85), (cf. ref. 32),

$$\langle l(t) \rangle \simeq \frac{\bar{l}(E)}{c\Gamma(1+\alpha)} t^\alpha. \tag{3.87}$$

Hence:

$$I(t) = \frac{ned\langle l \rangle}{dt} \simeq \frac{ne\bar{l}}{c\Gamma(\alpha)} t^{-(1-\alpha)}, \quad 0 < \alpha < 1, \tag{3.88}$$

the current is decreasing in time even in the absence of an absorbing boundary! We will first include the effects of such a boundary and then discuss the physical significance of the results we have obtained for $P(l, t)$, $\langle l \rangle$, and $I(t)$ for a disordered system.

3.6.2 Absorbing boundary in CTRW

It will suffice to consider the effect of a permanent trap on a one-dimensional RW for our problem. $P(l, t)$, defined in eqn. (3.78), is a pseudo one-dimensional propagator [26]; it is the probability of finding the carrier on the plane $\boldsymbol{l} \cdot \hat{\imath} = l$ if it started out on the plane $\boldsymbol{l} \cdot \hat{\imath} = l_0$. The absorbing plane is defined as $\boldsymbol{l} \cdot \hat{\imath} = N$, where N is the number of cells in the x-direction of the lattice (we are assuming periodic boundary conditions).

Let us denote the solution of eqn. (3.16) (now considered in one dimension) as $G(l, u)$, i.e.:

$$G(l, u) - \tilde{\psi}(u) \sum_{l'} \tilde{p}(l-l') G(l', u) = \delta_{l, l_0}, \tag{3.89}$$

and $\tilde{R}(l, u)$ the solution in the presence of the "trap" at $l = N$. It is straightforward to show that $\tilde{R}(l, u)$ must satisfy

$$\tilde{R}(l, u) = \tilde{\psi}(u) \sum_{l'} \tilde{p}(l-l') \tilde{R}(l', u) - \tilde{\psi}(u)[\tilde{p}(l-N) - \delta_{l, N}] \tilde{R}(N, u) + \delta_{l, l_0} \tag{3.90}$$

where we exclude the possibility of carrier hopping from N to any other site. The formal solution of eqn. (3.90) in terms of the Greens' function $G(l, u)$ is:

$$\tilde{R}(l, u) = \sum_{l'} G(l-l', u) \theta(l') \tag{3.91}$$

where

$$\theta(l) = \delta_{l, l_0} - \tilde{\psi}(u)[\tilde{p}(l-N) - \delta_{l, N}] \tilde{R}(N, u) \tag{3.92}$$

Inserting eqn. (3.92) for $\theta(l)$ into eqn. (3.91) and using the definition of $G(l, u)$ in eqn. (3.89) to affect the sum over l', we have,

$$\tilde{R}(l, u) = G(l - l_0, u) - [(1 - \tilde{\psi}(u)) G(l - N, u) - \delta_{l, N}] \tilde{R}(N, u) \qquad (3.93)$$

Now, we set $l = N$ in eqn. (3.93), solve for $\tilde{R}(N, u)$ and insert it back into eqn. (3.93), evaluated for $l \neq N$, to obtain:

$$\tilde{R}(N, u) = \frac{G(N - l_0, u)}{[1 - \tilde{\psi}(u)] G(0, u)} \qquad (3.94)$$

and:

$$\tilde{R}(l, u) = G(l - l_0, u) - G(l, u) \frac{G(N - l_0, u)}{G(0, u)}, \quad l \neq N, \qquad (3.95)$$

where we made use of the periodic condition $G(l - N, u) \equiv G(l, u)$.

Thus, the net effect of the absorbing boundary condition is to multiply $\mathcal{L}\{\langle l(t) \rangle\}$ in eqn. (3.85) by the factor $1 - G(N - l_0, u)/G(0, u)$. This factor has a simple interpretation: it eliminates all paths in the CTRW that go through $l = N$. The quantity:

$$\tilde{F}(N - l_0, u) \equiv G(N - l_0, u)/G(0, u) \qquad (3.96)$$

is the Laplace Transform of the first passage time distribution function $F(N - l_0, t)$ and it is discussed in refs. 26 and 28. It will suffice for our purpose to evaluate $\tilde{F}(N - l_0, u)$ in the limit of small u. It is shown in ref. 26 that:

$$\tilde{F}(N - l_0, u) \simeq \exp[-(N - l_0)(1 - \tilde{\psi}(u))/\bar{l}] \qquad (3.97)$$

for $u/W_M \ll 1$. We use the asymptotic expression for $\tilde{\psi}(u)$ in eqn. (3.86) to arrive at a compact form for $\tilde{I}(u)$, the Laplace Transform of $I(t) = ne \dfrac{d\langle l \rangle}{dt}$, $[\tilde{I}(u) = ne(u \mathcal{L}\{\langle l(t) \rangle\} - l_0)]$

$$\tilde{I}(u) \simeq ne\bar{l} \left[\frac{1 - \exp[-(N - l_0) cu^\alpha / \bar{l}]}{cu^\alpha} \right] \qquad (3.98)$$

Thus, for $(N - l_0) cu^\alpha / \bar{l} \gg 1$, one obtains the same result we derived above for $I(t)$ in eqn. (3.88). However,

$$\tilde{I}(u) \simeq ne \left[N - l_0 - 1/2 \frac{(N - l_0)^2}{\bar{l}} cu^\alpha + \ldots \right] \qquad (3.99)$$

for $(N - l_0) cu^\alpha / \bar{l} \ll 1$ and one can show [28, 31] that the $\tilde{I}(u)$ in eqn. (3.99) leads to

$$I(t) = \frac{ne}{2\bar{l}} (N - l_0)^2 c/[-\Gamma(-\alpha) t^{1 + \alpha}]. \qquad (3.100)$$

References pp. 114–115

Thus,

$$I(t) = \begin{cases} n e \bar{l}/[c\Gamma(\alpha)t^{1-\alpha}] & \dfrac{(N-l_0)}{\bar{l}t^\alpha}c \gg 1, \\ \dfrac{n e \bar{l}}{2}\dfrac{(N-l_0)^2}{\bar{l}}c/[-\Gamma(-\alpha)t^{1+\alpha}] & \dfrac{(N-l_0)}{\bar{l}t^\alpha}c \ll 1. \end{cases} \qquad (3.101)$$

The transition time which marks the change in the time dependence of $I(t)$:

$$t_T \simeq \left[\dfrac{(N-l_0)c}{\bar{l}(E)}\right]^{1/\alpha} \qquad (3.102)$$

will be shown to characterize the transit time for the carrier packet.

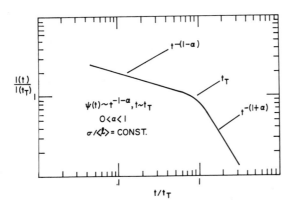

Fig. 3.12. A log(I)—log(t) plot indicating the current $I(t)$ associated with a packet of carriers moving, in an electric field, with a hopping-time distribution function $\psi(t) \simeq t^{-1-\alpha}$, $0 < \alpha < 1$, towards an absorbing barrier at the sample surface. (From Scher and Montroll [28].)

In Fig. 3.12, we exhibit a log(I)—log(t) plot of eqn. (3.101) and for contrast we show a normalized plot of $I(t)$ in Fig. 3.13 corresponding to the Gaussian packet motion in Fig. 3.10.

3.6.3 Qualitative discussion of dispersed transport and question of mobility

We can gain some physical understanding of the nature of the transport results we have obtained with the following pictures.

In Fig. 3.14, we illustrate the motion of a carrier hopping through a random array of sites. Each line represents a transition rate $W(r)$, eqn. (3.28). A heavy line symbolizes a longer (than average) hop. The carrier (hole) hops with greater relative probability in the direction of the applied field E, i.e. with the spatial asymmetry $\bar{l}(E)$.

At the initial time t_0, just after the incidence of the light flash and the

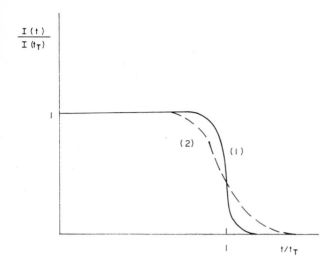

Fig. 3.13. Normalized current trace $I(t)/I(t_T)$ *versus* t/t_T expected for a propagating Gaussian packet. Curve (1) corresponds to the longer t_T and curve (2) corresponds to the shorter t_T. This figure illustrates the incompatibility of a Gaussian with the universality of $I(t)$. (From Scher and Montroll [28].)

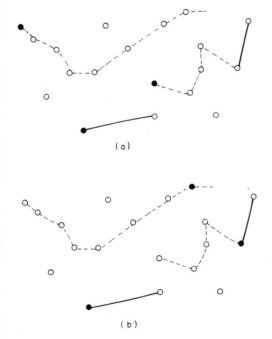

Fig. 3.14. Schematic diagram illustrating hopping in a disordered system. (a) At an initial time t_0, some carriers (●) are at "fast" hopping sites and some (○) at "slow" hopping sites. (b) At $t > t_0$, more carriers encounter the "slower" hopping situations, *i.e.* the net velocity of the packet is decreasing in time.

References pp. 114—115

charge separation, the carriers are distributed at a variety of site environments (Fig. 3.14(a)). Some of these local environments permit the immediate motion of the carrier (fast hops). On the time scale generated by the finite displacement of these "fast" carriers, a considerable fraction of the carriers remain near their region of formation. Hence the peak in $P(l, t)$ does not move (cf. Fig. 3.11). At a later time (Fig. 3.14(b)) a number of the "fast" carriers will eventually encounter a long hop and will therefore be immobilized. If the hopping time associated with these (relatively rare) long hops is of the order of the time it takes for the fast carriers to cross the sample, one will have a number of unusual properties of the propagating carrier packet.

The experimentally determined transit time t_T is associated with the motion of the small fraction of carriers that manage to avoid these long hops, i.e., the time for the forward tail of $P(l, t)$ to encounter the absorbing boundary. Hence, on the time scale of the experiment (which is of the order of t_T) a small percentage of the large number of intersite hopping events can have an inordinately large effect on the individual carrier transit time. This situation leads to an extensive dispersion of carrier transit times. Moreover, not only are there various numbers n_i of carriers moving at different effective velocities v_i, whereby the mean velocity is:

$$\bar{v} = \sum n_i v_i, \qquad (3.103)$$

but the n_i are functions of time. Eventually, every carrier will encounter at least one hopping time of the order of the laboratory time. Hence, $\bar{v}$ in eqn. (3.103) is a decreasing function of time, *i.e.*, the carrier is getting tired. The current $I(t)$, therefore, decreases in time [eqn. (3.88)] even before the carrier loss through absorption.

While it is possible, with some mechanism, to generate a fixed velocity absorption n_i to explain the universality property of the transient photocurrents $I(t)$, this is insufficient to account for other experimental facts, *e.g.*, the current decaying in time for $t < t_T$. When the first carriers hit the absorbing boundary the current decreases more rapidly in time, as the total number of carriers in transit are now being diminished. This type of behavior for $I(t)$ is shown on the log–log plot in Fig. 3.12. The break in the time decay of $I(t)$ is associated with the transit time t_T and this time is, of course, also associated with the motion of the fastest carriers, as indicated above.

Quantitatively, the change in the time decay of $I(t)$ is indicated in eqn. (3.101). The transition corresponds to

$$t^{-(1-\alpha)} \to t^{-(1+\alpha)}, \qquad (3.104)$$

and the sum of the slopes on a log–log plot add to -2, independent of α. The meaning of t_T in eqn. (3.102) can also be understood from the fact that at the transit time, the spatial mean $\langle l \rangle$ should be nearly equal to the thickness of the sample

$$\rho \langle l \rangle \simeq L, \tag{3.105}$$

where ρ is the average intersite separation.

Now, with eqn. (3.87) for $\langle l(t) \rangle$ and:

$$L \simeq (N - l_0)\rho, \tag{3.106}$$

one has:

$$\frac{\bar{l}(E) t^\alpha}{c\Gamma(1 + \alpha)} \equiv \frac{L}{\rho} \tag{3.107}$$

or

$$t_T \simeq \left(\frac{L}{\rho \bar{l}(E)}\right)^{1/\alpha}. \tag{3.108}$$

The transit time defined in eqn. (3.108) or eqn. (3.102), for $0 < \alpha < 1$, predicts an unusual superlinear thickness dependence.

In the same way t_T could have a superlinear dependence on E arising from an intrinsically linear response behavior $\bar{l}(E) \propto E$.

If one uses the conventional relation:

$$t_T^{-1} = \mu L/E \tag{3.109}$$

to define an effective drift mobility, μ, then, besides the possibility of an apparent E dependence of μ one would have to rationalize a thickness dependence of μ! It is obvious now that unless the mean $\langle l \rangle \propto t$, the idea of a μ depending solely on the properties of the material breaks down. For example, if one has $\langle l \rangle \propto t^\alpha$, $0 < \alpha < 1$, as in our case, one could write:

$$\langle l \rangle = (V_0/t^{1-\alpha}) t \tag{3.110}$$

and the effective μ would be time dependent. Thus, for this type of dispersive transport, the present theory establishes the limitations of the concept of a mobility. The theory identifies what property of the propagating packet is associated with the definition of the transit time. This transit time is related to the intrinsic-rate processes in the material as we shall see in the next section.

3.7 COMPARISON WITH EXPERIMENT ON HOPPING SYSTEMS

The most complete comparison of theory with experiment has been made in the As—Se chalcogenide glasses. This work is thoroughly reviewed in Chapter 7 of this book. The theory is in excellent agreement with experiment in As_2Se_3. The current shapes and the thickness and field dependence of the transit time are correlated in detail. There is convincing evidence that the dispersion of the carrier packet is proceeding via the stochastic dynamics we have discussed above. The more specific assignment of the stochastic process to hopping among localized states is highly plausible but yet not unambiguously proven. We still have not identified the origin of the localized states in the gap of As_2Se_3, much less controlled their density through fabrication techniques.

In this section, we shall compare the theory with the measurements made on a class of organic insulators where we have independent knowledge of hopping transport. The independent knowledge is gained through the control of the identified hopping sites in a system of molecularly doped polymers [32]. Studies by Gill [33] of the charge-transfer complexes of trinitrofluorenone [TNF] with polyvinyl carbazole [PVK] and TNF-doped polyesters, and more recently, by Mort, Pfister and Grammatica [34], on N—isopropyl carbazole [NIPC]-doped Lexan, have shown conclusively that transport occurs via hopping among the dopant molecules. An Arrhenius plot of the electron mobility, *i.e.* the inverse transit time, *versus* the average intermolecular separation of TNF in PVK is given in Fig. 8.10, page 319. A similar plot for hole transport in NIPC-doped Lexan is shown in Fig. 3.15. Both plots indicate that the transit time has an exponential dependence on the average dopant separation, ρ:

$$t_T^{-1} \propto \exp(-2\rho/\rho_0) \qquad (3.111)$$

where ρ_0 is a localization radius that is a measure of the extension of the electronic charge density outside the molecule. The slope of the line in Figs. 3.15 and 8.10 determine that ρ_0(NIPC) $\simeq 1.5$ Å, ρ_0(TNF) $\simeq 1.8$ Å; these are very reasonable values for atomic radii. This is strong evidence of hopping transport between the molecules controlled by the wave function overlap as shown for the transition rate in eqn. (3.28), with $R_0 \simeq \rho_0/2$. Moreover, it has been demonstrated in these systems that there is *no* measurable transport of the appropriate carrier when the dopant molecules are not present in the polymer. In Fig. 3.16, we have a log—log plot of $I(t)$ for 1:1 TNF:PVK. The current is seen to obey the scaling law for universal shape and the slopes of the dashed lines add to -2.0 in agreement with eqn. (3.104). The value of α is 0.8 for this composition. If one compares the $\psi(t)$ in eqn. (3.81) to the first-principles result in eqn. (3.34) we can write

$$\alpha \simeq \langle (\eta/3)(\ln\tau)^2 \rangle_{av}. \qquad (3.112)$$

Now using this relation in eqn. (3.112) with the *known* density of TNF and the *measured* $R_0 \simeq 0.9$ Å, which together yield a $\eta (\equiv 4\pi N_{TNF} R_0^3)$ of 10^{-2}, one obtains [*cf.* eqn. (3.28)]:

$$W_M \simeq 10^9 \text{ s}^{-1} \qquad (3.113)$$

a very reasonable value of W_M considering the large measured activation

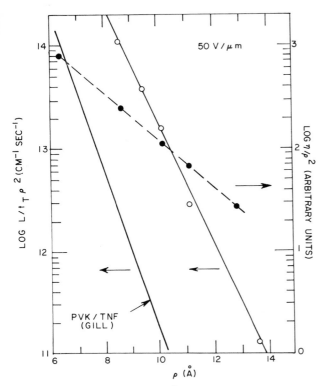

Fig. 3.15. Plot of Log $(L/t_T\rho^2)$ versus ρ. t_T was measured at a fixed bias field of 50 V/μm in samples containing different concentrations of NIPC. For comparison data for hole transport in PVK:TNF by Gill [33] are shown. The dashed line is a plot of Log (η/ρ^2) in arbitrary units versus ρ, where η is the photogeneration efficiency. The data for $\rho = 6.5$ Å correspond to PVK. (From Mort, et al. [34].)

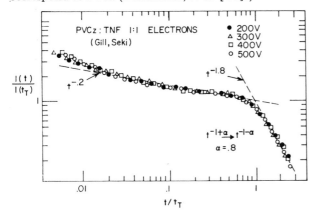

Fig. 3.16. A log(I)–log(t) plot for 1:1 TNF–PVK. The slopes of the dashed lines are −0.2, and −1.8, respectively. (From Gill, Chapter 8.)

References pp. 114–115

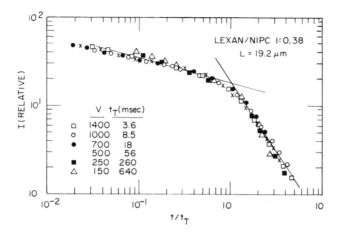

Fig. 3.17. Composite plots of hole transient currents, obtained at the bias voltage indicated, in logarithmic scales for a 1:0.38 weight ratio polycarbonate:NIPC sample. (From Mort et al. [34].)

energy $\mathcal{E}$ contained in W_M of $\mathcal{E} \lesssim 0.6\,\text{eV}$. The uncertainty in the value of $\mathcal{E}$ to put into W_M is associated with the field dependence of $\mathcal{E}$. In Fig. 3.17 we show a similar plot for the hole current $I(t)$ measured in NIPC-doped Lexan. Here, one can demonstrate the universal scaling law for the current shape for a transit time range of over two orders of magnitude. Again, the slopes of lines in Fig. 3.17 add to a value very close to -2.0 as predicted by the theory.

The NIPC loading in Lexan has been varied and the relation in eqn. (3.112) has been shown to give a very satisfactory correlation between the extent of dispersion and the density of the hopping sites [35], using the same value of W_M as determined for TNF:PVK.

One can also obtain valuable information about the photogeneration process in molecular solids with this system. The dashed line in Fig. 3.15 is a measure of the ρ dependence of the quantum efficiency [34]. The concentration dependence of the quantum efficiency, as in the case of charge transport, shows that overlap between dopant molecules plays an important role in the photogeneration process. This process will be discussed in the next section. From the slope, one obtains a value of 3.0 Å for the localization radius. This value, which is twice as large as ρ_0 (NIPC) quoted above for the transport data, is consistent, since photogeneration involves excited states. The hole transport involves ground state wave functions.

The organic molecularly-doped polymers are emerging as a prototype system for hopping transport, and the stochastic theory we have developed gives a very satisfactory explanation of the effects of disorder on the propagating carrier packet. One can now turn to a study of the microscopic aspects of the charge transfer step between a molecule and its ion in a system where one can vary the intermolecular separation. A study of these redox reactions of molecules has significance in many other areas of molecular physics and biophysics besides photoconductivity.

3.8 ONSAGER THEORY AND RELATION TO HOPPING TRANSPORT

As discussed in Section 3.1, photoexcitation in molecular solids does not immediately produce free carriers as it does in the familiar covalently-bonded semiconductors. In Fig. 3.18, we show a recent measurement of the optical absorption spectrum of crystalline sulfur [36]. The spectrum is dominated by exciton effects well into the wavelength region above threshold.

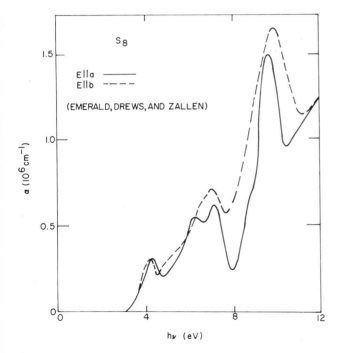

Fig. 3.18. Optical absorption spectra of orthorhombic sulfur, obtained by Kramers–Kronig analyses of ultraviolet reflectivity data. (From Emerald et al. [36].)

The photogeneration process in molecular solids involves the ionization of the local excitation. This local excitation is not necessarily an exciton. The process can be envisaged as follows: After the creation of the local

References pp. 114—115

excitation, the electron hops (or diffuses) to another molecular site, while losing the excess kinetic energy over the local potential energy (thermalization). The electron—hole pairs are now separated by an "initial" distance r_0 and then diffuse in their mutual Coulomb field, modified by the external field

$$U = \frac{-e^2}{4\pi\kappa\kappa_0 r} - eEr\cos\theta. \tag{3.114}$$

Onsager [37] solved the problem, in 1938, of the effect of an external field on the dissociation of two oppositely charged ions in a weak electrolyte. The Onsager theory involves the solution of an equation describing diffusion in a force field (the Smoluchowski equation):

$$\frac{\partial P}{\partial t} = D\nabla \cdot [\nabla P + P\nabla U/kT] \tag{3.115}$$

where $P(r, t)d^3r$ is the probability of finding a carrier in a volume d^3r, centered about r, at a time t, and D is the sum of the diffusion constants of the electron and hole $[D = kT(\mu_n + \mu_p)/e]$. Onsager solved for the steady-state ($\partial P/\partial t = 0$) case for a source at r_0 and sinks at the origin and infinity. The ratio of the stationary flow into $r = \infty$ to the source gives the probability, $f(r, \theta, E)$ that an ion pair thermalized with an initial separation r and at an angle θ with E will escape initial recombination. Onsager found:

$$f(r, \theta, E) = \exp(-A)\exp(-B) \sum_{n=0}^{\infty} \sum_{m=0}^{\infty} \frac{A^m}{m!} \frac{B^{m+n}}{(m+n)!} \tag{3.116}$$

where

$$A = e^2/4\pi\kappa\kappa_0 kTr, \quad B = (eEr/2kT)(1 + \cos\theta) \tag{3.117}$$

If ϕ_0 is defined as the efficiency of production of thermalized ion pairs per absorbed photon and $g(r, \theta)$ as the initial spatial distribution of separation of each ion pair, the overall photogeneration efficiency will be given by

$$\phi(E) = \phi_0 \int f(r, \theta, E) g(r, \theta) d^3r. \tag{3.118}$$

A reasonable assumption is that the initial distribution of thermalized pairs is an isotropic δ function,

$$g(r, \theta) = \delta(r - r_0)/4\pi r_0^2. \tag{3.119}$$

The Onsager theory has been applied to photogeneration in a number of molecular solids. The most complete comparison of the theory, as contained in eqns. (3.116)—(3.119), and experiment has been carried out, recently, by Pai and Enck [38] on amorphous selenium. The very satisfactory agreements they found are reviewed in this book in Chapter 7.

The Onsager theory is based on diffusion in a liquid. However, Pai and Enck fit their data (with 620-nm wavelength excitation) with $r_0 = 8.4$ Å, a distance of the order of the intermolecular spacing. At such small distances it is reasonable to expect that the carrier hops between sites. Therefore, we wish to show the connection between the Onsager Model and hopping transport [39].

Starting with eqn. (3.10), we shall indicate how one can obtain the Smoluchowski equation, eqn. (3.115). First, it is easy to show how the solution of eqn. (3.10) corresponds to solving a RW on a random medium. One simply takes the Laplace transform of eqn. (3.10) and algebraically regroups the terms to arrive at the RW equation:

$$\tilde{R}(s, u) - \delta_{s, s_0} = \sum_{s'} \tilde{\psi}[s, s', u|\{s_i\}] R(s', u) \tag{3.120}$$

where now:

$$\tilde{\psi}[s, s', u|\{s_i\}] \equiv W(s, s')\tilde{Q}(u|\{s_i\}), \tag{3.121}$$

$$\tilde{R}(s, u) \equiv \tilde{P}(s, u)/\tilde{Q}(u|\{s_i\}), \tag{3.122}$$

with $\tilde{Q}(u|\{s_i\})$ the Laplace transform of $Q(t|\{s_i\})$ [eqn. (3.26)], the probability to leave a site and $W(s, s')$, the transition rate between sites. The fixed configuration of the sites in this RW is emphasized by the designation $\{s_i\}$. A full discussion of the RW in a random medium, described by eqn. (3.120) and equivalent to the Master Equation [eqn. (3.10)], and its relation to the CTRW, can be found in Appendix B of ref. 10.

In deriving a (continuum) diffusion equation from the discrete case, one makes a Fokker—Planck type of expansion

$$P(r', t) \simeq P(r, t) + (r' - r) \cdot \nabla P(r, t) + \tfrac{1}{2}(r' - r)(r' - r):\nabla\nabla P(r, t) \tag{3.123}$$

The validity for such an expansion rests on the assumption that $P(r, t)$ is a smoothly varying function of r over a typical displacement (as provided by a hop or between scatterings). We insert the expression in eqn. (3.123) into the Master Equation and obtain

$$\frac{\partial P}{\partial t}(r, t) = \tfrac{1}{2}\sum_{r'} W(r, r')(r' - r)(r' - r):\nabla\nabla P(r, t) +$$

$$+ \sum_{r'} W(r, r')(r' - r) \cdot \nabla P(r, t) + \sum_{r'} [W(r, r') - W(r', r)] P(r, t). \tag{3.124}$$

If the transition rate is a symmetric function, as it is for an isotropic medium:

$$W(r, r') = W(r', r), \tag{3.125}$$

then the last term on the right-hand side of eqn. (3.124) vanishes as well as the second term (by symmetry) and we have the usual diffusion equation:

References pp. 114—115

$$\frac{\partial P}{\partial t}(r, t) = D \nabla^2 P(r, t), \qquad (3.126)$$

where:

$$D = \tfrac{1}{2} \sum_{r'} W(r, r')(r' - r)^2, \qquad (3.127)$$

the diffusion constant, which is assumed to be isotropic.

To incorporate a force field into the diffusive motion one must relax the condition in eqn. (3.125). The force field $\nabla U(r)$ must enter through the transition rates $W(r, r')$. We now use the more complete characterization of these transition rates, developed by Miller and Abrahams [13], rather than the expression cited in eqn. (3.28):

$$W(r, r') = F(|r' - r|)|U_{rr'}| \begin{cases} \bar{n}(|U_{rr'}|) & U(r) > U(r') \\ \bar{n}(|U_{rr'}|) + 1 & U(r) < U(r') \end{cases} \qquad (3.128)$$

where:

$$U_{rr'} \equiv U(r) - U(r'), \qquad (3.129)$$

and:

$$\bar{n}(U) = [\exp(U/kT) - 1]^{-1} \qquad (3.130)$$

is the familiar Bose–Einstein distribution function. The spherically symmetric part of the transition rate, $F(|r' - r|)$ contains the exponential dependence on the intersite separation. The carrier transition is accomplished by the absorption or emission of phonons. A more compact way of writing eqn. (3.128), to cover both cases, is

$$W(r, r') = F(|r' - r|) U_{rr'} \bar{n}(U_{rr'}). \qquad (3.131)$$

To derive the Smoluchowski equation, it is necessary to make two further approximations: (1) High temperature expansion of the Bose–Einstein factor:

$$(\exp(x) - 1)^{-1} = (x + 1/2 x^2 + \ldots)^{-1} \simeq x^{-1} - 1/2 + 0(x), \qquad (3.132)$$

which results in

$$W(r, r') = F(|r' - r|)(kT - 1/2 U_{rr'}). \qquad (3.133)$$

(2) A Taylor expansion of $U_{rr'}$ [the same as the one used for $P(r', t)$ in eqn. (3.123)]

$$U_{rr'} = U(r) - U(r') = -(r' - r) \cdot \nabla U(r) - 1/2(r' - r)(r' - r):\nabla\nabla U(r). \qquad (3.134)$$

One now inserts the expansion in eqn. (3.134) into the high temperature ($U_{rr'}/kT < 1$) form of the transition rate in eqn. (3.133) and substitutes the resulting expression into eqn. (3.124) to obtain the Smoluchowski equation, eqn. (3.115), where we have used:

$$\sum_{r'} F(|r'-r|)(r'-r) = 0 \qquad (3.135)$$

and:

$$D \equiv \frac{kT}{2} \sum_{r'} F(|r'-r|)(r'-r)(r'-r) \qquad (3.136)$$

the diffusion constant, which is isotropic (a scalar).

Hence, the Onsager model is related to the discrete hopping case when one assumes that the probability function $P(r, t)$ and the potential $U(r)$ are smoothly varying over a typical hopping displacement and that the corresponding potential differences $|U(r) - U(r')|$ are less than kT. One expects these approximations to be valid when the initial separation r_0 of the electron–hole pair is greater than the intermolecular separation. It is surprising that the Onsager solution continues to give a good description of the quantum efficiency in amorphous Se even down to $r_0 \simeq 8$ Å. The next step that should be taken to elucidate this point is a CTRW approach to discrete hopping in the presence of a Coulomb center and a d.c. bias field.

3.9 LANGEVIN RECOMBINATION [40]

If one examines the Smoluchowski equation in the steady-state limit and for zero external field one can easily solve for $n(r) \equiv P(r, \infty)$. One has, for the radial current:

$$\frac{I}{4\pi r^2} = D\left[\frac{dn(r)}{dr} + \frac{n(r)}{kT}\frac{dU(r)}{dr}\right] \qquad (3.137)$$

with $U(r)$ a central potential. One can rewrite eqn. (3.137):

$$\frac{I}{4\pi r^2} \exp[U(r)/kT] = D\frac{d}{dr}[n(r)\exp(U(r)/kT)] \qquad (3.138)$$

which can be integrated to give

$$n(r) = \left[\frac{I}{4\pi D}\int_a^r \frac{dz}{z^2}\exp[U(z)/kT] + C\right]\exp[-U(r)/kT]. \qquad (3.139)$$

The expression in eqn. (3.139) is general for a central force interaction $U(r)$ and can be used to solve for recombination, dissociation and the Onsager case (source at r_0, sinks at $r = 0$, $r = \infty$). The distinction between the cases lies in the boundary conditions employed. For recombination, we use the boundary conditions $n(a) = 0$ and $n(\infty) = n_0$, a constant average density. For a Coulomb potential $U(r) = -e^2/4\pi\kappa\kappa_0 r$, one can evaluate the integral in eqn. (3.139) and obtain

$$n_0 = \frac{I}{D}\frac{\kappa\kappa_0 kT}{e^2}[1 - \exp(-e^2/4\pi\kappa\kappa_0 kTa)] \qquad (3.140)$$

Inserting the Einstein relation $D = \mu kT/e$, and defining the recombination coefficient R as the current per unit density, I/n_0, we have:

$$R = \frac{\mu e}{\kappa\kappa_0}[1 - \exp(-e^2/4\pi\kappa\kappa_0 kTa)] \qquad (3.141)$$

which for $e^2/4\pi\kappa\kappa_0 kTa \gg 1$ becomes the well-known Langevin formula [41]. One could have obtained this last result simply by dropping the diffusion term (dn/dr) in eqn. (3.137). The criterion for dropping this term is now seen to be $e^2/4\pi\kappa\kappa_0 kTa \gg 1$.

The Onsager case involves the dissociation of a charge with its parent counter charge. The Langevin recombination is associated with a charge diffusing in the Coulomb well of any counter charge.

Langevin recombination has been well established as the mechanism for volume recombination in anthracene [42]. Double injection experiments in orthorhombic sulfur by Dolezalek and Spear [43] have also confirmed the Langevin value for the carrier recombination. For mobile carriers of both signs:

$$R = \frac{e}{\kappa\kappa_0}(\mu_p + \mu_n) \qquad (3.142)$$

and in sulfur $\mu_p \gg \mu_n$, the hole mobility is more than four orders of magnitude larger than the electron mobility. When a pulse of electrons was made to drift through a uniform distribution of mobile holes, the recombination coefficient was measured and found to be in numerical agreement with $R \simeq e\mu_p/\kappa\kappa_0$. However, because of the shorter trapping time of holes, one could arrange to have the electrons interacting with the *trapped holes*, i.e. $\mu_p = 0$. In the latter case, the measured recombination coefficient decreased by nearly five orders of magnitude and is consistent with $R \simeq e\mu_n/\kappa\kappa_0$. Thus, it is confirmed in single crystal sulfur that the recombination is a diffusion-controlled process.

3.10 CONCLUSION AND SUMMARY

The central theme of this chapter has been the important role played by localized electronic states in the transport, photogeneration and recombination in (disordered) molecular solids.

For the transport, we have considered the carrier hopping from one localized site to another. To describe this hopping among a random distribution of sites we have introduced the formalism of continuous time random walks (CTRW) and have shown how a computation of the basic probability function $P(s, t)$ in the CTRW can determine the a.c. conductivity $\sigma(\omega)$ and the transient photocurrent $I(t)$. In addition, we have demonstrated an intimate connection between $\sigma(\omega)$, for hopping carriers, and the time-dependent pair luminescence spectra that has been extensively studied in semiconductors. The Fourier spectral analysis of the time-dependent luminescence completely determines $\sigma(\omega)$. A number of the physical arguments used in the application of the CTRW to hopping in a random medium can be verified experimentally in the luminescence measurements. Pair radiative recombination corresponds to a single hop with the emitted photon as a signal that the hop has taken place.

In the band picture of transport, the carrier moves in the extended states and interacts with a localized electron state through trapping and release or through recombination. Thus, to describe carrier motion one needs a minimal set of parameters such as the microscopic mobility μ_0, the trapping time τ_n and the release time τ_r. As discussed in Section 3.1, these parameters can be determined from the matrix elements of the appropriate quantum transitions. In the case of hopping, the mobility, itself, can be considered as a sequence of trapping and release, and they are all determined from one transition rate $W(r)$ between the localized sites.

In most molecular solids, one has strong evidence that the photogeneration and recombination are diffusion-controlled processes (*cf.* Sections 3.8 and 3.9). These are processes where the carrier mean free path is short compared to the effective cross-sectional radius of the "trap". We have shown how the diffusion description is a continuum limit of discrete hopping motion and we have indicated how these processes can be determined by a direct consideration of hopping in a Coulomb well modified by an external field.

An implicit theme of this chapter concerns the need to use time-dependent methods to study photoconductivity in highly insulating molecular solids. Even in the case of photogeneration and recombination, the experimental technique involves the measurement of time-dependent currents [38, 43]. The Onsager model uses the steady-state solution of the Smoluchowski equation, however, the quantum yield is determined from time-of-flight experiments [38]. The procedure is valid because the time of observation of the quantum yield ($\gtrsim t_T$) is usually long compared to an intrinsic time of diffusion t_D in the Coulomb well. If one delays the application of the field after the incidence of the short light pulse, in a transient photocurrent measurement, by a time Δt and if $\Delta t \lesssim t_D$, then one has to use the time-dependent Onsager solution. The next step will be to determine the time-dependent Onsager solution on a discrete lattice as we have indicated before. Comparing the theoretical results to time-delay experiments will be a further check on the validity of the Onsager model.

In general, the time dependence of $I(t)$ has been an important probe of

References pp. 114—115

the nature of the carrier transport. The shape of $I(t)$ and its invariance to a change in transit time has been decisive in establishing the validity of the stochastic transport model of carrier motion. The basic dynamics is traced to the wide distribution of hopping times in a disordered system. It is this same wide distribution of hopping times that accounts for the behavior of $\sigma(\omega)$ in impurity-hopping conduction and the time-dependent luminescence spectra of impurity-pair recombination.

One can now use $I(t)$ measurements in, for example, molecularly doped polymers, in combination with the theoretical analyses described in this chapter, to probe the microscopics of molecular charge transfer.

REFERENCES

1 R.H. Bube, Photoconductivity of Solids, Wiley, New York, London, 1960.
2 A. Rose, Concepts in Photoconductivity and Allied Problems, Interscience Publishers, New York, 1963.
3 A.K. Jonscher, Principles of Semiconductor Operation, Wiley, New York, 1959.
4 S.H. Ryvkin, Photoelectric Effects in Semiconductors, Consultants Bureau, 1964.
5 L. Heijne, Physical principles of photoconductivity, Philips Tech. Rev., 25 (1963—64), 27 (1966) and 27 (1968).
6 Proceedings at the Enrico Fermi Summer School on the Optical Properties of Solids, J. Tauc (Ed.), Academic Press, New York, 1965.
7 J. Ziman, Electrons and Phonons, Oxford Univ. Press, London, 1960.
8 T.C. Arnoldussen, R.H. Bube, E.A. Fagen and S. Holmberg, J. Appl. Phys., 43 (1972) 1798.
9 R.A. Smith, Semiconductors, Cambridge Univ. Press, Cambridge, 1959; in, F. Seitz and D. Turnball (Eds.), Solid State Physics, Academic Press Inc., New York, 1955 *et. seq.*
10 H. Scher and M. Lax, Phys. Rev. B, 7 (1973) 4491.
11 M. Lax, Phys. Rev., 109 (1958) 1921.
12 M. Lax, unpublished.
13 A. Miller and E. Abrahams, Phys. Rev., 120 (1960) 745.
14 A.W. Montroll and G.H. Weiss, J. Math. Phys., 6 (1965) 167.
15 H. Scher and M. Lax, Phys. Rev. B, 7 (1973) 4502.
16 S.O. Rice, Bell Syst. Tech. J., 23 (1944) 282.
17 K. Colbow, Phys. Rev. A, 139 (1965) 274.
18 D.G. Thomas, J.J. Hopfield and W.M. Augustyniak, Phys. Rev. A, 140 (1965) 202.
19 R.C. Enck and A. Honig, Phys. Rev., 177 (1969) 1182.
20 M. Pollak and T.H. Geballe, Phys. Rev., 122 (1961) 1742.
21 A.I. Lakatos and M. Abkowitz, Phys. Rev. B, 3 (1971) 1791; M. Abkowitz, A. Lakatos and H. Scher, Phys. Rev. B, 9 (1974) 1813.
22 H. Scher, unpublished.
23 M. Pollak, J. Non-Cryst. Solids, 11 (1972) 1; V. Ambegaokar, B.I. Halperin and J.S. Langer, Phys. Rev. B, 4 (1971) 2612.
24 A.K. Jonscher, J. Phys. C, 6 (1973) L235.
25 M. Abkowitz, P. LeComber and W. Spear, Chelsea University Lectures 1975, unpublished, and Commun. Phys. (submitted).
26 E.W. Montroll and H. Scher, J. Stat. Phys., 9 (1973) 101.
27 H. Scher, in J. Stuke and W. Brenig (Eds.), Amorphous and Liquid Semiconductors, Taylor and Francis, London, 1974, p. 135.

28 J. Scher and E.W. Montroll, Phys. Rev. B, 12 (1975) 2455.
29 J.W. Cooley and J.W. Tukey, Math. Comput., 19 (1965) 297.
30 Handbook of Mathematical Functions, M. Abramowitz and I.A. Stegun (Eds.), U.S. GPO Washington D.C., 1964.
31 M. Shlesinger, J. Stat. Phys., 10 (1974) 421.
32 H. Hoegl, J. Phys. Chem., 69 (1965) 755.
33 W.D. Gill, J. Appl. Phys., 43 (1972) 5033.
34 J. Mort, G. Pfister and S. Grammatica, Solid State Commun., 18 (1976) 693.
35 G. Pfister, Personal Communication.
36 R.L. Emerald, R.E. Drew and R. Zallen, Phys. Rev. B, in press.
37 L. Onsager, Phys. Rev., 54 (1938) 554; J. Chem. Phys., 2 (1934) 599.
38 D.M. Pai and R.C. Enck, Phys. Rev. B, 11 (1975) 5163.
39 M. Lax and H. Scher, unpublished.
40 M. Silver and R.C. Jarnagin, Mol. Cryst., 3 (1968) 461.
41 P. Langevin, Ann. Chim. Phys., 28 (1903) 287, 433.
42 F.N. Coppage and R.G. Kepler, Mol. Cryst., 2 (1967) 231; M. Pope and J. Burgos, Mol. Cryst., 1 (1966) 395.
43 F.K. Dolezalek and W.E. Spear, J. Phys. Chem. Solids, 36 (1975) 819.

CHAPTER 4

COVALENT SEMICONDUCTORS

RICHARD H. BUBE

4.1. Introduction
4.2. Imperfection control of lifetime
4.3. Achieving maximum photoconductivity performance
4.4. Photo-Hall and photothermoelectric effects in crystals
4.5. Analysis of trapping
4.6. Photoadsorption effects
4.7. Photochemical changes
4.8. Photoconductivity in polycrystalline films
4.9. Heterojunctions

4.1 INTRODUCTION

If the title of covalent semiconductors is taken to include all inorganic materials in crystalline form, except for metals and strictly ionic compounds, the task of describing photoconductivity in these materials in the space allotted in this chapter is formidable indeed. The publications dealing with photoconductivity in such materials over only the last decade surely number several thousand. To make the task commensurate with the boundary conditions, we have therefore been forced to a number of simplifying procedures. First, we note the existence of a number of previous surveys and reviews on the same or a similar topic; these include books on photoconductivity itself by Bube [1] and Rose [2], proceedings of conferences on photoconductivity [3—5], books on photoelectronic processes in semiconductors by Ryvkin [6] and Vavilov [7], and review papers on photoconductivity and photoelectronic effects [8—13]. Second, relying on these publications for adequate treatment of the past, we concern ourselves only with work done in the last 10—15 years, although we do make some effort to include literature references to earlier key work. Third, and perhaps most unforgivable, even if most understandable, we limit ourselves to topics with which we have had some personal contact during recent years.

References pp. 147—153

4.2 IMPERFECTION CONTROL OF LIFETIME

The incorporation of imperfections into a semiconductor is normally expected to decrease the free carrier lifetime, and hence the photoconductivity, because of the additional recombination paths provided by the imperfections. Such behavior can be described by the Shockley—Read recombination model [14] or variations of that model [15]. More striking, however, is the occasional observation that the incorporation of specific types of imperfection actually results in increased free carrier lifetime for the majority carrier because of the effect they have on the occupancy of recombination centers and their own very small capture cross-section for majority carriers [16—27]. In the case of an n-type photoconductor, for example, such imperfections lie below the thermal equilibrium Fermi level, have a much smaller cross-section for the capture of electrons than of holes, and under photoexcitation effectively transfer their electrons to other recombination centers with larger cross-section for electrons; in this way, they present photoexcited electrons with a smaller density of hole-occupied large cross-section centers and hence increase the electron lifetime. Three other phenomena are associated with the presence of such sensitizing centers: superlinear photoconductivity in which the photoconductivity increases with a power of the light intensity greater than unity as the sensitizing centers change function from simple hole traps to recombination centers with increasing light intensity; thermal quenching of photoconductivity in which holes captured at sensitizing centers are thermally released to take part in recombination at other larger cross-section centers; and optical quenching of photoconductivity in which holes captured at sensitizing centers are optically released, by photoexcitation of electrons from the valence band, to take part in recombination at other larger cross-section centers.

The actual physical, chemical and crystallographic identity of the sensitizing centers in a particular material varies from well known in Ge and Si to still uncertain in spite of years of research in III—V and II—VI compounds. Examples of known sensitizing centers are Mn in Ge [28—34] and Zn in Si [35—48]. In both cases, the sensitizing center is an impurity which exists in the crystal in the dark in the doubly-negative state; photoexcitation results in a change to a singly-negative state which still offers a Coulomb repulsion for the capture of electrons and hence results in a small electron capture cross-section.

A recent investigation of Zn impurity in Si [47] provides an illustration of the methods used to obtain information about such sensitizing centers. Zinc impurity in Si has two energy levels: a double-acceptor level lying about 0.5 eV below the conduction band, and a single acceptor level lying about 0.31 eV above the valence band. It is when the zinc is in the double-acceptor state that it functions as a sensitizing center. The energy level location

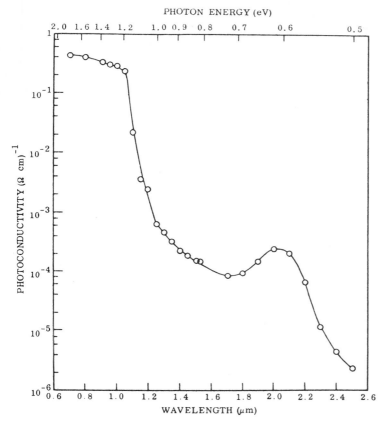

Fig. 4.1. Excitation spectrum for photoconductivity in high-resistivity n-type Zn-doped silicon at 80 K. (From Sklensky and Bube [47])

corresponding to this center can be determined in three independent ways: (a) the Hall effect analysis of conductivity as a function of temperature yields an electron ionization energy of 0.51 eV for the occupied Zn^{-2} center since the dark conductivity is associated with thermal excitation from this level to the conduction band; (b) the long wavelength cutoff of the photoconductivity excitation spectrum gives a value of 0.52 eV for the same energy, and (c) the long wavelength cutoff of the optical quenching spectrum gives a value of 0.58 eV for the height of the level above the valence band when unoccupied, corresponding to a depth below the conduction band of about 0.58 eV. The photoconductivity excitation spectrum at 80 K is given in Fig. 4.1; the minimum occurring at about 0.75 eV results from the simultaneous photoexcitation (occupied level to conduction band) and optical quenching (valence band to unoccupied level) caused by photons with energy greater than 0.58 eV.

References pp. 147—153

The density of sensitizing centers can be determined from the fact that the photoexcited carrier density saturates at high light intensities when the sensitizing centers are essentially all hole-occupied. Such an effect was observed for Zn impurity in Si, the actual density of electrons existing in the saturated condition being equal to the total density of sensitizing centers minus the density of electrons excited from sensitizing centers but captured by electron traps.

The capture cross-section of the sensitizing centers can be determined from the value of the electron density at which optical quenching of photoconductivity starts; it corresponds, in the simple model, to an equality between the rate of photoexcitation of electrons from the valence band to unoccupied sensitizing centers and the rate of capture of electrons from the valence band by unoccupied sensitizing centers. The electron capture cross-section associated with the singly-negative Zn center in Si was found to be about 10^{-20} cm^2 and temperature independent below 100 K, and to increase with an activation energy of 40 meV for temperatures between 100 and 200 K. The weak temperature dependence can probably be interpreted in terms of a thermally-assisted tunneling through the repulsive barrier surrounding the center.

The charge state of the sensitizing center can be determined through photo-Hall measurements of the change in the low-temperature mobility upon photoexcitation. As the sensitizing centers become unoccupied as the result of photoexcitation, their charge changes from -2 to -1, producing a corresponding increase in the mobility controlled by charged impurity scattering. Since the scattering cross-section varies as the square of the charge, this change in charge due to scattering represents a fourfold decrease in the scattering cross-section of the sensitizing centers. Measurements on Zn impurity in Si are consistent with this interpretation.

The lifetime of photoexcited electrons in Si with Zn impurity can be determined in three independent ways and checked for internal consistency: (a) from the steady-state photoconductivity and Hall effect, the lifetime τ can be calculated from $\Delta n = f\tau$, where f is the photoexcitation rate per unit volume per second; (b) from the decay of photoconductivity after the cessation of excitation in a regime where the decay is due to recombination at the centers of interest and is not affected by trap emptying; and (c) from the measured value of capture cross-section using optical quenching and a knowledge or assumption about the density of unoccupied centers, the lifetime can be calculated from $\tau = (NSv)^{-1}$, where N is the density of recombination centers, S is their electron capture cross-section, and v is the average thermal velocity of an electron. A comparison of lifetimes determined in these three ways for Zn impurity in Si is given in Fig. 4.2; over the temperature range of relevance, reasonable agreement is obtained.

Similar effects due to sensitizing centers have also been reported in GaAs into which Cu has been diffused [49, 50]. It is uncertain, however, whether

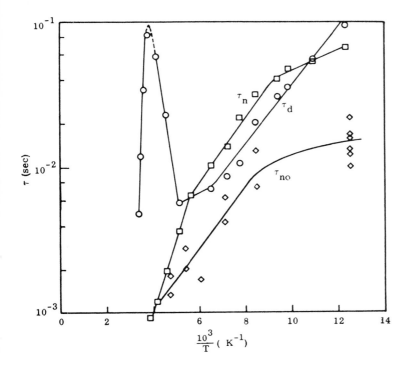

Fig. 4.2. Comparison of three methods for determining the electron lifetime as a function of temperature in high-resistivity n-type Zn-doped silicon. The curve marked with squares is the lifetime calculated from measurements of steady-state photoconductivity. The curve marked with circles is the decay time; above 200 K this curve is determined by trap emptying and does not represent the electron lifetime. The curve marked with triangles is the lifetime calculated from the independently measured capture cross-section of the Zn^- centers, assuming that all the Zn are in this state. (From Sklensky and Bube [47])

the sensitizing center level that results at 0.45 eV above the valence band is due simply to Cu impurity or to a complex involving Cu impurity. Sufficient Cu can be incorporated into n-type GaAs to lower the Fermi level below the middle of the gap and thus convert the conductivity to p-type; as long, however, as the Fermi level lies more than 0.45 eV above the valence band, the sensitizing effect of the center associated with that level is experienced and the photoconductivity is n-type even though the dark conductivity is p-type. The electron capture cross-section of these centers has been measured as a function of temperature from the optical quenching effect [51]; the cross-section is about 5×10^{-22} cm^2 and almost independent of temperature. A sample of InP with Cu impurity also showed sensitized photoconductivity behavior, and for this sample the capture cross-section was measured to be 4×10^{-21} cm^2 almost independent of temperature. Such temperature-independent cross-sections might be associated with neutral

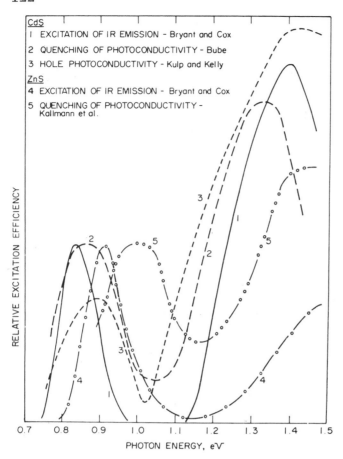

Fig. 4.3. Spectral dependence for optical quenching of n-type photoconductivity, excitation of infrared emission, and excitation of p-type photoconductivity in CdS and ZnS. (From refs. 66—71)

centers involving radiative transitions as the recombination process; some support for this hypothesis is derived from the location of energy levels associated with sensitizing centers and known imperfection-associated luminescence emission in these materials.

Sensitized photoconductivity effects have been observed in CdS and many other II—VI photoconductors almost from the beginning of research in this area [52]. Investigations include work on CdSe [53—55], ZnS [56—58], ZnSe [59—61], and ZnTe [62]. Quite similar results have been observed in other materials such as HgI_2 [63] and GaSe [64, 65]. In all these cases, the identity of the sensitizing center is unknown, although its electronic properties in many cases are completely reproducible and determinable. A long-standing debate has sought to establish whether the sensitizing centers in CdS and CdSe are basically associated with defects such as Cd vacancies, or are associated with residual Cu impurity. A number of processes have the same type of spectrum in ZnS and CdS as shown in Fig. 4.3; measurements of excitation of infrared emission, quenching of photoconductivity and excitation of hole conductivity all indicate a narrow band with maximum at about 0.9 eV and a broad band or band edge at about 1.1 eV [66—71]. It has been suggested [72] that the sensitizing center is an isolated substitutional Cu impurity on a Cd site, and that the narrow band with maximum at 0.9 eV corresponds to an internal d-shell transition in the Cu^{+2} d^9 configuration. Both excitation of infrared emission and of hole conductivity do appear to be associated with the Cu impurity, but optical quenching of photoconductivity (the only one of the three processes explicitly involving a sensitizing center) does not. The most sensitive CdS photoconductors are prepared with only a trace of halogen donors; incorporation of any density of Cu impurity serves only to decrease the electron lifetime. The characteristic optical quenching spectrum shown in Fig. 4.3 is found in the purest high photosensitivity CdS crystals, and does not change in shape for a wide variety of CdS crystals prepared under quite different conditions [11]. Evidence has been produced that the center responsible for the green luminescence in ZnS:Cu, almost certainly involving the Cu, is not the same as the sensitizing center for photoconductivity in that material [56]. Perhaps the most instructive results come from an investigation of luminescence and photoconductivity in ZnSe [60, 61]. In p-type ZnSe:Cu a specific optical absorption, optical quenching of n-type photoconductivity, and photoexcitation of p-type photoconductivity can all be identified directly with transitions involving the Cu^{+2} d^9 configuration. Cu is thereby shown to be the major sensitizing center for n-type photoconductivity in p-type ZnSe:Cu, with an electron capture cross-section of about 4×10^{-19} cm^2. However, investigation of the photoelectronic properties of ZnSe without Cu indicate that the sensitizing center for n-type photoconductivity in this n-type ZnSe lies closer to the valence band than the Cu level and has an electron capture cross-section of 3×10^{-22} cm^2. It can therefore be concluded that Cu is a sensitizing center for n-type photoconductivity, but that another sensitizing center also exists which in terms of its electron capture cross-section is some thousand times more effective.

Electron capture cross-sections have been measured for a variety of different types of photosensitive CdS crystals, yielding values of the order of 10^{-21} cm^2 [51]. Saturation of photoconductivity due to photoexcited emptying of sensitizing centers has been investigated in CdS and CdSSe solid solutions [73, 74]. High-intensity laser quenching of photoconductivity in CdS and GaAs was examined to see if there was any indication of Auger recombination competing with the normal recombination processes [75]; within the range of densities available, no evidence of Auger recombination

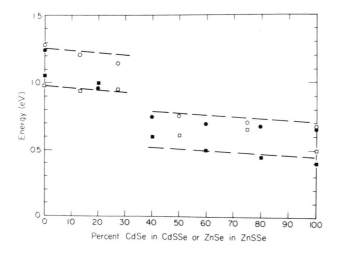

Fig. 4.4. Variation of the hole ionization energy of photoconductivity sensitizing centers in CdSSe and of Cu-associated luminescence centers in ZnSSe, as a function of the Se/S ratio. Open points are for CdSSe, solid points for ZnSSe; circles are optical ionization energies determined from optical quenching, squares are thermal ionization energies determined from thermal quenching. (From Bube [13])

was obtained, indicating that the appropriate Auger recombination coefficient for the sensitizing centers is less than 10^{-28} cm^6 s^{-1} and hence more than four orders of magnitude smaller than that calculated [76] for a hydrogen-like center.

The variation of the location of the energy level for the sensitizing center with composition in the CdSSe solid solution system was investigated using both thermal quenching and optical quenching to locate the level [77]. Strikingly similar results were obtained from luminescence measurements on ZnSSe:Cu:Cl powders [78]. The two sets of data are compared in Fig. 4.4; the similarity compounds the problem of distinguishing between Cu and defects as the sensitizing center in CdS and CdSe described above, but since the two types of imperfection do have approximately the same energy level it is perhaps not surprising that they have similar variations with composition. The thermal ionization energy for a hole from a sensitizing center in the CdSSe system is always less than the optical ionization energy, presumably because of the Franck—Condon principle. At the sulfide and selenide ends of the composition there are small variations in energy approximately the same as the band gap with composition; between 30 and 40 percent selenide, however, there is an almost abrupt transition from a sulfide-like ionization energy to a selenide-like ionization energy. The transition occurs approximately when on the average there are more centers with two selenium and one sulfur nearest neighbors than centers with two sulfur and one selenium nearest neighbors.

The effective thermal ionization energy of sensitizing centers in CdS can be decreased (*i.e.*, the critical temperature for rapid thermal quenching is reduced) by increasing the density of impurities. In a series of CdS:Ga:Cu powders, the effective hole ionization energy was reduced from 0.75 eV at 4×10^{17} cm^{-3} Cu and Ga impurities to 0.29 eV at 2×10^{20} cm^{-3} in a consistent manner satisfying the relationship $E = 0.92 - 9 \times 10^{-8} N_I^{1/3}$ eV, where N_I is the density of either Cu or Ga impurities [79]. In similar measurements on CdS:Cu:I single crystals, a similar decrease in effective hole ionization energy was observed from 1.17 to 0.55 eV with increasing density of impurities [80]. Below the critical temperature for thermal quenching, the electron lifetime in these materials is unaffected by the increasing impurity density. These results may indicate a genuine decrease in hole ionization energy of the sensitizing centers due to imperfection interactions, or they may instead be interpretable as the result of an increasing density of trapping states introduced with the higher impurity densities, these trapping states coming to dominate the recombination and cause the thermal quenching of sensitivity at lower temperatures [26, 27].

Unusual photoconductivity properties associated with Cl and Ga donors in CdTe crystals lead to some interesting speculations. These properties are quite distinguishable from the small capture cross-section centers described earlier and labeled double acceptors by Lorenz *et al.* [81]. Both Cl and Ga impurities were found to act as single donors in CdTe with a neutral (occupied) donor level that is degenerate in energy with the lowest conduction band minimum at Γ_1 [82, 83]. Photoconductivity resulting from photoionization of neutral Cl or Ga donors exhibits very long decay times. These decay times are strongly temperature dependent but are *independent* of carrier and donor densities for a wide range of values. The optical ionization energy of neutral Cl and Ga donors was measured from the photon energy necessary to excite electrons from neutral donors into the Γ_1 conduction minimum. For Cl, the thermal depth determined from the temperature dependence of decay time is 0.51 eV, and the optical depth obtained from analyzing the optical data in terms of Lucovsky's model [84] is 0.90 eV, in spite of the fact that electrically the level is at an energy degenerate with the conduction band minimum at Γ_1. For Ga, the corresponding values are 0.31 eV and 1.0 eV. In addition, the free carrier mobility in the Γ_1 minimum is found to be almost independent of the state of ionization of the Cl or Ga donors, changes with photoexcitation being less than 1/10 of that expected for the transformation of a singly-charged impurity to a neutral impurity; these results are in marked contrast to the effect of photoexcitation on the double acceptor imperfections in the same samples of CdTe for which changes in mobility of the expected magnitude were easily detectable. These observations, together with the results of changes in conductivity caused by hydrostatic pressure [82], suggest a resolution in terms of a model proposed by Paul [85, 86] to account for

References pp. 147—153

unusual donor behavior in GaSb. It is proposed that the Cl and Ga donor states are constructed of wave functions from the higher X_1 and L_1 minima, and have no overlap in k with the lowest minimum at Γ_1. Then photoionization of the neutral donors results from optical excitation of the bound electron to the higher band minimum followed by phonon scattering to the Γ_1 minimum. Recombination of free carriers with the ionized donor occurs only for electrons with sufficient energy to occupy conduction band states which overlap the donor ground state in k space.

4.3 ACHIEVING MAXIMUM PHOTOCONDUCTIVITY PERFORMANCE

When a photoconductor is being used as a light detector, its performance is judged by the dual criteria of photoconductivity gain and speed of response. The gain is increased by increasing the applied voltage, but in systems with high gain utilizing ohmic contacts, the injection of space-charge-limited current from the contacts limits the maximum gain that can be achieved [85]. The relationship between maximum gain and other materials parameters is given by $G_{max} = M(\tau_0/\tau_r)$, where M is a multiplying factor that can be larger than unity under special conditions of trap distribution, τ_0 is the observed response time, and τ_r is the dielectric relaxation time at the operating conditions. This relationship makes evident the tradeoff nature of maximum gain *versus* response time, in the usual manner in which the gain—bandwidth product is a constant. Improvement in the speed of response must result in decreases in the maximum gain obtainable unless some means are found to provide M values in excess of unity. Several attempts have been made to see what possibilities are available.

Nearly pure CdS single crystals were grown by a vapor phase method using extensive precautions to insure materials purity and absence of oxygen [86]. The random occurrence of photosensitive crystals could be made much more reproducible by the inclusion of a trace of iodine in the preparation. Typical spectral response curves for the pure and slightly iodine-doped CdS crystals are given in Fig. 4.5. Improvements in the speed of response were judged by comparison with a "standard" CdS:Cl:Cu crystal photoconductor, M factors were evaluated experimentally, and trap densities were measured using thermally stimulated conductivity. Trap densities were in the low 10^{14} cm^{-3} range, with the optimum case being one crystal with a total trap density of 5×10^{13} cm^{-3} and most of these traps were sufficiently shallow to empty very rapidly at room temperature. Typical of the best performance achieved for low-light illumination is that for one crystal which was 500 times faster on decay and 2000 times faster on rise, and showed an $M = 24$. Very high values of M up to over 500 were observed on certain acicular whisker crystals about 25 μm in diameter. Unfortunately all the best crystals were characterized by small size and high fragility.

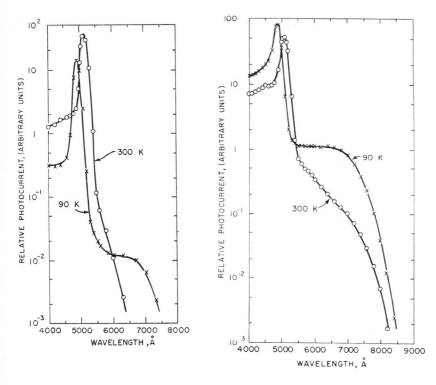

Fig. 4.5. Excitation spectra for CdS single crystals. Curves to the left are for a photosensitive crystal prepared under conditions of high purity. Curves to the right are for a photosensitive crystal prepared with a trace of iodine present during the crystal growth. The increase in extrinsic photoresponse due to the trace of iodine is evident. (From Bube and Barton [86])

At 90 K the speed of response of GaAs:Cu crystals prepared by diffusion of Cu is dominated by a set of trapping levels with depths of about 0.2—0.3 eV and densities of about 10^{14} cm^{-3} [50]. This relatively low trap density results in the GaAs:Cu being about an order of magnitude better than the "standard" CdS:Cl:Cu at room temperature as far as the speed of response at equal sensitivity is concerned.

Speed of response can be improved in highly photosensitive materials not only by decreasing the trap density but also by operating at such conditions of sensitizing center hole ionization energy and temperature that thermal quenching of sensitivity effectively occurs during the decay period. Such a process shortens the decay time, although it may actually lengthen the rise time. The impurity density effect on thermal quenching in CdS:I:Cu [80] crystals discussed above can be used to produce the conditions of enhanced decay rate at room temperature. In spite of the fact that the total density of traps in such a high-impurity material is large (greater than 10^{17} cm^{-3}), these

References pp. 147—153

traps have a monotonic distribution [87] in which the density decreases exponentially with depth so that the effective density at room temperature is several orders of magnitude less than the total. Improvements in speed of response at constant sensitivity compared to CdS:Cl:Cu were found by factors of 75 in decay and 4 in rise. The expected asymmetry in rise and decay times was observed, the rise time being about 40 times longer than the decay time for light intensities between 10^{-1} and 10^{-3} ft.-c.

4.4 PHOTO-HALL AND PHOTOTHERMOELECTRIC EFFECTS IN CRYSTALS

In order to use photoconductivity effects as an effective analytical tool in investigating electrical transport processes and the role of imperfections, it becomes desirable to be able to distinguish between the effects due to changes in carrier density and effects due to changes in carrier mobility. Two effects that can be conveniently used for this purpose are the Hall effect and the thermoelectric (Seebeck) effect. In a later section, we discuss the utility of these methods in providing an understanding of photoconductivity in polycrystalline films; in this section, we consider measurements of these effects in single crystals.

In the first section of this chapter we described how the charge state of sensitizing centers can be determined by measuring the change in charged-impurity scattering with photoexcitation. Such measurements had been made relatively early on doubly negative impurities such as Fe, Co, Ni and Mn in Ge [88]. The presence of such impurities in Ge and Si was also found to give rise to a change in conductivity type with increasing intrinsic excitation, detectable through measurements of the photo-Hall effect [34, 89, 90].

When photo-Hall techniques were used to measure the change in mobility with photoexcitation in high-resistivity photosensitive CdS, CdSe and GaAs crystals [50, 91—93], however, much larger changes in Hall mobility were observed than would have been expected for the change in scattering cross section of point defects. Since these abnormally large effects of photoexcitation on mobility can be decreased by raising the temperature of the crystal, it has been concluded that such effects are to be interpreted in terms of an inhomogeneous distribution of scattering centers producing space-charge regions surrounding microvolumes with differing Fermi level [94].

The photo-Hall mobility is particularly sensitive to two-carrier effects, and becomes zero if $p\mu_p^2 = n\mu_n^2$, where p and n are the hole and electron densities respectively, and μ_p and μ_n are the hole and electron mobilities respectively. A variety of two-carrier effects in CdS and CdSe [91, 93], GaAs [50, 95, 96], and GaP [97] have been investigated by means of the photo-Hall effect.

The specific problem of identifying defects introduced into GaAs by annealing provides a good illustration of the utility of the photo-Hall effect in such problems [95, 98, 99]. It was found that when GaAs crystals were annealed at temperatures between 500° and 700°C, depending on the nature of the original growth process, imperfections were introduced in densities as high as 10^{19} cm^{-3} that acted like traps for charge carriers. Three variations of the photo-Hall technique were used: (a) measurement of the photo-Hall mobility as a function of photoexcitation intensity at fixed temperatures; (b) measurement of the photoconductivity and photo-Hall mobility as a function of temperature at fixed photoexcitation intensity; and (c) measurement of the Hall effect of carriers thermally released from traps in a thermally-stimulated conductivity experiment. At each of four temperatures, the crystal was high-resistivity n-type in the dark, became p-type as the excitation intensity was increased, and finally became n-type at still higher intensities; this behavior could be interpreted in terms of a high density of traps for electrons lying just above the Fermi level in the dark. When the crystal was cooled under constant photoexcitation, the n-type Hall mobility increased rapidly, reached a maximum value, and then dropped off again as, at low temperatures, the photoconductivity abruptly became p-type; this behavior could be interpreted in terms of the existence of a hole trap somewhat shallower than the electron trap observed above, and also of shallow electron traps which remove electrons from the conduction band at low temperatures. These proposals can be checked by the direct measurement of the Hall effect during thermally-stimulated conductivity, with the results shown in Fig. 4.6. The first traps to empty at the lowest temperature are hole traps. Then the Hall mobility drops at the center of the thermally-stimulated conductivity peak to a minimum. This minimum may be associated with the emptying of an electron trap with slightly larger ionization energy than the hole trap; the existence of such traps was indicated by the low-temperature photoconductivity behavior described above. Except for this transient decrease, however, the measured hole mobility retains the same high value both before and after the shallow traps have emptied; this result suggests that the hole and electron traps exist in pairs, and are either neutral when un-ionized, or give small dipole scattering when ionized. After the principal low-temperature peak has been passed, there is an indication of electron trap emptying at the foot of the current peak for the deeper traps. Then strong evidence is obtained for the emptying of holes from the deep traps, the Hall effect finally becoming n-type at high temperatures as the dark conductivity dominates. The absence of a clearly n-type peak associated with a deep trap does not contradict the evidence for a high density of deep electron traps mentioned above, for these electron traps would serve as recombination centers for the holes freed from the slightly shallower hole traps and would not appear in the measured thermally-stimulated conductivity. The utility of the photo-Hall method is realized when it is recognized that the only information available before the temperature-dependent photo-Hall measurements was that there were two traps, one shallow and one deep,

References pp. 147—153

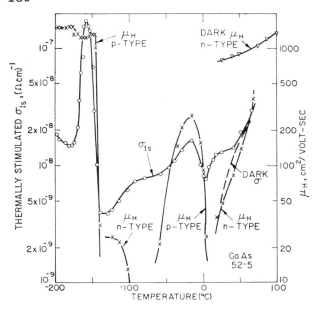

Fig. 4.6. Temperature dependence of thermally-stimulated conductivity and thermally-stimulated Hall effect for an annealed GaAs crystal. (From Bube and MacDonald [92])

of unknown type.

A number of investigations have been directed toward the properties of semi-insulating GaAs:O [96, 100—113] and GaAs:Cr [96, 114—138]. One of the most recent of these [96] involves a comprehensive investigation of the photoelectronic properties of n-type GaAs:O and GaAs:Cr single crystals through measurements of dark conductivity and Hall effect, optical absorption, photoluminescence, photoconductivity, and photo-Hall effect as a function of photon energy, light intensity and temperature, optical quenching of photoconductivity, and thermally-stimulated conductivity. Although the GaAs:O and GaAs:Cr crystals were grown by different methods and by different companies, almost all imperfection levels identified in GaAs:O were found also in GaAs:Cr, which is consistent with the mass spectrographic analysis of these crystals. At low temperatures, intrinsic photoconductivity is n-type and increases exponentially with $1/T$ for both GaAs:O and GaAs:Cr samples, but the slope of the $\ln \Delta n$ versus $1/T$ plot is different from sample to sample. This kind of result can be described successfully by a simple one-level Shockley—Read recombination model which predicts that the slope observed should be given by the difference between the recombination center level and the Fermi level. Applying the model to six different samples of GaAs:Cr with Fermi levels 0.605—0.638 eV below the conduction band at 300 K gave a consistent location for the recombination center level of 0.663 ± 0.003 eV below the conduction band. The same model applied to a GaAs:O crystal with Fermi level 0.580 eV below the conduction band gave a recombination center level 0.646 eV below the conduction band. There is every reason to assume that this is the same recombination center in GaAs:O and GaAs:Cr, and that it is associated with oxygen impurity. The location of the level by this method is quite close to that observed from the conductivity thermal activation energy (0.68 eV in GaAs:Cr) and from the low-energy threshold of the spectral response curve in GaAs:O and GaAs:Cr.

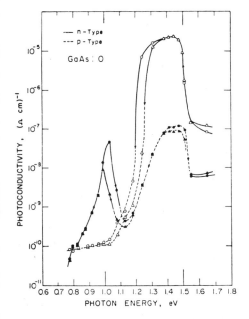
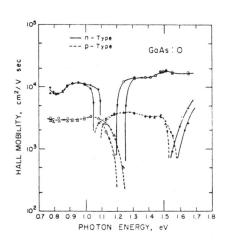

Fig. 4.7. Curves to the left are excitation spectra for photoconductivity in oxygen-doped GaAs at 82 K. Solid curves represent n-type photoconductivity, and dashed curves represent p-type photoconductivity. Arrows on the curves indicate that the curve that is higher at low photon energies is measured from low-to-high energies; the curve that is higher at high photon energies is measured from high-to-low energies. To show time effects, the value after 3 min is shown by circles, and after 15 min by triangles. The curves to the right are the corresponding variations of Hall mobility with photon energy. Exposure to radiation in the 1.0—1.3 eV range at 82 K transforms the photoconductivity permanently to p-type, until the crystal is heated above 120 K. (From Lin [96])

For both GaAs:O and GaAs:Cr at 82 K, a slow photoconductivity quenching process is found under extrinsic photoexcitation of $1.25 \leqslant h\nu \leqslant 1.0$ eV. This process produces a shift in the photoconductivity from n-type toward p-type; the p-type photoconductivity is persistent at 82 K even if the quenching radiation is removed and intrinsic photoexcitation is applied, but recovers abruptly to the n-type photoconductivity if the crystal is heated to

References pp. 147—153

above 105 K. Figure 4.7 illustrates the description of this phenomenon that is made possible by photo-Hall measurements. The persistent quenching can be explained in terms of increasing recombination flux through centers with a large cross-section for electrons after these centers have been emptied by photoexcitation placing their electrons into electron traps; recovery occurs only when the electrons are thermally released from the traps and return to the centers.

The use of the photothermoelectric effect for the separation of charge carrier density and mobility effects has certain advantages over the photo-Hall effect in spite of the power of the latter described in the above examples. The apparatus for the photothermoelectric effect can be much simpler, since it is in general easier to set up a small temperature gradient than to use a large and bulky magnet. The photothermoelectric effect is even more sensitive to two-carrier effects than the photo-Hall effect since the photothermoelectric effect goes to zero approximately when $p\mu_p = n\mu_n$. Finally, the photothermoelectric effect allows the determination of very small carrier mobility values since it is only the carrier density that determines the thermoelectric power, unlike the case of the Hall effect where the measured Hall voltage is proportional to the Hall mobility. The photothermoelectric effect was first described by Tauc for Ge [139, 140]; he developed a relation for the change in thermoelectric power in the small-signal photoconductivity regime using the concept of a steady-state Fermi level, and showed that general agreement was found between experiment and theory. The analysis of Tauc has been generalized by Van der Pauw and Polder [141], but in its first application to the large-signal case in experiments on CdS, even the wrong sign of the thermoelectric power was reported [142].

Recent developments of the photothermoelectric effect for use in the photoelectronic analysis of single-crystal semiconductors include the investigation of n- and p-type Si [143, 144], two-carrier effects in GaAs [123], and a variety of large-signal effects on CdS [145]. In the work on Si, it was shown that the electronic contribution to the thermoelectric power in small-signal transport can be adequately described by a simple model based on the steady-state Fermi level, and that photoexcitation decreases the phonon drag contribution to the thermoelectric power [146, 147]. In the work on CdS, an effort was made to establish in detail the validity of the photothermoelectric approach using a steady-state Fermi level in the large-signal regime with photoconductivity much larger than dark conductivity. Figure 4.8 illustrates the degree of agreement found between carrier densities determined by photo-Hall measurements and by photothermoelectric measurements on the same sample over a wide range of photoconductivities. In addition to these steady-state nonequilibrium situations, the photothermoelectric effect was also shown effective in transient measurements such as photoconductivity decay.

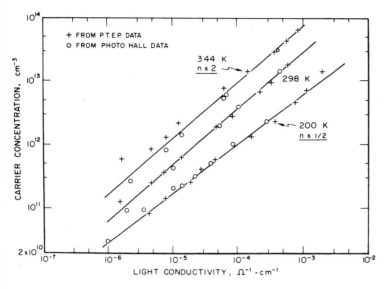

Fig. 4.8. Comparison of electron density calculated from independent measurements of photo-Hall and photothermoelectric effect as a function of light conductivity under white-light excitation for a high-resistivity photosensitive CdS crystal. The data for 200 K and 344 K are offset slightly in magnitude, as indicated, for clarity. (From Kwok and Bube [145])

4.5 ANALYSIS OF TRAPPING

There are a variety of methods available for determining the trap depth, density and capture cross-section of traps in semiconductors. These methods include the analysis of photoconductivity rise or decay curves [16, 148–153], thermally-stimulated conductivity (TSC) curves [96, 153–170], optically-stimulated conductivity [171, 172], dependence of space-charge-limited current on applied voltage [173–179], and electric-field ionization of traps [96, 169, 180, 181].

A careful investigation of the determination of trapping parameters in crystals like CdS [153] emphasizes that there is no simple formula that can be applied to all trapping analysis, guaranteed without further investigation to supply correct values of trap depth and cross-section. A close correlation is found between the quasi Fermi level and the trap depth as long as the traps may be considered to be in effective thermal equilibrium with the conduction band. A quasicontinuous distribution of such traps with depths lying between 0.1 and 0.7 eV were found in crystals of CdSSe. The depths of such traps are given approximately by $\mathcal{E}_t = 20\,kT_m$, where T_m is the temperature for TSC corresponding to these traps. The constant multiplier of kT_m can be shown to lie in the range between 15 and 26 for a wide variety of trapping mechanisms and initial conditions [170]. In these same crystals,

References pp. 147—153

however, a set of traps was also found with a depth of 0.73 eV and a cross-section of 10^{-14} cm^2 that emptied by a completely monomolecular process without retrapping of freed electrons, to which the same type of quasi-Fermi-level analysis was completely inapplicable since these traps are not in effective thermal equilibrium with the conduction band. The nature of these latter traps is discussed further in Section 4.7.

Of all the ways of determining trap depths by analysis of photoconductivity transients and TSC, perhaps the simplest, the most widely applicable, and the most reliable is the original method of decayed TSC curves introduced in principle by Garlick and Gibson in 1948 [155]. Even this method must be used with care, and TSC remains a very useful qualitative technique, difficult to make into a thoroughly reliable quantitative technique.

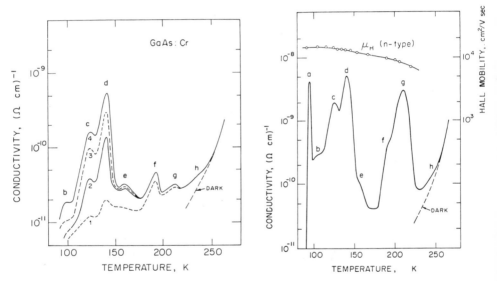

Fig. 4.9. Thermally-stimulated conductivity curves for high-resistivity GaAs:Cr (left) and GaAs:O (right). Letters indicate the identical peaks in both materials. Family of curves to the left shows the effect of varying initial excitation time, in seconds (1) 2, (2) 6, (3) 10, (4) 60 and 600. Plot to the right also shows variation of thermally-stimulated Hall mobility, indicating that electron traps are involved. (From Lin [96])

A particularly varied TSC structure is found in the examination of electron—hole pairs, depending on the conductivity type of the material. The The same sequence of seven or eight traps is found in a variety of high-resistivity GaAs crystals regardless of the specific preparation procedure. Figure 4.9 compares the TSC curves found for high-resistivity GaAs:O and GaAs:Cr; as discussed above, it is known that the GaAs:Cr also contains oxygen impurity. The Hall mobility during TSC is n-type and relatively temperature independent, indicating that electron traps are being measured. The trap depth can be calculated from the maximum TSC assuming a suitable model, the trap density can be estimated from the area under the TSC peak associated with the particular trap of interest, and the trap capture cross-section can be determined by measuring the rate of trap filling. Trap cross-sections are found to be of the order of 10^{-18} cm^2 for traps a, c and d, but traps f and g are found to have a cross-section of the order of 10^{-16} cm^2. A simple model to explain this series of discrete trapping states is to associate these traps to different spatial configurations of the same complex. For example, if the complex were considered to be a donor—acceptor pair, the different trapping states would correspond to different spatial separations between the donor and the acceptor.

Shallow traps in both CdS [169] and GaAs:O or GaAs:Cr [96] can be emptied by the application of a moderate electric field. The results obtained are consistent with a field emptying of Coulomb-attractive traps by a decrease in the trap depth and a decrease in the capture cross-section of traps by the electric field.

4.6 PHOTOADSORPTION EFFECTS

The properties of semiconductors may be changed radically by the adsorption of a gas on the surface of the material if that gas interacts electronically with the bulk. Such effects are observable on single crystals but they are naturally most pronounced on thin films. The adsorption may be either enhanced or reduced by the presence of photoexcitation creating electron—hole pairs, depending on the conductivity type of the material. The gas that has been investigated most extensively is oxygen, which changes from a physically adsorbed to a chemically adsorbed state by accepting an electron from the bulk material. Thus, the adsorption of oxygen on an n-type material reduces the conductivity and produces a depletion layer at the surface, whereas the adsorption of oxygen on a p-type material increases the conductivity and produces an accumulation layer at the surface. An extensive literature exists on the adsorption of oxygen on semiconductors of many different types: Ge [183—185], PbS [186—190], PbTe [191], BaO [192], ZnO [193—202], ZnS [203], CdS [153, 204—220], and CdSe [221, 222].

Mark [207, 219] has given a thorough report on the phenomenon of oxygen chemisorption on CdS crystals, both as to its kinetics and its energetics. He has located the energy level associated with chemisorbed oxygen at 0.91 eV below the conduction band. The changes in photoelectronic properties of thin platelets of CdS due to oxygen adsorption can be surprisingly large, amounting to seven to ten orders of magnitude in dark conductivity and to three to four orders of magnitude in electron lifetime [210]. When the depletion layer due to adsorbed oxygen extends across the entire platelet, the electrical properties of the CdS are dominated by the oxygen adsorption; in general, it would be expected that this would occur for platelets less than a hundred microns in thickness.

Sintered layers of CdS and CdSe are polycrystalline layers with connections between the individual particles formed during the sintering process [223]. They are particularly sensitive to oxygen adsorption effects, both as such adsorption affects the particles and especially as it affects the inter-particle connections [208, 209]. Such properties of sintered layer photoconductors as instability, nonreproducibility, variability through low-temperature annealing, and long-term drifts may all be attributed to the effects of adsorption or desorption of oxygen. Annealing such a layer in the light in oxygen decreases photoconductivity and dark conductivity, whereas annealing in the dark in oxygen or in either light or dark in helium increases both photoconductivity and dark conductivity. As in the case of the single crystal platelets described above, adsorbed oxygen decreases the photoconductivity lifetime by acting as a recombination center, capturing photoexcited holes through the depletion layer field at the surface and preventing these holes from being captured by smaller electron-capture-cross-section centers in the bulk. By using the Hall effect it can be shown that both the electron density and the electron mobility are reduced in these layers by the adsorption of oxygen [209]. The magnitudes of the effects are such that in all cases the mobility change due to adsorption contributes significantly to the total photoconductivity change.

Similar effects of oxygen adsorption have been observed on solution-sprayed films of CdS [212, 220, 224]. These are polycrystalline films with transport in the plane of the film dominated by potential barriers between grains; the photoconductivity in these films is described in a later section of this chapter. Chemisorption of oxygen at the film surface changes the electron density with a much smaller effect on mobility for films of one micron or larger thickness [216]. Large chemisorption effects on the mobility are observed with the solution-sprayed films, however, and these are attributed to the chemisorption of oxygen at the grain boundaries, which alters the barrier height and barrier width. Under photoexcitation, the adsorbed oxygen at the film surface and at the grain boundaries acts as additional recombination centers and reduces the photoexcited electron lifetime. The adsorbed oxygen also decreases the tunneling probability for electron transport through the intergrain barriers, a major transport mode at low temperatures [220].

Unlike most of the materials with which the effects of oxygen adsorption have been investigated, PbS is a *p*-type semiconductor when prepared as a photoconductor and therefore adsorption of oxygen increases the dark conductivity. The Hall effect was measured as a function of time during

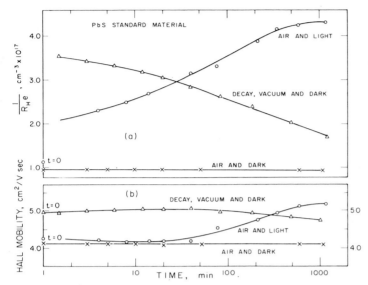

Fig. 4.10. Hole density ($1/R_H e$) and mobility as a function of time after a vacuum anneal of a photosensitive PbS layer, in air and light, and in air and dark. Also the same quantities as a function of time in dark and vacuum after the previous exposure to air and light. (From Espevik et al. [190])

photoadsorption of oxygen and as a function of time subsequently during desorption of oxygen in vacuum and dark, with the results shown in Fig. 4.10 [190]. The increase in conductivity associated with photo-adsorption of oxygen occurs through an increase in both free hole density and free hole mobility, although the time dependence of the two variations is not the same. When the sample is exposed to air and light the mobility does not change for the first 40 min; for longer times the increase in conductivity is contributed equally by increases in carrier density and mobility. Similar behavior in reverse occurs upon desorption of oxygen in vacuum and dark. These results are interpreted in terms of an initial photoadsorption of oxygen on the free surface of the layer, producing free holes and leaving the mobility unaffected; if time is allowed for this adsorbed oxygen to diffuse to a grain boundary, however, then its negative charge decreases the depletion layer width and tunneling-controlled mobility is increased.

4.7 PHOTOCHEMICAL CHANGES

One of the more exotic effects associated with CdS-like materials is that given the name of "photochemical changes". An excellent summary is given by Tscholl [225]. The effects manifest themselves in two general ways: (a) a slow decrease in photocurrent with constant photoexcitation

References pp. 147—153

if the temperature is above some lower limit [86, 226—231], and (b) a change in the thermally-stimulated-conductivity curve as the result of photoexcitation over certain temperature ranges [153, 232—237]. The generally favored photochemical model for this process involves the concept of agglomeration of defects to form a particular trap under the effects of photoexcitation, subsequent thermal emptying of this trap being associated with dispersal of the defects and hence with the destruction of the trap.

Some of the most striking data showing the nature of the effect are shown in Fig. 4.11, where CdS crystals with quite different background histories

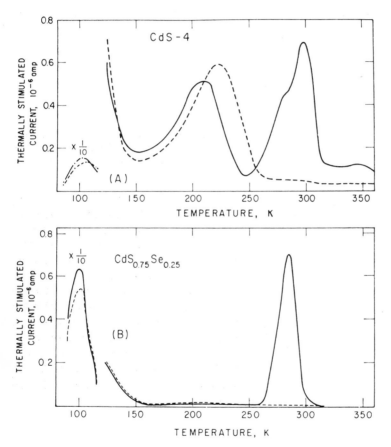

Fig. 4.11. Evidences for photochemical effects in thermally-stimulated conductivity of (a) a CdS crystal, previously irradiated with 240 keV electrons and annealed at $200°C$ under photoexcitation, and (b) a $CdS_{0.75}Se_{0.25}$ crystal. The solid curves indicate the values of thermally-stimulated conductivity observed after photoexcitation during cooling from high temperatures; the dashed curves indicate the values of thermally-stimulated conductivity after photoexcitation at low temperatures only. (From Im et al. [237])

show similar properties. The traps with a thermally-stimulated-conductivity (TSC) peak near 300 K have the following unique characteristics: (a) they can be filled by photoexcitation only if the temperature of the photoexcitation is sufficiently high, (b) the trap depth is about 0.8 eV, (c) the cross-section for electrons is of the order of 10^{-14} cm^2 as measured from the kinetics of trap emptying, and (d) in spite of such a large cross-section, retrapping is negligible and the kinetics associated with trap emptying is monomolecular. It was earlier advanced that these properties could be explained if the trap were a Coulomb-repulsive imperfection without invoking photochemical changes [153, 235]. Suppose the imperfection consisted of a trap level $\mathcal{E}_r$ below the bottom of the conduction band, surrounded by a potential barrier of maximum height $\mathcal{E}_b$. The activation energy for thermally freeing a trapped electron is $\mathcal{E}_t = \mathcal{E}_r + \mathcal{E}_b$, if excitation over the top of the barrier is necessary for freeing (a similar treatment can be given for tunneling through the barrier). Under thermal equilibrium conditions, $\nu \exp(-\mathcal{E}_t/kT) = n S_n^t v_n$, where ν is the attempt-to-escape frequency, n is the density of free electrons, S_n^t is the capture cross-section for capturing a free electron, and v_n is the thermal velocity of a free electron. If we write $\nu = N_c S_n^{t*} v_n$, thereby distinguishing between the effective cross-section S_n^{t*} for the escape process and the effective cross-section S_n^t for the capture process, $S_n^{t*} = S_n^t \exp(\mathcal{E}_b/kT)$. It is the value of $S_n^{t*} = 10^{-14}$ cm^2 that is determined from the decay kinetics, but this value may be some five orders of magnitude or more larger than S_n^t that determines the ability of the trap to fill and to retrap.

In spite of the appeal of this wholly electronic model, subsequent investigation of the details of the trap filling process [237] seems to make a photochemical reaction explanation necessary. The principal contribution to this argument comes from the dependence of trap filling, not only on temperature but also on wavelength. It was found that trap filling is most efficient for a band of wavelengths, the maximum of this band occurring at a wavelength larger than the wavelength for maximum photoconductivity. If a Coulomb-repulsive barrier model were adopted, it would follow that the optimum wavelength for trap filling should be either the wavelength corresponding to maximum photoconductivity or possibly a shorter wavelength to aid in surmounting the barrier. The photochemical reaction model proposes three steps: (a) dissociation of a neutral center into a pair of shallow metastable occupied electron and hole traps by photoexcitation (the most effective wavelength is about 200 Å longer than the most effective wavelength for exciting photoconductivity); (b) formation of a stable trap-pair by this dissociated center either by lattice relaxation or diffusion (temperature-dependent processes), producing traps with much larger depths; and (c) collapse of the center upon thermal emptying of an electron from the electron trap, which simultaneously releases the hole from the hole trap. Once the electron is excited to the conduction band, the trapping levels

References pp. 147—153

no longer exist, and retrapping cannot occur. Both the photoexciting wavelength (step a) and the photoexciting temperature (step b) have to be appropriate in order for the formation and filling of the trap.

Quite analogous effects have been observed in CdS single crystals with excess Cd donors and a density of Cu acceptors larger than that of the Cd donors [238—240]. Cu impurity is presumed to form two types of center in CdS with excess Cd donors: Type I centers are Cu-acceptor Cd-donor pair complexes; Type II centers are due to Cu impurity present in excess of the excess Cd concentration. It is observed that the higher the Cu concentration over the Cd concentration, the more rapid and larger are the decreases in electron lifetime upon photoexcitation. It is assumed that Type II centers are surrounded by some kind of potential barrier and are not photoelectronically active. Before photoexcitation, the Type I levels are occupied and form neutral complexes with the Cd-donor levels, which are assumed to have a low electron capture cross-section and a large hole capture cross-section. When Type I centers dominate, therefore, there is a high electron lifetime. Photoexcitation removes electrons from the Type I centers, placing them on the Type II centers and producing Type I' centers from the Type I centers. Since Type I' centers can capture an additional hole, the capture cross-section for free electrons is greatly increased and their lifetime is decreased. Some degree of heat treatment is required to free the electrons from the Type II centers and restore the system to its initial condition. Similar effects found in Cu_2S—CdS heterojunction photovoltaic cells as a result of diffusion of Cu into the CdS near the junction interface, produce marked degradation effects in the photovoltaic performance of the cell [241, 242].

4.8 PHOTOCONDUCTIVITY IN POLYCRYSTALLINE FILMS

In homogeneous single crystals, photoconductivity is almost always the result simply of an increase in the density of free carriers. There may be some small effect due to a change in the scattering and a consequent small change in the carrier mobility under photoexcitation (see Section 4.4), but this is always a secondary and small effect. In polycrystalline films, however, it is conceptually possible that the major effect of photoexcitation may be to decrease the potential barriers between grains and hence to increase the mobility without making a major change in the carrier density if this is initially high in the dark.

The use of chemically deposited PbS layers as infrared detectors (see also Chapter 10) has stimulated a more-or-less continuous investigation of their properties over the last 30 years [1, 243—245]. Throughout this investigation, a particular controversy has been sustained: Is the photoconductivity primarily due to an increase in the density of free carriers ("numbers model") or to an increase in the mobility of free carriers ("barrier modulation model")? Hall and photo-Hall techniques have been applied by several investigators in an effort to answer this question [246—251]. Woods [246] found no increase in Hall mobility with photoexcitation in PbS layers, but Bode [247] found no increase in carrier density with photoexcitation of PbSe layers.

A model for PbS films proposed by Petritz et al. [251] involved potential barriers between grains, but a photoconductivity that was determined primarily by recombination kinetics within grains and depended mostly on an increase in carrier density with photoexcitation. This model was carefully reinvestigated [190] using both photo-Hall and photothermoelectric effects, and was found to be substantially valid. The dark conductivity process is characterized by a small thermally-activated mobility associated with intergrain boundary potential barriers, which is essentially independent of the details of the deposition or the nature of the substrate [252]. In the higher temperature range, the activation energy is about 0.08 eV; in the lower temperature range, the apparent activation energy decreases to about 0.04 eV and probably corresponds to tunneling through the potential barrier. Carrier densities derived from Hall effect measurements were found to be equal to those derived from thermoelectric power measurements, thus supporting the conclusion that these densities corresponded to grain properties.

PbS layers prepared without added oxidant in the preparation were weakly photosensitive, corresponding to sensitizing centers with an energy level 0.13 eV below the conduction band. Layers prepared with added oxidant in the preparation were strongly photosensitive, corresponding to sensitizing centers with levels lying 0.22 eV below the conduction band. In all cases, photoconductivity resulted primarily from an increase in free hole density, with some small contribution from an increase in hole mobility, particularly at low temperatures where the primary effect is probably an increase in tunneling probability through the intergrain potential barriers. Figure 4.12 shows the temperature dependence of hole density and hole mobility for the sensitive PbS material in two states: one with considerable adsorbed oxygen and another with the oxygen desorbed by heat treating in vacuum. It is only at low temperatures in the initial high-conductivity state that the proportional change in mobility is comparable to the proportional change in hole density. As the layer is made less conducting by desorbing oxygen, the relative contribution of the change in mobility with photoexcitation decreases.

Many investigations have also been made of the nature of charge transport in polycrystalline films of CdS-like materials. Measurements of Hall effect on evaporated films [253—260] indicate a thermally-activated mobility with values of the activation energy between 0.07 and 0.2 eV. The applicability of the barrier model to these polycrystalline films is supported by the

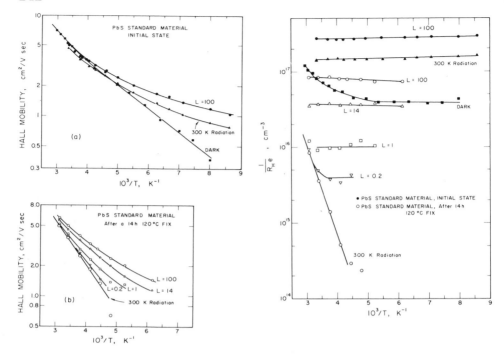

Fig. 4.12. Variation of hole density and mobility in photosensitive PbS layers with photoexcitation, in a film with adsorbed oxygen and a film free of adsorbed oxygen. Curves to the left show the temperature dependence of dark and photo-Hall mobility for the initial state (oxygen adsorbed) and after a 14 hour vacuum anneal at 120°C (oxygen desorbed). Curves to the right show the corresponding variation of hole density. Intensities L are in units of mW/cm^2. (From Espevik et al. [190])

field-effect measurements by Waxman et al. [257] which showed that the measured mobility activation energy could be decreased by the application of a field voltage. Other investigations have been carried out on CdS films prepared by sintering, chemical deposition and sputtering [261—266].

A detailed investigation of photoconductivity in solution-sprayed CdS films using the photothermoelectric effect [220] showed that in this system the principal effect of photoexcitation is to increase the electron mobility, with only a small effect on the electron density. Figure 4.13 shows typical variation of electron density and mobility with temperature for several different intensities of photoexcitation. The major effect of light in this case is to increase the tunneling probability through the intergrain potential barriers by decreasing the depletion layer widths adjacent to the barrier. Figure 4.13 shows that the effect of photoexcitation is to decrease the barrier height from 0.46 eV in the dark to 0.22 eV in the full light. If the mobility, as limited by tunneling through an ideal Schottky barrier, is

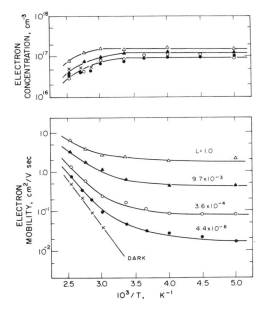

Fig. 4.13. Electron density and mobility as a function of temperature for a solution-sprayed CdS film for different photoexcitation intensities. $L = 1$ corresponds to white-light excitation of 50 mW/cm^2. (From Wu and Bube [220])

calculated as a function of barrier height, the observed change from a mobility of 0.015 cm^2/V.s at the lowest light level to 1.5 cm^2/V.s at the highest light level agrees very well with that theoretically predicted.

In measurements on polycrystalline films, the question may be raised whether or not standard analysis of Hall and thermoelectric power measurements can be applied to such an inhomogeneous system. Volger [248] proposed a geometric model for an inhomogeneous material consisting of conducting grains separated by more resistive intergranular material, and showed how the measured Hall effect could be affected by such a configuration. Basically, he argued that the Hall effect in such a system was effectively determined by the carrier density in the grains, whereas the resistivity was determined by the intergranular higher-resistivity regions; no conclusions can therefore be drawn about the mobility in either the grains or the intergranular regions. Davis and Greene [249] applied this model to the case of a space-charge inversion layer on PbS crystallites. Snowden and Portis [250] used a similar model to describe their results obtained as a function of the frequency of the applied electric field in the microwave range. The implications of the resistance network model were investigated further for the Hall effect [267] and for the thermoelectric effect [268]. The model was also extended to include the presence of potential barriers between the grains and intergranular regions of the film. These extensions of

References pp. 147—153

the resistance network model indicated that even the Hall coefficient could give carrier density values appreciably different from those of either grain or intergranular region, although the carrier density determined from thermoelectric power measurements was equal to that in the grains under all except the most extreme conditions. Criticisms of this model have been voiced by several authors [269—271], contending that the possibility that the Hall effect might not give the carrier density in the grains was due to an insufficient physical analysis of the model. Such criticisms might not be as applicable to the form of the model involving potential barriers between grain and intergranular regions. In any event, the best experimental test of the equivalence of Hall and thermoelectric power carrier densities in PbS films indicate that the Hall effect is measuring the carrier density in the grains in this case, as mentioned above. Unfortunately this result does not unambiguously resolve the question, for one of the possible cases in the resistance network model with potential barriers [268] also yields this same result.

An extreme case of a polycrystalline system is that associated with a photoconducting powder dispersed in an insulating "binder". In this case, it is likely that tunneling between photoconducting particles through a thin film of binder plays an important role in the conductivity process for low applied electric fields. Investigations on CdS and CdSe powders in such a system reveal three distinctive properties [272, 273]: (a) the photocurrent *versus* voltage curve is ohmic at very low and very high applied voltages, but involves processes in which the photocurrent increases with about the 2nd or 3rd power of the voltage in between; (b) for all voltages above the low-voltage ohmic region, there is a time constant associated with the application and removal of a voltage as well as with the normal application and removal of light; (c) it is possible for such a system to exhibit an on/off switching behavior through a hysteresis effect in the current—voltage dependence. If the power/binder system is conceived as a system of particles separated by potential barriers, then such distinctive powder properties as voltage-dependent sensitivity, polarization, voltage-associated time delays, and hysteresis can all be explained by the build-up of a positive space-charge near the barrier that decreases the barrier width and facilitates tunneling through the barrier. Such a build-up of positive space-charge is time consuming, strongly dependent on the direction of the electric field, and takes place effectively only at temperatures such that holes can be stably held at sensitizing centers in the material.

4.9 HETEROJUNCTIONS

Homojunctions, whether $p-n$ or $p-i-n$ or $n-p-n$, have long been used for photodetectors through a measurement of the photoconductivity induced in them by absorbed light [274]. Of increasing interest, however, are heterojunctions (see also Chapter 10) in which light can be directly incident on the junction if directed through the larger band gap material [275]. With increasing interest in conversion of solar energy into electricity, heterojunctions have taken on an even greater interest in recent years; with increasing concern for terrestrial utilization of solar energy, the direction of interest has turned toward systems that can readily be prepared in large-area, low-cost form.

One of the systems possibly suitable for such application is the Cu_2S—CdS heterojunction. This system has two principal advantages: (a) CdS can be deposited in thin film form by a variety of possible techniques including vacuum evaporation or solution spraying; and (b) junctions can be conveniently made by a "dipping process" by which a layer of Cu_2S is topotaxially formed on CdS by dipping the CdS into an aqueous solution of Cu^{+2} ions at a temperature between 75° and 100°C. This Cu_2S—CdS system also has several disadvantages: (a) Cu_2S is not really that exact composition but Cu_xS, with a number of different phases differing only slightly in Cu/S ratio but with appreciable differences in desirability for photovoltaic utilization; (b) the convenient dipping process also carries with it considerable complexity in the nature of the ion-exchange processes involved; (c) decomposition of the Cu_xS layer is a constant possibility either through interaction with the atmosphere or by Cu diffusion into the CdS; (d) electrochemical decomposition of the Cu_xS is possible if it finds itself in a region where the decomposition voltage is exceeded; and (e) diffusion of Cu into the CdS leads to a high-resistivity compensated layer in the CdS, the properties of which are extremely important in determining cell parameters.

Investigations into the properties of Cu_2S—CdS heterojunctions formed by dipping single crystal CdS have revealed a rich complexity of possible phenomena in these junctions [241, 242, 276—279]. Trapped charge near the interface in the CdS causes a persistent increase in junction capacitance and plays a significant role in determining carrier transport over the interface. Excitation of capacitance by short-wavelength illumination (photocapacitance) is accompanied by an enhancement of the long-wavelength photovoltaic response. Similarly, a decrease in photocapacitance by exposure to long-wavelength light (optical quenching) or to heat (thermal quenching) causes a decrease in response.

A long heat treatment of a single crystal Cu_2S—CdS junction causes a large decrease in response and an increase in the magnitude of the effects of enhancement and quenching. This decrease in response consists of two components: (a) a relatively small thermal component occurring on heat treatment in the dark; and (b) a much larger degradation caused by exposure to light at room temperature. By a short additional heat treatment above about 100°C, the cell can be completely restored to its condition before optical degradation with no change in depletion layer width (see Section 4.7).

The measured short-circuit current in a single crystal Cu_2S—CdS junction

References pp. 147—153

is the product of two independent processes: (a) the generation of electrons in the Cu_2S and their injection into the CdS (a process that is almost completely independent of heat treatment, degradation/restoration, or enhancement/quenching); and (b) the probability that an excited electron in the Cu_2S will make it safely through the CdS, which depends on the properties of the CdS:Cu layer near the junction interface that controls the rate of loss of injected electrons via tunneling recombination through interface states (a process that is affected by heat treatment, degradation/restoration, and enhancement/quenching).

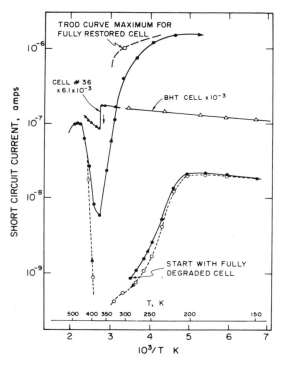

Fig. 4.14. A complete thermal cycle of short-circuit current excited by white light at an intensity of $1400\,\mu W/cm^2$ for a p-Cu_2S/n-CdS heterojunction photovoltaic cell, after long heat treatment. Additional curves on the figure are a before-heat-treatment curve (BHT), data for another cell (No. 36) obtained after only a short heat treatment at 150°C and normalized to fit the BHT data, and an indication of the current for a fully restored cell (called a TROD maximum on the figure; TROD stands for Thermally Restorable Optical Degradation). (From Fahrenbruch and Bube [242])

The wide range of short-circuit currents possible in such a system is illustrated by Fig. 4.14, which shows a complete thermal cycle of a heat-treated junction. The cycle starts at room temperature with the cell in the degraded—quenched state. As the temperature is increased, the cell goes to the degraded—enhanced state due to a decrease in the thermal quenching; little hysteresis occurs upon returning to 300 K. From 325 to 375 K, the short-circuit current drops to unmeasurably small values as the state changes from degraded—enhanced to degraded—quenched, and then increases as the temperature is raised as thermal restoration produces the restored—quenched state. As the temperature is decreased from 475 K, the enhancement/quenching ratio increases and the cell enters the restored—enhanced state with also a small effect due to optical degradation. Not shown in Fig. 4.14 is the fact that the cell in the restored state also shows little hysteresis between 150 and 300 K.

Other heterojunction systems based on CdS may avoid some of the complexities encountered with the Cu_2S—CdS system because of the properties of Cu_2S and because of diffusion of Cu into the CdS. A number of II—VI heterojunctions are possibilities for this purpose [280].

REFERENCES

1 R.H. Bube, Photoconductivity of Solids, Wiley, New York, 1960.
2 A. Rose, Concepts in Photoconductivity and Allied Problems, Wiley, New York, 1963.
3 R.G. Breckenridge, B.R. Russell and E.E. Hahn (Eds.), Photoconductivity Conf., Atlantic City, 1954, Wiley, New York, 1956.
4 H. Levinstein (Ed.), Photoconductivity, Pergamon Press, Oxford, 1962.
5 E.M. Pell (Ed.), Proc. Third Int. Conf. on Photoconductivity, Pergamon Press, Oxford, 1971.
6 S.M. Ryvkin, Photoelectric Effects in Semiconductors, Consultants Bureau, New York, 1964.
7 V.S. Vavilov, Effects of Radiation on Semiconductors, Consultants Bureau, New York, 1965.
8 R.H. Bube, Methods Exp. Phys., 6 (1959) 335.
9 R.H. Bube, Solid State Phys., 11 (1960) 223.
10 R.H. Bube, in S. Larach (Ed.), Photoelectronic Materials and Devices, van Nostrand, New York, 1965 p. 100.
11 R.H. Bube, in M. Aven and J.S. Prener (Eds.), Physics and Chemistry of II—VI Compounds, North-Holland, 1967, p. 658.
12 R.H. Bube, in R.K. Willardson and A.C. Beer (Eds.), Semiconductors and Semimetals, Academic Press, New York, 1967, p. 461.
13 R.H. Bube, in H. Eyring, D. Henderson, and W. Jost (Eds.), Physical Chemistry: Vol. X, Solid State, Academic Press, New York, 1970, p.515.
14 W. Shockley and W.T. Read, Phys. Rev., 87 (1952) 835.
15 J.S. Blakemore, Semiconductor Statistics, Pergamon Press, Oxford, 1962.
16 A. Rose, RCA Rev., 12 (1951) 362.
17 A. Rose, Proc. IRE, 43 (1955) 1850.
18 A. Rose, Phys. Rev., 97 (1955) 322.
19 A. Rose, Prog. Semicond., 2 (1957) 109.
20 G. Wlerick, C. R. Acad. Sci., 238 (1954) 2514.
21 H.J. Dirksen and O.W. Memelink, Appl. Sci. Res. Sect. B, 4 (1954) 205.
22 M. Schön, Physica, 20 (1954) 930.

23 C.A. Duboc, Br. J. Appl. Phys. Suppl., 4 (1955) 107.
24 R.H. Bube, J. Phys. Chem. Solids, 1 (1957) 234.
25 H.A. Klasens, J. Phys. Chem. Solids, 7 (1958) 175.
26 F. Cardon and R.H. Bube, J. Appl. Phys., 35 (1964) 3344.
27 G.A. Dussel and R.H. Bube, J. Appl. Phys., 37 (1966) 13.
28 R. Newman, H.H. Woodbury and W.W. Tyler, Phys. Rev., 102 (1956) 613.
29 W.W. Tyler and H.H. Woodbury, Phys. Rev., 96 (1954) 874.
30 R. Newman and W.W. Tyler, Phys. Rev., 96 (1954) 882.
31 W.W. Tyler, R. Newman and H.H. Woodbury, Phys. Rev., 97 (1955) 669.
32 W.W. Tyler, R. Newman and H.H. Woodbury, Phys. Rev., 98 (1955) 461.
33 R. Newman, Phys. Rev., 94 (1954) 278.
34 H.H. Woodbury and W.W. Tyler, Phys. Rev., 100 (1955) 659.
35 C.S. Fuller and F.J. Morin, Phys. Rev., 105 (1957) 379.
36 B.V. Kornilov, Sov. Phys. Solid State, 7 (1965) 1446.
37 B.V. Kornilov and S.E. Gorskii, Sov. Phys. Semicond., 2 (1968) 216.
38 K.D. Glinchuk, A.D. Denisova and N.M. Litovchenko, Sov. Phys. Solid State, 5 (1964) 1412.
39 B.V. Kornilov, Sov. Phys. Solid State, 8 (1966) 157.
40 B.V. Kornilov, Sov. Phys. Solid State, 7 (1966) 2794.
41 K.D. Glinchuk and N.M. Litovchenko, Sov. Phys. Solid State, 5 (1964) 2197.
42 B.V. Kornilov, Sov. Phys. Solid State, 6 (1965) 2982.
43 B.V. Kornilov, Sov. Phys. Solid State, 5 (1964) 2420.
44 Yu. I. Zavadskii and B.V. Kornilov, Phys. Status Solidi, 42 (1970) 617.
45 Yu. I. Zavadskii and B.V. Kornilov, Sov. Phys. Semicond., 5 (1971) 56.
46 E.E. Loebner and T.J. Diesel, Tech. Conf. on Preparation and Properties of Electron. Mater. Control Radiative Processes, Boston, 1966.
47 A.F. Sklensky and R.H. Bube, Phys. Rev., 6 (1972) 1328.
48 S. Rabie and N. Rumin, J. Appl. Phys., 45 (1974) 3988.
49 J. Blanc, R.H. Bube and F.D. Rosi, Proc. Int. Conf. on Semicond., Prague, 1960, p. 936.
50 J. Blanc, R.H. Bube and H.E. MacDonald, J. Appl. Phys., 32 (1961) 1666.
51 R.H. Bube and F. Cardon, J. Appl. Phys., 35 (1964) 2712.
52 R.H. Bube and S.M. Thomsen, J. Chem. Phys., 23 (1955) 15.
53 R.H. Bube, in R.G. Breckenridge, B.R. Russell and E.E. Hahn (Eds.), Photoconductivity Conf., Atlantic City, 1954, Wiley, New York, 1956, p. 575.
54 R.H. Bube and L.A. Barton, J. Chem. Phys., 29 (1958) 128.
55 A.L. Robinson and R.H. Bube, J. Appl. Phys., 42 (1971) 5280.
56 C.S. Kang, P.B.P. Phipps and R.H. Bube, Phys. Rev., 156 (1967) 998.
57 G.H. Blount, A.C. Sanderson and R.H. Bube, J. Appl. Phys., 38 (1967) 4409.
58 G.H. Blount, P.B.P. Phipps and R.H. Bube, J. Appl. Phys., 38 (1967) 4450.
59 R.H. Bube and E.L. Lind, Phys. Rev., 110 (1958) 1040.
60 G.B. Stringfellow and R.H. Bube, Phys. Rev., 171 (1968) 903.
61 G.B. Stringfellow and R.H. Bube, J. Appl. Phys., 39 (1968) 3657.
62 R.H. Bube and E.L. Lind, Phys. Rev., 105 (1957) 1711.
63 R.H. Bube, Phys. Rev., 106 (1957) 703.
64 R.H. Bube and E.L. Lind, Phys. Rev., 115 (1959) 1159.
65 R.H. Bube and E.L. Lind, Phys. Rev., 119 (1960) 1535.
66 F.J. Bryant and A.F.J. Cox, Br. J. Appl. Phys., 16 (1965) 463.
67 F.J. Bryant and A.F.J. Cox, Proc. Phys. Soc. (London), 87 (1966) 551.
68 R.H. Bube, Phys. Rev., 99 (1955) 1105.
69 B.A. Kulp and R.H. Kelley, J. Appl. Phys., 32 (1961) 1290.
70 H. Kallman, B. Kramer and A. Perlmutter, Phys. Rev., 99 (1955) 391.

71 S.O. Hemila and R.H. Bube, J. Appl. Phys., 38 (1967) 5258.
72 I. Broser, in M. Aven and J.S. Prener (Eds.), Physics and Chemistry of II—VI Compounds, North-Holland, 1967, p. 533.
73 R.H. Bube, J. Appl. Phys., 31 (1960) 1301.
74 R.H. Bube and C.-T. Ho, J. Appl. Phys., 37 (1966) 4132.
75 J. Saura and R.H. Bube, J. Appl. Phys., 36 (1965) 3660.
76 E. Burstein, G. Picus and N. Sclar, in R.G. Breckenridge, B.R. Russell and E.E. Hahn (Eds.), Photoconductivity Conf., Atlantic City, 1954, Wiley, New York., 1956, p. 353.
77 R.H. Bube, J. Appl. Phys., 35 (1964) 586.
78 F.F. Morehead, J. Phys. Chem. Solids, 24 (1963) 37.
79 R.H. Bube and A.B. Dreeben, Phys. Rev., 115 (1959) 1578.
80 R.H. Bube, E.L. Lind and A.B. Dreeben, Phys. Rev., 128 (1962) 532.
81 M.R. Lorenz, B. Segall and H.H. Woodbury, Phys. Rev., 134 (1964) 751.
82 G.W. Iseler, J.A. Kafalas, A.J. Strauss, H.F. MacMillan and R.H. Bube, Solid State Commun., 10 (1972) 619.
83 H.F. MacMillan, Ph.D. Thesis, Stanford Univ., 1972.
84 G. Lucovsky, Solid State Commun., 3 (1965) 299.
85 A. Rose and M.A. Lampert, RCA Rev., 20 (1959) 57; Phys. Rev., 113 (1959) 1227.
86 R.H. Bube and L.A. Barton, RCA Rev., 20 (1959) 564.
87 R.H. Bube, J. Appl. Phys., 32 (1961) 1621.
88 W.W. Tyler and H.H. Woodbury, Phys. Rev., 102 (1956) 647.
89 W.W. Tyler and R. Newman, Phys. Rev., 98 (1955) 961.
90 R.O. Carlson, Phys. Rev., 104 (1956) 937.
91 R.H. Bube and H.E. MacDonald, Phys. Rev., 121 (1961) 473.
92 R.H. Bube and H.E. MacDonald, Phys. Rev., 128 (1962) 2062.
93 R.H. Bube, H.E. MacDonald and J. Blanc, J. Phys. Chem. Solids, 22 (1961) 173.
94 L.R. Weisberg, J. Appl. Phys., 33 (1962) 1817.
95 R.H. Bube and H.E. MacDonald, Phys. Rev., 128 (1962) 2071.
96 Alice L. Lin, Ph.D. Thesis, Stanford Univ., 1974; J. Appl. Phys., 47 (1976) 1852, 1859.
97 D.L. Bowman, J. Appl. Phys., 38 (1967) 568.
98 J. Blanc, R.H. Bube and L.R. Weisberg, Phys. Rev. Lett., 128 (1962) 532.
99 J. Blanc, R.H. Bube and L.R. Weisberg, J. Phys. Chem. Solids, 25 (1964) 225.
100 C.H. Gooch, C. Hilsum and B.R. Holeman, J. Appl. Phys., 32 (1961) 2069.
101 R.W. Haisty, E.W. Mehal and R. Stratton, J. Phys. Chem. Solids, 23 (1962) 829.
102 J.F. Woods and N.G. Ainslie, J. Appl. Phys., 34 (1963) 1469.
103 W.J. Turner, G.D. Pettit and N.G. Ainslie, J. Appl. Phys., 34 (1963) 3274.
104 G. Day, IEEE Trans. Electron Devices, 13 (1966) 88.
105 J.M. Woodall and J.F. Woods, Solid State Commun., 4 (1966) 33.
106 F. Huth, Phys. Status Solidi, 31 (1969) K119.
107 Yu. V. Vorob'ev, Yu. I. Karkhanin and O.V. Tretyak, Phys. Status Solidi, 36 (1969) 499.
108 Y. Tokumam, Jpn. J. Appl. Phys., 9 (1970) 95.
109 Y. Zohta, Appl. Phys. Lett., 17 (1970) 284.
110 N.M. Kolchanova, Ga. N. Talalakin and E.A. Dretova, Sov. Phys. Semicond., 4 (1970) 174.
111 G.A. Aleksandrova, Yu. I. Zavadskii, B.V. Kornilov and I.M. Skvortsov, Sov. Phys. Semicond., 6 (1973) 1170.
112 E. Omelianovski and R.H. Bube, personal communication; see ref. 96.
113 A.P. Jonath and R.H. Bube, Surf. Sci., 37 (1973) 167.
114 G.R. Cronin and R.W. Haisty, J. Electrochem. Soc., 111 (1964) 874.
115 W.J. Turner and G.D. Pettit, Bull. Am. Phys. Soc., 9 (1964) 269.
116 C.E. Jones and A.R. Hilton, J. Electrochem. Soc., 113 (1966) 504.

117   E.W. Williams and D.M. Blacknall, Trans. Metall. Soc. AIME, 239 (1967) 387.
118   R.F. Broom, J. Appl. Phys., 38 (1969) 3483.
119   D.R. Heath, P.R. Selway and C.C. Tooke, Br. J. Appl. Phys., 1 (1968) 29.
120   G.A. Allen, Br. J. Appl. Phys., 1 (1968) 593.
121   T. Inoue and M. Ohyama, Jpn. J. Appl. Phys., 8 (1969) 1362.
122   G.A. Egiazaryan, V.I. Murygin, V.S. Rubin and V.I. Stafeev, Sov. Phys. Semicond., 3 (1970) 1389.
123   J.G. Harper, H.E. Matthews and R.H. Bube, J. Appl. Phys., 41 (1970) 3182.
124   O.V. Tretyak, Sov. Phys. Semicond., 4 (1970) 517.
125   T. Inoue and M. Ohyama, Solid State Commun., 8 (1970) 1309.
126   N.M. Kolchanova, D.N. Nasledov, M.A. Mirdzhalilova and V.Yu. Ibrugimov, Sov. Phys. Semicond., 4 (1970) 294.
127   E.M. Omel'yanovskii, L.Ya. Pervova, E.P. Rashevskaya and V.I. Fistal', Sov. Phys. Semicond., 5 (1971) 484.
128   B.V. Kornilov, V.A. Vil'kotskii, G.V. Aleksandrova and G.N. Tereshko, Sov. Phys. Semicond., 5 (1971) 119.
129   V.M. Gontar', G.A. Egiazaryan, V.S. Rubin, V.I. Murygin and V.I. Stafeev, Sov. Phys. Semicond., 5 (1971) 1939.
130   A.A. Gutkin, M.B. Kagan, D.N. Nasledov, B.A. Kholev and T.A. Shaposhnikova, Sov. Phys. Semicond., 5 (1971) 1006.
131   A.T. Gorelenok, B.V. Tsarenkov and N.G. Chiabrishvili, Sov. Phys. Semicond., 5 (1971) 95.
132   D.S. Domanevskii and V.D. Tkachev, Sov. Phys. Semicond., 4 (1971) 1790.
133   T. Nishino and T. Yanagida, Jpn. J. Appl. Phys., 11 (1972) 1221.
134   G.P. Peka and Y.I. Karkhauin, Sov. Phys. Semicond., 6 (1972) 261.
135   S.S. Li and C.I. Huang, J. Appl. Phys., 43 (1972) 1757.
136   A.A. Gutkin, A.A. Lebedev, G.N. Talalakin and T.A. Shaposhnikova, Sov. Phys. Semicond., 6 (1972) 928.
137   D. Bois and P. Pinard, Jpn. J. Appl. Phys., 12 (1973) 936.
138   A.A. Gutkin, A.A. Lebedev, R.K. Radu and G.N. Talalakin, Sov. Phys. Semicond., 6 (1973) 1674.
139   J. Tauc, Czech. J. Phys., 5 (1955) 528.
140   J. Tauc, Photo and Thermoelectric Effect in Semiconductors, Pergamon, London, 1962.
141   L.J. Van der Pauw and D. Polder, J. Electron., 2 (1956) 239.
142   F.L. Weichman and R. Lomnes, Bull. Am. Phys. Soc., 12 (1967) 657.
143   J.G. Harper, H.E. Matthews and R.H. Bube, J. Appl. Phys., 41 (1970) 765.
144   H. Kwok and R.H. Bube, J. Appl. Phys., 41 (1970) 5026.
145   H. Kwok and R.H. Bube, J. Appl. Phys., 44 (1973) 138.
146   C. Herring, Semiconductors and Phosphors, Interscience, New York, 1958.
147   T.H. Geballe and G.W. Hull, Phys. Rev., 98 (1955) 940.
148   H.B. Devore, RCA Rev., 20 (1959) 79.
149   E.A. Niekisch, Ann. Phys., 15 (1955) 279, 288.
150   E.A. Niekisch, Z. Phys., 161 (1961) 38.
151   E.A. Niekisch, Z. Phys. Chem. (Frankfurt), 217 (1961) 110.
152   R.H. Bube, J. Appl. Phys., 35 (1964) 3067.
153   R.H. Bube, G.A. Dussel, C.-T. Ho and L.D. Miller, J. Appl. Phys., 37 (1966) 21.
154   J.T. Randall and M.H.F. Wilkins, Proc. R. Soc. London, Ser. A, 184 (1945) 366.
155   G.F.J. Garlick and A.F. Gibson, Proc. Phys. Soc. London, Sect. A, 60 (1948) 574.
156   L.I. Grossweiner, J. Appl. Phys., 24 (1953) 1306.
157   A.H. Booth, Can. J. Chem., 32 (1954) 214.
158   R.H. Bube, J. Chem. Phys., 23 (1955) 18.

159 W. Hoogenstraaten, Philips Res. Rep., 13 (1958) 515.
160 K.W. Böer, S. Oberländer and J. Voigt, Ann. Phys. Leipzig, 2 (1958) 136.
161 A. Halperin and A.A. Braner, Phys. Rev., 117 (1960) 408.
162 R.R. Haering and E.N. Adams, Phys. Rev., 117 (1960) 451.
163 P.N. Keating, Proc. Phys. Soc. London, 78 (1961) 1408.
164 A. Bohun, Czech. J. Phys., 4 (1954) 91.
165 C. Haake, J. Opt. Soc. Am., 47 (1957) 649.
166 Ch.B. Luschik, Dokl. Akad. Nauk. SSSR, 101 (1955) 641.
167 H.J. Dittfeld and J. Voigt, Phys. Status Solidi, 3 (1963) 1941.
168 K.H. Nicholas and J. Woods, Br. J. Appl. Phys., 15 (1964) 783.
169 G.A. Dussel and R.H. Bube, J. Appl. Phys., 37 (1966) 2797.
170 G.A. Dussel and R.H. Bube, Phys. Rev., 155 (1967) 764.
171 J. Lambe, Phys. Rev., 98 (1955) 985.
172 R.H. Bube, Phys. Rev., 101 (1956) 1668.
173 R.W. Smith and A. Rose, Phys. Rev., 97 (1955) 1531.
174 A. Rose, Phys. Rev., 97 (1955) 1538.
175 M.A. Lampert, Phys. Rev., 103 (1956) 1648.
176 M.A. Lampert, A. Rose and R.W. Smith, J. Phys. Chem. Solids, 8 (1959) 464.
177 R.W. Smith, RCA Rev., 20 (1959) 69.
178 R.H. Bube, J. Appl. Phys., 33 (1962) 1733.
179 C. Manfredotti, A. Rizzo, L. Vasanelli, S. Galassini and L. Ruggiero, J. Appl. Phys., 44 (1973) 5463.
180 K.W. Böer and U. Kümmel, Ann. Phys., 6 (1957) 393.
181 K.W. Böer and U. Kümmel, Z. Naturforsch. Teil A, 13 (1958) 698.
182 R.W. Haisty, Appl. Phys. Lett., 10 (1967) 31.
183 W.H. Brattain and J. Bardeen, Bell Syst. Tech. J., 32 (1953) 1.
184 J. Bardeen and S.R. Morrison, Physica, 20 (1954) 873.
185 E.N. Clarke, Sylvania Technol., 7 (1954) 102.
186 H. Hintenberger, Z. Naturforsch., 1 (1946) 13.
187 H.T. Minden, J. Chem. Phys., 23 (1955) 1948.
188 M. Smollett and R.G. Pratt, Proc. Phys. Soc. London Sect. B, 68 (1955) 390.
189 R.J. Ryerson and R.H. Bube, J. Appl. Phys., 41 (1970) 5355.
190 S. Espevik, C.Wu and R.H. Bube, J. Appl. Phys., 42 (1971) 3513.
191 D.E. Bode and H. Levinstein, Phys. Rev., 96 (1954) 259.
192 M. Sakamoto, S. Kobayashi and S. Ishii, Phys. Rev., 98 (1955) 552.
193 G. Heiland, Z. Phys., 138 (1954) 459; ibid. 142 (1955) 415.
194 S.M. Ryvkin, Zh. Tekh. Fiz., 22 (1952) 1930.
195 E.K. Putseiko and A.N. Terenin, Dokl. Akad. Nauk. SSSR, 101 (1955) 645.
196 E. Mollwo, in R.G. Breckenridge, B.R. Russell and E.E. Hahn (Eds.), Photoconductivity Conf., Atlantic City, 1954, Wiley, New York, 1956, p. 509.
197 P.H. Miller, in R.G. Breckenridge, B.R. Russell and E.E. Hahn (Eds.), Photoconductivity Conf., Atlantic City, 1954, Wiley, New York, 1956, p. 287.
198 D.A. Melnick, J. Chem. Phys., 26 (1957) 1136.
199 D.B. Melved, J. Chem. Phys., 28 (1958) 870; J. Phys. Chem. Solids, 20 (1961) 255.
200 G. Heiland, E. Mollwo and F. Stöckmann, Solid State Phys., 8 (1959) 268.
201 B. Hoffman and E. Mollwo, Z. Angew. Phys., 14 (1962) 734.
202 W. Ruppel, H.J. Gerritsen and A. Rose, Helv. Phys. Acta, 30 (1957) 495.
203 A. Kobayashi and J. Kawaji, J. Phys. Soc. Jpn, 10 (1955) 270; ibid 11 (1956) 369; J. Chem. Phys., 24 (1956) 907.
204 R.H. Bube, Phys. Rev., 83 (1951) 393.
205 S.H. Liebson, J. Chem. Phys., 23 (1955) 1732.
206 R. Williams, J. Phys. Chem. Solids, 23 (1962) 1057.

207 P. Mark, J. Phys. Chem. Solids, 25 (1964) 911; ibid. 26 (1965) 959; ibid. 26 (1965) 1767.
208 H. Shear, E.A. Hilton and R.H. Bube, J. Electrochem. Soc., 112 (1965) 997.
209 A.L. Robinson and R.H. Bube, J. Electrochem. Soc., 112 (1965) 1002.
210 R.H. Bube, J. Electrochem. Soc., 113 (1966) 793.
211 K.J. Haas, D.C. Fox and M.J. Katz, J. Phys. Chem. Solids, 26 (1965) 1779.
212 F.B. Micheletti and P. Mark, Appl. Phys. Lett., 10 (1967) 136.
213 A. Waxman, Solid-State Electron., 9 (1966) 303.
214 D.M. Hughes and G. Carter, Phys. Status Solidi, 25 (1968) 449.
215 J.P. Lerge and S. Martinuzzi, Phys. Status Solidi A, 1 (1970) 689.
216 P.A. Thomas, C. Sebenne and M. Balkanski, Rev. Phys. Appl., 5 (1970) 683.
217 F.B. Micheletti and P. Mark, J. Appl. Phys., 39 (1968) 5274.
218 R.H. Bube, J. Appl. Phys., 34 (1963) 3309.
219 S.B. Roy, W.R. Bottoms and P. Mark, Surf. Sci., 28 (1971) 517.
220 C. Wu and R.H. Bube, J. Appl. Phys., 45 (1974) 648.
221 R.H. Bube, J. Chem. Phys., 27 (1957) 496.
222 G.A. Somorjai, J. Phys. Chem. Solids, 24 (1963) 175.
223 S.M. Thomsen and R.H. Bube, Rev. Sci. Instrum., 26 (1955) 664.
224 R.R. Chamberlin and H.S. Skarman, J. Electrochem. Soc., 113 (1966) 86; Solid-State Electron., 9 (1966) 819.
225 E. Tscholl, Philips Res. Rep. Suppl., 6 (1968) 1.
226 R.H. Bube, J. Chem. Phys., 30 (1959) 266.
227 K.W. Böer, Physica, 20 (1954) 1103.
228 K.W. Böer, E. Borchardt and W. Borchardt, Z. Phys. Chem., 203 (1954) 145.
229 K.W. Böer, W. Borchardt and S. Oberländer, Z. Phys. Chem., 210 (1959) 218.
230 W. Borchardt, Phys. Status Solidi, 1 (1961) K52.
231 W. Borchardt, Phys. Status Solidi, 2 (1962) 1575.
232 J. Woods and D.A. Wright, Solid State Phys. Electron. Telecommun. Proc. Int. Conf., Brussels, 1958, 1960, p. 880.
233 K.H. Nicholas and J. Woods, Br. J. Appl. Phys., 15 (1964) 783.
234 J. Woods and K.H. Nicholas, Br. J. Appl. Phys., 15 (1964) 1361.
235 A.P. Trofimenko, G.A. Fedorus and M.K. Sheinkman, Sov. Phys. Solid State, 5 (1964) 1316.
236 N.E. Korsunskaya, I.V. Markevich and M.K. Sheinkman, Phys. Status Solidi, 13 (1966) 25.
237 H.B. Im, H.E. Matthews and R.H. Bube, J. Appl. Phys., 41 (1970) 2581.
238 S. Kanev, V. Stojanov and V. Sekerdzijski, Acta Phys. Pol., 25 (1964) 3.
239 S. Kanev, V. Sekerdzijski and V. Stojanov, C. R. Acad. Bulg. Sci., 16 (1963) 7.
240 S. Kanev, V. Stojanov and M. Lakova, C.R. Acad. Bulg. Sci., 22 (1969) 863.
241 S.K. Kanev, A.L. Fahrenbruch and R.H. Bube, Appl. Phys. Lett., 19 (1971) 459.
242 A.L. Fahrenbruch and R.H. Bube, J. Appl. Phys., 45 (1974) 1264.
243 G. Brükmann, Kolloid Z., 65 (1933) 148.
244 R.A. Smith, Adv. Phys., 2 (1953) 321.
245 R.A. Smith, Semiconductors, Cambridge Univ. Press, Cambridge, England, 1959 pp. 414—432.
246 J.F. Woods, Phys. Rev., 106 (1957) 235.
247 D.E. Bode, Proc. Natl. Electron. Conf., Chicago, 19 (1963) 630.
248 J. Volger, Phys. Rev., 79 (1950) 1023.
249 J.L. Davis and R.F. Greene, Appl. Phys. Lett., 11 (1967) 227.
250 D.P. Snowden and A.M. Portis, Phys. Rev., 120 (1960) 1983.
251 R.L. Petritz, F.L. Lummis, H.E. Sorrows and J.F. Woods, Semiconductor Surface Physics, Univ. of Pennsylvania Press, Philadelphia, Pa., 1957, p. 229.

252 E.-H. Lee and R.H. Bube, J. Appl. Phys., 43 (1973) 4259.
253 L.L. Kazmerski, W.B. Berry and C.W. Allen, J. Appl. Phys., 43 (1972) 3515.
254 R.S. Muller and B.G. Watkins, Proc. IEEE, 52 (1964) 425.
255 R.G. Mankarious, Solid-State Electron., 7 (1964) 702.
256 I.A. Karpovich and B.N. Zvonkov, Sov. Phys. Solid State, 6 (1965) 2714.
257 A. Waxman, V.E. Heinrich, F.V. Shallcross, H. Borkan and P.K. Weimer, J. Appl. Phys., 36 (1965) 168.
258 F.V. Shallcross, Trans. Am. Inst. Min. Metall. Pet. Eng., 236 (1966) 309.
259 C.A. Neugebauer, J. Appl. Phys., 39 (1968) 3177.
260 W.P. Bleha, W.H. Hartman, R.L. Jimenez and R.N. Peacock, J. Vac. Sci. Technol., 7 (1970) 135.
261 D.A. Cusano, in M. Aven and J.P. Prener (Eds.), Physics and Chemistry of II—VI Compounds, North-Holland, Amsterdam, 1967, pp. 710—719.
262 B. Ray, II—VI Compounds, Pergamon, Oxford, 1969, pp. 41—44.
263 D.B. Fraster and H. Mechior, J. Appl. Phys., 43 (1972) 3120.
264 I. Lagnado and M. Lichtensteiger, J. Vac. Sci. Technol., 7 (1970) 318.
265 R.R. Chamberlin and H.S. Skarman, J. Electrochem. Soc., 113 (1966) 86.
266 R.R. Chamberlin and H.S. Skarman, Solid-State Electron., 9 (1966) 819.
267 R.H. Bube, Appl. Phys. Lett., 13 (1969) 136; ibid., 14 (1969) 84.
268 G.H. Blount, R.H. Bube and A.L. Robinson, J. Appl. Phys., 41 (1970) 2190.
269 J. Heleskivi and T. Salo, J. Appl. Phys., 43 (1972) 740.
270 R.H. Bube, J. Appl. Phys., 43 (1972) 742.
271 K. Lipskis, A. Sakalas and J. Viscakas, Phys. Status Solidi A, 4 (1971) K217.
272 R.H. Bube, J. Appl. Phys., 31 (1960) 2239.
273 F.M. Nicoll, RCA Rev., 19 (1958) 77.
274 R.H. Bube, Trans. Metall. Soc. AIME, 239 (1967) 291.
275 A.G. Milnes and D.L. Feucht, Heterojunctions and Metal-Semiconductor Junctions, Academic Press, New York, 1972.
276 W.D. Gill and R.H. Bube, J. Appl. Phys., 41 (1970) 1694.
277 W.D. Gill and R.H. Bube, J. Appl. Phys., 41 (1970) 3679.
278 P.F. Lindquist and R.H. Bube, J. Appl. Phys., 43 (1972) 2839.
279 P.F. Lindquist and R.H. Bube, J. Electrochem. Soc., 119 (1972) 936.
280 A.L. Fahrenbruch, V. Vasilchenko, F. Buch, K. Mitchell and R.H. Bube, Appl. Phys. Lett., 25 (1974) 609.

CHAPTER 5

MOLECULAR CRYSTALS

HIROO INOKUCHI and YUSEI MARUYAMA

5.1. Introduction
5.2. Photogeneration processes
    5.2.1. The "*via* exciton" mechanisms
    5.2.2. Direct production of electron-hole pairs
    5.2.3. Photoinjection from electrodes
    5.2.4. Photo-detrapping of trapped charges
5.3. Trapping and recombination processes
    5.3.1. Trapping and electric field dependence of photocurrent
    5.3.2. Charge-carrier recombination and temperature dependence of photocurrent
5.4. Photo-carrier transport processes
    5.4.1. Charge-carrier mobility: experimental and phenomenological
    5.4.2. Charge-carrier mobility: theoretical
5.5. Related phenomena
    5.5.1. Photovoltaic effect
    5.5.2. Sensitized photoconduction
    5.5.3. Photoconduction in amorphous organic solids
    5.5.4. External photoelectric effect: photoemission phenomena

---

5.1 INTRODUCTION

Solids constructed from molecules (or atoms of the inert gases) that interact by van der Waals forces are called molecular crystals. The energy of these interactions is very small in comparison with the binding energy of the electrons in the molecules. Some typical molecular crystals are those formed by the aromatic anisotropic molecules, naphthalene, anthracene, naphthacene, and so on. Photoconductivity in organic molecular crystals was first observed for anthracene in 1906 [1]. After about a forty-year interval, photoconductivity in organic molecular crystals was studied again from a new point of view — the typical organic semiconductors [2]. Experimental and also theoretical studies on the electrical properties of organic molecular crystals and their complexes have been carried out intensively in the last decade. There are two fundamental processes in the photoconduction phenomena; the first is the generation of charge carriers and the second is their transport. In the studies on photo-carrier generation mechanisms, spectral response of the photocurrents has been examined and the creation of exciton states has

References pp. 182—184

been realized to be very crucial for the carrier generation processes, of which details and also other important processes will be elucidated in Section 5.2. Trapping and recombination effects are also important for understanding photoconduction phenomena. Information is obtained by measurements of the electric field and temperature dependence of the photocurrents, some results of which will be presented in Section 5.3.

The epoch-making step in the field of studies on the electrical properties of organic molecular crystals was the charge carrier drift mobility measurements in anthracene single crystals by LeBlanc [3a] and Kepler [3b] in 1960. The drift mobility ratios in the crystal axis directions observed by Kepler were compared with the overlap integrals of π—electron orbitals between relevant molecules by Murrell [4], and then LeBlanc [5] also calculated the transfer (or resonance) energies between molecules and, using a tight-binding-approximation, estimated the mobility ratios. A rather comprehensive treatment concerning the transport properties of anthracene crystals was made by Friedman [6], in which he made an interesting prediction for the anomaly in sign and magnitude of the Hall mobility in anthracene crystals. Several attempts to observe this predicted behavior have not been successful. A recent approach to an understanding of the conduction mechanism in organic molecular crystals by Munn and Siebrand [7] seems to be most relevant and comprehensive, the essential part of which is summarized in Section 5.4.2. Finally in Section 5.5, we present briefly some related phenomena, such as the photovoltaic effect, sensitized photoconduction, and the photoemission effect.

5.2 PHOTOGENERATION PROCESSES

Many mechanisms for charge carrier generation by light in a molecular crystal have been reported so far. Of importance are the following: the "via excitons" mechanisms; direct production of electron—hole pairs (intrinsic bulk photoconduction); and injection from an electrode. In the following paragraphs, each of these mechanisms is elucidated and reviewed.

5.2.1 The "via Exciton" mechanisms

(a) Exciton states

In a molecular crystal which is characterized as a translationally-symmetric periodic system, the electronic excited state of any particular molecule cannot be an eigen state of the crystal. All of the molecules in the crystal are identical, and hence the eigen crystal excited state should be represented as the set of N (the total number of molecules) orthonormalized functions, in each of which a different molecule is excited. These functions are identified by the values of the wave vector k and the molecular excited state. In a system having no intermolecular interaction, all the states differing in the values of k have an identical energy. However, if we introduce the intermolecular coupling such as the case for real crystals, this degeneracy is removed. In crystals of large dimensions, the consecutive values of k differ little from one another, and hence the N excited states form a quasicontinuous band of excited states of the crystal, consisting of N sublevels. Each of these excited states, pertaining to a definite value of k, is a collective excited state of the whole crystal. Such elementary excitations were first discussed by Frenkel [8], and termed excitons. The concept of excitons for organic molecular crystals has been developed extensively by Davydov [9]. Thus, the excitation of a crystal corresponding to an exciton state is distributed throughout the crystal, rather than concentrated in one molecule. On the other hand, the states of a crystal corresponding to the excitation of some special region of the crystal are represented by wave packets, $i.e.$, linear combinations of exciton functions. These wave packets are moving in the crystal and their group velocities are determined by the interaction energy between molecules, roughly equal to the exciton bandwidth. Hence, the exciton bandwidth $\Delta\mathcal{E}$ gives the energy transfer time τ' between nearest-neighbour molecules, and also it gives a measure of the energy migration time τ through the use of the uncertainty principle argument, $\tau \simeq h/\Delta\mathcal{E}$. For the singlet exciton in an anthracene crystal, the splitting $\Delta\mathcal{E}$ is of the order of 200—300 cm^{-1}, making τ about 1.3×10^{-14} s. At room temperature, the singlet and also triplet ($\tau \simeq 4 \times 10^{-14}$ s) exciton are both consistent with the hopping model in which the exciton is scattered at every molecular site. At low temperature, there is evidence that the coherent length of the singlet exciton exceeds the lattice spacing. The intermolecular coupling is reduced to the transition dipole—dipole interaction, and this value also determines the Davydov splitting [9]. In Fig. 5.1, absorption spectra of the crystal state of anthracene and also the solution state are shown [10], where one notices

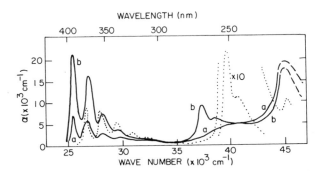

Fig. 5.1. Absorption spectrum of anthracene crystal (solid line) using a and b polarized light, and of anthracene dissolved in iso-octane (dotted line). (From Bree and Lyons [10])

References pp. 182—184

the similarity in the vibrational spectrum in the presence of the crystal spectral red shift.

(b) Exciton—surface interaction

Singlet or triplet excitons can react with surfaces; presumably adsorbed oxygen, impurities or defects and/or electrodes. For most crystals, the diffusion length of the exciton, l, is too small ($\simeq$ 300 Å for the singlet exciton in anthracene [11]) and the absorption coefficient α too large for any exciton to reach large distances from the surface. As a result of exciton annihilation at the surface, a photocurrent is produced that is linear with light intensity since the annihilation process is a monomolecular reaction. A detailed analysis of the annihilation quantum efficiency was carried out by Mulder [11], which showed that it is directly proportional to α. Figure 5.2 shows [11] a comparison of the spectral dependence of α with that of electrical conductivity, illustrating the close correlation between α and the photoconductivity.

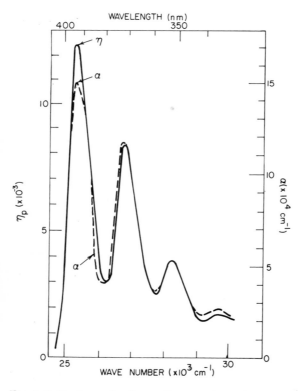

Fig. 5.2. Photoconduction action spectrum (η_p) measured on an anthracene crystal doped with naphthacene compared with absorption spectrum (α). η_p is a quantum efficiency for carrier generation. (From Mulder [11])

Both singlet and triplet excitons also dissociate at metal electrodes, resulting in carrier injection when energetic conditions are satisfied. At the interface between crystalline anthracene and a metal contact a $\pi\pi^*$ singlet exciton is able to dissociate to yield a free hole within the crystal which can be detected by photocurrent measurements. The essential process is transfer of the excited electron through a surface barrier to empty metal states above the Fermi level [12]. The quantum yield for generation of free carriers is of the order of 10^{-2}. In contrast, an injection of a hot metal electron or hole into the conduction or valence band of the crystal occurs with a quantum yield of the order 10^{-6} or less, as evidenced by internal photoemission studies [13].

Vaubel et al. [14] studied the reaction of singlet excitons at an anthracene/aluminum interface. They measured the rate constant k_q for energy transfer from an anthracene singlet exciton to an aluminum contact, varying the transfer distance d by inserting fatty acid monolayers between crystal surface and metal. They arrived at the following conclusions:

(1) k_q varies as d^{-3}. This agrees with Kuhn's theory for energy transfer from a donor to an absorbing layer if the layer thickness is large enough compared with the donor—acceptor distance d. Since this theory is derived for dipole—dipole coupling, the result indicates that this type of coupling provides the dominating channel for non-radiative energy transfer between singlet excitons and free metal electrons.

(2) It is found that $k_q = Ad^{-3}$ where $A = 1 \times 10^{-9}\,\text{cm}^3\text{s}^{-1}$ and the critical distance is $d_0 = 62 \pm 4$ Å.

(3) The thickness l_0 of the surface quenching zone in a metal-covered anthracene crystal in the absence of a fatty acid spacer, can be estimated to be $l_0 \lesssim 20$ Å. This means that the singlet exciton will be quenched when it reaches the second or third molecular layer adjacent to a metal contact. The triplet excitons are also involved in the surface photo-carrier generation in anthracene crystals [15]. The quenching length l_0 for the reaction of triplet exciton in a chloranil crystal was found to be $l_0 \simeq 1.2$ Å. This means that the triplet exciton must actually strike the electrode in order to be quenched.

(c) Exciton—exciton interaction

Singlet and triplet excitons interacting with themselves or with each other are capable of producing carrier pairs if the energy of the two excitons exceeds the band gap energy $\mathcal{E}_g$ which can be estimated from the intrinsic photoconductivity responses (see Section 5.2.2). Choi and Rice [16] first pointed out theoretically the possibility of charge carrier generation by an exciton—exciton interaction mechanism and computed the bimolecular rate constant to be $2.6 \times 10^{-12}\,\text{cm}^3\text{s}^{-1}$. Silver et al. [17] verified their theoretical work by observing a photocurrent in anthracene which increased with the square of the light intensity. In Silver's experiment only long-wavelength (415—455 nm) weakly absorbed light was used, and hence carriers are generated approximately uniformly throughout the bulk of the crystal. The bimolecular rate constant was found to be 5×10^{-12} cm^3 s^{-1}, which is in remarkably good agreement with theory. Hasegawa and Yoshimura [18], and Kepler [19] reported that photo-carriers can be produced in an anthracene crystal by the interaction of singlet excitons induced by a Q-switched ruby laser. Hasegawa and Yoshimura found that the photocurrent increases with the fourth power of the laser light intensity in the low-intensity region (around 6.8×10^{25} photons cm^{-2} s^{-1}) in contrast to the square dependence in the high-intensity region ($\sim 6.8 \times 10^{26}$ photons cm^{-2} s^{-1}). The rate of increase of charge carriers is:

$$dn/dt = \beta N^2 - \gamma n^2, \tag{5.1}$$

where n is the density of conduction carriers, β and γ are the singlet–singlet interaction and recombination constant, respectively, and N is the density of singlet excitons. Since the singlet exciton is produced by two-photon absorption, density N is:

$$N = KF^2, \tag{5.2}$$

where F is the laser intensity (photons cm^{-2} s^{-1}), K a constant (s^2 cm/photon). Integrating eqn. (5.1), one directly obtains the total number of carriers, n_T, produced in the crystal during the laser duration Δt

$$n_T = (\beta/\gamma)^{1/2} KF^2 \tanh [(\beta\gamma)^{1/2} (\Delta t) KF^2]. \tag{5.3}$$

If the argument of the hyperbolic tangent is larger than 1.5 (high-intensity case),

$$n_T \approx (\beta/\gamma)^{1/2} KF^2 \propto F^2. \tag{5.4}$$

In the low-intensity case, that is,

$$(\beta\gamma)^{1/2} (\Delta t) KF^2 \leqslant 0.5, \quad n_T \approx \beta(\Delta t) K^2 F^4. \tag{5.5}$$

Using appropriate values for parameters, Hasegawa and Yoshimura estimated $1.5 \times 10^{-7} < \gamma < 6 \times 10^{-7}$ cm^3 s^{-1}.

Kepler [20] pointed out the possibility of photo-ionization of excitons and experimentally showed that the photocurrent increased with the third power of the laser light intensity. This dependence may be elucidated by the one-photon ionization process of excitons which are created by two-photon absorption. Bergmann et al. [21] provided a direct evidence for bimolecular annihilation of singlet excitons in crystalline anthracene by the study of the radiative decay of two-photon excited states. Their directly observed rate constant for singlet–singlet annihilation, γ_s, is $(4 \pm 3) \times 10^{-8}$ cm^3 s^{-1} which is higher by about three orders of magnitude than the results reported for photocurrent generation in this system. Bergmann et al. explained this discrepancy as follows. The free carrier density for the photoconductivity experiment is lower than the collision-ionization yield due to electron–hole geminate recombination in a nonequilibrium state.

(d) Interaction of excitons with free and trapped charges

Singlet or triplet excitons can react with trapped charges yielding free carriers and a ground state crystal. By monitoring the decrease in fluorescence efficiency that accompanies the introduction of free charge, the rate constant, $\gamma_{S,t}$, for the interaction can be estimated. For triplet excitons, $\gamma_{T,t}$ is determined by monitoring the triplet exciton concentration by means of the delayed fluorescence as displayed by the triplet—triplet (T—T) fusion; for anthracene $\gamma_{S,t} \simeq 10^{-8}\,\mathrm{cm}^3\,\mathrm{s}^{-1}$ [22], $\gamma_{T,t} \simeq 2 \times 10^{-11}\,\mathrm{cm}^3\,\mathrm{s}^{-1}$ [23]. Excitons are also known to react with free charges. For anthracene, the details of the interaction in case of triplet excitons have been studied by Wakayama and Williams [24], and they found $\gamma_{T,f} \simeq 10^{-9}\,\mathrm{cm}^3\,\mathrm{s}^{-1}$. Trapped charges are detrapped to form free carriers by direct absorption of a photon. For phthalocyanine crystals, Day and Price [25] observed a long wavelength photoconduction response which was ascribed to a detrapping effect of trapped charges at oxygen impurities. Injected charges into a crystalline anthracene can fill up the traps in the crystal. These trapped charges may react with a photon directly or excitons yielding free carriers [26].

5.2.2 Direct production of electron—hole pairs

By pulsed light excitation in the region of 250—300 nm, Castro and Hornig [27] observed intrinsic electron—hole pair production in the bulk of an anthracene crystal. Kearns [28] also measured the spectral responses of the photoconduction for an anthracene crystal very carefully and obtained almost the same results as Castro and Hornig. The observed spectra of Kearns are shown in Fig. 5.3.

The pair production process is distinguished from the exciton induced surface production as follows:

(1) The action spectrum of pair production will not resemble the absorption spectrum of the crystals whereas the action spectrum for single carrier production will;

(2) the quantum efficiency for hole and electron generation will be the same and will be independent of the nature of the electrodes.

The long wavelength absorption threshold lies at about 3.9 eV, and this is the generally accepted value of the band gap, $\mathcal{E}_g$, of anthracene crystal. The quantum efficiency of pair production is quite low ($\sim 10^{-4}$), but this is mainly due to geminate recombination. According to Lyons, $\mathcal{E}_g$ is given by:

$$\mathcal{E}_g = 2I_c - I_g - A_g,$$

where I_c is the ionization energy of crystal, I_g the ionization energy of a free molecule and A_g its electron affinity. For anthracene, using the values of $I_g = 7.4\,\mathrm{eV}$, $I_c = 5.9\,\mathrm{eV}$, and $A_g = 0.58\,\mathrm{eV}$, $\mathcal{E}_g$ may be predicted as 3.8 eV [29] which is in quite good agreement with the above-mentioned intrinsic photoconductivity threshold, 3.9 eV. In the case of naphthacene, $\mathcal{E}_g$ may be

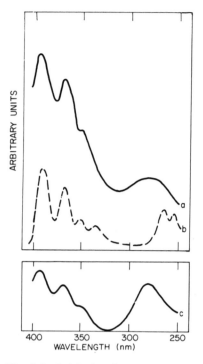

Fig. 5.3. Intrinsic photoconductivity spectra of crystalline anthracene. (a) photocurrent for holes, (b) absorption spectrum, and (c) photocurrent for electrons. (From Kearns [28])

estimated to be 3.0 eV and the same value was observed by Geacintov and Pope [30].

A striking aspect of the intrinsic photoconductivity action spectrum is shown in Fig. 5.4. [30]. The peaks in this action spectrum seem to be correlated with those that appear in the action spectrum of the external photoelectric effect and have been attributed to the excitation of electrons from lower-lying valence states into the conduction band. Silver and Sharma [31] analysed the observations of Castro and Hornig, and Kearns, in terms of band-to-band transition to an almost free electron band above the lowest narrow tight-binding band. From a consideration of the spectral response of the photocurrent, they concluded that auto-ionization of the Frenkel exciton levels at these energies is shown to be unimportant. However, by comparing the external photoelectric effect produced by a single photon process with that produced by an exciton—exciton process, Pope [32] was led to the conclusion that the internal intrinsic carrier generation process is that of auto-ionization at least for excitation energies up to 6.3 eV. Theoretical arguments for the auto-ionization process were made by Choi [33a] and Jortner [33b] in an attempt to understand this discrepancy in the carrier

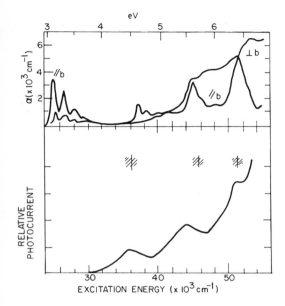

Fig. 5.4. Anthracene: relative number of electrons/photon as a function of photon energy, and absorption spectrum, $\|b$ and $\perp b$ crystal axes. Shaded regions: energy sequence of the density of states peaks in valence band estimated from kinetic energy distribution of external photoelectrons. (From Geacintov and Pope [30])

generation process. Recently, Baessler and Killesreiter [34] have analysed the auto-ionization data, taking the excess energy dissipation balance of the excited state into consideration, and have determined the precise band-gap energy for anthracene and naphthacene to be 4.00 ± 0.02 eV and 3.11 ± 0.03 eV, respectively.

5.2.3 Photoinjection from electrodes

There exists the possibility for the occurrence of electron or hole photoemission from an optically excited electrode into a crystal. This was reported by Pope and Kallmann [35] who used optically excited I_3^-(aq.) electrodes to inject holes into anthracene. Applying this method, Vaubel and Baessler [36] determined the band-gap energy for crystalline anthracene by measuring the threshold for the photoemission of electrons and holes. They obtained $\mathscr{E}_g = 3.9 \pm 0.1$ eV, taking into consideration an image force effect. Photoemission from a metal electrode was found by Williams and Dresner [37] who studied hole emission from a gold metal electrode into anthracene. They obtained a sequence of resonances that could be correlated with the vibrational modes in the optical absorption spectrum and interpreted this structure as a whole series of narrow conduction bands. Vaubel and Baessler

References pp. 182—184

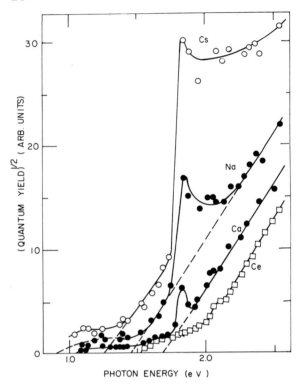

Fig. 5.5. Spectral response curve for the negative photoemission current for various contact metals. (From Caywood [38])

[36] also concluded that there was a broader electron conduction band (~ 0.5 eV wide) beginning about 0.55 eV above the edge of the lower band. Caywood [38] critically reviewed the experimental evidence and supported the above mentioned interpretation. Typical spectral response curves for photoemission quantum yield for various metal contacts are shown in Fig. 5.5 [38].

However, a serious question arises concerning this photoemission effect as pointed out by Many *et al.* [26, 39]. After a critical examination of the available data and their own carefully obtained experimental data, they insisted that two processes could partly or completely mask photoemission from metal contacts into anthracene; these are carrier photoexcitation out of surface traps which are occupied under thermal equilibrium, and photoexcitation out of bulk traps which are continuously being filled by carrier injection from the contact under study. The first process, *i.e.* hole photoexcitation out of surface traps, is greatly influenced by surface treatments, but even under the most optimal treatment the spectral yield is similar, both in magnitude and threshold, to yield curves obtained by several workers and attributed to photoemission. Further, they obtained almost the same response curve for the blocking electrode contact. As to the second process, they suggested that with contacts of alkali and alkaline earth metals (electron injectors) and gold (hole injectors) the reported yield curves are probably associated with photo-enhanced space-charge-limited currents. A more elaborate experiment from a quite different point of view, such as tunneling spectroscopy, should be made to achieve a full understanding of this problem.

5.2.4 Photo-detrapping of trapped charges

All organic crystals contain chemical or physical traps. Using modern techniques of purification, it is often possible to reduce the level of chemical impurities to 1 part in 10^7. Many kinds of crystal defect may act as physical traps, of which the density is estimated to be 10^{15}—10^{17} cm^{-3} [40]. Trapped carriers may be detrapped either optically, thermally, or by means of electric field. The optical excitation energy can be provided by direct absorption of a photon or by interaction with an exciton. Singlet or triplet excitons can react with trapped charges involving an ionization process. The bimolecular rate constants for the singlet-trap and triplet-trap are known to be $\gamma_{S,t} \simeq 10^{-8}$ cm^3 s^{-1} [22] and $\gamma_{T,t} \simeq 3 \times 10^{-11}$ cm^3 s^{-1} [23], respectively.

5.3 TRAPPING AND RECOMBINATION PROCESSES

5.3.1 Trapping and electric field dependence of photocurrent

The steady flow of a unipolar current through a real insulator containing bulk trapping sites and provided with an injecting contact has been divided into several regimes. At low voltages, there is an ohmic regime in which bulk and thermal generation of carriers predominates; this is followed at higher voltages by a space-charge-limited current (SCLC) regime during which the traps are filled by injected carriers, and at a particular voltage called V_{TFL}, the current rises steeply signalling the completion of the trap-filled process (traps filled limit, TFL). At sufficiently high voltages, the contact becomes depleted, and the current becomes independent of this applied field as saturation is reached. However, in real molecular crystals there are quite different physical mechanisms underlying the observed current voltage (J—V) response. For example, in the work of Campos [41] on naphthalene crystals, it is shown that the values obtained with simple SCLC theory for the density and depth of traps do not agree with the values obtained by the dependence with temperature of the factor θ (θ: the ratio of free electrons to trapped electrons). A typical result observed for the currents as a function of voltage is shown in Fig. 5.6 [41]. From this figure, several regions can be identified;

References pp. 182—184

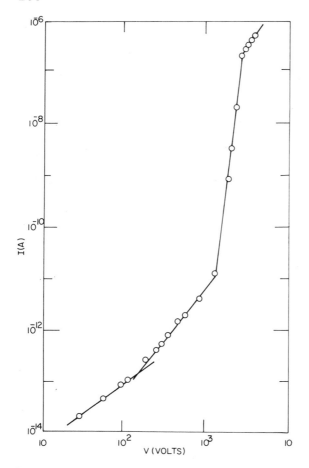

Fig. 5.6. Current as a function of voltage at room temperature for a naphthalene single crystal. (From Campos [41])

the first is ohmic, and this is followed by a square-law region ($J = (9/8)\mu\kappa\kappa_0 V^2/L^3$), and then a large-slope region of the J–V curve may be seen and, finally, a quadratic region of magnitude 10^4 smaller than a trap-free SCL current. This behaviour is characteristic of multiple discrete trapping levels. The disagreement of the theory seems to indicate that the fast increase of the current is not due to the traps filling. These traps are well characterized by a trap density of 10^{16}–$10^{17}\,\mathrm{cm}^{-3}$, a mean depth of $\sim 0.65\,\mathrm{eV}$, a capture cross-section of $10^{-17}\,\mathrm{cm}^2$, and a frequency of escape of $10^9\,\mathrm{s}^{-1}$, respectively. Photocurrent characteristics in anthracene crystals originating from exciton decay at metal (silver) and aqueous electrolytic electrodes show markedly different behaviour at field strengths in the range 10^3–$10^5\,\mathrm{V.cm}^{-1}$ [42]. This is attributed to the influence of the image force on charge separation at the metal electrode contrary to electrolyte contacts where the image force is negligible due to the slow orientation polarization of water as compared to the hopping frequency of injected charge carriers.

Recently, Chance and Braun [43] made a very careful study of the electric field dependence of hole and electron quantum yields. For electric fields up to 2×10^4 V.cm^{-1} applied normal to the *ab* plane of *virgin* anthracene crystals, both the hole and electron photo-carrier quantum yields increase linearly with field. Field-dependence plots exhibit a slope to intercept ratio, S/I, of 3.1×10^{-5} cm.V^{-1}, a value which is in quantitative agreement with that predicted by Onsager's theory of geminate recombination, 3.4×10^{-5} cm.V^{-1}, and this agreement seems to exclude the possibility for a Poole—Frenkel mechanism proposed by Geacintov and Pope [44]. Chance and Braun also showed that in *non-virgin* crystals, sharply reduced photo-carrier quantum yields at low fields are found to result from recombination of free carriers with oppositely charged trapped carriers left behind in the excitation region from previous experiments. Besides the SCLC, the most commonly used technique for determining the concentration and energy distribution of traps within the forbidden band is that of thermally-stimulated currents (TSC). Reucroft *et al.* [45] studied the carrier trapping characteristics of anthracene, chrysene, and naphthacene crystals using this TSC technique and also the space-charge-limited trap-limited current (SCLTLC) technique, comparing the variations of the trap characteristics in accordance with the different crystal conditions which are mainly due to differences of the crystal-growth methods. Silinsh [40] also studied the physical nature of traps in anthracene crystals.

5.3.2 Carrier recombination and temperature dependence of photocurrent

Kepler and Coppage [46] studied the generation and recombination of electrons and holes in anthracene crystals by X-rays from a 600-kV pulsed X-ray source. They found that the average energy, W, deposited per electron has a mean free path λ, the probability that it will travel a distance nation effects become important and electron—hole recombination coefficient was found to be $(3 \pm 1) \times 10^{-6}$ cm^3 s^{-1}, which was in good agreement with a theoretical value calculated on the basis of a very small mean free path for electrons and holes (less than 100 Å), and the large value of W was also explained in terms of this model. Further, they extended these considerations to carrier generation by relatively low-energy photon processes such as a band-to-band transition, and concluded that the very low quantum efficiency for free carrier generation in such processes could be explained, and that the commonly observed exponential temperature dependence of photoconductivity might also be explained by the same mechanism. If an electron has a mean free path λ, the probability that it will travel a distance r_0 without suffering a collision is $\exp(-r_0/\lambda)$, where r_0 is the critical

References pp. 182—184

distance which is defined by $(3/2)kT = e^2/\kappa\kappa_0 r_0$ ($\kappa\kappa_0$ is the dielectric constant of the material). Therefore, it appears that only the fraction $\exp(-r_0/\lambda)$ of those electron—hole pairs produced by low-energy photons will appear as free carriers. Since r_0 is proportional to $1/T$, $\exp(-r_0/\lambda)$ would look like $\exp(-T_0/T)$ provided λ was relatively independent of temperature, where T_0 is a characteristic temperature, being equal to $2e^2/3\kappa\kappa_0 k\lambda$. If we use the experimental fact that $kT_0 \simeq 0.16$ eV, it is found that, at room temperature, r_0/λ is approximately 6: a very reasonable value. Schott [47] gave a more detailed study on the mechanism of carrier-generation in electron-bombarded crystalline anthracene, and showed that the energy needed to create an electron—hole pair by low-energy electron-bombardment was less than 75 eV. He also reported the results of characteristic losses of 41 keV electrons in anthracene. The dominant loss, in the range 14—25 eV, is a solid-state effect, tentatively assigned to collective oscillations of the π-electrons.

5.4 PHOTO-CARRIER TRANSPORT PROCESSES

5.4.1 Charge-carrier mobility: experimental and phenomenological

(a) Drift mobility measurements

In order to elucidate the electronic charge transport mechanisms in organic molecular crystals, the charge-carrier mobility measurement has been a very useful tool since the experimental works by Kepler and LeBlanc [3]. If a small number of carriers are introduced into the system they will move in a constant electric field and will stop moving when they reach the metal—insulator interface. These conditions suggest the simplest and most direct method of determining the mobility of the carriers, the drift transient experiment. A typical experimental set-up is shown in Fig. 5.7 [48]. The crystal is sandwiched between two plane-parallel electrodes, at least one of which may be transparent to the exciting light. If the carriers are created at one surface by illuminating with strongly absorbed light for a time duration which is short compared with the carrier transit time, then the carriers will travel in a sheet across the crystal causing a constant current in the external circuit until the sheet hits the opposite interface. The drift mobility, μ, for those particular carriers, can be determined from the transit time, t_T:

$$\mu = \frac{L}{t_T E},$$

where L is the crystal thickness and E is the electric field strength.

Examples are given in Fig. 5.8, where about 1 μs pulse of strongly absorbed *UV* light created a sheet of electrons in anthracene at one surface. In the photocurrent pulse and its integral form, one sees a ramp (*i.e.*, constant

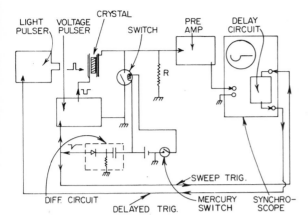

Fig. 5.7. Block diagram of experimental arrangement for drift mobility measurement. (From Maruyama and Inokuchi [48])

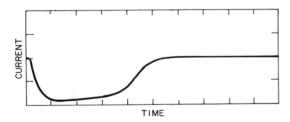

Fig. 5.8. Typical transient photocurrent for electrons. (abscissa: 10 μs/division).

current) as the sheet encounters the opposite face of the crystal. The slight curvature of the ramp is caused by the deep trapping of some of the carriers on the way across and the broadening of the sheet by simple diffusion. A rather long tail of the current pulse may also be caused by the broadening of the sheet and/or the detrapping effect from some shallow-trapping states. The mobility of electrons in the c'-direction of an anthracene crystal is determined to be 0.2—0.3 cm^2 V^{-1} s^{-1} from Fig. 5.8, and by reversing the polarity of the field holes can be drawn across in the same fashion (a mobility of 0.05—0.1 cm^2 V^{-1} s^{-1} was observed for holes). The temperature dependence of the mobilities for electrons and holes along each crystal axis was measured by Kepler [49], Nakada and Ishihara [50], and Fourny and Delacote [51]. All the results are shown in Fig. 5.9, from which one may imply that in the direction of the a or b axis a coherent (band-type) motion of charge-carriers is predominant and in the c'-direction (perpendicular to the ab plane) a thermally-activated incoherent (hopping-type) motion appears to be the case. A more detailed discussion will be given later in Section 5.4.2.

Kepler [49] and Kajiwara et al. [52] reported the behaviour of carrier

References pp. 182—184

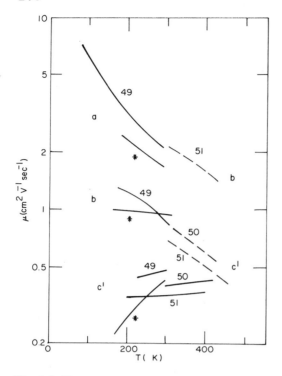

Fig. 5.9. Temperature dependence of carrier mobilities in anthracene. The solid lines are electron mobilities in the directions indicated on the left of the lines. The broken lines are hole mobilities in the direction indicated on the right. The lines are labeled with the number of the reference from which they were taken (*: G.T. Pott and D.F. Williams, J. Chem. Phys., 51 (1969) 1901).

mobilities in anthracene crystals under high pressure (up to 7.4 kbar). As shown in Fig. 5.10 [52], the electron mobilities along the a and b axes increased with the pressure applied; the values of the drift mobility at 7 kbar were 1.7 times as large as those at atmospheric pressure. The one along the c' axis was, however, nearly constant. In perylene single crystals, a rather low value of electron mobility ($\sim 0.02 \, \mathrm{cm^2 \, V^{-1} \, s^{-1}}$) was reported by Maruyama et al. [53], who explained the result by taking a fluctuation of transfer integrals and a stabilization of electron energy due to the excimer-like excess electron state into consideration. Quite recently Kamura et al. [54] have carried out the drift mobility measurement using a surface-type of cell and suggested the possibility for measuring the drift mobility in the direction parallel to a crystal surface. In usual simple organic crystals, the drift velocities are proportional to applied electric field up to $\sim 10^5 \, \mathrm{V.cm^{-1}}$. However, the drift mobility in a naphthacene thin crystal was reported to be inversely proportional to the electric field (10^2—$10^3 \, \mathrm{V.cm^{-1}}$) [55]. In amorphous photoconductive polymers (PVK and its complexes), the drift

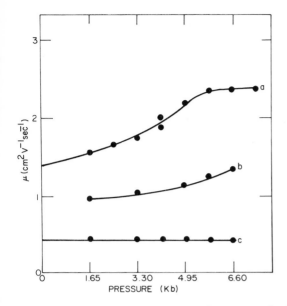

Fig. 5.10. Electron mobility of anthracene under high pressure. (a) along a-axis, (b) along b-axis, and (c) along c'-axis (From Kajiwara et al. [52])

mobilities are known to be usually dependent on the electric field. A more detailed discussion on this point will be found in Chapter 8. The values of charge-carrier drift mobilities in simple organic crystals observed so far are summarized in Table 5.1. The first theoretical approach to the mobility in a molecular crystal was made by LeBlanc [56] who calculated the transfer integrals between molecules in an anthracene crystal. Using the tight-binding-approximation, LeBlanc evaluated the electron and hole bandwidth to be about 0.02 eV and also estimated the mobility ratios in each crystal axis direction. The transfer integral calculation was improved by Katz et al. [57], and afterwards by Silbey et al. [58], taking into consideration the vibrational wavefunction overlaps between molecules.

(b) Hall mobility measurements

In the drift mobility measurement, the presence of a shallow trap may modulate the value of the drift mobility as was observed by Hoesterey and Letson [59], who doped anthracene with various amounts of naphthacene (a shallow trap for holes). The transit time in the presence of a shallow trap may involve a mean trap-residence time and a mean trapping time, and hence the observed drift mobility might not correspond with the microscopic mobility. To avoid this difficulty, a Hall mobility measurement can be tried. The great virtue of measuring the Hall mobility is that it ignores any trapping the carriers might undergo, and gives only the intrinsic mobility. Moreover, as was pointed out by Friedman [60], the sign, temperature dependence,

References pp. 182—184

TABLE 5.1

Room temperature drift mobilities in organic crystals[a]

Substances	Hole			Electron			References
	μ_a	μ_b	$\mu_{c'}$	μ_a	μ_b	$\mu_{c'}$	
Anthracene	1.0	2.0	0.98	1.7	1.0	0.54	[3a]
			0.8			0.4	[3b]
			0.8			0.4	[50]
			0.1			0.3	[61c]
Naphthalene	0.9	1.4	0.4	0.7	0.7	0.4	[b]
Naphthacene			1—0.1[c]				[55]
			0.35[d]				[e]
Pyrene			1.2			3.0	[f]
			$\sim 10^{-4}$			$\sim 10^{-4}$	[g]
Perylene			0.02			0.017	[53]
Tetrabenzo-perylene		$\sim 0.8^h$			$\sim 0.8^h$		[54]
m-Terphenyl			10^{-5}				[i]
Stilbene			10^{-3}				[i]
p-Quaterphenyl			$\sim 10^{-4}$				[j]

[a] Units in $cm^2 V^{-1} s^{-1}$.
[b] M. Silver et al., J. Chem. Phys., 38 (1963) 3030.
[c] The mobility is dependent on the electric field; 1—0.1 $cm^2 V^{-1} s^{-1}$ for 10^2—10^3 V.cm^{-1}.
[d] Observed at 100°C.
[e] O.H. LeBlanc, J. Chem. Phys., 37 (1962) 916.
[f] K. Ohki et al., Bull. Chem. Soc. Jpn., 36 (1963) 1512.
[g] A. Suzuki, personal communication.
[h] Parallel to a—b plane.
[i] R. Raman and S.P. McGlynn, J. Chem. Phys., 40 (1964) 515.
[j] A. Lipiński et al., Mol. Cryst. Liq. Cryst., 13 (1971) 381.

and magnitude of the Hall mobility compared with the intrinsic drift mobility provide important clues concerning the mechanism of transport. Many investigators have attempted to solve this problem by the measurement of the Hall mobility. Unfortunately, the Hall angle has proved difficult to measure in cases where μ is less than about 5 $cm^2 V^{-1} s^{-1}$. The large inconsistencies in the reported Hall mobilities for anthracene, ranging from 0.8 to 200 $cm^2 V^{-1} s^{-1}$ [61], indicate that reliable measurements for mobilities in this range pose formidable experimental problems. An apparatus for measuring the Hall mobility of photo-excited carriers with blocking contacts used by Maruyama and Inokuchi is shown in Fig. 5.11 [61c]. The values of Hall mobilities observed so far are summarized in Table 5.2.

Very recently, Burland [62] has succeeded in observing the cyclotron resonance absorption of photo-injected holes in anthracene crystals, and obtained the result of $m_p^* = 11 m_n$, where m_p^* is the effective mass of a hole and m_n is the free electron mass.

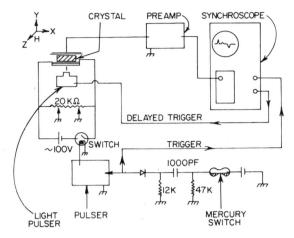

Fig. 5.11. Block diagram of experimental arrangement for photo-Hall effect measurement. (From Maruyama and Inokuchi [61c])

TABLE 5.2

Room-temperature Hall mobility in anthracene[a]

Orientation	E[b] B	a c'	b c'	a b	c' b	b a	c' a	other
Hole	References [61a] [61b] [61c] [61e] [61d] c d	 −1.5 0.8	 −3.1 2.5	 7.6 <1.0	 7.0 <1.0	 −8.6 0.7	 −3.8 	>~30 −35 ± 15 25−200(c') 5.2($\perp c'$) 1.4(c') 0.2−2.0(c')
Electron	[61f] [61a] [61g] [61f]	 9 	8−20 7 >20	 4 	 <|2| 	 <|2| 		 50

[a] Units in $cm^2 V^{-1} s^{-1}$.
[b] The notation E, a; B, c' means that the electric field is applied along a and the magnetic field along c'.
[c] R. Pethig and K. Morgan, Nature, 214 (1967) 266.
[d] T. Tombs, Ph.D. Thesis, Princeton University, 1968, unpublished.

5.4.2 Charge-carrier mobility: theoretical

As was pointed out earlier, band calculations by the tight-binding-approximation have been made rather intensively for anthracene, and they indicate

References pp. 182—184

that the bandwidths of excess electrons and holes are narrow, at most 0.2 eV and more likely less. It became obvious that electron—phonon interactions have an appreciable effect. The small polaron theory (Yamashita and Kurosawa [63], and Holstein [64]) indicates that when the effect of the local energy of the charge carrier by the vibrational disturbance is sufficiently large, the translational coherence of the perturbed band states is lost and the carrier then hops incoherently between localized states. This has been applied to anthracene mainly with respect to the acoustic lattice phonons (Glarum [65], and Glaeser and Berry [66]). Gosar and Choi [67] studied the effect of the fluctuations of the polarization energy and the transfer integrals on the excess electron and hole motion in anthracene crystals using the Kubo linear-response theory and the Wannier representation. Maruyama et al. [53] have applied this theory to the calculation of the drift mobility in perylene crystals taking into account the fluctuation of transfer integrals and a stabilization of electron energy due to the excimer-like excess electron state.

The recent work of Munn and Siebrand [7a, b] seems to have established a basis by which the fundamental processes governing charge-carrier transport in organic molecular crystals can now be understood. They have noted that in the localized hopping mode the interaction with the intramolecular vibrational modes can be significant. The interaction with the out-of-plane bending modes, which involves quadratic interaction with respect to the vibrational displacement, is particularly dominant. They considered three extreme limits for which their perturbation method could be applied. These are the slow phonon coherent limit, the slow phonon hopping limit and the slow electron hopping limit. The slow phonon coherent limit corresponds to the conventional band mode in which the carriers are quasifree and the mobility decreases with temperature as various scattering processes increase. In the latter two limits, the electron bandwidth is much smaller than the electron—phonon coupling energy and the carrier transport is in the hopping mode. One being the slow phonon hopping limit where the electron exchange energies (the transfer integrals) are large compared with phonon dispersion energies, and since the electron is strongly coupled to the phonons its rate of transfer is limited by the rate of phonon transfer [7a]. The other is the slow electron hopping limit where the electron exchange energies are small compared with phonon dispersion energies and transfer of the electron between adjacent molecules is the rate-determining step. The mutual relations among three parameters, ω_1, J, and ω_2, where ω_1 is the vibrational coupling parameter between adjacent molecules, J the electronic coupling between adjacent molecules, and ω_2 the electron—phonon coupling, are as follows (ω_0 is the phonon frequency):

(a) the slow phonon coherent limit; $J \gg \omega_2^2/\omega_0 \gg \omega_1^2/\omega_0$,

(b) the slow phonon hopping limit; $\omega_2^2/\omega_0 \gg J \gg \omega_1^2/\omega_0$,

(c) the slow electron hopping limit; $\omega_2^2/\omega_0 \gg \omega_1^2/\omega_0 \gg J$.

These three limiting cases were calculated for a linear chain model using these parameters involved which corresponded to a phonon bandwidth of about 0.001 eV and a zero temperature electron—phonon binding energy of about 0.012 eV. The results of these model computations as a function of the electron exchange integral were used to interpolate the mobility in the regions intermediate to these limiting cases, as shown in Fig. 5.12 [67]. The

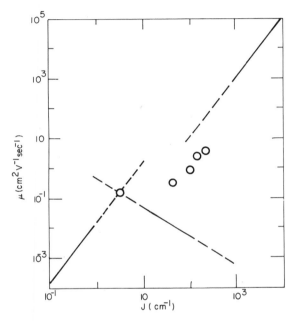

Fig. 5.12. Mobility as a function of the intermolecular electronic coupling J. The solid lines represent, in order of increasing J, the slow-electron hopping, slow-phonon hopping, and slow-phonon coherent limits. The broken lines are simple extrapolations. The open circles are the mobility contributions deduced from the experimental data for anthracene. (From Gosar and Choi [67])

model calculations and the values of intermolecular electronic couplings calculated by Glaeser and Berry [66] (including the polarization effect) were used to determine the dominant contributions to carrier drift mobilities in anthracene self-consistently from the experimental data, assuming that carrier mobilities in three-dimensional lattices consisted of contributions from mobilities along nearest neighbour directions. The results indicated that the transport of electrons in the c'-direction either by slow phonon hopping or slow electron hopping accounts for the observed magnitude, temperature dependence, and deuterium effect [68]. The other diagonal mobility components for electrons and holes were attributed to transport which is either

References pp. 182—184

coherent or intermediate between coherent and hopping transport in the slow phonon limit. Using this theory, Munn and Siebrand [69] also calculated the Hall mobilities. General expressions were derived for the Hall mobilities in these three different limits. It was shown that the Hall mobility may be anomalous in sign in each of these limits, depending on the signs of the intermolecular exchange integrals. The results of model calculations were compared with experimental Hall mobilities for anthracene by Korn et al. [61e] and fairly good agreement has been found. Further, they calculated the electrical conductivity of aromatic hydrocarbon crystals using linear-response theory and the fundamental processes governing charge-carrier transport in these crystals appeared to be understood [70].

5.5 RELATED PHENOMENA

5.5.1 Photovoltaic effect

The photovoltaic effect in organic crystals was first reported by Kallman and Pope [71], who studied this effect in anthracene, using aqueous NaCl electrodes. With highly absorbed 365 nm light (3.15 eV) a photovoltage of about 200 mV developed across the crystal. This effect is produced by the non-uniform optical excitation of the crystal, producing a gradient of exciton-injected carrier of only one sign, namely holes. This phenomenon is called the Dember effect. Using a metal electrode evaporated on the organic thin film, Inokuchi et al. [72] observed the photovoltaic effect due to the contact potential difference between the metal—crystal interface, that is, of surface origin. Geacintov et al. [73] studied the photocurrent and photovoltage of naphthacene single crystals with aqueous electrodes as a function of excitation wavelength in the 220—560 nm region. At a wavelength longer than 410 nm, the photocurrent was due to the injection of holes at the illuminated electrode. A bulk-generated (electron) photocurrent was shown to be produced with excitation energies in excess of 3 eV, *i.e.*, at a wavelength less than 410 nm. The photovoltaic effect in naphthacene has been also studied by Fang et al. [74], who found a reversal of the polarity of the photovoltaic current in a certain wavelength region, and interpreted this effect as the consequence of the photovoltaic transport equation. Recently, Tsubomura et al. [75] have also examined the time and spectral dependence of the photovoltage in naphthacene thin film, using a chopped excitation light source. They observed fast and slow components in time which were opposite in polarity, and discussed the effect in terms of the singlet and triplet exciton migration and trapping.

An interesting approach to photovoltaic phenomena from a somewhat different point of view has been made in the field of surface chemistry in the studies of the physicochemical properties of bimolecular lipid layers by Tien [76]. An ultra-thin bimolecular layer (40—130 Å) is held at the small hole separating electrolyte solutions which take a role of the electrode or sensitizer. This film shows a photovoltaic effect, photoconduction and photosensitized conduction, which could be analyzed by analogy to organic semiconductors. These studies may be developed in relation to the problem on the action of living membranes. More substantial theoretical and experimental work should be done in the future on the photovoltaic effect in organic crystals containing traps.

5.5.2 Sensitized photoconduction

As was stated in Section 5.2, the photoconduction of molecular crystals is often determined by the surface properties of the crystals. The photo-carrier injection processes in the case of an anthracene single crystal with an appropriate dye, *e.g.* rhodamine B, adsorbed on its surface are rather well understood. When either the adsorbed dye is excited in the visible or the anthracene in the near u.v., the occurrence of the following processes can be demonstrated experimentally:
(a) singlet—singlet-energy transfer,
(b) triplet—triplet-energy transfer,
(c) charge transfer quenching of excited singlet states,
(d) formation of triplet states by recombination of radical ions,
(e) triplet—triplet-annihilation.
The exceptional behaviour of this system has its origin in the relative energies of the excited states of the anthracene crystal A and the adsorbed dye D (Fig. 5.13) [77]. In Fig. 5.14 the most important reactions are shown, which occur when either the dye or the anthracene single crystal is excited.

In sensitization experiments, an electrolytic contact arrangement developed by Kallmann and Pope [78] has been used and one of the aqueous electrolyte solution is a dilute solution of a dye. The dye is partly adsorbed on the surface of the crystal. Upon excitation with visible light [Fig. 5.14 (2)] a sensitized hole injection current i_+ is found. The excitation spectrum of the photocurrent, $i_+(\tilde{\nu})$, is very similar to the absorption spectrum of the dye in homogeneous solution, $\epsilon(\tilde{\nu})$. Upon excitation with u.v. light, the photocurrent is increased by the adsorption of the dye, but the excitation spectrum $i_+(\tilde{\nu})$ is still parallel to the absorption spectrum of the anthracene crystal, $\alpha(\tilde{\nu})$. In both cases, the adsorbed dye may be considered as the primary electron acceptor, which is necessary for the surface generation of holes. Several investigators have studied more detailed reaction schemes for the sensitized generation of holes. This interesting work has been reviewed recently by Nickel [77] and references are given.

References pp. 182—184

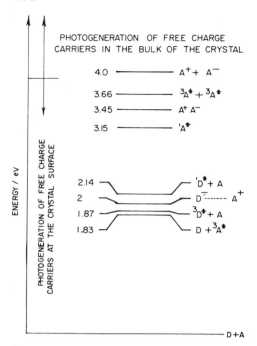

Fig. 5.13. Energies of excited states of crystalline anthracene (A) and of rhodamine B (D). For completeness the energies of the postulated CT-exciton and of internal ionization of crystalline anthracene are included. (From Nickel [77])

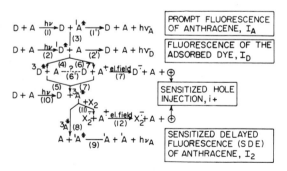

Fig. 5.14. Reactions following the excitation of an anthracene single crystal A or of a dye D, adsorbed at the surface of A. X_2 is a strong oxidant. (1), (2), (10) correspond to excitation at $\lambda \leqslant 420$ nm, 500 nm $< \lambda < 600$ nm, $\lambda = 633$ nm. Measurable quantities are shown in frames. (From Nickel [77])

5.5.3 Photoconduction in amorphous organic solids

The main features of photoconduction in amorphous solids will be presented in Chapter 7 of this book. However, a recent topic on the photoconduction in amorphous organic simple-molecular solids will be described here.

Using a low-temperature evaporation technique, Maruyama and Iwasaki [79] prepared amorphous organic thin films (naphthacene, perylene, coronene, violanthrene-A, and quaterrylene), and observed the optical [79a] and the electrical [79b] properties of these films compared with those of crystalline-state (well-oriented) films. They noted, for the optical properties of amorphous films, the following: (a) the disappearance of a Davydov splitting (naphthacene, Fig. 5.15), (b) the reduction of the intensity of a charge

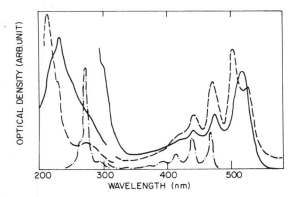

Fig. 5.15. Absorption spectra of naphthacene. ———: amorphous film; -----: crystalline film; —·—: solution.

transfer band (perylene), and (c) the presence of a characteristic band for the amorphous state (violanthrene-A and quaterrylene). The photoconduction action spectra of amorphous naphthacene and violanthrene-A films also showed a very interesting correlation with their structures (Fig. 5.16 and Fig. 5.17). Although the photoconductivity spectral responses of the crystal-like films were parallel to their absorption spectra, those of the amorphous films were quite different from their absorption spectra, which correlate well with the absorption spectra of crystal-like films. This fact may be interpreted as follows. The excited states corresponding to the absorption peaks in amorphous films may be so localized in nature that they do not migrate in the solid and consequently no charge-carriers can be created, and only the exciton absorption at the small crystalline portions in the "amorphous" films release charge carriers.

References pp. 182—184

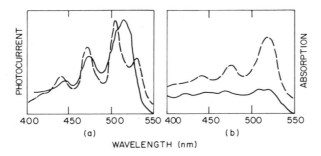

Fig. 5.16. Photoconductivity spectral responses of naphthacene films; ———: photocurrent; -----: absorption (a) Room-temperature evaporated crystal-like film. (b) Low-temperature evaporated amorphous film.

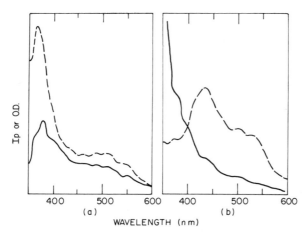

Fig. 5.17. Photoconductivity spectral responses of violanthrene-A films; ———: photocurrent; -----: absorption. (a) High-temperature evaporated crystal-like film. (b) Low-temperature evaporated amorphous film.

5.5.4 External photoelectric effect: photoemission phenomena

The absorption of a quantum of sufficient energy by an organic crystal will result in the ejection of an electron. In anthracene, this energy is about 5.9 eV. The external photoelectron emission process can be considered to involve three processes. The first is the production of the superexcited state by the absorption of a photon of requisite energy and its subsequent ionization; the second is the movement of the ejected electrons in all directions in the crystal, during which the electron becomes thermalized; finally, the electron that has travelled in the direction of the surface must also overcome any surface potential barriers created by impurities and/or by the ambient atmosphere. Instead of the ionization process due to direct transitions from

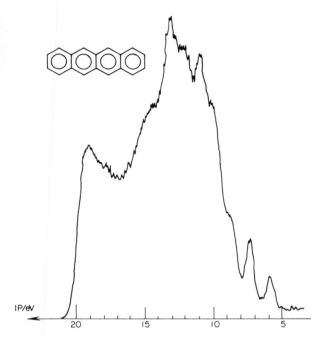

Fig. 5.18. Kinetic energy distribution curve for naphthacene under an illumination of 21.22 eV He resonance line.

the valence band to the conduction band in an inorganic semiconductor, electron—hole pair creation most likely occurs as a result of an autoionization of a superexcited molecular state in an organic crystal. The measurement of the distribution of the kinetic energies of the emitted electrons can provide information on the bound states of electronic excitation involved in the ionization process. From this distribution, it is possible to determine whether the energy that is available to the ejected electron can be used for the excitation of the positive ion that remains. The kinetic energy distribution curve for a naphthacene thin film obtained by the use of a 21.22 eV (584 Å) He resonance line source is shown in Fig. 5.18 [80]. From these curves, the first ionization potential of naphthacene crystal can be determined to be 5.40 eV and the sequence of the peaks are considered to correspond to the lower valence state of this crystal. These low-lying valence band states were also compared with the intrinsic photoconductivity spectral peaks and a relevant elucidation was provided by Geacintov and Pope (Fig. 5.4) [30]. The thermalization distance for the ejected electron is close to 100 Å almost without regard for the initial energy of the electron in the range < 20 eV. In the energy loss process that cannot excite a molecular vibration, the sub-vibrational excitation may be the main loss mechanism and it has been given as 0.003 eV per collision by Kochi et al. [81]. Multi-quantum processes for

TABLE 5.3

Ionization potentials, electron affinities, predicted band gaps (eV) for some organic molecules and crystals

Compound	I_g	A_g	I_c	E_c^a	$\mathcal{E}_g^b$	$\mathcal{E}_g$(obs.)
Benzene	9.24^c	−0.95				
Naphthalene	8.15^c	−0.2	6.76^d	5.4	5.6	5.2
Anthracene	7.47^c	0.58	5.70^e	3.9	3.3	3.9
Naphthacene	7.04^c	0.88	5.28^e	3.5	2.6	3.0
Phenanthrene	7.86^c	0.31	6.45^d	5.0	4.7	
Pyrene	7.41^c	0.50	5.83^e	4.3	3.8	4.1
Perylene	7.00^c	1.3	5.37^e	3.7	2.4	
Pentacene	6.74^c	1.2	4.85^e	3.0	1.8	

^{a} $E_c = 2I_c - I_g$
^{b} $\mathcal{E}_g = 2I_c - I_g - A_g$
^{c} R. Boschi, E. Clar and W. Schmidt, J. Chem. Phys., 60 (1974) 4406.
^{d} L.E. Lyons and G.C. Morris, J. Chem. Soc., (1960) 5192.
^{e} T. Hirooka, Ph.D. Thesis, University of Tokyo, 1973.

the external photoelectric effect have also been observed. In anthracene crystals, a singlet—singlet excitons annihilation can result in an external electron emission [82], and in naphthacene crystals, a CT—CT excitons annihilation process may take part in this effect [83]. The available data of ionization potentials and the electron affinities for gas and solid, respectively, are shown in Table 5.3.

REFERENCES

1 A. Pochettino, Atti Accad. Naz. Lincei Cl. Sci. Fis. Mat. Nat. Rend., 15 (1906) 355.
2 H. Inokuchi, Bull. Chem. Soc. Jpn., 27 (1954) 22.
3(a) O.H. LeBlanc, J. Chem. Phys., 33 (1960) 626.
3(b) R.G. Kepler, Phys. Rev., 119 (1960) 1226.
4 J.N. Murrell, Mol. Phys., 4 (1961) 205.
5 O.H. LeBlanc, J. Chem. Phys., 35 (1961) 1275.
6 L. Friedman, Phys. Rev. A, 133 (1964) 1668.
7(a) R.W. Munn and W. Siebrand, J. Chem. Phys., 52 (1970) 47.
7(b) R.W. Munn and W. Siebrand, ibid., 52 (1970) 6391.
8 J. Frenkel, Phys. Rev., 37 (1931) 17, 1276.
9 A.S. Davydov, Sov. Phys. Usp., 530 (1964) 145.
10 A. Bree and L.E. Lyons, J. Chem. Soc., (1959) 2662.
11 B.J. Mulder, Phys. Res. Rep. Suppl. No. 4 (1968).
12 H. Killesreiter and H. Baessler, Chem. Phys. Lett., 11 (1971) 411.
13 H. Baessler, N. Riehl and G. Vaubel, Mol. Cryst. Liq. Cryst., 9 (1969) 249.
14 G. Vaubel, H. Baessler and D. Möbius, Chem. Phys. Lett., 10 (1971) 334.
15 S.Z. Weisz, J. Menendez, L.F. Rojas and S. Dellonte, Mol. Cryst. Liq. Cryst., 24 (1973) 45.

16 S. Choi and S.A. Rice, J. Chem. Phys., 38 (1963) 366.
17 M. Silver, D. Olness, M. Swicord and R.C. Jarnagin, Phys. Rev. Lett., 10 (1963) 12.
18 K. Hasegawa and S. Yoshimura, Phys. Rev. Lett., 14 (1965) 689.
19 R.G. Kepler, J. Chem. Phys., 40 (1964) 1173.
20 R.G. Kepler, Phys. Rev. Lett., 18 (1967) 951.
21 A. Bergmann, M. Levine and J. Jortner, Phys. Rev. Lett., 18 (1967) 593.
22 M. Schott and J. Berrihar, Mol. Cryst. Liq. Cryst., 20 (1973) 13.
23 V. Ern, H. Bouchriha, J. Fourny and G. Delacote, Solid State Commun., 9 (1971) 1201.
24 N. Wakayama and D.F. Williams, J. Chem. Phys., 57 (1972) 1770.
25 P. Day and M.G. Price, J. Chem. Soc. A, (1969) 236.
26 A. Many, J. Levinson, and I. Teucher, Mol. Cryst., 5 (1969) 121.
27 G. Castro and J.F. Hornig, J. Chem. Phys., 42 (1965) 1459.
28 D.R. Kearns, J. Chem. Phys., 45 (1966) 3966.
29 D.M. Hanson, Crit. Rev. Solid State Sci., 3 (1973) 243.
30 N.E. Geacintov and M. Pope, J. Chem. Phys., 50 (1969) 814.
31 M. Silver and R. Sharma, J. Chem. Phys., 46 (1967) 692.
32 M. Pope, J. Chem. Phys., 47 (1967) 2197.
33(a) S.I. Choi, Phys. Rev. Lett., 19 (1967) 358.
33(b) J. Jortner, Phys. Rev. Lett., 20 (1967) 244.
34 H. Baessler and H. Killesreiter, Mol. Cryst. Liq. Cryst., 24 (1973) 21.
35 M. Pope and H. Kallmann, Symp. on Electrical Conduction in Organic Solids, Interscience, New York, 1960.
36 G. Vaubel and H. Baessler, Phys. Status Solidi, 26 (1968) 599.
37 R. Williams and J. Dresner, J. Chem. Phys., 46 (1967) 2133.
38 J.M. Caywood, Mol. Cryst. Liq. Cryst., 12 (1970) 1.
39 J. Levinson, Z. Burnshtein and A. Many, Mol. Cryst. Liq. Cryst., 26 (1974) 329.
40 J. Sworakowski, Mol. Cryst. Liq. Cryst., 11 (1970) 1.
    E.A. Silinsh, Phys. Status Solidi, 3 (1970) 817.
41 M. Campos, Mol. Cryst. Liq. Cryst., 18 (1972) 105.
42 M.E. Michel-Beyerle, W. Harengel, R. Haberkorn and J. Kinder, Mol. Cryst. Liq. Cryst., 25 (1974) 323.
43 R.R. Chance and C.L. Braun, J. Chem. Phys., 59 (1973) 2269.
44 N.E. Geacintov and M. Pope, Proc. 3rd Int. Photoconductivity Conf., Stanford, Calif., 1969, Pergamon, Oxford, 1971, p. 289.
45 P.J. Reucroft, F.D. Mullins and E.E. Hillman, Mol. Cryst. Liq. Cryst., 23 (1973) 179.
46 R.G. Kepler and F.N. Coppage, Phys. Rev., 151 (1966) 610.
47 M. Schott, Mol. Cryst., 5 (1969) 229.
48 Y. Maruyama and H. Inokuchi, Bull. Chem. Soc. Jpn., 40 (1967) 2073.
49 R.G. Kepler, Organic Semiconductors, Macmillan, New York, 1962, p.1.
50 I. Nakada and Y. Ishihara, J. Phys. Soc. Jpn., 19 (1964) 695.
51 J. Fourny and G. Delacote, J. Chem. Phys., 50 (1969) 1028.
52 T. Kajiwara, H. Inokuchi and S. Minomura, Bull. Chem. Soc. Jpn., 40 (1967) 1055.
53 Y. Maruyama, T. Kobayashi, H. Inokuchi and S. Iwashima, Mol. Cryst. Liq. Cryst., 20 (1973) 373.
54 Y. Kamura, H. Inokuchi and Y. Maruyama, Chem. Lett., (1974) 301.
55 J. Kondrasiuk and A. Szymanski, Mol. Cryst. Liq. Cryst., 18 (1972) 379.
56 O.H. LeBlanc, J. Chem. Phys., 35 (1961) 1275.
57 J.L. Katz, S.A. Rice, S. Choi and J. Jortner, J. Chem. Phys., 39 (1963) 1683.
58 R. Silbey, J. Jortner, S.A. Rice and M.T. Vala, J. Chem. Phys., 42 (1965) 733.
59 D.C. Hoesterey and G.M. Letson, J. Phys. Chem. Solids, 24 (1963) 1609.
60 L. Friedman, Phys. Rev., 133 (1964) 1668.
    L. Friedman and T. Holstein, Ann. Phys., 21 (1963) 494.

61(a) J. Dresner, Phys. Rev., 143 (1966) 558.
61(b) G. Delacote and M. Schott, Solid State Commun., 4 (1966) 177.
61(c) Y. Maruyama and H. Inokuchi, Bull. Chem. Soc. Jpn., 40 (1967) 2073.
61(d) G.C. Smith, Phys. Rev., 185 (1969) 1133.
61(e) A.I. Korn, R.A. Arndt and A.C. Damask, Phys. Rev., 186 (1969) 938.
61(f) J. Dresner, J. Chem. Phys., 52 (1970) 6343.
61(g) M. Schadt and D.F. Williams, Phys. Status Solidi, 39 (1970) 223.
62   D.M. Burland, Phys. Rev. Lett., 33 (1974) 833.
63   J. Yamashita and T. Kurosawa, J. Phys. Chem. Solids, 5 (1958) 34.
64   T. Holstein, Ann. Phys. (N.Y.), 8 (1959) 325, 343.
65   S.H. Glarum, J. Phys. Chem. Solids, 24 (1963) 1577.
66   R.M. Glaeser and R.S. Berry, J. Chem. Phys., 44 (1966) 3797.
67   P. Gosar and S. Choi, Phys. Rev., 150 (1966) 529.
68   W. Mey, T.J. Sonnonstine, D.L. Morel and A.M. Herman, J. Chem. Phys., 58 (1973) 2542.
69   R.W. Munn and W. Siebrand, J. Chem. Phys., 53 (1970) 3343.
70   R.W. Munn and W. Siebrand, Discuss. Faraday Soc., 51 (1971) 17.
71   H. Kallmann and M. Pope, J. Chem. Phys., 30 (1959) 585.
72   H. Inokuchi, Y. Maruyama and H. Akamatu, Bull. Chem. Soc. Jpn., 34 (1961) 1093.
     Y. Maruyama, H. Inokuchi and Y. Harada, Bull. Chem. Soc. Jpn., 36 (1963) 1193.
73   N. Geacintov, M. Pope and H. Kallmann, J. Chem. Phys., 45 (1966) 2639.
74   P.H. Fang, A. Golubovic and N.A. Dimond, Jpn. J. Appl. Phys., 11 (1972) 1298.
     P.H. Fang, Jpn. J. Appl. Phys., 12 (1973) 536.
75   H. Tsubomura, personal communication.
76   H.T. Tien, Techniques of Surface and Colloid Chemistry and Physics, Vol. 1, 1972, p. 109.
     H.T. Tien, Surface and Colloid Science, Vol. 4, 1971, p. 361.
77   B. Nickel, Mol. Cryst. Liq. Cryst., 18 (1972) 227.
78   H. Kallmann and M. Pope, Rev. Sci. Instrum., 30 (1959) 44.
79(a) Y. Maruyama and N. Iwasaki, Chem. Phys. Lett., 24 (1974) 26.
79(b) Y. Maruyama and N. Iwasaki, J. Non-Cryst. Solids, 16 (1974) 399.
80   K. Seki, Y Harada, K. Ohno and H. Inokuchi, Bull. Chem. Soc. Jpn., 47 (1974) 1608.
81   M. Kochi, Y. Harada, T. Hirooka and H. Inokuchi, Bull. Chem. Soc. Jpn., 43 (1970) 2690.
82   M. Pope, H. Kallmann and J. Gianchino, J. Chem. Phys., 42 (1965) 2540.
83   M. Pope, J. Burgos and J. Giachino, J. Chem. Phys., 43 (1965) 3367.

CHAPTER 6

AMORPHOUS TETRAHEDRALLY BONDED SOLIDS

W.E. SPEAR and P.G. LECOMBER

6.1. Introduction
6.2. Photoconductivity in glow discharge silicon
   6.2.1. Specimen preparation and survey of electronic properties
   6.2.2. Photoconductivity at room temperature
   6.2.3. The temperature dependence of the photoconductivity
   6.2.4. Interpretation and discussion
   6.2.5. The recombination process
   6.2.6. Photoluminescence
6.3. Photoconductivity in evaporated a-Si
6.4. Photoconductivity in a-Ge
   6.4.1. Evaporated films
   6.4.2. Electrolytic a-Ge
6.5. Photoconductivity in other tetrahedrally bonded amorphous materials
6.6. Concluding remarks

6.1 INTRODUCTION

The extensive work on amorphous silicon and germanium (a-Si, a-Ge) and other tetrahedrally bonded materials has shown that the electronic properties of these non-crystalline solids depend critically on the method of preparation and also on the detailed experimental conditions during the specimen deposition. For instance, a-Si prepared by two different techniques may well differ in its electrical conductivity by eight orders of magnitude and the optical absorption coefficient at photon energies below the optical gap may differ by a factor of two or more. Another remarkable example is provided by the strong dependence of the properties of evaporated films on the rate of deposition. Thus a four-fold increase in this rate raises the density of unsaturated spins in a-Si by about two orders of magnitude [1].

These effects, which are not generally met in the other amorphous materials discussed in this book (such as Se or the chalcogenides), have complicated the meaningful comparison of results from different laboratories and have proved a serious obstacle in the detailed understanding of the electronic properties of amorphous tetrahedral solids. The main problem arises through the defect structure of these materials, which is extremely sensitive to preparation

References pp. 213—214

conditions and at the same time has a determining influence on the electronic properties. It is likely that the defects are mainly associated with unsaturated bonds in the random network, situated on internal surfaces of vacancy clusters. Strong support for such a model comes, for instance, from the work of Brodsky *et al.* [2, 3] who correlated the spin density with the electrical and optical properties of their evaporated or sputtered specimens.

The photoconductivity in a-Si, Ge and other tetrahedral materials produced by vacuum evaporation or sputtering is normally small and has received far less attention in the literature than the photoconductive properties of Se or the chalcogenide glasses. The main reason for the small response appears to be the pronounced defect structure which leads to a high overall level of the density of states in the mobility gap and consequently to short recombination lifetimes of the photogenerated carriers. Although it has been possible, by carefully controlled evaporation techniques and subsequent annealing, to improve the photoresponse of a-Si [4, 5] and Ge [6] it is difficult to draw reasonably unambiguous conclusions from the results at present available. This work, as well as measurements on some amorphous III—V compounds will be discussed in Sections 6.3—6.5 of this chapter.

Most of the review is concerned with a detailed discussion of photoconductive and related properties in a-Si specimens prepared by the decomposition of silane gas (SiH_4) in a radio-frequency glow discharge. This method was developed by Chittick *et al.* [7] and produced a-Si films with a high photoresponse [7, 8]. The discussion, in the following section, of photoconductivity in glow discharge Si and its spectral, temperature, and intensity dependence is based on the recent work of Spear and his collaborators [9, 10]. Its primary aim has been to correlate and interpret the photoconductive results in the light of detailed information on the transport properties, obtained previously from drift mobility and conductivity measurements [11, 12]. In addition, the main features of the density of state distribution in glow discharge specimens have been determined from field effect measurements [13, 14]. This aspect is of particular interest, because it has made it possible to relate the recombination process to particular features of the density of state distribution.

6.2 PHOTOCONDUCTIVITY IN GLOW DISCHARGE SILICON

6.2.1 Specimen preparation and survey of electronic properties

In the preparation of the specimens, silane was passed at a constant rate through a quartz reaction tube. The substrate, a 20-mm-diameter Vitreosil or Corning glass (7059) disc, was held in the centre of the tube on a molybdenum hearth. This was surrounded on the outside by an r.f. coil, powered from a 5 MHz generator, which produced the glow discharge. An electric heater

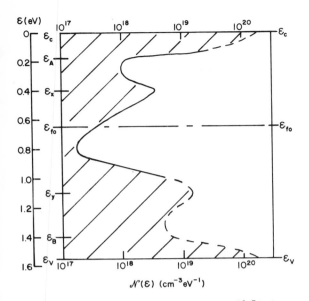

Fig. 6.1. Density of states distribution, $\mathcal{N}(\mathcal{E})$, for an a-Si specimen deposited by the glow discharge technique on a substrate held at a temperature $T_d \gtrsim 500$K. The energy is measured from $\mathcal{E}_c$. Full lines: $\mathcal{N}(\mathcal{E})$ obtained from field effect measurements; broken lines: based on other information. (From Madan et al. [13] and Spear [14])

controlled the temperature of the substrate during deposition. This quantity, denoted by T_d in the following, is an important parameter, as it largely determines the electronic properties of the glow discharge specimens. Most of the photoconductive experiments were carried out on gap cells, using aluminium or gold surface electrodes, separated by 1.5 mm. Several sandwich cells, with molybdenum bottom electrodes, were also investigated and led to substantially similar results. A Keithley 602 Electrometer was used for the measurement of the steady photocurrent; applied fields lay between 4 and 8 kV cm^{-1} and ohmic behaviour was found in most cases.

Before turning to the photoconductivity, it will be useful to summarise our present understanding of the electronic properties of glow discharge silicon. Some of the most relevant information has come from recent field effect experiments [13] which have led to the distribution of localized states throughout an appreciable part of the mobility gap. Figure 6.1 shows the distribution function $\mathcal{N}(\mathcal{E})$ (i.e. the number of states per unit volume per unit energy range at energy $\mathcal{E}$) for specimens deposited at a relatively high T_d, between 500 and 600K. The energy is normalised to the onset of the extended electron states at $\mathcal{E}_c$. The solid line represents $\mathcal{N}(\mathcal{E})$ obtained directly from the field effect experiments, whereas the remaining parts of the curve have been deduced from other information.

The localized tail states, between $\mathcal{E}_c$ and $\mathcal{E}_A$ and also between $\mathcal{E}_v$ and $\mathcal{E}_B$,

References pp. 213—214

are thought to arise from the lack of long-range order in the solid. The distribution in the central region, between $\mathscr{E}_A$ and $\mathscr{E}_B$, depends markedly on T_d and is associated with reasonably well-defined structural defects, probably unsaturated bonds in vacancy complexes. The field effect experiments show that in a low T_d specimen ($\sim$ 300K) the overall density of these gap states may be up to two orders of magnitude higher than in specimens prepared between 500 and 600K.

$\mathscr{N}(\mathscr{E})$ possesses a well-defined minimum approximately in the centre of the mobility gap. The position of the thermal equilibrium Fermi level, $\mathscr{E}_{f0}$, is determined by the defect structure and thus depends on T_d. As shown in Fig. 6.1, $\mathscr{E}_c - \mathscr{E}_{f0} \simeq 0.6$ eV for high T_d specimens. With decreasing T_d, $\mathscr{E}_{f0}$ moves towards the maximum at $\mathscr{E}_y$ and $\mathscr{E}_c - \mathscr{E}_{f0} \simeq 0.8$ eV for $T_d \simeq 350$K.

The electronic transport properties of the specimens are also largely determined by the defect structure. The correlation of conductivity and drift mobility results have led to the following information [11, 12]:

(a) At temperatures above about 250K the current in specimens prepared at $T_d \gtrsim 350$K is predominantly carried by electrons in extended states above $\mathscr{E}_c$. The electron mobility μ_c in these states is about $10\,\mathrm{cm}^2\,\mathrm{s}^{-1}\,\mathrm{V}^{-1}$. The drift mobility measurements indicate that excess electrons interact through trapping and thermal release with localized states lying at $\mathscr{E}_A$, 0.18 eV below $\mathscr{E}_c$ in the rapidly decreasing density of the electron tail states. The latter are substantially in thermal equilibrium with the extended states at all times during an electron transit, and a closer analysis of the transport [15] indicates that a local maximum in the density of occupied states does, in fact, lie close to $\mathscr{E}_A$.

(b) Below 250K, a change in transport mechanism occurs. Electrons now propagate by phonon-assisted hopping through states around $\mathscr{E}_A$ although, with decreasing temperature, the predominant hopping paths lie progressively closer to the Fermi level.

(c) The dark conductivity and its temperature dependence are largely determined by T_d. In specimens prepared at $T_d \lesssim 350$K (i.e. $\mathscr{E}_c - \mathscr{E}_{f0} \gtrsim 0.8$ eV), the current is carried predominantly by holes. These propagate by phonon-assisted hopping in states around the maximum at $\mathscr{E}_y$ (Fig. 6.1).

(d) The energy gap between extended electron and hole states is estimated as $\mathscr{E}_c - \mathscr{E}_v \gtrsim 1.55$ eV from transport experiments.

6.2.2 Photoconductivity at room temperature

This investigation extends over glow discharge specimens, deposited between 300 and 600K. Table 6.1 is a summary of relevant specimen data: the thickness d, the deposition temperature, electrical constants such as the room temperature conductivity, σ_{RT}, and the gradient of the $\ln \sigma$ versus $1/T$ curves, $\mathscr{E}_1^\sigma = (\mathscr{E}_c - \mathscr{E}_f)_0$. The remaining columns list several photoconductive properties which will be referred to in the following.

TABLE 6.1

Summary of specimen data

Specimen	T_d(K)	d(μm)	$\sigma_{RT}(\Omega cm)^{-1}$	$\mathcal{E}_1^\sigma$(eV)	$\mathcal{E}_1^{opt}$(eV)	$\mathcal{E}_0$(eV)
1	600	1.0	2.5×10^{-9}	0.66	0.60	1.64
2	600	3.9	1×10^{-8}	0.68	0.65	1.43
3	600	2.1	1.4×10^{-8}	0.70	0.63	1.46
4	550	1.0	1×10^{-9}	0.63	0.58	1.50
5	500	2.3	3.7×10^{-10}	0.70	0.62	1.56
6	500	0.8	1×10^{-9}	0.79	0.68	1.55
7	490	4.0	2.5×10^{-10}	0.72	0.60	1.53
7S	490	4.0	10^{-11}	—	0.65	1.57
8	440	1.7	1.5×10^{-10}	0.70	0.65	1.53
9S	420	0.8	10^{-11}	—	—	1.57
10	380	3.4	4×10^{-12}	0.95	0.80	1.57
11	370	0.6	10^{-11}	0.80	—	1.58
12	350	1.1	2×10^{-12}	0.80	0.90	1.55
13	300	4.3	2×10^{-11}	0.3^a	0.95	1.55
14E	550	0.1	10^{-12}	0.3^a	—	—

a-Si specimens 1—13, prepared by glow discharge; 14E, evaporated; 7S and 9S sandwich cells, all others gap cell geometry. T_d deposition temperature; d specimen thickness; σ_{RT} conductivity at room temperature; $\mathcal{E}_1^\sigma = (\mathcal{E}_c - \mathcal{E}_f)_0$, position of Fermi level from conductivity measurements, $\mathcal{E}_1^{opt}$ and $\mathcal{E}_0$ obtained from extrapolation in Fig. 6.3.

^{a} These energies do not represent $(\mathcal{E}_c - \mathcal{E}_f)_0$.

The photocurrent i_p, defined as the difference between the current measured with and without illumination, is given by the following expression:

$$i_p = eN_0(1-R)[1-\exp(-\alpha d)]\eta\tau/t_T \qquad (6.1)$$

where $N_0(1-R)$ is the number of photons per second falling onto the specimen, corrected for surface reflection. η describes the efficiency of the generation process; it is defined as the number of electron—hole pairs produced per absorbed photon. τ/t_T is the ratio of the recombination lifetime and the carrier transit time for electrons in states above $\mathcal{E}_c$. The latter is given by $t_T = L/E\mu_c$ where L is the appropriate drift distance. In general, τ/t_T will be the sum of an electron and a hole contribution. However, on the basis of the transport results summarised in Section 6.2.1, we should like to suggest that the steady state photocurrent at room temperature is carried predominantly by electrons in the extended states, so that τ/t_T will refer to the generated electrons. This applies without doubt to specimens prepared at the higher T_d and further evidence for this is presented in the following.

In all room temperature experiments it was found that with the incident light intensities used ($\gtrsim 5 \times 10^{12}$ photons cm^{-2} s^{-1}) the photocurrent was almost proportional to the intensity throughout the spectral range ($i_p \propto F^{0.9}$ approximately).

References pp. 213—214

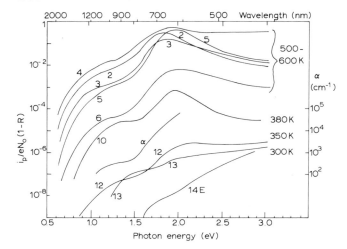

Fig. 6.2. Spectral dependence of photoresponse in glow discharge Si specimens prepared at the deposition temperatures indicated. Curve 14E refers to a specimen evaporated onto a substrate at 550K. The ordinate represents the number of charge carriers flowing around the circuit per photon entering the specimen. α is the absorption coefficient for a 500K specimen. (From Loveland *et al.* [9])

Figure 6.2 shows the steady state photoresponse as a function of photon energy measured on a number of gap cell specimens deposited at the T_d values indicated. The ordinate plotted here is $i_p/eN_0(1-R)$, representing the number of charge carriers flowing round the circuit for each photon entering the specimen. The figure also includes, for comparison, the spectral dependence of the absorption coefficient α for a 500K specimen (see right-hand ordinate). The following features are of interest:

(a) The photoresponse of the group of high temperature specimens approaches unity at photon energies between 1.8 and 2 eV, for the values of the applied field used. With higher fields, unity could be exceeded, indicating that carriers are replenished at the electrodes.

(b) Specimens prepared below about 500K exhibit a drastic decrease in photoresponse at all wavelengths. A limiting case is the evaporated specimen, 14E, which was deposited on a substrate at 550K.

(c) The photoresponse disappears below about 0.6 eV.

(d) Practically all the curves show a shoulder between 1.1 and 1.3 eV, followed by a rapid rise in photoresponse above about 1.5 eV.

These last features appear to be closely similar to those of the absorption curve. There exists no marked displacement between photoconductivity and absorption edges, such as has been found for instance in amorphous Se [16]. For most specimens $\alpha d \lesssim 0.4$ at photon energies below 1.7 eV. Thus, in the weakly absorbed region, eqn. (6.1) becomes

$$i_p/eN_0(1-R) \simeq \alpha d \eta \tau / t_T. \tag{6.2}$$

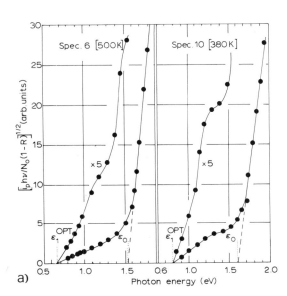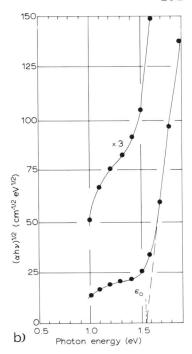

Fig. 6.3. Determination of the optical gap $\mathscr{E}_0$ of glow discharge Si. (a) from the photoconductivity measurements, which also give a value of $\mathscr{E}_1^{opt}$; (b) from absorption measurements. (From Loveland et al. [9])

Equation (6.2) suggests that the photoconductive results of Fig. 6.2 could be used as a sensitive measure of absorption between 1.7 and 0.6 eV, which would be unaffected by the experimental uncertainties involved in the direct determination of α in the weakly-absorbing region.

This suggestion hinges on the assumption that $\eta\tau/t_T$ remains substantially independent of photon energy. A test, using some of the experimentally determined photoconductivity and absorption data in eqn. (6.2) showed that $\eta\tau/t_T$ generally decreased between 1.6 and 1 eV [9]. In spite of this dependence, it seemed worthwhile to use the photoconductive data as an approximate measure of α, particularly in extrapolations involving only a relatively small range of photon energies. Theoretical work on the optical properties of non-crystalline solids has indicated that in analysing absorption data, $(\alpha h\nu)^{1/2}$ should be a meaningful independent variable. This is based on the analysis of Tauc [17] which led to $(\alpha h\nu)^{1/2}$ proportional to $(h\nu - \mathscr{E}_0)$, if parabolic bands are assumed. The constant $\mathscr{E}_0$ can be used to define an optical gap. A relation of the same form has been derived by Davis and Mott [18], who assumed that the density of states at the band edges are linear functions of energy.

References pp. 213—214

In Fig. 6.3(a), the corresponding quantity from the photoconductivity results, $[i_p h\nu/N_0(1-R)]^{1/2}$, is plotted against photon energy in the low absorption region for two typical samples, prepared at 500 and 380K respectively. Most of the specimens have been analysed in this way and the results are summarised in Table 6.1. It can be seen that two meaningful extrapolations can be made. From the well-defined straight portion of the curves we obtain values of $\mathscr{E}_0$ lying essentially between 1.5 eV and 1.6 eV for most specimens, irrespective of T_d. Secondly, the "tail" of the curves extrapolates to values of $\mathscr{E}_1^{opt}$ between 0.6 eV and 0.8 eV, which is compared in Table 6.1 with values of $(\mathscr{E}_c - \mathscr{E}_f)_0$ obtained from conductivity measurements on the same specimens. The expanded parts of these curves ($\times$ 5) in Fig. 6.3(a) also show the presence of the shoulder near 1.2 eV, referred to above in connection with Fig. 6.2. It is of interest to compare these deductions from the photoconductivity results with those from absorption measurements. Figure 6.3(b) is a plot of $(\alpha h\nu)^{1/2}$ against the photon energy for a high T_d specimen. $\mathscr{E}_0 = $ 1.55 eV in agreement with the photoconductivity results, but $\mathscr{E}_1^{opt}$ cannot be determined with any certainty because of the large experimental error in absorption measurements on these specimens [9] at photon energies below about 1 eV.

As the transport properties are known, it is possible to calculate the transit time t_T of photogenerated electrons in the extended states and, from eqn. (6.1), determine their lifetime τ in these states. This has been done at a photon energy of 1.8 eV, corresponding approximately to the maximum of the curves in Fig. 6.2. The τ values so obtained are plotted in Fig. 6.4 against the deposition temperature.

It is evident that the photosensitivity reaches its maximum value at $T_d \simeq 500$K and then remains essentially constant for specimens deposited above this temperature. For $T_d < 500$K, the sensitivity decreases rapidly by orders of magnitude, probably as a direct result of the decreasing recombination lifetime.

To conclude this section, the photoconductive results will now be discussed on the basis of the transport properties and the density of state distribution. Previous work, summarised in Section 6.2, led to the conclusion that on glow discharge films ($T_d \gtrsim 350$K) the transport of excess carriers takes place predominantly in the extended electron states. Further evidence in support of this interpretation is provided by the satisfactory agreement in Table 6.1 between the photoconductive threshold $\mathscr{E}_1^{opt}$ and the position of the Fermi level $(\mathscr{E}_c - \mathscr{E}_f)_0$, deduced from conductivity measurements on the same specimens. It implies that the onset of photoconductivity coincides with transitions from filled states at $\mathscr{E}_f$ into extended states at $\mathscr{E}_c$. In comparing the two energies it must be borne in mind that $(\mathscr{E}_c - \mathscr{E}_f)_0$ refers to the extrapolated position of the Fermi level at $T = 0$, whereas $\mathscr{E}_1^{opt}$ is measured at room temperature and should therefore be compared with $(\mathscr{E}_c - \mathscr{E}_f)_{RT} = (\mathscr{E}_c - \mathscr{E}_f)_0 - \delta T$. With a reasonable value of the temperature coefficient δ,

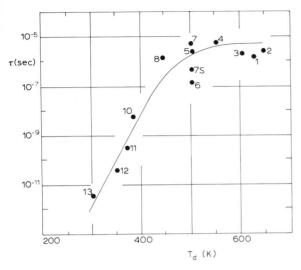

Fig. 6.4. Recombination lifetime of photogenerated carriers as a function of deposition temperature for a number of glow discharge specimens. All measurements were made in gap cell geometry, with the exception of 7S, which refers to sandwich cell geometry. (From Loveland et al. [9])

$\delta T \simeq 0.06$ eV. Also, $\mathcal{E}_1^{opt}$ is observed under photoexcitation and the increased electron density is described by a quasi-Fermi level, lying slightly closer to $\mathcal{E}_c$ than the thermal equilibrium energy, $\mathcal{E}_f$. With the relatively low light levels used in the experiments, this shift is estimated to be somewhat smaller than δT. Both effects are in the same direction and we would expect $\mathcal{E}_1^{opt}$ to be smaller than $(\mathcal{E}_c - \mathcal{E}_f)_0$ by up to 0.1 eV. This is in fact borne out by the values in Table 6.1.

We differ basically, however, from Spicer et al. [19] who suggested that the rise in absorption, observed in evaporated silicon films between 0.6 and 0.7 eV (i.e. close to our $\mathcal{E}_1^{opt}$), marks the onset of intrinsic absorption (see also Section 6.3).

On the basis of the above evidence, we shall regard the extended states above $\mathcal{E}_c$ as the final states of all transitions which contribute to a measurable steady photocurrent at room temperature. An interesting case is presented by specimen 13 (Table 6.1) which was prepared at a low T_d ($\simeq 300$K). The activation energy from conductivity measurements is 0.3 eV and the sign of the field effect shows that the dark current is carried by holes. The small pre-exponential factor of the conductivity ($\sigma_0 \simeq 10^{-4}$ ohm^{-1} cm^{-1}) implies further that the transport is by phonon-assisted hopping. Suppose that these holes propagate through the localized state distribution at an energy $\mathcal{E}_y$, where $\mathcal{E}_c - \mathcal{E}_y \simeq \mathcal{E}_c - \mathcal{E}_f + 0.3 - W_h$, and W_h is the hole hopping energy. For the same specimen, the photoconductive threshold is $\mathcal{E}_1^{opt} = \mathcal{E}_c - \mathcal{E}_f = 0.95$ eV, which means that under external excitation the transport of the

References pp. 213—214

photogenerated electrons at $\mathcal{E}_c$ predominates over the hole transport at $\mathcal{E}_y$. From transport experiments [12], it is suggested that $W_h \simeq 0.09$ eV, so that $\mathcal{E}_c - \mathcal{E}_y \simeq 1.16$ eV.

This estimate, and particularly the shoulder between 1.1 and 1.3 eV in most of the photoconductivity curves, point towards a prominent maximum in the density of states spectrum around an energy $\mathcal{E}_y$ as has been indicated in Fig. 6.1. Further evidence for this feature comes from absorption measurements on evaporated silicon films [20, 21]. There is no doubt that the localized states at $\mathcal{E}_y$ provide a hole conduction path in low T_d glow discharge specimens and also, we believe, in most evaporated silicon films [12, 14].

The results of Table 6.1 lead to a value for the optical gap $\mathcal{E}_0$ which is consistent with estimates for $\mathcal{E}_c - \mathcal{E}_v$ from the conductivity measurements. With a few exceptions, $\mathcal{E}_0$ lies between 1.5 and 1.6 eV. Very similar values were obtained by Chittick [8]. It is difficult to decide on the basis of Table 6.1 whether $\mathcal{E}_0$ depends on T_d or not.

The value of τ, between 10^{-5} and 10^{-6} s deduced from the experimental results for high temperature specimens, is remarkably long for a non-crystalline material and compares with carrier lifetimes in sensitive crystalline photoconductors. This property of amorphous silicon films prepared by the glow discharge technique, together with reasonably fast transient response and stable behaviour, suggests that these films may prove useful in photoconducting devices.

It should be emphasized that the above value of τ represents the lifetime of electrons in extended states; it is terminated by transitions into lower-lying states which lead to recombination rather than thermal re-excitation back to $\mathcal{E}_c$. In the steady state, photogenerated electrons involved in transport are essentially distributed between the extended states and the tail states around $\mathcal{E}_A$ (Fig. 6.1). The latter are in thermal equilibrium with electrons at $\mathcal{E}_c$ and n_c/n_A is given by eqn. (6.8) in Section 6.2.4. With the constants derived from the field effect data, $n_c/n_A \simeq 10^{-2}$ at room temperature, so that only about 1% of the generated carriers are mobile above $\mathcal{E}_c$. However, in a transient experiment, such as the decay of the photocurrent after removing the light, a response time of about $10^2 \tau$ would be expected, because the trapped carriers in the "reservoir" at $\mathcal{E}_A$ will have to be removed through the recombination channel. In fact, transient experiments on glow discharge films [9] give rise and decay times of the order of 10^{-3} s, in reasonable agreement with the above estimate.

The rapid decrease of τ in Fig. 6.4 must be connected with the overall increase in $\mathcal{N}(\mathcal{E})$ on lowering T_d, which is clearly shown by the field effect data [13]. In addition, $\mathcal{E}_f$ moves towards $\mathcal{E}_v$ (Table 6.1) into regions of larger state density, resulting probably in a further increase in the density of recombination centres. The virtual absence of photoconductivity in most evaporated (and sputtered) specimens, shown for instance by specimen 14E in Fig. 6.2, points to an even higher density of recombination centres.

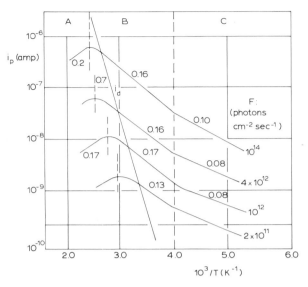

Fig. 6.5. Photocurrent in a-Si plotted against the inverse temperature for a specimen prepared at $T_d \simeq 550K$. The measurements were made at a photon energy of 2 eV at the stated intensities. The numbers along the curves are the activation energies in eV. i_d denotes the dark current. (From Spear et al. [10])

6.2.3 The temperature dependence of the photoconductivity

Interesting information on the fundamental aspects of photoconductivity, particularly on the carrier recombination process, can be obtained by studying the temperature dependence of i_p. In the following experiments on glow discharge silicon [10] the excitation is now confined to a single photon energy in the region of maximum photoresponse (Fig. 6.2). Light from a quartz-iodine lamp was passed through a band pass filter, peaked at about 2 eV; by means of neutral density filters the photoconductivity could be investigated over a reasonably wide intensity range.

Figure 6.5 shows the photocurrent between 500K and 100K as a function of the inverse temperature for a specimen deposited at $T_d \simeq 550K$. The four curves refer to different incident light intensities, ranging from 10^{14} to 2×10^{11} photons cm^{-2} s^{-1}, and the line marked i_d represents the temperature dependence of the dark current; the numbers along the curves denote the activation energies in eV. Clearly, the general features of Fig. 6.5, in particular the photoconductivity maximum, closely resemble the corresponding chalcogenide results (see Chapter 7). It will be useful for the purpose of discussion to label the temperature ranges as shown in the figure. Range A denotes the high-temperature region, $T > T_{max}$, where T_{max} is the temperature corresponding to the photoconductive maximum. Because $i_p \ll i_d$, with the latter varying rapidly with T, reasonably accurate measurements in this

References pp. 213—214

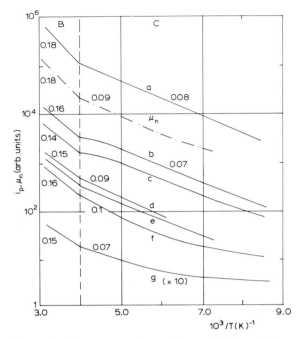

Fig. 6.6. Temperature dependence of the photocurrent in the low-temperature range for specimens prepared at different T_d. Photon energy 2 eV, except for (e) obtained with 1.2 eV photons. Numbers along the curves are the activation energies in eV. (a): $T_d \simeq 500$K, high F; μ_n: electron drift mobility; (b), (c) and (d): different specimens, $T_d \simeq 500$K, low F; (e): photon energy 1.2 eV, high F; (f): $T_d \simeq 350$K; (g): $T_d \simeq 320$K, note that this curve has been scaled × 10. (From Spear et al. [10])

temperature range were difficult to obtain. The available data suggest an activation energy between 0.17 and 0.20 eV.

In common with the chalcogenide results, the maximum lies on the high-temperature side of the dark current line, which it approaches with increasing light intensity. Temperature range B extends from $T_{\max}$ to about 250K, where in most of the curves a well-defined change in gradient is observed; the temperature range below 250K will be denoted by C.

Figure 6.6 shows in more detail some of the results below $T_{\max}$ obtained with high and low T_d specimens and at widely differing intensities. A typical electron drift mobility curve, μ_n, has been included in the diagram and the close similarity in the temperature dependence of i_p and μ_n will be a significant point in the discussion. The main features of Fig. 6.6 may be summarised as follows:

(a) In range B, activation energies between 0.18 and 0.14 eV are found. There appears to be no systematic dependence of the activation energy on T_d (curves (a), (f) and (g)) or on intensity (curves (a) and (c)).

(b) At about 250K, the activation energy of most i_p curves (and also of

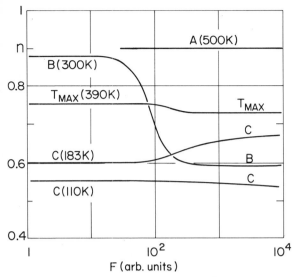

Fig. 6.7. Exponent n in relation $i_p \propto F^n$ plotted against intensity F in different temperature ranges for a high T_d specimen (550K). (From Spear et al. [10])

μ_n) changes fairly abruptly to an average value of 0.08 eV. At low light levels the flat transition region shown by curves (b) and (c) is sometimes observed.

(c) In range C, a gradual decrease in the activation energy is found at the lowest temperatures. This applies particularly to the low T_d specimens (curves (f) and (g)). High T_d samples, generally show an almost constant activation energy between 250 and 100K.

(d) There is no difference in the i_p curves obtained with photon energies of 2 eV and of 1.2 eV (curve (e)).

Finally, Fig. 6.7 summarises the intensity dependence of the photoconductivity in the various temperature ranges for a high T_d specimen (550K). It is found that within experimental error $\sigma_p \propto F^n$, and Fig. 6.7 shows n as a function of F. At high temperatures (range A) an almost linear dependence with $n \simeq 0.9$ was observed throughout. The same was found at room temperature (range B) at the lower intensities, where $i_p \lesssim i_d$, which corresponds to the experimental conditions used in the previous section.

6.2.4 Interpretation and discussion

In view of the general similarity between the temperature dependence of photoconductivity in a-Si and the chalcogenides, it will be useful to begin by considering some of the relevant features of the analysis used in the case of the chalcogenide work.

Arnoldussen et al. [22], Main and Owen [23] and recently Simmons and Taylor [24] have put forward and evaluated fairly elaborate models to

References pp. 213—214

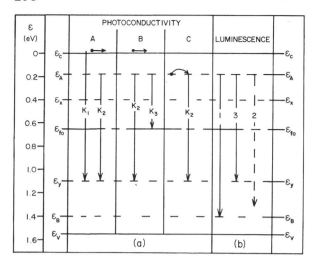

Fig. 6.8. Schematic energy level diagram including the main features of the density of state distribution of Fig. 6.1 for a-Si. (a) illustrates the recombination transitions for the temperature ranges A, B and C discussed in this section. (b) shows a possible assignment of the photoluminescent transitions (Section 6.2.6).

account for the observed activation energies and intensity dependences in the various temperature ranges. An interesting conclusion can be drawn from this work, namely that at least two levels of localized states (in addition to the extended electron and hole states) have to be incorporated for a general interpretation of the observed features. The suggested models do not seem to depend critically on the details of the density of states distribution in the neighbourhood of the two recombination levels. In fact, Main and Owen as well as Simmons and Taylor treat these states as discrete energy levels, representative of fairly narrow bands of localized states. Arnoldussen et al. adopt a somewhat different approach, also discussed in Chapter 7, by placing the two levels of recombination centres in regions of rapidly decreasing densities of localized states below $\mathcal{E}_c$ and above $\mathcal{E}_v$ respectively. The above authors also make the important point that a consistent explanation of the experimental results is possible only if transitions between localized states are included in the model in addition to those between extended and localized states.

In the following discussion of the a-Si results, we shall derive simple expressions for the recombination kinetics which, however, take proper account of the known transport properties of the material. On turning back to the density of state distribution in Fig. 6.1, it is tempting to suggest that localized states in the region of $\mathcal{E}_A$, $\mathcal{E}_x$ or $\mathcal{E}_y$ may be predominantly involved in the recombination process. We shall now explore this possibility in the case of high T_d specimens, using the schematic energy level diagram shown in Fig. 6.8(a).

(a) Temperature range A ($T > T_{\max}$)

At temperatures above the photoconductive maximum, $i_p \ll i_d$ and the occupancies of the various states are practically the same as in thermal equilibrium. The quasi Fermi levels for electrons and holes coincide with the thermal equilibrium Fermi level, $\mathscr{E}_{f0}$. We know from the results of Section 6.2.2 and from drift mobility measurements, that in high T_d specimens the predominant transport (above 250K) is by electrons in extended states at $\mathscr{E}_c$. The photoconductivity is therefore:

$$\sigma_p = n_c e \mu_c, \tag{6.3}$$

with an electron mobility μ_c of 5–10 cm^2 s^{-1} V^{-1}. Consider first an electron at $\mathscr{E}_c$ recombining with trapped holes at $\mathscr{E}_y$. In the steady state:

$$f = K_1 n_c p_y, \tag{6.4}$$

where f is the generation rate (cm^{-3} s^{-1}) and K_1 the recombination constant. But p_y is given by the thermal equilibrium value, independent of light intensity:

$$p_y = N_y \exp[-(\mathscr{E}_{f0} - \mathscr{E}_y)/kT], \tag{6.5}$$

where N_y is the density (cm^{-3}) of the centres at $\mathscr{E}_y$.

From eqns. (6.3), (6.4) and (6.5) we find that

$$\sigma_p = \frac{ef\mu_c}{N_y K_1} \exp(\mathscr{E}_{f0} - \mathscr{E}_y)/kT. \tag{6.6}$$

According to our density of state distribution we would then expect a gradient of 0.4 to 0.5 eV, whereas the experimental value lies around 0.2 eV.

If, on the other hand, the recombination transition goes from $\mathscr{E}_A$ to $\mathscr{E}_y$, then eqn. (6.4) has to be replaced by

$$f = K_2 n_A p_y. \tag{6.7}$$

Also, in thermal equilibrium n_A and n_c are related by:

$$\frac{n_c}{n_A} = \frac{N_c}{N_A} \exp\left\{\left[-(\mathscr{E}_c - \mathscr{E}_A)\right]/kT\right\}, \tag{6.8}$$

where $N_A \simeq \mathscr{N}(\mathscr{E}_A) kT$.

With eqn. (6.8), one obtains in this case

$$\sigma_p = \frac{efN_c}{K_2 N_y N_A} \mu_c \exp[(\mathscr{E}_{f0} - \mathscr{E}_y) - (\mathscr{E}_c - \mathscr{E}_A)]/kT. \tag{6.9}$$

Equation (6.9) now predicts a gradient of about 0.2 eV and a linear intensity

References pp. 213–214

dependence of i_p, both of which are in approximate agreement with the experimental results. It is, therefore, suggested that recombination from the electron tail states at $\mathscr{E}_A$ to the hole traps near $\mathscr{E}_y$ is the most likely process in range A.

(b) Temperature range B

In this range, the photocurrent curves of Fig. 6.6 as well as the drift mobility curves, lead to activation energies of approximately $\mathscr{E}_c - \mathscr{E}_A$. Also, at higher levels of illumination, $i_p \propto F^{0.6}$ (Fig. 6.7). Although the transitions between $\mathscr{E}_A$ and $\mathscr{E}_y$ still appear to be the predominant recombination path at higher intensities, the conditions are now basically different from those described previously. The photogenerated density of carriers at $\mathscr{E}_c$ (and $\mathscr{E}_v$) is far larger than that in the dark and, as a consequence, the quasi Fermi levels (and the "demarcation levels" [25]) will move towards $\mathscr{E}_A$ and $\mathscr{E}_y$. States enclosed by the quasi Fermi levels will now become depopulated and the charge neutrality condition tends towards $n_A \simeq p_y$. The fact that states near $\mathscr{E}_A$ and $\mathscr{E}_y$ become relatively more populated at higher light intensities, makes the suggested recombination path from $\mathscr{E}_A$ to $\mathscr{E}_y$ a likely one. Here, we assume, in common with the other two-level models, that the localized negative and positive charge densities are largely concentrated at $\mathscr{E}_A$ and $\mathscr{E}_y$. In view of the probably quite appreciable density of trapped negative charge near $\mathscr{E}_x$ (see Fig. 6.1) this assumption may well be an over-simplification.

If the above neutrality condition is used instead of eqn. (6.5), now no longer applicable, then eqns. (6.3), (6.7) and (6.8) lead to:

$$\sigma_p = e \left(\frac{f}{K_2}\right)^{1/2} \mu_c \frac{N_c}{N_A} \exp\left[-(\mathscr{E}_c - \mathscr{E}_A)/kT\right] \qquad (6.10)$$

which predicts the correct activation energy and approximately the correct intensity dependence. It can easily be shown that recombination between carriers at $\mathscr{E}_c$ and $\mathscr{E}_y$ would lead to a gradient of $\frac{1}{2}(\mathscr{E}_c - \mathscr{E}_A)$ at high intensities [23, 24], in disagreement with the experimental results.

It now remains to explain the fact that the photocurrent tends towards a linear intensity dependence at lower illumination levels. Following Arnoldussen et al. [22], we would suggest that at low light intensities conditions are still reasonably close to thermal equilibrium, but that compared with region A the lower temperature has decreased the occupation at $\mathscr{E}_A$ and $\mathscr{E}_y$, whilst leaving the population around $\mathscr{E}_{f0}$ relatively unchanged. Recombination from $\mathscr{E}_A$ through states at the Fermi level will now become important. If the density of states involved in such a transition is $N_f \simeq \mathscr{N}_f kT$, then the photoconductivity will be dominated by the term:

$$\sigma_p = \frac{ef}{K_3 N_F} \frac{N_c}{N_A} \mu_c \exp\left[-(\mathscr{E}_c - \mathscr{E}_A)/kT\right] \qquad (6.11)$$

which now gives a linear intensity dependence but the same activation energy as in eqn. (6.10).

As mentioned earlier, the room temperature results in Section 6.2.2 were obtained under the conditions described by eqn. (6.11). If we compare this model with eqn. (6.1), we find that τ, the lifetime of electrons in the extended states, is related to K_3 by

$$\tau = \left(\frac{1}{K_3 N_f}\right)\left(\frac{n_c}{n_A}\right)$$

Or, referring to the discussion at the end of Section 6.2.2, the recombination lifetime of electrons at $\mathscr{E}_A$, the initial state for the transition, will be $1/K_3 N_f$.

Figure 6.7 shows that the exponent in the intensity dependence changes at 300K from 0.9 to 0.6 with increasing F whereas, in measurements on the chalcogenides, it generally approaches more closely to the predicted values of 1 and 0.5. The reason could be that in the present case the distinction between the suggested recombination paths ($\mathscr{E}_A \to \mathscr{E}_y$, or $\mathscr{E}_A \to \mathscr{E}_{f0} \to \mathscr{E}_y$) is less sharply defined as a function of intensity. For instance, in the low intensity regime, 20% of bi-molecular recombinations could account for the observed $F^{0.9}$ dependence; similarly, the presence of about 10% of mono-molecular transitions would lead to the $F^{0.6}$ dependence at high illumination levels. This last situation could arise under our experimental conditions where the intensity varies rapidly with depth below the specimen surface.

(c) Temperature range C

With decreasing temperature, a change of gradient is observed around 250K in all the i_p versus $1/T$ curves of Fig. 6.6. Once again, the same feature is shown by the corresponding drift mobility curves. In our previous work [11, 12] this was interpreted as signifying a change in the predominant transport mechanism from extended state conduction at $\mathscr{E}_c$ to electron hopping conduction around $\mathscr{E}_A$. The reason for this is that with decreasing T the probability of thermal release, P_{th}, of an electron localized at $\mathscr{E}_A$ will drop rapidly so that below a critical temperature it will become more probable that the electron hops to a neighbouring site with phonon assistance. When $P_{hop} > P_{th}$, the predominant current path will shift from $\mathscr{E}_c$ to $\mathscr{E}_A$. The hopping mobility is given by:

$$\mu_H = (\mu_H)_0 \exp(-W/kT), \tag{6.12}$$

where the hopping energy, W, is about 0.09 eV from drift mobility measurements [12].

The change in conduction mechanism implies that eqn. (6.3) takes the form:

$$\sigma_p = n_A e \mu_H, \tag{6.13}$$

and eqn. (6.10) becomes

$$\sigma_p = e\left(\frac{f}{K_2}\right)^{1/2} (\mu_H)_0 \exp(-W/kT). \tag{6.14}$$

The steady state photoconductivity below about 250K is therefore controlled by the hopping activation energy of electrons near $\mathcal{E}_A$. This arises because in the case of glow discharge silicon the observed photocurrent flows through localized states which represent also the initial state in the recombination process. There does not appear to be any comparable evidence for this from the chalcogenide results. It should therefore be emphasised that the close correlation between the temperature dependence of i_p and the electron drift mobility found here in regions B and C lends strong support to the kinetic model of Fig. 6.8 and implies that the latter is consistent with the interpretation of the transport and optical data. It is also significant in this connection that the i_p curves retain their typical characteristics if the incident photon energy is reduced to 1.2 eV, when electrons near $\mathcal{E}_y$ are excited to $\mathcal{E}_c$.

A curious effect shown by the two low intensity curves in Fig. 6.6 is the flat portion just below 250K where the change in conduction mechanism takes place. The effect has not been observed under transient conditions in the drift mobility measurements and must therefore be associated with the steady state, most likely with the recombination process.

It is apparent from the results in Figs. 6.2 and 6.6 that the magnitude of the photoconductivity throughout the temperature range depends critically on the deposition temperature. The photoresponse in low T_d specimens can be four to five orders of magnitude lower than that in a good specimen prepared between 500 and 600K. The reason for this is undoubtedly the large increase in the overall level of $\mathcal{N}(\mathcal{E})$ within the mobility gap [13]. In this case, it seems likely that other recombination paths below $\mathcal{E}_A$ will develop and act in parallel to those considered in Fig. 6.8. The analysis presented in this section is therefore restricted to high T_d specimens.

6.2.5 The recombination process

Although the interpretation given above can account for the experimental results in a satisfactory manner, it provides only limited insight into the details of the recombination process. It has been shown that a prominent recombination path exists between initial states at $\mathcal{E}_A$ and final states at $\mathcal{E}_y$ which has been formally described by the recombination constant K_2, defined by eqn. (6.7). As $\mathcal{E}_A - \mathcal{E}_y \simeq 0.95$ eV, which amounts to ten or more phonon energies in silicon, the question arises as to how this energy is dissipated. One possibility is by radiative recombination, but photoluminescence experiments on glow discharge silicon (Section 6.2.6) show that the intensity of the emitted radiation decreases exponentially above about 100K and is no longer observable at 200K. This mechanism would therefore be an unlikely explanation for most of the temperature range used in the photoconductivity experiments. One has to conclude that the energy is dissipated by multiphonon emission.

The experimental results in ranges B and C lead with eqns. (6.10) and (6.14) respectively to values of K_2 around 10^{-10} cm^3 s^{-1}. The question arises: can we account theoretically for such a value? In the following, consider, for instance, region C at a sufficiently high temperature where the hopping probability P_{hop} still exceeds the probability of a multiphonon emission P_{ph} from $\mathscr{E}_A$ to $\mathscr{E}_y$. We set:

$$K_2 \simeq P_{ph} V \qquad (6.15)$$

where V is the maximum volume in which recombination between a localized electron and hole is likely to occur. If the centres are charged when occupied, then $V \simeq r_c^3$ where r_c represents the Coulomb radius. At 200K this is about 10^{-6} cm for silicon. We set $V \lesssim 10^{-18}$ cm^3 because the volume would be appreciably smaller if one of the centres were neutral on trapping a carrier. With this estimate, eqn. (6.15) then requires that $P_{ph} \gtrsim 10^8$ s^{-1}.

The difficulty in proceeding further arises from the lack of an adequate theory for multiphonon transitions between localized states in an amorphous semiconductor. In a recent paper [26] Mott et al. have adapted the treatment of Englman and Jortner [27], which had been applied originally to the case of large molecules.

An alternative approach has been taken by Robertson and Friedman [28], based on a one-dimensional molecular crystal model [29]. The results obtained from both these calculations are in substantial agreement. Suppose that on a configurational diagram the energy difference between the minima corresponding to the ground and excited states of a recombination centre is $\Delta\mathscr{E}$. The number of photons emitted in the transition will then be $(\Delta\mathscr{E} - \mathscr{E}_s)/\hbar\omega$, where $\mathscr{E}_s$ is the Stokes shift. In the weak coupling limit, believed to be applicable to a-Si [26], the predicted non-radiative transition probability is of the form:

$$P_{ph} \simeq C \exp\left[-\gamma \frac{\Delta\mathscr{E}}{\hbar\omega}\right], \qquad (6.16)$$

with $\gamma = \ln(\Delta\mathscr{E}/\mathscr{E}_s) - 1 + \mathscr{E}_s/\Delta\mathscr{E}$. Although $\mathscr{E}_A$ is unlikely to be an excited state of $\mathscr{E}_y$, eqn. (6.16) should give an order of magnitude estimate of P_{ph}. In the following, we shall take a phonon energy $\hbar\omega = 0.07$ eV, corresponding to the maximum in the Raman spectrum for a-Si [30], and also set $C \simeq \omega/2\pi = 1.7 \times 10^{13}$ s^{-1}. Equation (6.16) then leads to the experimentally determined value of $P_{ph} \simeq 10^8$ s^{-1} with $\mathscr{E}_s = 0.18$ eV. A Stokes shift of about 20% cannot be excluded on the basis of a comparison between the features of the luminescent spectrum (Section 6.2.6) and the density of state distribution in Fig. 6.1. In principle, eqn. (6.16) can, therefore, account for the non-radiative recombination from $\mathcal{E}_A$ to $\mathcal{E}_y$, but in view of the uncertainties and assumptions involved in the analysis this is by no means conclusive.

Mott [26] suggested that some of the recombination centres at $\mathcal{E}_y$ may contain a hydrogen atom, incorporated during the decomposition of SiH_4. The presence of the light atom in the structure will give rise to more energetic local phonon modes. In this case, eqn. (6.16) could give the required P_{ph} in the absence of any Stokes shift.

The possibility of recombination through a defect centre [10] is worth considering. It has been suggested [14] that the two prominent peaks in $\mathcal{N}(\mathcal{E})$ at $\mathcal{E}_x$ and $\mathcal{E}_y$ (see Fig. 6.1) may be associated (at least partly) with defect states formed by the interaction of dangling bonds in multivacancy complexes. It should be recalled that this type of defect, in the form of divacancies, is very stable in irradiated crystalline silicon. In the amorphous phase there is convincing evidence, particularly from e.s.r. measurements [2], for the existence of unsaturated bonds probably situated on internal surfaces of the vacancies.

Suppose that, as a first step in the recombination, an electron at $\mathcal{E}_A$ drops into the empty upper state ($\mathcal{E}_x$) of the defect by a multiphonon transition. The electron is now localized in a region which is structurally less "rigid" than the random network of saturated silicon bonds. The structural distortion caused by the presence of the electron will lower the energy of the localized state and, in this way, most of the energy $\mathcal{E}_x - \mathcal{E}_y$ could be dissipated. The mechanism envisaged is thus similar to the self-localization of the charge carrier when a small polaron is formed [29]. Again, it is not possible on the basis of the present data to decide whether such a recombination process occurs, or if it does, whether it could compete with the direct multiphonon recombination discussed earlier.

6.2.6 Photoluminescence

The recent results of Engemann and Fischer [31, 32] on the radiative recombination in glow discharge silicon are of considerable interest in the present context and will be briefly discussed in this section.

In these experiments, a 0.5 W Krypton laser was used to excite the luminescence. Figure 6.9 shows the emission spectra at three different temperatures using a specimen prepared at $T_d \simeq 450K$. The spectra consist of a broad band with features at 1.23 eV, 1.13 eV and 0.92 eV, which will be referred to as 1, 2 and 3, respectively. It can be seen that both the overall intensity and the intensity distribution within the band are dependent on temperature. As far as the former is concerned, it has already been mentioned that photoluminescence becomes observable below 200K, increases rapidly with decreasing T and reaches a constant level of intensity below about 80K. The changes in the relative intensities of the three peaks are clearly shown in Fig.

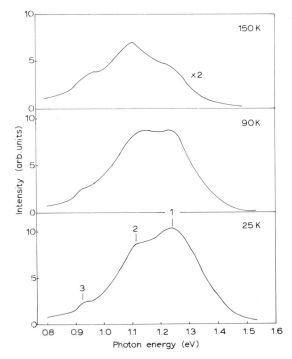

Fig. 6.9. Photoluminescence spectra at three temperatures for a glow discharge Si specimen prepared at $T_d \simeq 450$K. The features marked 1, 2 and 3 are discussed in the text. (From Engemann and Fischer [31])

6.9. With rising temperature, the low energy peaks 2 and 3 become the dominant structure.

It is encouraging to find that the main features of the luminescent spectrum appear to be reasonably consistent with the density of states distribution of Fig. 6.1. A possible interpretation of the radiative transitions associated with peaks 1, 2 and 3 is shown in Fig. 6.8(b). With $\mathscr{E}_c - \mathscr{E}_v \gtrsim 1.55$ eV, as indicated by the photoconductive and transport results, transition 1 (1.23 eV) takes place almost certainly between the localized tail states of the conduction and valence bands. If the initial state lies near $\mathscr{E}_A$, where $\mathscr{E}_c - \mathscr{E}_A \simeq 0.18$ eV, then $\mathscr{E}_B$ should be at about the same energy above $\mathscr{E}_v$. Transition 3 (0.93 eV) matches almost exactly $\mathscr{E}_A - \mathscr{E}_y$, which, according to the photoconductivity results (Section 6.2.4), represents the main recombination transition above 200K. Transition 2 (1.13 eV) is more difficult to assign, largely because little is known about the density of state distribution between $\mathscr{E}_y$ and the valence band. It could possibly be associated with a transition from $\mathscr{E}_A$ to states below $\mathscr{E}_y$ [32].

The dependence of the overall luminescent intensity (below 80K) on the deposition temperature [32] is interesting, because the graph shows a close

References pp. 213—214

resemblance to Fig. 6.4 of τ versus T_d. The luminescent intensity saturates for $T_d > 450K$, showing that in specimens with long τ the radiative recombination process reaches its optimum efficiency. In low T_d specimens, on the other hand, with a higher level of $\mathcal{N}(\mathscr{E})$ and therefore small τ, non-radiative recombination competes strongly with the photoluminescence. No luminescence could be observed in evaporated silicon specimens which, as in photoconductivity, represent a case of extremely low τ.

6.3 PHOTOCONDUCTIVITY IN EVAPORATED a-Si

In the previous sections, it has been pointed out that the photoresponse in evaporated silicon films is generally many orders of magnitude smaller than in glow discharge specimens. The curve for specimen 14E in Fig. 6.2 represents a fairly typical example and similar results have been obtained for evaporated germanium.

However, as pointed out in the introduction to this chapter, the electronic properties of evaporated tetrahedral materials depend strongly on the deposition conditions and particularly on the deposition rate [1]. Donovan and his collaborators at the Michelson Laboratory have investigated the preparation conditions for optimum photoresponse. They succeeded in producing evaporated silicon [4] and also germanium [6] specimens (to be discussed in the next section) which showed appreciably higher photosensitivity than that found by previous authors. The experimental conditions were as follows: a source to substrate distance of 42 cm, a deposition rate of 5 Å s^{-1} or less, a pressure of $\leqslant 10^{-6}$ torr during evaporation. For optimum response, the films were then annealed at 350°C for one hour.

Figure 6.10 shows the spectral response at 300K and 77K obtained by Fischer and Donovan [4] for a silicon film, about 0.2 μm thick, produced under the above conditions. The ordinate of the graph is $\Delta\sigma/\sigma F$, where $\Delta\sigma$ is the photoconductivity, σ the dark conductivity, and F the number of photons incident per unit area per second. The spectral dependence of the photoconductivity is remarkably similar to the curves in Fig. 6.2 for glow discharge silicon. In particular, one recognises the rapid rise in i_p between 1.4 and 1.6 eV and the flattening out of the response at higher photon energies. The shoulder between 1.1 and 1.3 eV is clearly shown, although the authors ascribe it to an interference effect. This appears highly unlikely, because the same feature is shown consistently in Fig. 6.2 for specimens varying in thickness between 0.6 and 4 μm, and also appears in the absorption measurements of different authors (see Section 6.2.2).

Fischer and Donovan confirm the proportionality of the photoconductivity and absorption curves implied by eqn. (6.2) for the weakly absorbed region. In their work they use photoconductivity as a sensitive measure of α at low photon energies as discussed in Section 6.2.2 for glow discharge specimens.

The quantitative comparison of the results shown in Figs. 6.10 and 6.2 is complicated by the different ordinates. However, the photosensitivity is determined by the product $\eta\mu\tau$ [*cf.* eqn. (6.1)], which is about 2×10^{-7} cm^2 V^{-1} for the annealed, evaporated films. For a high temperature glow discharge specimen, $\eta\mu\tau$ approaches 10^{-4} cm^2 V^{-1}, so that they are two to three orders of magnitude more photosensitive than the evaporated films. The increase in $\Delta\sigma/\sigma F$ at 77K shown in Fig. 6.10 is largely due to the decrease in the dark conductivity.

In their discussion, Fischer and Donovan [4] suggest that the photocurrent, both at 300K and 77K, is carried by electrons in the extended states. Whereas this may be the case at room temperature, it would seem hardly possible at 77K and we would suggest that hopping transport is a more likely interpretation for the observed photoconductivity in this temperature range.

6.4 PHOTOCONDUCTIVITY IN *a*-Ge

In the previous sections of this chapter, we have presented extensive data on *a*-Si and shown that the photoconductivity and other properties of this material can be understood in terms of a relatively simple model. The photoresponse in *a*-Ge is much smaller than in glow discharge *a*-Si and, as a consequence, the published information is far more limited and less well understood. In the following, we shall summarise the data available for specimens prepared by vacuum evaporation and by an electrolytic process. Although *a*-Ge films can also be prepared by the glow discharge decomposition of germane, photoconductivity in such specimens is small and has not been investigated in any detail.

6.4.1 Evaporated films

The first results were published by Clark [33] who could only measure photoresponse at 4.2K. At this temperature, i_p peaked at a photon energy of about 1.5 eV. Grigorovici *et al.* [34] measured the spectral response between liquid nitrogen temperatures and room temperature. Their curves disagreed with the data of Clark in that they observed a broad maximum at about 1.1 eV or 1.2 eV. Grigorovici *et al.* attributed this peak to transitions between extended valence and conduction band states.

The most extensive measurements have been made by Donovan and his collaborators[5, 6, 35, 36] on specimens evaporated under the same conditions as their silicon films (Section 6.3). Again the data were used to extend the investigation of the absorption towards lower photon energies. We shall confine ourselves in the following mainly to the most recent paper [6], because the authors themselves expressed considerable doubt about the information contained in their previous publications on the subject.

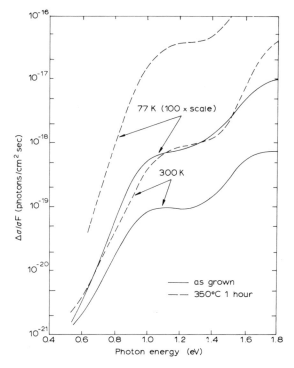

Fig. 6.10. The spectral dependence of the photoresponse at 300K and 77K for two films of evaporated a-Si. The solid lines are for an as-deposited film and the dotted lines refer to a similar sample after annealing at 350°C for one hour. (From Fischer and Donovan [4])

Figure 6.11 shows the spectral dependence of the photoconductivity per incident photon, $\Delta\sigma/F$ for 800 Å films deposited in UHV and standard vacuum ($\sim 10^{-6}$ torr) and measured at 77K. All the curves were normalized to a value of $\Delta\sigma/F = 0.5$ at a photon energy of 1.5 eV. The triangles and squares are data for UHV films deposited at substrate temperatures T_s of 25°C and 180°C respectively, and measured *in situ*. The circles are for a film deposited in standard vacuum and annealed after air exposure to 350°C. The threshold region at low photon energies was not as sharp as previously reported [36]. The results were similar whether the films were deposited in ultra-high vacuum or in standard vacuum and whether the samples were annealed either before or after air exposure. The authors concluded that the effect of annealing was to shift the threshold to higher photon energies but no sharpening of the edge was observed.

In their earlier papers, these authors reported a time constant associated with their photoresponse of the order of microseconds. They identified the threshold of the photoresponse (and absorption) with transitions between

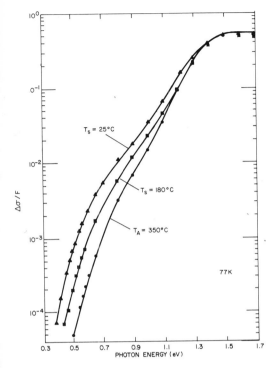

Fig. 6.11. The spectral dependence of the photoconductivity per incident photon for evaporated a-Ge films. All the curves were normalized to a value of $\Delta\sigma/F = 0.5$ at a photon energy of 1.5 eV. The triangles and squares are data for UHV films deposited at substrate temperatures T_s of 25°C and 180°C respectively, and measured *in situ*. The circles are for a film deposited in standard vacuum and annealed after air exposure to 350°C. (From Donovan et al. [6])

extended states in the valence and conduction bands. In their recent work [6], the time constant of the photoresponse is reported as milliseconds and the room temperature signal is associated with a bolometric effect. Only below about 77K do the authors suggest that the bolometric effects are negligible and that they are observing a true photoconductive signal. In view of these complications, the interpretation of the photoconductivity in their samples must presumably remain uncertain at present.

The photoconductivity in evaporated a-Ge has also been measured by Vescan and Croitoru [37]. These authors report a small photoconductive response ($\eta\mu\tau \sim 10^{-10}$ cm^2 V^{-1}) which is much less than the dark conductivity. However, they were able to measure its temperature and intensity dependence and to separate out a bolometric effect observed above room temperature. One other paper has been published on photoconductivity in evaporated a-Ge films by Burke and Clark [38] who observed a residual conductivity after the removal of the illumination in the temperature range

References pp. 213—214

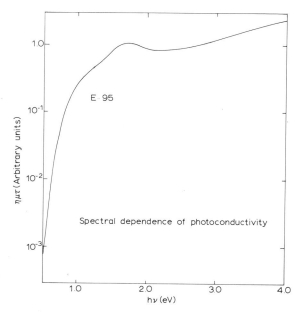

Fig. 6.12. The spectral dependence of the product $\eta\mu\tau$, for a film of electrolytic a-Ge. (From Camphausen et al. [41])

from 77K to 120K. This effect, which has also been found in single crystals of CdS [39] and polycrystalline CdSe [40] is not understood.

6.4.2 Electrolytic a-Ge

Camphausen et al. [41] deposited these films onto copper substrates at about 55°C from a solution of $GeCl_4$ and propylene glycol. The samples, formed at a rate of about 8 Å s^{-1} and between 5 µm and 50 µm thick, were removed from the copper substrate before measurements were made. The spectral response of $\eta\mu\tau$ was measured up to photon energies of 4.0 eV and is shown in Fig. 6.12.

Camphausen et al. interpret their data as follows. At the lowest photon energies, carriers are excited from states below $\mathscr{E}_f$ to states in the localized part of the conduction band tail. For higher energies, it is suggested that excitation is increasingly into the extended states. The voltage dependence of $\eta\mu\tau$ lends support to this interpretation. Above photon energies of about 1.2 eV, the photocurrent varied linearly with applied voltage and Camphausen et al. assume that most of the carriers were excited above the mobility edge. Below 1.2 eV, the photocurrent varied superlinearly with field and this was associated with field-assisted emission of electrons from the localized final states of the optical transition.

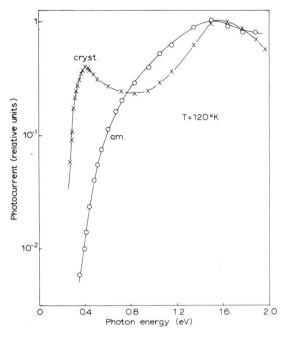

Fig. 6.13. The spectral dependence of the photocurrent in amorphous and polycrystalline InSb. The photocurrents are normalized to the same magnitude at the maximum near 1.5 eV. At this energy, the response in the amorphous film is smaller by about five orders of magnitude than that for the crystal. (From Eckenbach et al. [42])

6.5 PHOTOCONDUCTIVITY IN OTHER TETRAHEDRALLY BONDED AMORPHOUS MATERIALS

Of these materials, photoconductivity has been reported in a-InSb [42], a-GaP [43], a-GaAs [37, 44] and a-GaAs + 16% Ge [37]. The experimental data are generally limited by the relatively small response of the samples. The spectral dependence of the photoconductivity in a-InSb is shown in Fig. 6.13 and compared with that for the corresponding polycrystalline material. At the maximum near 1.5 eV, the signal in the amorphous case is smaller by about five orders of magnitude than that for the crystal. The other obvious difference is that the amorphous edge is far more gradual and does not exhibit the structure observed in the crystal near 0.4 eV. The authors conclude that a large density of states in the gap reduces the excess carrier lifetime in the amorphous state to values well below that found in the crystal.

Measurements of the photocurrent per absorbed photon in a-GaP have recently been reported by Yates et al. [43]. This quantity rises rapidly up to photon energies of about 1 eV and then increases more gradually towards the limit of measurement of 3 eV. The magnitude of the response is such

References pp. 213—214

that $\eta\mu\tau$ has a maximum value of about 5×10^{-11} cm^2 V^{-1} at room temperature. From simultaneous absorption measurements these authors conclude that in a-GaP there is a narrow range of weakly-absorbing tail states, in which the mobility decreases rapidly below $\mathscr{E}_c$ or above $\mathscr{E}_v$.

Botilă et al. [44] have measured the a.c.-photoconductivity of a-GaAs as a function of frequency, photon energy and temperature. The signal is small under all experimental conditions. The authors suggest that the observed response represents a super-position of at least three components, one of which is associated with heating of the sample due to the large incident light intensities used. The presence of a slow component is taken by the authors to indicate the existence of a high density of localized states in the gap.

Finally, similar measurements have also been reported on a-GaAs and a-GaAs + 16% Ge by Vescan and Croitoru [37]. In both materials, the photoresponse at low temperature increases with illumination, following a power law dependence with an exponent slightly greater than one. The temperature dependence of the photoconductivity is similar in form to that shown in Fig. 6.5 for a-Si with the maximum in the photocurrent at about 250K. At the lowest temperatures investigated (about 130K), the response approaches asymptotically a constant value.

6.6 CONCLUDING REMARKS

An attempt has been made in this chapter to review and correlate the photoconductive properties of tetrahedrally bonded amorphous materials. It is astonishing to find that except for a-Si and Ge we seem to know very little about photoconductivity in these solids. This applies to the amorphous III—V compounds and particularly to the group of tetrahedral glasses (e.g. CdGe$_x$As$_2$, ZnSi$_x$P$_2$). Reference to the previous section suggests that the main reason for the lack of information and perhaps interest lies in the extremely small photoresponse of these materials, which severely restricts the type and range of any investigation.

However, it has been stressed in the foregoing that the defect structure of tetrahedral amorphous materials, and consequently the lifetime of excess carriers, is critically dependent on the method of preparation. It may thus be possible, as has been done in the case of evaporated Si and Ge, similarly to improve the photoresponse of a wider range of these solids through careful attention to the preparation technique or through the development of new methods which produce a lower level of localized gap states.

The glow discharge technique has achieved this to a remarkable extent in the case of a-Si. The high photoresponse of specimens deposited at $T_d > 500$K allowed an investigation of photoconductivity over a wide range in several important variables. At the same time, the low level of the localized state density made field effect, drift mobility and luminescent measurements possible. It is largely through correlation with results of this kind that photoconductivity studies can make a really valuable contribution to our understanding of the non-crystalline state.

REFERENCES

1   P.G. LeComber, R.J. Loveland, W.E. Spear and R.A. Vaughan, in J. Stuke and W. Brenig (Eds.), Amorphous and Liquid Semiconductors, Taylor and Francis Ltd., London, 1974, p. 245.
2   M.H. Brodsky and R.S. Title, Phys. Rev. Lett., 23 (1969) 581.
3   M.H. Brodsky, R.S. Title, K. Weiser and G.D. Pettit, Phys. Rev. B, 1 (1970) 2632.
4   J.E. Fischer and T.M. Donovan, J. Non-Cryst. Solids, 8—10 (1972) 202.
5   J.E. Fischer, Thin Solid Films, 17 (1973) 223.
6   T.M. Donovan, M.L. Knotek and J.E. Fischer, in J. Stuke and W. Brenig (Eds.), Amorphous and Liquid Semiconductors, Taylor and Francis Ltd., London, 1974, p. 549.
7   R.C. Chittick, J.H. Alexander and H.F. Sterling, J. Electrochem. Soc., 116 (1969) 77.
8   R.C. Chittick, J. Non-Cryst. Solids, 3 (1970) 255.
9   R.J. Loveland, W.E. Spear and A. Al-Sharbaty, J. Non-Cryst. Solids, 13 (1973/74) 55.
10 W.E. Spear, R.J. Loveland and A. Al-Sharbaty, J. Non-Cryst. Solids, 15 (1974) 410.
11 P.G. LeComber and W.E. Spear, Phys. Rev. Lett., 25 (1970) 509.
12 P.G. LeComber, A. Madan and W.E. Spear, J. Non-Cryst. Solids, 11 (1972) 219.
13 A. Madan, P.G. LeComber and W.E. Spear, J. Non-Cryst. Solids, 20 (1976) 239. For a summary of the results see [14].
14 W.E. Spear, in J. Stuke and W. Brenig (Eds.), Amorphous and Liquid Semiconductors, Taylor and Francis Ltd., London, 1974, p. 1.
15 W.E. Spear and P.G. LeComber, J. Non-Cryst. Solids, 8—10 (1972) 727.
16 N.F. Mott and E.A. Davis, Electronic Processes in Non-Crystalline Materials, Clarendon Press, Oxford, 1971, p. 235 ff.
17 J. Tauc, in F. Abeles (Ed.), Optical Properties of Solids, North-Holland, Amsterdam, 1970.
18 E.A. Davis and N.F. Mott, Philos. Mag., 22 (1970) 903;
    See also N.F. Mott and E.A. Davis, Electronic Processes in Non-Crystalline Materials, Clarendon Press, Oxford, 1971, p. 248 ff.
19 W.E. Spicer, T.M. Donovan and J.E. Fischer, J. Non-Cryst. Solids, 8—10 (1972) 122.
20 R. Grigorovici and A. Vancu, Thin Solid Films, 2 (1968) 105.
21 A. Lewis, Phys. Rev. Lett., 29 (1972) 1555.
22 T.C. Arnoldussen, R.H. Bube, E.A. Fagen and S. Holmberg, J. Appl. Phys., 43 (1972) 1798.
23 C. Main and A.E. Owen, in P.G. LeComber and J. Mort (Eds.), Electronic and Structural Properties of Amorphous Semiconductors, Academic Press, London, 1973, p. 527.
24 J.G. Simmons and G.W. Taylor, J. Phys. C, 7 (1974) 3051.
25 A. Rose, Concepts in Photoconductivity and Allied Problems, Interscience Publishers, New York, 1963.
26 N.F. Mott, E.A. Davis and R.A. Street, Philos. Mag., 32 (1975) 961, and personal communication.
27 R. Englman and J. Jortner, Mol. Phys., 18 (1970) 145.

28 N. Robertson and L. Friedman, Philos. Mag., 33 (1976) 753.
29 See, for instance, D. Emin, in P.G. LeComber and J. Mort (Eds.), Electronic and Structural Properties of Amorphous Semiconductors, Academic Press, London, 1973, p. 261.
30 J.E. Smith, Jr., M.H. Brodsky, B.L. Crowder and M.I. Nathan, J. Non-Cryst. Solids, 8—10 (1972) 179.
31 D. Engemann and R. Fischer, in J. Stuke and W. Brenig (Eds.), Amorphous and Liquid Semiconductors, Taylor and Francis Ltd., London, 1974, p. 947.
32 D. Engemann and R. Fischer, Proc. 12th Intern. Conf. on the Physics of Semiconductors, Stuttgart, 1974, p. 1042.
33 A.H. Clark, Phys. Rev., 154 (1967) 750.
34 R. Grigorovici, N. Croitoru and A. Dévényi, Phys. Status Solidi, 23 (1967) 627.
35 J.E. Fischer and T.M. Donovan, Opt. Commun., 3 (1971) 116.
36 M.L. Knotek and T.M. Donovan, Phys. Rev. Lett., 30 (1973) 652.
37 L. Vescan and N. Croitoru, Conf. Amorphous Semiconductors, 1974, Reinhardsbrunn, East Germany, Akad. Wiss. der DDR.
38 T.J. Burke and A.H. Clark, Thin Solid Films, 10 (1972) 413.
39 D.E. Brodie and P.C. Eastman, Can. J. Phys., 43 (1965) 969.
40 I.A. Karpovich, B.N. Zvonkov and M.A. Rizakhanov, Sov. Phys. — Solid State, 12 (1971) 1773.
41 D.L. Camphausen, G.A.N. Connell and W. Paul, J. Non-Cryst. Solids, 8—10 (1972) 223.
42 W. Eckenbach, W. Fuhs and J. Stuke, J. Non-Cryst. Solids, 5 (1971) 264.
43 D.A. Yates, C.M. Penchina and J.E. Davey, in J. Stuke and W. Brenig (Eds.), Amorphous and Liquid Semiconductors, Taylor and Francis Ltd., London, 1974, p. 617.
44 T. Botilă, N. Croitoru, Gr. Ioanid, T. Stoica and L. Vescan, in M.H. Brodsky, S. Kirkpatrick and D. Weaire (Eds.), Tetrahedrally Bonded Amorphous Semiconductors, Am. Inst. Phys., New York, 1974, p. 33.

CHAPTER 7

AMORPHOUS CHALCOGENIDES

R.C. ENCK and G. PFISTER

7.1 Introduction
7.2 Photogeneration efficiency
   7.2.1 Amorphous selenium
   7.2.2 Amorphous As_2Se_3 and As_2S_3
7.3 Steady-state photoconductivity
   7.3.1 Introduction
   7.3.2 Temperature dependence
   7.3.3 Decay time
   7.3.4 Radiative recombination
   7.3.5 Summary
7.4 Electronic transport properties
   7.4.1 Introduction
   7.4.2 Time-of-flight experiment
   7.4.3 Hole transport in amorphous As_2Se_3
   7.4.4 Amorphous Se
   7.4.5 Effect of impurities on charge transport in amorphous Se
   7.4.6 Hole transport in amorphous As_2S_3 and amorphous As_2Te_3
   7.4.7 ac Conduction
   7.4.8 The mobility edge

7.1 INTRODUCTION

The subject of photoconductivity in the amorphous chalcogenides divides naturally into three parts: photogeneration, steady-state photoconductivity, and electronic transport. Photogeneration, covered in Section 7.2, has been extensively studied using transient techniques, particularly in amorphous selenium, where field-dependent quantum efficiencies are observed. This material has become a model system for demonstrating the applicability of the Onsager theory to photogeneration in low mobility materials. Steady-state photoconductivity, covered in Section 7.3, has been useful primarily in determining the free-carrier recombination mechanisms and the density of states in the forbidden gap. Recent work in this area has suggested that in many of these materials there are sharp discontinuities in the density of localized states in the forbidden gap and that carrier recombination is dominated by transitions between localized states. Electronic transport, covered in Section 7.4, has generally been studied by transient drift mobility experiments, while photo a.c. conductivity has recently evolved as an additional tool. Amorphous selenium and amorphous As_2Se_3 have been extensively studied and have been considered to be model systems for trap-controlled band-like transport, and a stochastic transport process with a wide distribution of event times, respectively. Recent work, however, casts some doubt on the fundamental difference of the transport mechanism in these two materials.

Each section of this chapter is essentially self-contained with a separate introduction which outlines its contents. It should be possible, therefore, for the reader to understand each section independently.

7.2 PHOTOGENERATION EFFICIENCY

The ability to time-resolve the drift of a thin sheet of photogenerated charge carriers in amorphous Se, amorphous As_2Se_3, and amorphous As_2S_3 allows the photogeneration efficiency in these materials to be determined directly under appropriate experimental conditions. Carriers are photogenerated in a thin surface region of a plate-like sample by strongly-absorbed light. A uniform electric field parallel to the thin dimension of the sample draws all of the photogenerated carriers of appropriate sign out of the photogeneration region and ideally causes them to drift completely across the sample with some unique drift mobility. Under these conditions, the height of the transient current pulse directly gives the photogeneration efficiency. If free-carrier recombination in the generation region is important, the photoinjection efficiency will be smaller than the generation region into the bulk of the sample. In this section, the terms "quantum efficiency", "photogeneration efficiency", and "photoinjection efficiency" will be used interchangeably to describe the results of this measurement. The exact meaning will be clear from the associated discussion. (See Chapter 2 for further details.) These ideal conditions are approached most closely for amorphous selenium (a-Se) and most of our discussion will be concerned with this material, which has been extensively investigated and in which the photogeneration process is substantially understood. Also included will be a discussion of the more complex cases of a-As_2Se_3 and a-As_2S_3, where the relatively small amount of data available suggests photogeneration effects similar to those observed in a-Se.

The most striking features of photogeneration in the amorphous chalcogenides are the strong dependences of the photogeneration efficiency on photon energy, temperature, and the applied electric field. The strong dependence on photon energy continues for energies substantially greater than the absorption edge. These characteristics are not present in conventional semiconductors, but are observed to some extent in a wide range of low mobility materials such as molecular crystals (S [1], anthracene [2], As_2S_3 [3]), organic polymers (PVK [4], PVK—TNF [5]), and liquids [6, 7] (hydrocarbons, argon, oxygen) in addition to the amorphous chalcogenides (Se [8], As_2Se_3 [9], As_2S_3 [10]).

It is likely that the short mean free path for free carriers characteristic of these materials is responsible for these features of the photogeneration. In this picture, a photoexcited carrier pair shares an excess kinetic energy $[h\nu - \mathcal{E}_g + \mathcal{E}_c(r)]$, where $h\nu$ is the photon energy, $\mathcal{E}_g$ is the band gap, r is the pair separation, and $\mathcal{E}_c(r) = e^2/(4\pi\kappa\kappa_0 r)$ is the Coulomb energy of the pair. The photogeneration efficiency cannot approach unity unless the carriers, before they thermalize, diffuse apart a distance $\gtrsim r_c$ such that $e^2/(4\pi\kappa\kappa_0 r_c) \simeq kT$. Otherwise, geminate recombination (that is, recombination of the hole and electron produced by a single photon before the two carriers escape the influence of their mutual Coulomb field) is more likely than creation of a free carrier pair. Since the rate of energy loss for a hot carrier is expected to be high due to the short mean free path, it is unlikely that thermalization will occur at separations greater than r_c unless a substantial amount of excess kinetic energy is present. This results in a strong dependence of photogeneration efficiency on photon energy for photons with energy greater than the band gap. The picture also predicts strong effects of temperature and external electric fields.

A number of different models have been used to explain the details of photogeneration or carrier yield in a-Se and the other low mobility materials previously mentioned. These models include exciton absorption, absorption by localized molecular excitations, the Poole—Frenkel effect, and the Onsager theory. The Onsager theory seemed to offer the best possibility for a complete explanation of the data, but it was generally applied in a very approximate form and in most cases the amount of data available was insufficient to come to a definite conclusion about the applicability of the theory.

Recent work with a-Se, however, has shown that the Onsager theory fits the photogeneration data extremely well over a very wide range of photon energy, electric field strength, and temperature. This work will therefore be discussed in some detail, not only because of its direct importance to the amorphous chalcogenides, but also because of its implications for the applicability of Onsager theory to the whole class of low mobility materials.

7.2.1 Amorphous selenium

(i) Sample preparation and measurement techniques

Photogeneration measurements on a-Se have usually been carried out on films prepared by vacuum evaporation of high purity vitreous selenium onto conducting substrates held at about 55°C. Substantially higher substrate temperatures induce crystallization, while lower temperatures result in increased bulk deep trapping which makes reliable photogeneration measurements difficult. Except when otherwise noted, we will be concerned with samples prepared in this way. Some work on bulk-quenched glass will also be discussed.

In addition to the electroded time-resolved transient conductivity measurements discussed in detail elsewhere in this book, xerographic discharge measurements have also been used to determine photogeneration quantum efficiency [11]. The major advantage of the xerographic discharge approach is the ability to apply electric fields of more than 6×10^5 V/cm while electroded samples can suffer breakdowns at fields greater than about 1.5×10^5 V/cm. With the xerographic discharge technique the sample is corona-charged with the same polarity ions as the mobile charge to be observed. There must, of course, be negligible charge injection from this corona contact on the time scale of the measurement (that is, it must form a blocking contact). The surface voltage on the film is measured by a transparent probe capacitatively coupled to the surface. The sample is then illuminated by a constant intensity of strongly absorbed light and the rate of change of surface potential is measured.

Assuming that the light intensity is sufficiently low that the change in surface potential during a carrier transit time is small compared to the surface potential (emission-limited current case), the rate of discharge of this film in the presence of bulk trapping can be shown to be [11]:

$$\frac{dE}{dt}\bigg|_{t=0} = \frac{\eta F e}{\kappa \kappa_0} \frac{\mu E \tau}{L} (1 - \exp(-L/\mu E \tau)) \tag{7.1}$$

where η is the photoinjection quantum efficiency [QE], F is the light intensity per unit area in photons/second, e is the electronic charge, $\kappa \kappa_0$ is the sample dielectric constant, μ is the carrier drift mobility, τ is the carrier bulk deep trapping lifetime, L is the sample thickness, V is the surface potential, and $E(= V/L)$ is the electric field in the sample (assumed uniform). When bulk trapping is negligible (that is, for $\mu \tau E \gg L$), this equation can be written as

$$\frac{dE}{dt} = \frac{\eta F e}{\kappa \kappa_0} \quad \text{or} \quad \eta = \frac{\kappa \kappa_0 (dE/dt)}{Fe}. \tag{7.2}$$

For strong bulk trapping ($\mu \tau E \ll L$) we have

$$\frac{dE}{dt}\bigg|_{t=0} = \frac{\eta F e}{\kappa \kappa_0} \frac{\mu E \tau}{L}. \tag{7.3}$$

The presence of a thickness dependence in the rate of change of electric field at constant field for a given light intensity thus serves to indicate the presence of a bulk trapping limitation. In the absence of such a thickness dependence, the photoinjection quantum efficiency can be determined from the rate of change of electric field without detailed knowledge of any microscopic parameter. In fact, the existence of a unique mobility is not necessary as long as some average of the carrier transit times is small enough to satisfy the emission-limited current condition. To minimize the cumulative effect of trapped space charge, the usual practice is to use only the initial $(dE/dt)E_0$ to determine the photoinjection efficiency at the initial charging field E_0. Efficiencies at other fields are determined by charging to other initial potentials.

(ii) Electric field and photon energy dependence of QE for holes

Photoinjection quantum efficiencies for holes in a-Se measured using the xerographic discharge technique are shown in Fig. 7.1 as a function of the

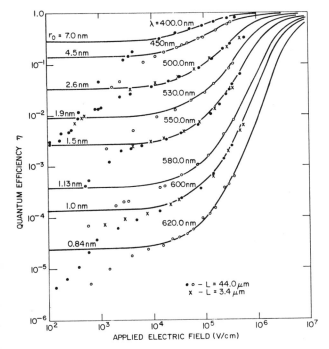

Fig. 7.1. The circles and crosses show the experimental quantum efficiency of photoinjection of holes in a-Se *versus* applied electric field with the wavelength of the exciting radiation as the parameter. Included are data on films of two different thicknesses. The solid lines are the theoretical Onsager dissociation efficiencies for $\phi_0 = 1$ and for initial separations r_0 indicated in the figure. (From Pai and Enck [8].)

applied electric field and for eight different exciting wavelengths and two sample thicknesses [8]. The room temperature absorption coefficient for a-Se is shown as a function of photon energy in Fig. 7.2. Photoinjection quantum efficiency for a-Se as a function of photon energy is also plotted

References pp. 297—302

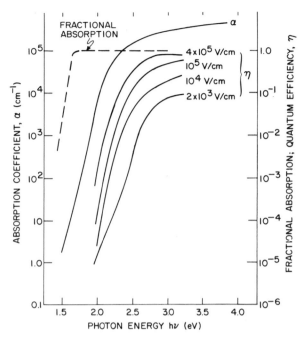

Fig. 7.2. Absorption coefficient, quantum efficiency for holes, and fractional absorption for 50-μm-thick a-Se versus exciting photon energy. Applied electric field is the parameter for the quantum efficiency curves. (From Pai [12].)

for four values of the applied electric field [12]. Comparing this to Fig. 7.1, it is seen that, for all exciting wavelengths in Fig. 7.1, the absorption depth of the light is a small fraction of the sample thickness. It should be noted that, for fields larger than about 2×10^3 V/cm, the two samples of different thickness shown in Fig. 7.1 have identical efficiencies within experimental error. This shows [see eqns. (7.1), (7.2), and (7.3)] that bulk trapping is negligible in this field range. The discrepancy between theory and experiment for low fields ($\lesssim 5 \times 10^3$ V/cm) will be shown to be due largely to surface recombination and trapping. Reduction of the low field QE has also been observed due to perturbation of the electric field in the generation region caused by small amounts of trapped space charge. Strong sample-to-sample variations and sample history effects have been observed at low fields [12, 13].

Several features characterize the sensitivity curves of Fig. 7.1:

(1) Since there is no significant bulk trapping and the light is always strongly absorbed compared to the sample thickness, the shapes of the curves directly display the quantum efficiency of photoinjection of holes from the photogeneration region into the bulk of the sample.

(2) The shape of the electric field dependence of QE for the

short-wavelength region of the spectrum is very different from that in the long wavelength region. In the blue region, QE varies approximately linearly with electric field for fields less than 10^4 V/cm, changing to sublinear ($E^{0.3}$—$E^{0.5}$) for larger fields. In the red region of the spectrum, the QE rises sharply for fields greater than about 10^5 V/cm and tends to saturate in the intermediate field region. A decrease from the saturation value is observed at lower fields and the field at which the decrease occurs is substantially lower in the red region of the spectrum than in the blue region. This drop-off has been observed to occur at higher fields in some instances but this could be due to surface effects and/or trapped space charge caused by inadequate resting of the sample.

(3) At a constant electric field, the QE can vary from essentially unity to a very small value as the wavelength varies, even though absorption of the light is complete in all cases.

The initial mechanism proposed to explain field-dependent sensitivity in a-Se was a carrier-range limitation (bulk trapping). When this was shown to be inappropriate as a result of the thickness independence of QE [14], a number of other models were considered in addition to the field dependence of the photogeneration process itself.

As a result of the apparent "gap" between the photogeneration edge and the absorption edge (see Fig. 7.2) it was proposed that the total absorption was the result of two independent absorption processes [15, 16]. One process, presumably ending in a localized molecular state, always had zero QE for producing free carriers and only produced significant absorption near the absorption edge. The other always had unity QE for producing free carriers and produced monotonically increasing absorption starting somewhat above the absorption edge. The observation of the sharp rise in QE for the high field and long wavelength, observed in Fig. 7.1, requires the introduction of a field dependence of one of these dual absorption edges for which there seems to be little published evidence. This, together with the impression that may be derived from Figs. 7.1 and 7.2, *i.e.* that no gap at all would appear if the electric field were high enough, suggests that this theory may not be applicable to photogeneration in a-Se.

Bimolecular recombination in the photogeneration region was rejected as an explanation for the field-dependent QE because the QE was found to be independent of light intensity for constant-energy light pulses with widths ranging from about 1 s to 4 ns [17]. Monomolecular recombination was rejected as an explanation because the large density of recombination centers necessary in the generation region to explain the field-dependent QE must necessarily act as efficient traps for either holes or electrons in the transport region. Since no such traps are observed, monomolecular recombination is not important [17]. A distribution of traps in the generation region different from that in the bulk cannot solve this problem since the identities of these two regions change very substantially as the absorption constant changes with exciting wavelength.

References pp. 297—302

It was therefore proposed that the field-dependent photoinjection was actually due to a field-dependent photogeneration process in which the carrier pair produced by a single photon was initially bound by their mutual coulomb attraction and subsequently either recombined or became a pair of free carriers through a field-aided thermal dissociation process [17, 18]. This process was described either by a theory due to Frenkel [17—20] or by a theory due to Onsager [8, 21].

Before investigating the application of these theories to the data, the portions of experimental results that are apparently inconsistent with both theories will be discussed further. For any theory of this type, the QE must become independent of electric field below that field which can give a bound carrier, at most, an additional energy of kT (or alternatively, lower the barrier to escape by kT). For a-Se, this field is $\sim 10^4$ V/cm at room temperature [17, 22]. However, we see that instead of a saturation in QE below this field, the QE drops at least linearly with electric field.

Suggested explanations for this discrepancy include failure of isotropic media theory to describe the random-walk process in a locally anisotropic material such as selenium [22], monomolecular bulk recombination in the generation region [23], and surface recombination [24]. None of these predict results in satisfactory agreement with the low-field QE data although the tendency for the break to the low-field linear region to move to lower fields as the wavelength increases is in qualitative agreement with the surface recombination model. This inconclusive result is probably to be expected, considering the previously mentioned difficulties inherent in low-field measurements.

Recent two-photon photogeneration measurements have resolved this question in favor of the surface recombination model [24]. In this experiment, an electroded sample is illuminated with a short, intense pulse of 1.17 eV photons. The sample is essentially transparent at this energy, but there is a small cross-section for two-photon absorption where two photons are absorbed essentially simultaneously and a carrier pair is produced with twice the incident photon energy, or 2.34 eV. Due to the small absorption coefficient for two-photon absorption, the carrier pairs are produced uniformly throughout the sample. This would result in a triangular current pulse, in agreement with experiment, where the height of the triangle is proportional to the photogeneration. Figure 7.3 shows the electric field dependence of the relative QE of two-photon photogeneration for a-Se. The low-field saturation predicted by a field-aided thermal dissociation mechanism for photogeneration is clearly seen in contrast with the low-field drop-off of the 2.34 eV single-photon QE data also shown in the figure. If surface recombination is the dominant process at low fields, it can be shown that the field at which the break to the linear dependence occurs would vary inversely with optical absorption depth [17, 22, 25]. This would require the linear field dependence to occur at $\lesssim 10^2$ V/cm for weakly-absorbed light in Fig.

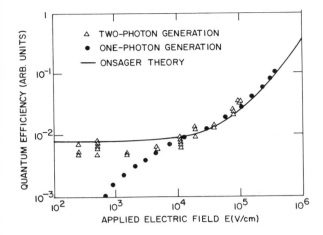

Fig. 7.3. Quantum efficiency *versus* applied electric field for holes in *a*-Se. Experimental results for one-photon generation ($h\nu = 2.34$ eV) and two-photon generation ($h\nu = 1.17$ eV) are shown. Also included are the QE values calculated from Onsager theory with $\phi_0 = 1$ and $r_0 = 1.9$ nm. (From Enck [22] and Pai and Enck [8].)

7.3, in agreement with the data. The other suggested mechanisms would cause the break to occur at $\gtrsim 10^4$ V/cm for weakly-absorbed light in contradiction to the experimental results. Surface recombination and/or surface trapping are thus seen to be the cause of the low-field drop in apparent photogeneration efficiency in *a*-Se.

This conclusion was subsequently confirmed and one-photon photogeneration and recombination at low fields was studied directly using "delayed voltage" measurements [26]. In this technique, an adjustable electric field (the generation field) is applied to the sample during and after the light pulse until a strong electric field (the collection field) is applied a time t_d (the delay time) after the light pulse. The collection field should be large enough to prevent further recombination or deep trapping, ensuring that all the free carriers present at t_d after the light pulse are collected. Further discussion of this measurement technique will be found in Chapters 2 and 9.

The amount of collected positive charge *versus* generation field for a delayed-voltage experiment in *a*-Se is plotted in Fig. 7.4. The exciting light was a 4-ns pulse at 337 nm. Delay time was 200 ns and collection field was 5×10^4 V/cm. The QE observed at 10^4 V/cm generation field was the same as that observed in a constant-field transient photoconductivity experiment. However, the saturation observed at lower fields in Fig. 7.4 corresponds to a much higher quantum efficiency than the constant-field case (see Fig. 7.1). The low-field QE increases further as the delay time is decreased from 200 ns, while this decrease has no effect on the QE at 10^4 V/cm. This shows that the drop in apparent QE at low field in Fig. 7.1 is not due to a drop in photogeneration efficiency but to recombination and trapping losses near

References pp. 297–302

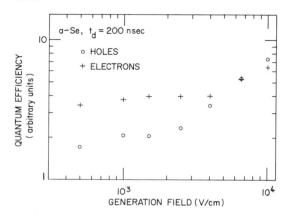

Fig. 7.4. Quantum efficiency *versus* electric field in a-Se for 337-nm light. The electric field is increased to 7×10^4 V/cm 200 ns after the light flash. (From Enck [26].)

the surface during transport. (Bulk losses are ruled out by the thickness independence of the QE.) The results shown in Fig. 7.4 are independent of the magnitude of the collection field for collection fields greater than about 10^4 V/cm, indicating that the field-dependent generation process is completed in a time less than t_d.

Hole collection efficiency as a function of delay time when the generation field is kept constant at zero and the collection field is 5×10^4 V/cm is shown by the points in Fig. 7.5. Bimolecular recombination is ruled out as the cause of this decrease of collection efficiency with delay time since the efficiency is independent of light intensity at all delay times. The predicted linear dependence of the inverse of the collection efficiency on delay time [1] is also not seen. Normal monomolecular recombination in the generation region is ruled out by the non-exponential dependence on delay time. The slow dependence on delay seen in Fig. 7.5, approaching square root at long times, suggests that diffusion may be the process limiting carrier loss in this experiment.

The solid line shown in Fig. 7.5 is the result of a calculation of the fraction of holes remaining as a function of delay time, assuming that the only loss mechanism is recombination or deep trapping at the sample surface of photogenerated free carriers in a field-free semi-infinite sample. The initial carrier concentration is assumed to decrease exponentially with distance from the surface and the diffusion velocity is derived from the drift mobility using the Einstein relation. An effective surface recombination velocity of 150 cm/s was used. This is reduced from the actual surface recombination velocity by the fraction of time the carriers spend in the shallow traps which determine the drift mobility and from which it is assumed they cannot recombine or become deep trapped. This excellent agreement is further proof that the apparent drop in QE at low fields is due to surface

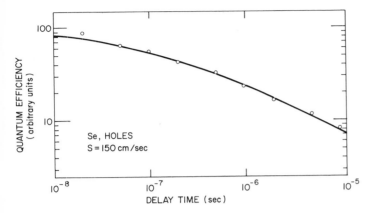

Fig. 7.5. Quantum efficiency for holes *versus* time delay between 337-nm light flash and application of electric field (5×10^4 V/cm) in a 10-μm-thick a-Se film. The solid line is the result of a calculation assuming exponential light absorption and carrier diffusion in a semi-infinite sample with a surface recombination velocity of 150 cm/s. (From Enck [26].)

recombination and that the actual QE saturates at low fields as predicted by field-aided thermal dissociation models of photogeneration.

As previously indicated, the only theories that satisfactorily explain the field dependence, photon energy dependence, and temperature dependence of photogeneration in a-Se describe the photogeneration process in terms of field-aided thermal dissociation of a photoexcited hole—electron pair bound by their mutual coulomb attraction. Until the recent application of the Onsager theory to this model, it was described in the framework of the Poole—Frenkel theory [20]. In its initial form [17, 18], the Poole—Frenkel approach assumed that the photoexcited hole—electron pair has an initial separation such that the carriers experience a binding energy $\mathcal{E}_0$ due to their mutual coulomb attraction. In the presence of an external electric field E this barrier to dissociation is lowered by the amount $\beta E^{1/2}$, so that the probability for dissociation P_d is proportional to:

$$P_d \propto \exp[-(\mathcal{E}_0 - \beta E^{1/2})/kT] \tag{7.4}$$

where

$$\beta = (e^3/\pi\kappa\kappa_0)^{1/2} \tag{7.5}$$

Assuming that $\eta \propto P_d$, this model gave a satisfactory qualitative description of the functional form of the field dependence of photogeneration at long exciting wavelengths, but was completely unsatisfactory at short wavelengths. In addition, $\mathcal{E}_0$ and β were found to be photon-energy dependent and the measured magnitude of β, even at low photon energies where β is largest and approximately independent of photon energy, is about half that calculated from eqn. (7.5) using a relative dielectric constant of 6.3 for a-Se.

References pp. 297—302

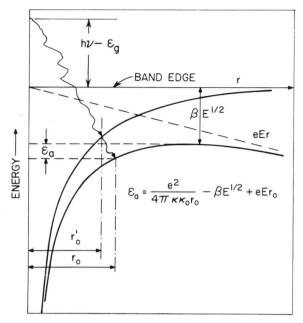

Fig. 7.6. Schematic of electron—hole pair separation in an applied electric field. (From Knights and Davis [19].)

This simple model was modified by the inclusion of a recombination lifetime τ_r (assumed to be independent of photon energy and temperature) to compete with the ionization lifetime τ_i:

$$\tau_i = 1/\nu_e P_d \tag{7.6}$$

where ν_e is an attempt-to-escape frequency [17]. The resulting rate equation leads to the following expression for the QE [17]

$$\eta = 1/(1 + \tau_i/\tau_r). \tag{7.7}$$

This correctly predicts the saturation in QE observed at high fields and high photon energies. Other modifications, based on models for the thermalization of the photoexcited hot carriers, were proposed to explain the photon-energy dependence of the activation energy $\mathcal{E}_0$ [19].

A schematic of the electron—hole pair separation based on this picture is shown in Fig. 7.6. The two solid curves show the local potential with and without an external field. The dashed line shows the potential due to the external field. The carrier of interest is photoexcited to an energy $(h\nu - \mathcal{E}_g)$ above the band edge, where $\mathcal{E}_g$ is the band gap energy. The jagged line shows the thermalization of this carrier during which it loses the excess kinetic energy $(h\nu - \mathcal{E}_g - \mathcal{E}_c)$ supplied by the photon, ending with a separation r_0 between the hole and electron. ($\mathcal{E}_c$ is the coulomb energy of the carrier pair.)

Since the mean free path in a-Se is approximately an interatomic spacing, the motion during this thermalization can be expected to be diffusive. The diffusion constant D is assumed to be independent of the carrier energy. Due to the relaxation of k-conservation for inelastic scattering in amorphous materials, it is also expected that the loss rate will reach its maximum value of a phonon frequency ν_{ph} times the phonon energy $h\nu_{ph}$. The attempt-to-escape frequency ν_e is taken to be the reciprocal of the time taken for the carriers to diffuse apart by a distance equal to the coulomb capture radius r_c (taken as the separation at which the coulomb energy equals $2kT$), so that

$$\nu_e^{-1} = \frac{r_c^2}{D} = \frac{e^4}{(4\pi)^2(\kappa\kappa_0)^2 4(kT)^2 D}.$$

With these assumptions, eqn. (7.7) can be written as

$$\eta = \left[1 + \frac{e^4}{4\tau_r(4\pi)^2(\kappa\kappa_0)^2(kT)^2 D} \exp\left(\frac{\mathcal{E}_a}{kT}\right)\right]^{-1}, \qquad (7.8)$$

where:

$$\mathcal{E}_a = \frac{e^2}{4\pi\kappa\kappa_0 r_0} - \beta E^{1/2} + eEr_0 \qquad (7.9)$$

and r_0 is defined by

$$\frac{h\nu_{ph}^2}{D} r_0^2 = (h\nu - \mathcal{E}_g) + \frac{e^2}{4\pi\kappa\kappa_0 r_0} + eEr_0. \qquad (7.10)$$

These expressions have been evaluated numerically for various values of the parameters D, ν_{ph}, τ_r, and $\mathcal{E}_g$, using a value of β equal to half the theoretical value [19]. For low fields, the calculation was modified using a one-dimensional random walk theory to account for the experimentally observed linear drop in QE. The results of this calculation using reasonable values of the parameters give field dependences, photon energy dependences, and temperature dependences which are qualitatively similar to those observed experimentally, suggesting that the model is basically correct. However, agreement with experiment is not observed over any substantial range of the experimental parameters, indicating that substantial changes may be necessary in the theory.

The characteristics of this model, which may require refinement to improve agreement with experimental results, include the one-dimensional nature of the theory, the observation of an incorrect value for β, the assumption that the recombination lifetime is independent of photon energy and temperature, and the introduction of a detailed model of thermalization using a constant value for D, the diffusion constant, to determine r_0. Some early work was done on three-dimensional Poole—Frenkel theory [27], but

References pp. 297—302

it has not been developed further. It appears that three-dimensional effects are most important at low fields, since escape at high fields, where barrier lowering is important, is mostly along the direction of the field and thus approximately one-dimensional. The magnitude of β is one of the basic predictions of the Poole—Frenkel theory and the substantial disagreement between the theoretical and observed values casts doubt on the applicability of the theory in its present form to these experiments.

The assumption that τ_r is constant is open to serious question since τ_r should be a function of r_0 which is a strong function of photon energy. The recombination lifetime must also be a function of temperature through the diffusion velocity. In order to take these matters into account, a recombination model must be developed which takes into account the coulomb nature of the potential barrier and the microscopic motion of the carriers.

Determining r_0 from a model for the thermalization process as part of the calculation for QE requires that both thermalization and subsequent separation or recombination be properly modeled before satisfactory agreement with experiment can be achieved. Of these two processes, thermalization, involving the properties of hot carriers, is least well understood. The thermalization calculation also requires the introduction of some additional interrelated adjustable parameters whose values are likely to be difficult to determine uniquely. This suggests that r_0 be used as a parameter in the calculation of the separation—recombination process. If r_0 can be uniquely determined as a function of photon energy and temperature, these results can then be used to develop a model for thermalization. This is the approach taken in the Onsager theory.

(iii) Onsager theory

The theory of geminate recombination (or initial recombination) can be reduced to the problem of Brownian motion in the presence of the coulomb attraction and the applied electric field. The Onsager approach is to solve the equation of Brownian motion for the two carriers under the influence of the coulomb field and the applied field [21]. Ionization and recombination were calculated by separately considering two stationary flow problems, one with a source at the origin and a sink at $r = \infty$, the other with a source at infinity and a sink at the origin. Using the solutions to these two extreme boundary-condition problems, Onsager derived a relation for the probability $p(r, \theta, E)$ that an ion pair thermalized with an initial separation r and at an angle θ with the electric field direction will escape geminate recombination.

If ϕ_0 is the efficiency of production of thermalized ion (or hole—electron) pairs per absorbed photon (which is the maximum possible QE), and $p(r, \theta, E)$ is the probability of dissociation of the thermalized pairs as a function of electric field and initial separation, and $g(r, \theta)$ is the initial distribution of separations between the thermalized ions of each ion pair, the overall QE will be given by:

$$\phi(E) = \phi_0 \int p(r, \theta, E) g(r, \theta) \mathrm{d}^3 r, \tag{7.11}$$

where ϕ_0 is assumed to be independent of the applied electric field. (Further details of Onsager theory can be found in Chapter 3 and ref. 21.) Using the assumption that the initial distribution of thermalized pairs is isotropic so that:

$$g(r, \theta) = \frac{1}{4\pi r_0^2} \delta(r - r_0), \tag{7.12}$$

where r_0 is a characteristic thermalization length, the integration in eqn. (7.11) can be carried out. The resulting expression for escape efficiency (or QE) is given by [8]:

$$\phi(r_0, E) = \phi_0 \frac{kT}{eEr_0} \exp(-A) \exp\left(-\frac{eEr_0}{kT}\right) \sum_{m=0}^{\infty} \frac{A^m}{m!} \sum_{n=0}^{\infty} \sum_{l=m+n+1}^{\infty} \left(\frac{eEr_0}{kT}\right)^l \frac{1}{l!}, \tag{7.13}$$

where $A = e^2/4\pi\kappa\kappa_0 kT r_0$. If a critical Onsager distance r_c is defined as that distance at which the coulomb energy is equal to kT

$$r_c = e^2/4\pi\kappa\kappa_0 kT. \tag{7.14}$$

The first few terms of eqn. (7.13) can then be written as

$$\phi(r_0, E, T) = \phi_0 \exp\left[-\frac{r_c(T)}{r_0}\right] \left\{ 1 + \left(\frac{e}{kT}\right) \frac{1}{2!} r_c E + \left(\frac{e}{kT}\right)^2 \frac{1}{3!} r_c \left(\frac{r_c}{2} - r_0\right) E^2 \right.$$
$$\left. + \left(\frac{e}{kT}\right)^3 \frac{1}{4!} r_c \left(r_0^2 - r_0 r_c + \frac{r_c^2}{6}\right) E^3 + \ldots \right\}. \tag{7.15}$$

The expression for generation efficiency given in eqn. (7.13) has been evaluated numerically for a-Se using a relative dielectric constant of 6.3 [8]. The results at room temperature for the best choice of ϕ_0 and r_0 are shown by the solid lines in Fig. 7.1. The appropriate r_0 for each photon energy is shown in the figure. The thermalization efficiency ϕ_0 is assumed to be independent of photon energy, electric field, and temperature. It is, thus, merely a multiplicative constant which can move the entire family of curves rigidly parallel to the vertical axis, but which cannot change their shape or relative magnitude. Within the experimental accuracy of the absolute efficiency measurement ($\sim \pm 10\%$), $\phi_0 = 1$. Once ϕ_0 is chosen, the choice of a particular r_0 determines both the shape of the QE versus electric field curve and the absolute values of the QE on that curve. Therefore, the shapes and relative magnitudes of the QE versus electric field curves at different photon energies cannot be fitted independently and r_0 is determined uniquely at each photon energy. Since the variations in shape and magnitude of the QE curves in Fig.

References pp. 297—302

7.1 are very large, the excellent fit with the one adjustable parameter r_0 indicates that the Onsager theory accurately describes photogeneration in a-Se and that the thermalization distances r_0 are physically meaningful. As discussed in an earlier section, the apparent disagreement between theory and experiment at low fields disappears when proper account is taken of surface recombination effects.

(iv) Temperature dependence of thermalization distance

The values of ϕ_0 and r_0 have been determined at a number of temperatures between room temperature and 223K by fitting the family of measured QE *versus* electric field curves for different photon energies at each temperature [8]. The value of ϕ_0 was found to be approximately unity at each temperature. Table 7.1 shows the temperature dependence of r_0 for several photon

TABLE 7.1

Temperature dependence of the initial separation r_0 of charge carriers for several excitation wavelengths in a-Se[a]

Excitation wavelength, λ (nm)	Initial separation r_0 (nm)				
	$T = 294$K	$T = 280$K	$T = 268$K	$T = 257$K	$T = 223$K
400	7.0	7.0	—	7.0	7.0
450	5.3	5.4	—	5.2	5.2
480	4.0	—	4.19	3.97	3.6
500	3.21	3.33	3.28	3.07	3.0
520	2.52	2.4	2.56	2.45	2.39
540	2.02	2.04	2.05	1.92	1.97
560	1.44	1.38	—	1.38	1.36
580	1.23	1.22	1.23	1.19	—

[a] These results were obtained using electroded samples. (From Pai and Enck [8]).

energies. The values of r_0 are almost independent of temperature, decreasing on the average about five percent for a temperature decrease from 294K to 223K.

Though precise experimental data have not been available, there has been some discussion about the expected form of the temperature dependence of r_0 when the Onsager theory of dissociation is applied to explain QE data in anthracene and carrier yield data in liquid hydrocarbons. In a crystalline material such as anthracene, the temperature dependence of the mean free path resulting from the temperature dependence of thermally-induced fluctuations in the lattice density causes the diffusion constant D to have an inverse temperature dependence. The thermalization distance, which is proportional to $\sqrt{D}$, is expected to have $T^{-0.5}$ dependence [28], which would give an *increase* of 15% in r_0 at 223K over 294K, rather than the *decrease* of

5% actually observed in a-Se. In an amorphous material, the mean free paths are very small and determined by the disorder of the material rather than by thermally-induced fluctuations in the atomic positions, so that the approximate temperature independence is not surprising. In liquid hydrocarbons, r_0 is predicted to decrease slowly as T increases due to the change in the amount of excess energy that must be lost by a hot carrier to achieve thermalization due to variation in kT with temperature [29]. This also gives the wrong sign for the variation of r_0 with temperature in a-Se. In the other materials mentioned, there are no accurate experimental determinations of r_0 to compare with theory.

The most likely cause of the bulk of the observed variation of r_0 with temperature in a-Se is the variation of the band gap with temperature, reported to be $\sim -7 \times 10^{-4}$ eV/K [19]. As the temperature decreases, the band gap becomes wider and a carrier excited by a photon of a particular energy must lose a smaller amount of excess kinetic energy to thermalize, thus causing a decrease in r_0. Using the slope of the curve showing r_0 as a function of photon energy (see Fig. 7.8) to determine the decrease in r_0 for the 50 meV increase in band gap in going from 294K to 223K, the calculated decreases in r_0 for the exciting wavelengths shown in Table 7.1 range from 5 to 10 percent, averaging about 8 percent. The counteracting increase in r_0 over the same temperature range due to the increase in necessary energy loss to achieve thermalization caused by the decrease in kT (~ 6 meV) is about 1 percent when calculated by the same method used for the band gap variation. This is in satisfactory agreement with the measured average decrease of 5 percent shown in the table.

(v) Temperature dependence of quantum efficiency

The activation energy of the photogeneration process has generally been determined by extrapolating the linear portions of QE *versus* $\sqrt{E}$ plots for a number of temperatures back to zero field and thus finding a zero-field activation energy [19]. This has then been compared with the prediction of Poole—Frenkel which gives a simple activated dependence at zero field (which is identical to the zero-field activation energy for Onsager theory).

For the Onsager theory, experiment and theory are most meaningfully compared at an electric field for which the QE can accurately be measured directly since the theoretical QE is not linear on a $\sqrt{E}$ plot [8]. Figure 7.7 shows both the theoretical and experimental values of the activation energy as a function of exciting photon energy for an applied electric field of 7×10^4 V/cm. The theoretical values are calculated numerically from eqn. (7.13), assuming r_0 to be independent of temperature. The agreement between theory and experiment is quite good except for the one point corresponding to the lowest photon energy. At such a low photon energy, r_0 is small (~ 1.2 nm), corresponding to only a few interatomic distances, and this error may indicate the beginning of the breakdown of the assumption of diffusive motion at such small separations.

References pp. 297—302

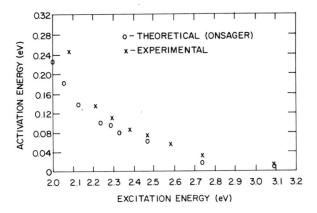

Fig. 7.7 Comparison of the theoretical (Onsager) and experimental values of the activation energy of quantum efficiency for holes in a-Se as a function of photon energy at an applied electric field of 7×10^4 V/cm. (From Pai and Enck [8].)

(vi) Photon energy dependence of thermalization distance

The variation of the thermalization distance r_0 with exciting photon energy is plotted in Fig. 7.8. The assumption that the variation in r_0 with the

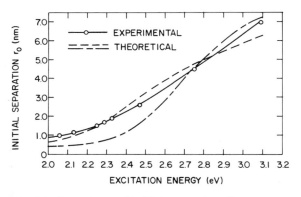

Fig. 7.8. Experimental initial separation distances as a function of photon energy for a-Se. Figure also shows expected variations of initial distances using simple models. (From Pai and Enck [8].)

energy of the exciting radiation occurs as a result of the variation of the initial kinetic energy of the carriers upon excitation is a reasonable one and has been applied for explaining photogeneration in liquid hydrocarbons and anthracene. In the treatment of thermalization in the application of Poole—Frenkel theory to a-Se, the motion of the hot carrier was assumed to be diffusive so that the thermalization distance $r_0 = \sqrt{Dt_0}$, where t_0 is the thermalization time. The rate of energy loss was assumed independent of excess energy and equal to $h\nu_{ph}^2$, where ν_{ph} is a typical phonon frequency. Assuming a band gap of $\mathcal{E}_g$, the amount of energy that must be lost to achieve thermalization is $h\nu - \mathcal{E}_g + e^2/4\pi\kappa\kappa_0 r_0 + eEr_0$.

Values of r_0 from eqn. (7.10) as functions of photon energy are plotted in Fig. 7.8 for two sets of the parameters ν_{ph}^2/D and $\mathcal{E}_g$. The dashed curve results when $\nu_{ph} = 1.34 \times 10^{13}\sqrt{D}$/second and $\mathcal{E}_g = 2.36$ eV. The second broken curve results when $\nu_{ph} = 9.3 \times 10^{12}\sqrt{D}$/second and $\mathcal{E}_g = 2.60$ eV. The fit is qualitatively correct, but indicates that a better theory for the motion of hot carriers may be necessary to adequately explain the experimental results.

(vii) Electron quantum efficiency

The Onsager theory of dissociation predicts equal quantum efficiency of generation for both holes and electrons, since the escape of the hole from the coulomb field of the electron also implies the escape of the electron from the influence of the hole. However, the bulk deep-trapping lifetime of electrons is very sensitive to the presence of certain impurities (see Section 7.4.5); hence, extreme care must be exercised to ensure that the electron injection efficiencies are not affected by bulk trapping. Figure 7.9 shows the comparison between injection efficiencies of electrons and holes at room temperature and at exciting wavelengths of 450 and 540 nm [8]. The

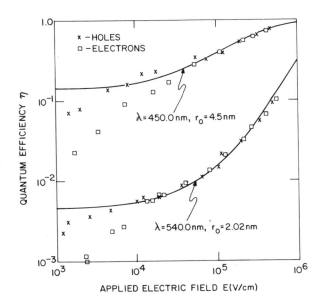

Fig. 7.9. Comparison of photoinjection efficiencies of holes and electrons for a-Se as functions of the applied electric field at exciting wavelengths of 450 nm and 540 nm. The solid lines are the Onsager dissociation efficiencies for $\phi_0 = 1$, $r_0 = 4.5$ nm and 2.02 nm. (From Pai and Enck [8].)

References pp. 297—302

theoretical Onsager dissociation efficiencies for appropriate r_0 are also shown in the figure. The hole and electron quantum efficiencies of injection are the same in the region where the measured injection efficiencies are equal to those predicted by the Onsager theory of dissociation. They are different, however, in the low-field region where the generated carriers diffuse back to the surface and are subject to loss by surface recombination. The nonequality between the electron and hole injection efficiency is probably due to difference in trapping and recombination kinetics of the surface for electrons and holes. One other interesting feature of Fig. 7.9 is that the departure from the predicted Onsager efficiency occurs at a higher field for 450-nm excitation. This observation is consistent with the picture of surface effects causing the low-field drop-off, since the absorption coefficient is larger for 450-nm excitation and the free-carrier generation takes place closer to the surface.

(viii) Discussion

In the literature, the validity of applying the Onsager theory of dissociation to photogeneration in materials such as anthracene [2], liquid hydrocarbons [30], and crystalline As_2S_3 [3] is often checked by consideration of the slope-to-intercept ratio. This is done by taking the low-field approximation to eqn. (7.15) given by

$$\phi(r_0, E, T) \simeq \phi_0 \exp\left(-r_c(T)/r_0\right)\{1 + [(e/kT)\tfrac{1}{2}r_c E]\}. \tag{7.16}$$

The carrier yield *versus* electric field plot in this region is linear and should have a slope-to-intercept ratio of $(e/kT)\tfrac{1}{2}r_c$. Good agreement with the experimentally measured slope-to-intercept ratio has recently been found in the case of anthracene [2]. Using a dielectric constant of 6.3 for amorphous selenium, it is seen from eqn. (7.15) that the E^2 term has a value of 10% of the E term at a field of 1.7×10^4 V/cm; hence, the approximation of eqn. (7.16) is valid for QE data points at fields below $\sim 10^4$ V/cm. However, as can be seen from Fig. 7.1, the fit between the measured injection efficiency for amorphous Se and the Onsager expression deteriorates below 10^4 V/cm, and the slope-to-intercept test will not be useful until better data are available at low fields.

The mechanism of field dependence of carrier yield from energetic electrons in liquids is relatively complex and a number of different initial distributions (*e.g.*, Gaussian, exponential, power law) of thermalized carriers have been required to fit the experimental results [6, 7]. This results in a decreased dependence of $\phi(E)$ on E compared to the delta function distribution [6] and provides additional arbitrary fitting parameters. In the charge transfer complex of polyvinylcarbazole and 2,4,7—trinitro—9—fluorenone (PVK—TNF) [5] and in PVK [4], the limited range of thermalization distances and fields for which QE data are available have made convincing, quantitative comparison between experiment and theory difficult.

In amorphous selenium, however, there is excellent quantitative agreement with Onsager theory using a delta function initial distribution over very wide ranges of thermalization distance (or photon energy) and electric field. This material thus provides an ideal model system for the application of Onsager theory and the success of the theory in a-Se should encourage its use in other low mobility materials.

(ix) Vitreous selenium

Photogeneration has also been studied in vitreous selenium samples prepared by quenching liquid selenium [31]. Excitation with light of photon energy greater than the band gap gives results similar to those obtained for evaporated a-Se. However, light of lower photon energy which is weakly absorbed produces carrier pairs uniformly throughout the sample and the photogeneration efficiency for this process is reported to be independent of electric field and temperature. If a sample has never been exposed to air or light of photon energy greater than the band gap, only this bulk generation process is observed for weakly-absorbed light. After exposure, however, a damaged layer a few μm thick is formed at the surface and a field- and temperature-dependent component of photogeneration is observed at the surface. It is not known whether this source of carriers is photogeneration in the damaged region or photoemission from the electrodes. The resistivity of the damaged region is $\sim 10^{14}$ ohm.cm while the bulk resistivity is $\sim 10^{10}$ ohm.cm.

The spectral distribution of the QE for this material is shown in Fig. 7.10. A constant times the QE is plotted in the low energy peak since the absorption coefficient of the sample was unknown in this region. The relative dependence of the QE on field and temperature for several photon energies is shown in Fig. 7.11. Curve 4 in Fig. 7.11 was taken at the photon energy of the long wavelength peak shown in Fig. 7.10 and is essentially independent of electric field and temperature.

The field independence of the QE is not a convincing argument for a fundamentally different photogeneration process since bulk generation in evaporated a-Se has been shown to be essentially field independent over the field range shown for curve 4 (see Figs. 7.3 and 7.4). (The field dependences of Curves 1 and 2 are probably due to surface recombination.) However, the temperature independence of the QE indicates that no thermal dissociation process is important and that a carrier pair bound by their coulomb attraction is not formed.

The most likely reason for this is that the mean free path for a hot carrier in vitreous selenium may be much larger than in evaporated a-Se. The thermalization distance would thus be greater than the critical radius at which the coulomb energy is equal to $\sim kT$ and the thermalized carriers would be free. Since vitreous selenium presumably has a much higher molecular weight than evaporated a-Se and is thus likely to be in a more ordered state

References pp. 297—302

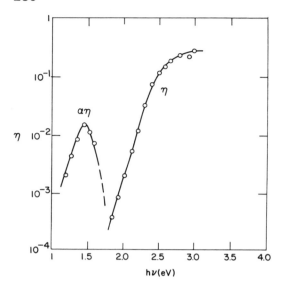

Fig. 7.10. The spectral distribution of the quantum efficiency η and the product of the quantum efficiency and the absorption coefficient α in vitreous Se. (From Juska et al. [31].)

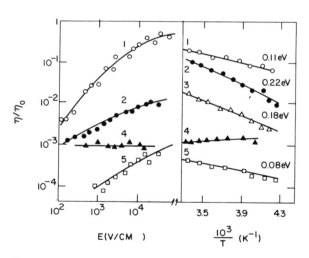

Fig. 7.11. The relative dependence of the quantum efficiency of holes in vitreous Se on field and temperature for the photon energies: ○ 3 eV, ● 2.24 eV, △ 2 eV, ▲ the spectral maximum, □ 1.38 eV (photoemission). (η_0 is chosen for each curve in an arbitrary and independent way.) (From Juska et al. [31].)

[32—34], a substantially larger mean free path is not unreasonable. This higher molecular weight, more ordered state is presumably chemically or photochemically degraded by exposure to energetic photons or oxygen.

7.2.2 Amorphous As_2Se_3 and As_2S_3

Field-dependent photogeneration has been reported in both a-As_2Se_3 [9] and a-As_2S_3 [10]. In a-As_2Se_3, xerographic discharge and electroded time-resolved current drift techniques both gave approximately $\eta \propto E^{0.5}$ above about 4×10^4 V/cm and $\eta \propto E$ below that field for 400-nm illumination [9] (similar to that originally reported for a-Se). More recent results [35] for photoinjection efficiency as a function of field, with exciting wavelength as the parameter, are shown in Fig. 7.12. At 650 nm the light is absorbed in a small fraction of the thickness of the 48-μm-thick sample ($\alpha \gtrsim 10^4$ cm^{-1}). At 700 nm the light penetrates the sample substantially ($\alpha \lesssim 10^3$ cm^{-1}) [36] (see ref. 40, p. 244). The field and wavelength dependence of QE is similar to, but much weaker than, that observed in a-Se over a similar range of absorption coefficients (related to excess kinetic energy) and electric fields (compare Fig. 7.1). This is due partly to the higher dielectric constant ($\kappa \approx 9$) of

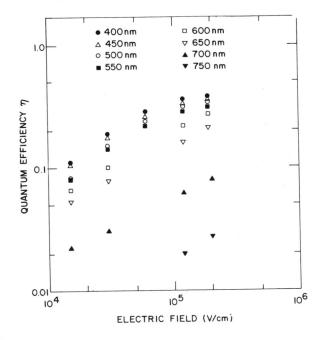

Fig. 7.12. Absolute quantum efficiency of photoinjection of holes in a-As_2Se_3 versus electric field with exciting wavelength as parameter. Sample thickness is 48 μm. (From Scharfe [35].)

References pp. 297—302

As$_2$Se$_3$ and the resulting lower value of r_c. However, the relatively high value of QE, even for relatively weakly-absorbed light (and thus very little excess kinetic energy), suggests that the thermalization process may be significantly different in a-As$_2$Se$_3$ and a-Se.

In a-As$_2$S$_3$ the carrier (hole) range is so short that fields high enough to make the hole range greater than the sample thickness can only be achieved in xerographic measurements. Field-dependent QE for 400-nm light were observed which fit the expression [$A \exp(\beta E^{1/2}/kT)$] ($\beta \simeq 1 \times 10^{-24}$ (Jm$^{1/2}$/V$^{1/2}$)) for electric fields $\gtrsim 3 \times 10^5$ V/cm and for sample thicknesses $\lesssim 10\,\mu$m. The maximum QE observed was ~ 0.6 [10]. Very similar results are obtained for a-Se at high fields with 450-nm light [18].

Only a limited amount of data is available on the variation of QE with exciting wavelength for a-As$_2$S$_3$, but the QE does decrease monotonically with decreasing photon energy [10], even for very high absorption coefficients ($\alpha > 10^5$ cm^{-1}), just as expected for an Onsager-type field-dependent photogeneration process.

7.3 STEADY-STATE PHOTOCONDUCTIVITY

7.3.1 Introduction

The general topic of steady-state photoconductivity has been covered in a number of earlier works [37, 38, 39] which provide good introductions to this field. Photoconductivity in amorphous chalcogenide materials has been discussed extensively in a recent book concerning the electronic properties of noncrystalline materials [40] and in a recent review article concerning photoelectric phenomena in amorphous chalcogenide semiconductors [41]. Excellent reviews of internal photoemission from metals into chalcogenides [42] and of the electronic properties of chalcogenides [43] are also available. We will, therefore, concern ourselves primarily with relatively recent developments and refer the reader to the sources cited for detailed discussions of earlier work. The focus of much of the recent work in this field has been the very similar temperature dependence of photoconductivity observed in a wide range of amorphous chalcogenide materials together with some additional experiments designed to shed further light on the mechanisms responsible for this characteristic temperature dependence. This work will be discussed in the following section, while other aspects of photoconductivity in these materials will be covered in the remainder of the introductory section.

All of the photoconductivity measurements to be discussed use the coplanar geometry for electrodes, allowing the entire sample area between electrodes to be uniformly illuminated over any plane parallel to the electrodes, independent of sample thickness and whether the light is weakly or strongly absorbed. Injecting contacts are also assumed. (Effects due to blocking contacts and nonuniform illumination in the "sandwich" geometry are discussed in the review article cited [41].)

The shape of the spectral distribution of photoconductivity in amorphous chalcogenides shows substantial differences from material to material and, in some cases, from sample to sample of the same material. In many materials, particularly when they are in the form of bulk-quenched glasses, the photocurrent per incident photon peaks at a photon energy near the absorption edge and falls at both higher and lower photon energy. Materials in which this has been observed include bulk glasses of the As_2Se_3—As_2Te_3 [41, 44], Se—As—Te [41], and Tl_2Se—As_2Te_3 [41] systems, evaporated amorphous films of the systems Sb_2S_2—Bi_2S_3 [41] and Ge—S—As—Te [45] and of As_2S_3 [41], and the bulk glasses As_2S_3 [41] and As_2Se_3 [41, 46—48]. A similar, although much more rapid, drop in photoconductivity with photon energy above the band gap is observed in crystalline materials and has been shown to be due to surface recombination effects [49]. This drop in photosensitivity at high photon energies was observed to disappear in thin amorphous films of Ge—S—As—Te when care was taken to remove surface contaminants [50], indicating that surface recombination can be important in amorphous chalcogenides. Additional evidence for the importance of surface recombination under suitable circumstances can be found in the low-field photogeneration experiments in amorphous selenium [24, 26] (see Section 7.2.1). In As_2Se_3 glass, however, the drop at high photon energies was shown to be due primarily to the sublinear dependence of the photocurrent on light intensity in this spectral region [46]. In general, the importance of surface recombination in many of the bulk glasses is unknown.

In most thin ($\sim 1~\mu m$) evaporated films of amorphous chalcogenides, however, the photocurrent per incident photon remains relatively constant for photon energies greater than the band gap, indicating that the surface recombination properties are similar to those of the bulk. The existence of a constant photosensitivity over a substantial wavelength range has been used as evidence that the quantum efficiency is unity in this spectral region [44, 50]. Examples of this behavior include GeTe [51], As—Se—Te [44, 52], Te_2AsSi [52], As_2Se_3 [52, 53], and Ge—Si—As—Te [50].

These results indicate that the relative importance of surface effects on photoconductivity in amorphous chalcogenides is much less than in crystalline materials. This is probably due to a combination of two circumstances, the relative importance of which will vary from sample to sample. First, since the degree of order in the bulk of an amorphous material is much less than in a crystalline material, the additional disorder represented by the discontinuity at the surface should be relatively less important. Second, the diffusion length in amorphous materials will be substantially less than in crystalline materials due to the low mobilities. (See the discussion of thermalization lengths in Section 7.2.1.) This will cause the influence of the surface

References pp. 297—302

to be restricted to a much thinner layer than in crystalline materials. The apparently greater sensitivity of the spectral dependence of photoconductivity to surface effects in bulk glasses than in evaporated films is probably related to the first circumstance noted. Evaporated films generally have less local order than bulk glasses due to the smaller atomic and molecular mobility during formation and to the larger probability of exposure to agents that may damage the molecular structure (for example, light, oxygen) during evaporation. This latter effect has been demonstrated to be important with respect to the photoconducting properties of amorphous selenium, where a damaged layer which appears identical to the bulk of the evaporated film can be formed on the surface of the bulk glass by exposure to energetic photons or oxygen [31]. (See also refs. 32, 33, 34.)

The drop in photosensitivity for photon energies less than the absorption edge is usually ascribed entirely to the decrease in optical absorption and the photoconductivity is frequently described as having the same spectral dependence as the optical absorption coefficient. (The optical absorption coefficient is usually exponential (Urbach rule) over most of the region in which it is rapidly changing near the absorption edge in these chalcogenide materials.) In order to check these ideas, however, the absolute quantum efficiency must be compared with the absorptance, not the absorption coefficient (see Fig. 7.2). This has been done for a number of As—Te bulk glasses [54] and it was determined that the drop in quantum efficiency occurs at a photon energy about 0.1 eV higher than the drop in optical absorptance. (The possibility that this drop was a mobility effect due to a change in transport from extended states to hopping was excluded by the magnitude, temperature dependence, and photon energy dependence of the photocurrent.) The drop in quantum efficiency per absorbed photon was ascribed to the formation of excitons by absorptions between pairs of localized states in band tails which extend ~ 0.1 eV from the band edges. The excitons are converted, with less than unity efficiency, to free carriers in the extended states by strong random microfields [54, 55]. These microfields have an r.m.s. magnitude of $\sim 2 \times 10^6$ V/cm and are attributed to density or composition fluctuations. Due to the weak temperature dependence of the effect, the dissociation of the exciton is thought to be due to field-assisted tunnelling.

Field-dependent photogeneration, as seen in amorphous selenium (see Section 7.2.1), would not be observable in the As—Te glasses for this model, since the largest achievable external field is substantially smaller than the microfields already present. This suggests that the local fields are substantially lower in the elemental amorphous chalcogenides, which have primarily positional disorder, than in the multicomponent glasses, which have compositional disorder as well as positional disorder.

7.3.2 Temperature dependence

The temperature dependences of the steady-state photoconductivity $\Delta\sigma$ in a wide variety of chalcogenide materials (As_2Te_3 [56—60], As_2SeTe_2 [44, 61, 62], As_2Se_3 [47, 48, 56], Ge—Se—As—Te [45, 50], GeTe [51, 63], $Ge_{15}Te_{81}Sb_2S_2$ [64, 65], $Si_{11}Ge_{11}As_{35}P_3Te_{40}$ [64, 65], $Ge_{16}As_{35}Te_{28}S_{21}$ [64, 65], Ge_3SeTe_2 [44], (Ge—Pb—Sn)Te [66], Se_2Te_2As [67], As—Te system [54], As—Te—Ge system [68], theory [69—72]) as well as in amorphous silicon deposited from silane [73] (see Chapter 6) exhibit essentially identical behavior. For constant light intensity the photoconductivity is a maximum at some characteristic temperature T_{max} which marks the transition between two regimes of recombination. At temperatures lower than T_{max} (LT regime), bimolecular recombination dominates at sufficiently high light intensities giving rise to a square root dependence of photocurrent on light intensity. At higher temperatures (HT regime) the dependence of photocurrent on light intensity is linear, indicating that monomolecular recombination is dominant. The photoconductivity $\Delta\sigma$ is always less than the dark conductivity σ_d in the HT regime.

For sufficiently low light intensities a linear dependence of $\Delta\sigma$ on light intensity has usually been observed over the entire temperature range. Both above and below T_{max} the photocurrent depends exponentially on $1/T$. When an exponential dependence of a quantity on inverse temperature is observed, the activation energy $\mathcal{E}$ is defined by writing the temperature dependence as $\exp(-\mathcal{E}/kT)$. The activation energy in the LT regime is $\mathcal{E}_{ph}^{LT}$ (>0), and in the HT regime is $\mathcal{E}_{ph}^{HT}$ (<0). At very low temperatures (VLT regime) the photocurrent depends linearly on light intensity and either becomes temperature-independent or varies exponentially with $1/T$ [47, 48, 56—58, 73]. In the latter case, the activation energy $\mathcal{E}_{ph}^{VLT}$ (>0) is smaller in absolute magnitude than the activation energy in the LT regime.

A typical plot of photoconductivity *versus* $1/T$ for these materials is shown in Fig. 7.13 with intensity as a parameter. The dark conductivity is also included and is observed to vary exponentially with $1/T$ with an activation energy $\mathcal{E}_\sigma$ [$\sigma_d = \sigma_0 \exp(-\mathcal{E}_\sigma/kT)$]. When $\Delta\sigma$ varies as the square root of light intensity in the LT region, the value of T_{max} must shift to lower temperatures with decreasing light intensity. The apparent activation energy for this shift can be shown to be $\mathcal{E}_{max} = \mathcal{E}_{ph}^{HT} + 2\mathcal{E}_{ph}^{LT}$. At lower light intensities, where $\Delta\sigma$ is linearly dependent on light intensity in the LT region, the value of T_{max} is independent of light intensity. The complementary data of $\Delta\sigma$ *versus* generation rate G for another typical chalcogenide film is shown in Fig. 7.14 (see Fig. 7.16 for $\Delta\sigma$ *versus* $1/T$ for this material). The linear and square root intensity dependences in the three temperature regimes are clearly illustrated.

The temperature dependence of $\Delta\sigma$, particularly the exponential regions above and below T_{max}, places severe restrictions on the possible energy

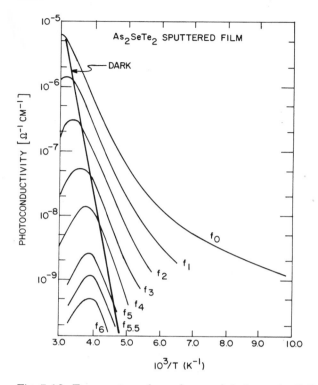

Fig. 7.13. Temperature dependence of dark conductivity and photoconductivity for different photoexcitation intensities in thin-film $As_2(Se_{1/3}Te_{2/3})_3$. White-light illumination was used with a peak wavelength at 1.1 μm. The incident photon flux is $10^{18}\,cm^{-2}\,s^{-1}$ evaluated at 1.1 μm. The subscripts on the f's represent approximate optical densities of neutral-density filters. Actual values of fractional transmission are $f_0 = 1$, $f_1 = 0.13$, $f_2 = 1.45 \times 10^{-2}$, $f_3 = 1.66 \times 10^{-3}$, $f_4 = 2.97 \times 10^{-4}$, $f_5 = 2.02 \times 10^{-5}$, $f_{5.5} = 7.35 \times 10^{-6}$, $f_6 = 3.15 \times 10^{-6}$. (From Arnoldussen et al. [44].)

distribution of recombination centers in these materials. Continuous distributions of recombination centers (*e.g.* uniform or exponentially decreasing from the band edges) all predict a maximum in the temperature dependence but are incapable of producing the two exponential regions and the square root light intensity dependence in the LT region [64, 65, 69, 71]. A "steep" exponential distribution (one which falls off more rapidly than the Boltzmann occupation function) is the most successful, predicting the HT exponential region and the square root light intensity dependence [65]. This suggests that the density of effective recombination centers must be restricted to bands of energies $\mathcal{E}_v^*$ above the valence band mobility edge and $(\mathcal{E}_c - \mathcal{E}_c^*)$ below the conduction band mobility edge, where $\mathcal{E}_c$ is the energy of the conduction band mobility edge. Additional recombination centers may be necessary near the middle of the gap to explain the VLT results and the LT, low light intensity results [47, 48, 65].

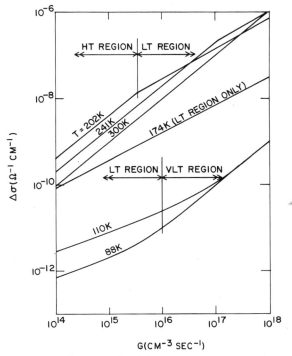

Fig. 7.14. Photoconductivity $\Delta\sigma$ versus generation rate for $Se_{40}Te_{40}As_{20}$ glass with temperature as parameter. (From Taylor and Simmons [67].)

A model for the effective recombination center density incorporating these features is shown in Fig. 7.15. Conduction is usually assumed to occur in band or nonlocalized (B) states above the conduction mobility edge or below the valence mobility edge. In the spirit of the Cohen, Fritzsche, and Ovshinsky (CFO) model of amorphous materials [74], the localized (L) states extending into the gap from the conduction band have generally been identified as conduction-band-like (C) states, neutral when empty of electrons. Those nearer the valence band are regarded as valence-band-like (V) states, neutral when filled with electrons. Energies are measured from the valence mobility edge and the conduction band mobility edge is located at $\mathcal{E}_c$. Thus $\mathcal{E}_c = \mathcal{E}_g$, the thermal band gap. The recombination transitions considered are shown in Fig. 7.15(b). Transitions 1 and 4 correspond to the normal band—local (B—L) processes considered in a Shockley—Read [75] analysis of semiconductor statistics. Transitions 2 and 3 correspond to transitions between pairs of localized states (L—L transitions). Consideration of these L—L processes has been found essential, in at least some amorphous chalcogenides, to consistently explain the experimental results.

The critical features of the above model are the sharp, discontinuous

References pp. 297—302

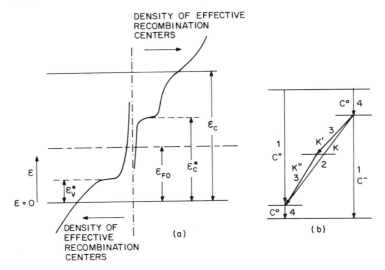

Fig. 7.15. (a) Schematic energy-level diagram for the proposed model. Localized states extend only into the mobility gap from the conduction and valence edges. With increasing energy above the valence edge, the density of effective recombination centers (see text) decreases sharply as the energy exceeds $\mathcal{E}_v^*$ for V states, and increases sharply as the energy exceeds $\mathcal{E}_c^*$ for C states. The equilibrium Fermi level lies $\mathcal{E}_{F0}$ above the valence edge; since the conductivity is p-type, $\mathcal{E}_{F0} \leq \frac{1}{2} \mathcal{E}_g$. (b) Typical transitions between individual states in the distribution shown in (a). (From Arnoldussen et al. [65].)

changes in effective recombination center density at $\mathcal{E}_v^*$ and $\mathcal{E}_c^*$. A distribution of recombination centers with these features can arise in a number of ways, such as (1) the distribution of L states can be smoothly varying but the recombination coefficients may vary in such a way that the product of the density of L states and the recombination coefficient (which gives the effective density of recombination centers) changes abruptly at $\mathcal{E}_c^*$ and $\mathcal{E}_v^*$; (2) the L states may be largely confined to the regions between $\mathcal{E}_v^*$ and the valence mobility edge and $\mathcal{E}_c^*$ and the conduction mobility edge, except perhaps for some states near $\mathcal{E}_{F0}$, the Fermi level at absolute zero; (3) the density of L states may peak at $\mathcal{E}_v^*$ and $\mathcal{E}_c^*$ giving effectively discrete recombination levels at $\mathcal{E}_v^*$ and $\mathcal{E}_c^*$. Possibility (1) has been developed further in connection with this model [61, 62] and the concept of a "recombination gap" corresponding to ($\mathcal{E}_c^* - \mathcal{E}_v^*$) has been proposed. Possibility (3) has generally been discounted for amorphous materials, but exactly this type of distribution of states has been observed in amorphous silicon, where the peaks in the density of states have been identified with structural defects [76, 77] (see Chapter 6).

Using the distribution of effective recombination centers shown in Fig. 7.15 with suitable relative values of the recombination coefficients and assuming that the states at $\mathcal{E}_v^*$ and $\mathcal{E}_c^*$ are in approximate thermal equilibrium with the nearest band (except in the VLT region), with almost all of

TABLE 7.2

Theoretical activation energies of photoconductivity and photoconductivity-decay time for the various recombination paths and transport mechanisms under consideration

Recombination (transport)	$\mathcal{E}_{ph}^{HT}$	$\mathcal{E}_{ph}^{LT}$	$\mathcal{E}_{\tau}^{HT}$	$\mathcal{E}_{\tau}^{LT}$
B–L (band)	$-(\mathcal{E}_c^* - \mathcal{E}_{F0}) + \mathcal{E}_\mu$	$\tfrac{1}{2}\mathcal{E}_v^* + \mathcal{E}_\mu$	$-(\mathcal{E}_c^* + \mathcal{E}_v^* - \mathcal{E}_{F0})$	$-\mathcal{E}_v^*/2$
L–L (band)	$-(\mathcal{E}_c^* - \mathcal{E}_v^* - \mathcal{E}_{F0}) + \mathcal{E}_\mu$	$\mathcal{E}_v^* + \mathcal{E}_\mu$	$-(\mathcal{E}_c^* - \mathcal{E}_{F0})$	zero
L–L (hopping)	$-(\mathcal{E}_c^* - \mathcal{E}_{F0}) + \mathcal{E}_\mu$	$\mathcal{E}_\mu$	$-(\mathcal{E}_c^* - \mathcal{E}_{F0})$	zero

the excess carriers in the localized states, expressions for the steady-state photoconductivity $\Delta\sigma$ and the decay time constant τ from the steady state when the light is turned off can be calculated [44, 57–61, 64, 65, 72]. Expressions for the thermal activation energies determined from these calculations for the high-temperature photoconductivity ($\mathcal{E}_{ph}^{HT}$), low-temperature photoconductivity ($\mathcal{E}_{ph}^{LT}$), high-temperature photoconductivity decay time constant ($\mathcal{E}_{\tau}^{HT}$), and low-temperature photoconductivity decay time constant ($\mathcal{E}_{\tau}^{LT}$) are shown in Table 7.2. The rows of the table are labeled with the assumed recombination paths and, in parentheses, transport mechanisms. It is assumed that the materials are p-type ($\mathcal{E}_{F0} \leq \tfrac{1}{2}\mathcal{E}_c$, $\mathcal{E}_v^* \leq (\mathcal{E}_c - \mathcal{E}_c^*)$ for approximately equal hole and electron mobilities) and that conduction takes place through extended states below the valence mobility edge, except that, in the last row of the table, hopping conduction through states at $\mathcal{E}_v^*$ is assumed. Activation energies for systems where only band–local (B–L, transitions 1 and 4 of Fig. 7.15) recombination transitions are allowed and for systems where local–local (L–L, transitions 2 and 3 of Fig. 7.15) recombination transitions are allowed in addition (for both extended-state and hopping conduction) are included in the table. The activation energy of the mobility $\mathcal{E}_\mu$ [$\mu = \mu_0 \exp(-\mathcal{E}_\mu/kT)$] has been found to be non-zero for some of the chalcogenide glasses [59, 60, 64, 65]. In the last row of the table, $\mathcal{E}_\mu$ represents the hopping activation energy. In the other rows, the mechanism has not yet been identified. In all cases, the mobility activation energy merely adds an additional multiplicative term $\exp[-\mathcal{E}_\mu/kT]$ to the temperature dependence of the steady-state photoconductivity.

For the model shown in Fig. 7.15 and for temperatures such that, in the dark, most thermally-excited carriers are in states near $\mathcal{E}_c^*$ and $\mathcal{E}_v^*$, so that the population of these states control charge neutrality, the following relationship between $\mathcal{E}_c^*$, $\mathcal{E}_v^*$, and $\mathcal{E}_{F0}$ holds

$$(\mathcal{E}_c^* + \mathcal{E}_v^*) = 2\mathcal{E}_{F0} + kT \ln\left(\frac{N_v^*}{N_c^*}\right) \tag{7.17}$$

where N_c^* and N_v^* are the densities of states at $\mathscr{E}_c^*$ and $\mathscr{E}_v^*$ respectively. In some of the papers referred to in connection with Table 7.2, this relationship, with $N_c^* = N_v^*$, has been used to rewrite the activation energies, making them appear different from those in the table.

In the HT region, the thermal population of all levels is large compared to the excess population due to the light. The temperature dependence of the recombination lifetime τ_0 is thus determined by the thermal population $\exp[-(\mathscr{E}_c^* - \mathscr{E}_{F0})/kT]$ of the recombination transition's terminal state $\mathscr{E}_c^*$. This is also the temperature dependence of the photoconductivity except in the L—L (band) case where the population of the non-transport states at $\mathscr{E}_v^*$ is determined by τ_0 and an additional temperature dependence of the carriers in the valence band is added by the Boltzmann distribution factor. Since it is determined by thermal populations, τ_0 is intensity independent and monomolecular recombination kinetics dominate.

For high light intensities in the LT region, the excess light-induced carrier population is large compared to the thermal population. By charge neutrality, the carrier populations n^* and p^* (equal to the excess carrier populations Δn^* and Δp^*) of the states at $\mathscr{E}_c^*$ and $\mathscr{E}_v^*$ must be equal (since only a very small fraction of the carriers are in the bands). In the two L—L cases, this results in bimolecular recombination statistics and temperature-independent values of p^* and τ_0. In the L—L (band) cases, a temperature dependence of the photoconductivity arises due to the Boltzmann distribution factor for the population of the band transport states. For B—L (band) transitions, the recombination process directly determines the population $p = \Delta p$ of the valence band, while Δp^*, which equals Δn^* and therefore determines τ_0, is related to Δp by the Boltzmann distribution factor. This results in a thermal activation energy for τ_0 and Δp of $\mathscr{E}_v^*/2$.

As the light intensity is lowered in the LT region, a transition to monomolecular recombination kinetics with an unchanged value of $\mathscr{E}_{\text{ph}}^{\text{LT}}$ is sometimes observed [64, 65, 67]. This is explained in the L—L case by assuming a density of L states near $\mathscr{E}_{F0}$ allowing transitions of Type 3 in Fig. 7.15 [65]. The recombination coefficients K' and K'' for these transitions are small compared to that for the Type 2 transitions (K), but when the combination of low light intensity and low temperature causes the populations at $\mathscr{E}_v^*$ and $\mathscr{E}_c^*$ to become small compared with the populations at $\mathscr{E}_{F0}$, the Type 3 transitions can dominate. These transitions give approximately monomolecular recombination kinetics and an approximately temperature independent τ_0 which causes p^* to be independent of temperature. The temperature dependence of $\Delta\sigma$ comes from the Boltzmann distribution factor for carriers in the transport band. These L states at $\mathscr{E}_{F0}$ can also explain the approximate temperature independence of the activation energy of the dark conductivity without requiring approximately equal densities of states at $\mathscr{E}_c^*$ and $\mathscr{E}_v^*$.

For B—L (band) transitions [59, 60], or L—L (hopping) transitions with hopping transport [67, 72], the question of a linear light intensity dependence at low intensities in the LT region has not been addressed. For the L—L (hopping) case, an approach similar to that discussed above should be adequate. For the B—L (band) case, there is some indication from the small amount of data available in this region that the transition to a linear light intensity dependence can be handled without postulating additional states [67].

In the VLT region, the electron and hole demarcation levels are closer to the conduction and valence mobility edges than almost all of the L states. Therefore, the populations of the L states are determined entirely by the recombination statistics and τ_0 for carriers in the bands is independent of temperature and light intensity. For band transport, this leads to temperature-independent photoconductivity which has a linear light intensity dependence. Hopping transport through L states under these circumstances would result in a thermally-activated mobility with activation energy $\mathcal{E}_\mu$.

The models introduced in Fig. 7.15 and Table 7.2 predict some simple relationships between the photoconductivity activation energies, the dark-conductivity activation energy, and the slope, $\mathcal{E}_s$, of the thermoelectric power versus T^{-1} plot. Since the thermoelectric power measures the number of free carriers, while the conductivity is additionally affected by the mobility, we have for extended-state transport through the valence band (assuming one carrier transport)

$$\mathcal{E}_s = \mathcal{E}_{FO} \tag{7.18}$$

$$\mathcal{E}_\sigma = \mathcal{E}_{FO} + \mathcal{E}_\mu = \mathcal{E}_s + \mathcal{E}_\mu. \tag{7.19}$$

Similarly, for hopping transport through the L states at $\mathcal{E}_v^*$ we have

$$\mathcal{E}_s = \mathcal{E}_{FO} - \mathcal{E}_v^* \tag{7.20}$$

$$\mathcal{E}_\sigma = \mathcal{E}_{FO} - \mathcal{E}_v^* + \mathcal{E}_\mu = \mathcal{E}_s + \mathcal{E}_\mu. \tag{7.21}$$

If, for a given material, the recombination mechanisms and transport mechanisms are the same for the HT and LT regions (that is, the photoconductivity activation energies belong to the same row of Table 7.2), it follows from the activation energies of Table 7.2 and eqns. (7.17), (7.19), and (7.21) that

$$2\mathcal{E}_{ph}^{LT} - \mathcal{E}_{ph}^{LT} = \mathcal{E}_\sigma + kT \ln\left(\frac{N_v^*}{N_c^*}\right). \tag{7.22}$$

If, however, the recombination path changes from B—L (band) in the HT region to L—L (band) in the LT region the relationship becomes

$$\mathcal{E}_{ph}^{LT} - \mathcal{E}_{ph}^{HT} = \mathcal{E}_\sigma - \mathcal{E}_\mu + kT \ln\left(\frac{N_v^*}{N_c^*}\right) \tag{7.23a}$$

References pp. 297—302

$$= \mathcal{E}_s + kT \ln \frac{N_v^*}{N_c^*} . \qquad (7.23b)$$

(In considering eqns. (7.22) and (7.23), we will assume $N_v^* = N_c^*$ unless they are otherwise specified.) When the L—L (hopping) mechanism is dominant on both sides of T_{max}, eqns. (7.22) and (7.23) will *both* be satisfied and

$$\mathcal{E}_{ph}^{LT} = \mathcal{E}_\mu . \qquad (7.24)$$

Thus, the temperature dependences of $\Delta\sigma$ and σ alone will usually not determine the recombination or transport mechanism. In most cases, additional measurements, such as thermoelectric power, field effect, and temperature dependence of photoconductivity decay time are necessary.

The currently available activation energies of photoconductivity, dark conductivity, thermoelectric power, and photoconductivity decay time of the various amorphous chalcogenides are shown in Table 7.3, which also includes ($\mathcal{E}_{ph}^{LT} - \mathcal{E}_{ph}^{HT}$) and ($2\mathcal{E}_{ph}^{LT} - \mathcal{E}_{ph}^{HT}$). The theory requires

$$(\mathcal{E}_{ph}^{LT} - \mathcal{E}_{ph}^{HT}) \lesssim \mathcal{E}_\sigma \lesssim (2\mathcal{E}_{ph}^{LT} - \mathcal{E}_{ph}^{HT}). \qquad (7.25)$$

The intermediate values of $\mathcal{E}_\sigma$ can only occur if there is a change in recombination mode as the temperature passes through T_{max} and if $\mathcal{E}_\mu \neq 0$.

Equation (7.25) is satisfied in Table 7.3 except for the last six materials from ref. 68 and, perhaps, As_2Se_3. For the first six lines of Table 7.3 (and probably, within experimental error, also for line 7), eqn. (7.22) is satisfied with $N_v^* \approx N_c^*$. For the material of line 6 (As_2Te_3), which is the only one of the first seven for which $\mathcal{E}_s$ is available, eqn. (7.23) is also satisfied, suggesting that L—L recombination and hopping transport dominate. This could also be the case for the other five materials, or extended-state transport could dominate with no change in recombination path on passing through T_{max}. According to the theory, the remaining materials that satisfy eqn. (7.25) (found in the next nine lines of Table 7.3) experience a change of recombination path [probably between B—L (band) and L—L (band)] on passing through T_{max}. This group includes materials which require $\mathcal{E}_\mu \approx 0$ and also materials which require $\mathcal{E}_\mu \neq 0$ to satisfy eqn. (7.22) or (7.23a). In the two cases where $\mathcal{E}_s$ is reported, eqn. (7.23b) is satisfied within experimental error with $N_v^* = N_c^*$.

These results have been used with the models discussed to calculate parameters such as $\mathcal{E}_\mu$, $\mathcal{E}_v^*$, $\mathcal{E}_c^*$, $\mathcal{E}_{FO}$, N_v, N_v^*, N_c^* and the various recombination constants for the materials studied [44, 57—59, 61, 64—67]. Usually $N_v^* = N_c^*$ is assumed for simplicity, although the case where $N_v^* \neq N_c^*$ has been considered [44, 64, 65]. Typical values of N_v^* are of the order of 10^{19} cm^{-3} eV. Theoretical curves of photocurrent *versus* $1/T$ computed from the B—L (band) model and the calculated parameters for $Si_{40}Te_{40}As_{20}$ [67] are shown by solid lines in Fig. 7.16. (In this reference, the value of $\mathcal{E}_{ph}^{LT}$ is erroneously written as $\mathcal{E}_v^*$ rather than $\mathcal{E}_v^*/2$. However, the correct value is

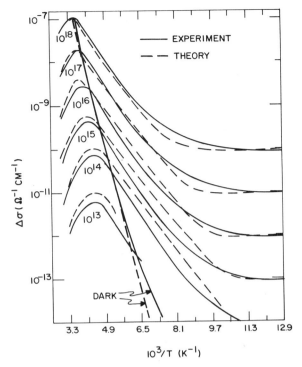

Fig. 7.16. Comparison of experiment and band—local recombination theory in $Se_{40}Te_{40}As_{20}$ glass for photoconductivity $\Delta\sigma$ versus T^{-1} with generation rate G as parameter. (From Taylor and Simmons [67].)

used in the calculation.) The dashed lines represent the experimental values [67] and indicate good agreement between theory and experiment over the entire temperature range with one set of parameters.

For the six materials in Table 7.3 which do not satisfy eqn. (7.25), $\mathcal{E}_\sigma \gtrsim 0.45\,\text{eV}$. It has been suggested [68] that for values of $\mathcal{E}_\sigma$ greater than about 0.4—0.5 eV localization becomes so severe in the shallow states that L—L recombination becomes insignificant and the dominant recombination process occurs from the band states through mid-gap recombination centers. However, there are a number of materials listed with $\mathcal{E}_\sigma > 0.5\,\text{eV}$ (including Te [78] with $\mathcal{E}_\sigma = 0.76\,\text{eV}$) which do satisfy eqn. (7.25). Whether a material is successfully described by one of the simple models associated with Fig. 7.15 is probably largely determined by the density and nature of deep states present since their effect is generally not considered in these models. Although the magnitude of the band gap may be expected to influence the formation of deep states, sample composition and preparation would be expected to be at least as important.

References pp. 297—302

250

TABLE 7.3

Experimental activation energies of photoconductivity and photoconductivity-decay time for the amorphous chalcogenides*

Composition	$\mathcal{E}_{ph}^{LT}$ (eV)	$-\mathcal{E}_{ph}^{HT}$ (eV)	$\mathcal{E}_\sigma$ (eV)	$\mathcal{E}_s$ (eV)	$\mathcal{E}_{ph}^{LT} - \mathcal{E}_{ph}^{HT}$ (eV)	$2\mathcal{E}_{ph}^{LT} - \mathcal{E}_{ph}^{HT}$ (eV)	$-\mathcal{E}_\tau^{HT}$ (eV)	$\mathcal{E}_\tau^{LT}$ (eV)
2(As$_2$Te$_3$)—As$_2$Se$_3$ [62]	0.18	0.13	0.47		0.31	0.49		
Ge(PbSn)Te glasses [66]	0.10	0.17	0.35		0.27	0.37		
	0.14	0.15	0.41		0.29	0.43		
	0.13	0.12	0.37		0.25	0.38		
Si$_{40}$Te$_{40}$As$_{20}$ [67]	0.135	0.21	≈0.48		0.34	0.48		
As$_2$Te$_3$ [59, 60]	0.14	0.16	0.44	0.30	0.30	0.44	0.26	0
Te [78]	0.24	0.34	0.76		0.58	0.82		
As$_2$Te$_3$ [56—58]	0.16 / 0.11[b]	0.21	0.42 / 0.26[b]		0.37	0.53	0.55	0.04
Ge$_{15}$Te$_{81}$Sb$_2$S$_2$ [65]	0.18	0.16	0.44	0.33	0.34	0.52		
Si$_{11}$Ge$_{11}$As$_{35}$P$_3$Te$_{40}$	0.20	0.19	0.44		0.39	0.59		
Ge$_{16}$As$_{35}$Te$_{28}$S$_{21}$	0.30	0.27[a]	0.57		0.57	0.87		
2(As$_2$Te$_3$)—As$_2$Se$_3$ [44]	0.25	0.20	0.55	0.41	0.45	0.70	0.08	
2(GeTe$_2$)—GeSe$_2$	0.19	0.38	0.66		0.57	0.76		
As$_{20}$Te$_{80}$ [68]	0.17	0.26	0.41		0.43	0.60		
Ge$_{20}$Te$_{80}$	0.21	0.28	0.47		0.49	0.70		
Si$_{20}$Te$_{80}$	0.18	0.24	0.49		0.42	0.60		
Ge$_{21}$Te$_{49}$Se$_{30}$	0.25		0.54					
As$_{21}$Te$_{49}$Se$_{30}$	0.24		0.52					
As$_5$Te$_{59}$Se$_{36}$	0.24	0.40	0.49		0.64	0.88		
As$_{10}$Te$_{56}$Se$_{34}$	0.23	0.45	0.48		0.67	0.91		
As$_{38}$Te$_{38}$Se$_{24}$	0.29	0.5	0.57		0.79	1.08		

TABLE 7.3 (continued)

Composition	ε_{ph}^{LT} (eV)	$-\varepsilon_{ph}^{HT}$ (eV)	ε_σ (eV)	ε_s (eV)	$\varepsilon_{ph}^{LT} - \varepsilon_{ph}^{HT}$ (eV)	$2\varepsilon_{ph}^{LT} - \varepsilon_{ph}^{HT}$ (eV)	$-\varepsilon_\tau^{HT}$ (eV)	ε_τ^{LT} (eV)
$As_5Te_{54}Se_{41}$	0.23	0.5	0.5		0.73	0.96		
$As_{10}Te_{51}Se_{39}$	0.21	0.5	0.46		0.71	0.92		
$As_{20}Te_{45}Se_{35}$	0.21	0.5	0.44		0.71	0.92		
As_2Se_3 [47, 48]	0.39		0.86				0.40	

[a] Not measured directly but inferred from measured ε_{ph}^{LT} and measured shift of T_{max} with temperature.
[b] There are two exponential regions for temperature below T_{max}. These values describe the behavior at the lowest measured temperatures.

* N.B. An additional extensive summary of energy parameters for photoconductivity in chalcogenides has recently been published by R.T.S. Shiah and R.H. Bube, J. Appl. Phys., 47 (1976) 2005.

7.3.3 Decay time

Additional evidence to determine the correct recombination path can be derived from the temperature dependence of the photoconductivity decay time. For L—L (band) or L—L (hopping) recombination [57—59]:

$$\mathcal{E}_\tau^{HT} = \mathcal{E}_{ph}^{HT} - \mathcal{E}_{ph}^{LT}, \tag{7.26}$$

while for B—L recombination [57, 58]

$$\mathcal{E}_\tau^{HT} = \mathcal{E}_{ph}^{HT} - 2\mathcal{E}_{ph}^{LT} + \mathcal{E}_\mu. \tag{7.27}$$

With respect to the two studies of As_2Te_3 reported in Table 7.3, $\mathcal{E}_\tau^{HT}$ for one [56—58] is approximately described by eqn. (7.27) while the other [59] is approximately described by eqn. (7.26). The corresponding recombination paths are the same as those deduced from the temperature dependence of the steady-state photoconductivity reported in these studies [56—59] (see discussion of temperature dependences in previous section), suggesting significant differences in recombination paths and transport mechanisms in the HT region for As_2Te_3 samples made by these two groups.

The value of $\mathcal{E}_\tau^{LT}$ in this material has been found to be zero [56—59] (corresponding to L—L (band) or L—L (hopping) recombination transport) or slightly positive [56—58] (≈ 0.04 eV) at lower temperatures (corresponding most closely to L—L (band) or L—L (hopping) recombination transport, since $\mathcal{E}_\tau^{LT} < 0$ for B—L (band) recombination). These results, plus the fact that $\mathcal{E}_{ph}^{LT}$ is essentially the same in these two studies, suggest that the recombination-transport mechanisms are the same in the LT region in the two sets of samples. A slightly positive value of $\mathcal{E}_\tau^{LT}$ has also been reported [44] for $2(As_2Te_3)$—As_2Se_3. The difference from the expected value of zero was ascribed to the effects of emptying of traps in the energy interval kT above the quasi-Fermi level for holes. If the release time of these traps is small compared to τ_0 (α-type traps [79]) for L—L (band) or L—L (hopping) recombination, τ is increased by approximately the ratio of the carrier density in the traps to the carrier density at $\mathcal{E}_v^*$. For B—L (band) recombination, the carrier density in the valence band replaces the density at $\mathcal{E}_v^*$ in this ratio. A Fermi-level analysis [80] then allows the L state distribution in the forbidden gap to be determined over a limited energy range. The results of this analysis [44] agree reasonably well with trap distributions determined from the photo-field-effect measurements on the same samples [44]. A density of states of the order of 10^{19} cm^{-3} eV^{-1} is indicated for this material, decreasing by about an order of magnitude for energies 0.1 eV closer to the valence edge.

Detailed analysis of the shapes of the photocurrent decay transients in the binary tellurium glasses $Te_{80}X_{20}$ (X = As, Ge, and Si) in the HT region have indicated that the decay time is determined by thermal release of minority carriers from traps to the conduction band [68]. The decays are composed of two or three distinct exponential portions, each of which is ascribed to the effects of a fairly discrete trapping level whose energy is determined from the temperature dependence of the associated portion of the photocurrent decay. It is argued that recombination proceeds through mid-gap recombination centers since the observed decay time constant is almost one order of magnitude shorter than the recombination lifetime predicted by the principle of detailed balance using only the shallow C and V states of the model in Fig. 7.15 [68]. Similar information on minority-carrier trapping levels in As_2Se_3 was obtained by analysis of the initial fast portion of the photocurrent decay in the part of the LT region near T_{max} [47, 48].

7.3.4 Radiative recombination

Further insight into the characteristics of L states in these materials should be available from an analysis of recombination radiation measurements. Radiative recombination has been reported for amorphous As_2Se_3–As_2Te_3 [81—84], As_2Se_3 [82—87], As_2S_3 [83—86, 88], $2As_2Te_3$–As_2Se_3 [89], and $(As_2Se_3)_x$–$(As_2Te_3)_{1-x}$ [84], but only in the case of $2As_2Te_3$–As_2Se_3 [61, 89] were photoconductivity, dark conductivity, and photoluminescence measured on the same samples. Typical values for the photoluminescence photon energy range from approximately 0.5—0.7 $\mathcal{E}_g$. The radiative transition has thus been viewed as either a B—L or L—L transition. An analysis [62] of some of this data using a simple version of the model of Fig. 7.15 found excellent agreement in $2As_2Te_3$–As_2Se_3 between the photoluminescence photon energy and ($\mathcal{E}_c^* - \mathcal{E}_v^*$) derived from photoconductivity experiments, thereby providing support for the L—L recombination model. For As_2Se_3, the agreement was not as good. Optical quenching and enhancement of photoluminescence by infrared light with photon energies $\sim \mathcal{E}_g/2$ have been reported [85] in amorphous As_2Se_3 and $(As_2Se_3)_x$–$(As_2Te_3)_{1-x}$ where $x \gtrsim 0.4$. These effects were interpreted in terms of a B—L recombination model. A limitation in comparing photoluminescence and photoconductivity results is that the photoluminescence experiments must typically be done at lower temperatures than the photoconductivity experiments and different recombination mechanisms may dominate in these two temperature regimes.

7.3.5 Summary

The results described in this section indicate that recombination in most of the chalcogenide glasses studied can generally be successfully explained in terms of the B—L (band) and/or L—L (band) [or L—L (hopping)] recombination models of Fig. 7.15 using shallow C and V states as recombination levels. Inclusion of a small number of mid-gap recombination levels may be necessary to explain the linear light intensity dependence at low light levels

References pp. 297—302

in the LT region. The correct recombination path cannot be determined from the temperature dependences of the steady-state photoconductivity and dark conductivity alone, although a change in recombination path as the temperature passes through T_{max} has been demonstrated in some materials. Measurements on some samples of the temperature dependence of the thermoelectric power, photoconductivity decay time, field effect, or photoluminescence have narrowed the choice of recombination paths and provided additional information on recombination levels and majority and minority carrier traps. Each of the recombination-transport paths of Table 7.2 was found to be dominant for some samples in some temperature ranges. Considerable differences in the dominant recombination mechanism were observed for different materials with some evidence, in one material, for different recombination paths for the same material prepared by different investigators.

The models discussed above have been found to be inappropriate for some chalcogenide glasses, probably due to the presence of large numbers of deep recombination centers and/or minority carrier traps. A more complete understanding of the densities and characteristics of the L states and the nature of the recombination processes in most of the chalcogenide glasses discussed now awaits the application of the complete set of measurements available to particular samples of the materials or to sets of samples prepared in identical, reproducible ways. An excellent example of this approach is described in Chapter 6 for amorphous silicon produced from silane.

7.4 ELECTRONIC TRANSPORT PROPERTIES

7.4.1 Introduction

This part of the chapter will be concerned with transport mechanisms of electronic charge generated by light. Results of dark-conductivity experiments will only be discussed as they bear on the transport of photogenerated carriers. Because of the vast number of chalcogenide glasses, including elemental glasses and binary, tertiary and quarternary alloys, a discussion of their electronic transport necessarily will have to be limited to a few representative examples. As prototype systems, we choose two extensively studied systems, namely the elemental glass amorphous selenium and the binary glass amorphous As_2Se_3. These two glasses exhibit, at least at room temperatures, electrical properties so different that their discussion will encompass most aspects of our present understanding of electronic transport in amorphous solids and, in particular, chalcogenide glasses. The main portion of the remainder of the chapter will, therefore, be devoted to these two prototype chalcogenides.

One of the most widely used methods to explore electronic transport in high-resistivity materials is the time-of-flight technique (Chapter 2) which because of its importance will be briefly discussed in the next section. Emphasis will be on its application to our model chalcogenides. The discussion of the transport properties will make extensive use of results from this type of experiment. More recently, photogenerated a.c.-conductivity has evolved as a valuable tool to study certain aspects of transport, and Section 7.4.7 will give a brief account of pertinent results.

Drift experiments on a-Se have commonly been interpreted using conventional semiconductor concepts such as extended states, drift mobility and shallow and deep traps. Similar studies on a-As_2Se_3 revealed drastically different transport properties and the concepts used to analyze the data for a-Se were challenged. Specifically, the meaning of the drift mobility parameter as derived from time-of-flight experiments was questioned. The significantly different transport properties of the two chalcogenides under discussion are best illustrated by some examples for hole transport. The rectangular shape of the room temperature transient signals in a-Se evidences little dispersion of the drifting hole sheet and, consequently, transit times are well determined. The hole drift mobility, $\sim 0.14 \, cm^2/V.s$, is thermally activated with $\sim 0.25 \, eV$ and electric field and hydrostatic pressure independent. Hole lifetimes with respect to deep traps are typically $10-50 \, \mu s$. In contrast, hole pulse shapes in a-As_2Se_3 strongly deviate from the ideal rectangular shape due to a wide dispersion of the carrier sheet in transit. This appears to be a feature in common with an increasing number of amorphous organic polymeric solids (see Chapter 8). A typical mobility for holes at a field of $\sim 10^5 \, V/cm$ is $\sim 10^{-5} \, cm^2/V.s$. It is thermally activated with $\sim 0.58 \, eV$ and dependent upon electric field, sample thickness (!) and hydrostatic pressure. Hole lifetimes with respect to deep traps can be of the order of a second.

In spite of these unusual properties, it appears that, at present, the understanding of (hole) transient conductivity in a-As_2Se_3 is more satisfactory than in a-Se although the elemental glass has been a subject of much earlier intensive investigation, starting with Spear's work in 1957. A large body of experimental information on a-As_2Se_3 can be explained in a consistent manner in terms of a theoretical model which conceives electronic transport as a stochastic process with a wide distribution of event times extending beyond the time range of experimental observation. This model successfully explains charge transport in a-As_2Se_3 without invoking extended states and thus differs distinctively from the multiple-trapping transport mechanism generally proposed for a-Se in which the carriers transit the sample in extended states, while being in rapid communication with a set of shallow traps below (above) the conduction (valence) band. Recent experiments on hole transient conductivity in a-Se, however, indicate that, at low temperatures, a stochastic process similar to that observed in a-As_2Se_3 occurs, but there is no indication that the transition from the well-defined high temperature transport to the dispersive low-temperature transport is paralleled by a significant change of the mechanism of carrier motion. Hence, in deviation from earlier interpretations, extended states again might not be involved. In view of these recent developments, we deviate from a chronological presentation and first discuss a-As_2Se_3 in Section 7.4.3 before we describe the results on a-Se in Section 7.4.4.

Electronic transport in chalcogenides is relatively insensitive to preparatory conditions and results from different laboratories obtained from liquid-quenched or evaporated materials are in reasonable agreement. This remarkable feature, which contrasts with the findings in tetrahedrally bonded amorphous solids (Chapter 6), will be exemplified by hole transport in a-Se in Section 7.4.4. Significant amounts of impurities are necessary ($> 0.1\%$) to influence the various transport parameters. By doping a-Se samples with increasing amounts of As, the progressive transition of the well defined transport in a-Se into the dispersive transport in a-As_2Se_3 has been extensively studied. Several recent papers on this subject will be summarized in Section 7.4.5. Experiments related to transport in a-As_2Se_3 and a-As_2Te_3 will be described in Section 7.4.6. Finally, Section 7.4.8 examines our present evidence for extended-state transport in some representative chalcogenides.

7.4.2 Time-of-flight experiment

(i) Experimental details

A general description of the time-of-flight technique has been presented in Chapter 2 to which we refer for references. In the following paragraphs, this technique will briefly be discussed in relation to its application to chalcogenides.

Drift experiments are most easily performed by injecting a thin unipolar carrier sheet at one end of the sample and measuring the time t_T (transit time) it takes for the sheet of charge to drift across the sample under the influence of a d.c. bias field. In chalcogenides, the sheet of charge is conveniently generated by a pulse of light (of duration much less than t_T) absorbed close to the surface of the specimen. The sample configuration typically consists of a film, of 1—100 μm thickness, evaporated onto a substrate electrode and covered with a suitable semitransparent top electrode to form a sandwich structure. Some experiments, like those under hydrostatic pressure, require unsupported films. These can easily be obtained by flexing the substrate onto which the amorphous film was evaporated and flaking off pieces of the glass. Time-of-flight experiments have also been reported using thin specimens obtained by polishing glass quenched from the melt. Since the absorption coefficient of chalcogenides is typically $\gtrsim 10^5 \, \text{cm}^{-1}$, the requirement for the absorption to occur within a fraction of the sample thickness can easily be established by choosing the wavelength of the incident light. The dielectric relaxation times in representative chalcogenides is much longer than typical transit times t_T, even at moderate electric fields

($\lesssim 1$ V/μm), so that charge screening is negligible and electrons and holes photogenerated at the surface can easily be separated. Hence, by choosing the polarity of the driving field, electron and hole transport can be studied separately. We will restrict our discussion to the small signal case in which the total injected charge σ per unit area is much less than CV and, therefore, the electric field is, to a good approximation, uniform across the film. V is the applied voltage and $C = \kappa \kappa_0/L$ is the sample capacitance per unit area. For experiments performed under space-charge conditions we refer to Chapter 2. In the small signal case the drifting charge sheet σ induces onto the electrodes a charge which is proportional to σ and the distance the sheet has moved. If the time constant of the external circuit is short compared to the transit time, $RC \ll t_T$, the d.c. voltage source is able to maintain a constant field across the sample and the drifting carrier sheet induces in the external circuit a constant current $I = \sigma v_d A/L$. A is the electrode area and v_d is the average drift velocity of the carrier sheet (current mode). The transit time is readily identified by the rapid drop of the current occurring when the carrier sheet strikes the back electrode. If $RC \gg t_T$, the drift of the carrier sheet is observed as a linear decrease of the voltage across the film (voltage mode). Our discussion of pulse shapes in the following paragraphs will be restricted to measurements performed in the current mode.

Deviations from the ideal current shape are expected if carriers in transit are lost in deep traps. In this event, the current decays for $t < t_T$ and the transit time becomes less well defined. Furthermore, in this case, the current signal exhibits an additional field dependence because the number of carriers transiting the sample depends upon the ratio of carrier lifetime and t_T. A conventional Hecht-type analysis of the current — taking into account a possible field-dependent generation mechanism — yields the average drift distance per unit field (Schubweg) and the deep trapping lifetime. It will be shown below that current signals in a-As_2Se_3 and at low temperatures in a-Se significantly deviate from the ideal rectangular pulse shape. However, the origin of this deviation is not attributable to carrier loss of the Hecht type but lies in the detailed mechanism of charge transport.

(ii) *Overview of current shapes*

The dominant characteristics of the type of current traces commonly observed in chalcogenides will be outlined in this section. A more detailed analysis of the current signals will be presented in conjunction with the discussion of hole transport in a-As_2Se_3.

The ideal case of a rectangular-shaped current signal is closely approached by transients of both carriers in a-Se at room temperature [13, 17, 32, 90—95]. Figure 7.17 shows such an example for hole drift. The transit time t_T is easily identified on this trace. The slight rounding of the signal around t_T has been attributed to a dispersion in carrier arrival times as a result of statistical fluctuations of the individual events characterizing the transport.

References pp. 297—302

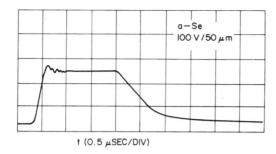

Fig. 7.17. Hole transient current signal in a-Se at room temperature. (From Marshall and Owen [95].)

At room temperature, the dispersion increases about as $t_T^{1/2}$, which is characteristic of a carrier sheet whose broadening can be described by Gaussian statistics [32]. This feature is not observed for hole transport in $a\text{-As}_2\text{Se}_3$. For $a\text{-As}_2\text{Se}_3$, the current signal is often found to be entirely featureless, as shown in Fig. 7.18(a), and no characteristic time t_T can be identified on the current trace [9, 96—102]. Pulse shapes of this kind are not only observed in some chalcogenides but also in an increasing number of organic polymeric solids. Also, a featureless current signal of the type shown in Fig. 7.18(a) is characteristic of a-Se at low temperatures [94, 95, 103]. In fact, current traces for hole drift in $a\text{-As}_2\text{Se}_3$ at room temperature and in a-Se at low temperatures are very similar.

In Fig. 7.18(b), we replotted the current trace of Fig. 7.18(a) in logarithmic scales. In these units, the current signal appears essentially as two straight lines of different negative slopes, thus indicating an algebraic time dependence of the current. The intersection of the tangents to the initial and final part of the current signal clearly defines a characteristic time t_T. A theoretical paper by Scher and Montroll [104] shows that an algebraically-decaying current signal of the type shown in Fig. 7.18(b) is expected if the carrier transport is governed by a stochastic process with a wide distribution of event times. Such a process would, for instance, be hopping among localized states. The predictions of the Scher and Montroll theory, which have been verified in great detail for hole transport in $a\text{-As}_2\text{Se}_3$ [98], will be discussed in Section 7.4.2 and 7.4.3 (compare also Chapter 3). For the present purpose, it suffices to remark that the current trace which is apparently featureless using linear scales can readily be analyzed using logarithmic scales and that the time t_T indicated in Fig. 7.18(b) is a meaningful parameter characteristic of bulk transport. It is important to point out that not only the fiduciary time t_T but also the shape of the current signal yields essential information on the transport mechanism.

In Fig. 7.18(c), we illustrate a current trace with a shape intermediate between the two extreme cases shown in Figs. 7.17 and 7.18(a). This shape

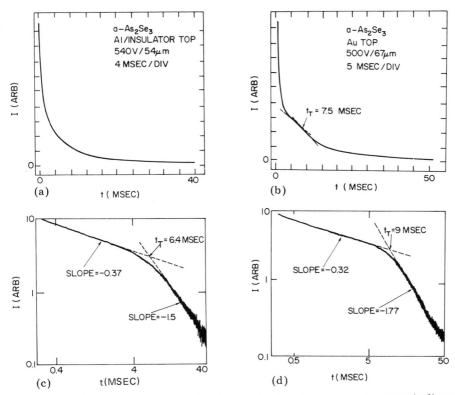

Fig. 7.18. Hole transient current signals in a-As_2Se_3 at room temperature in linear scales (a, c) and logarithmic scales (b, d), respectively. (From Pfister and Scher [98].)

is representative for a large number of a-As_2Se_3 films provided with a gold top electrode. From traces of this type, a time t_T characterizing the transport can be defined by the intersection of the tangents to the plateau and initial part of the tail region [9, 101].

The shape of the current trace shown in Fig. 7.18(c) has been the subject of wide speculation. It was generally agreed that the long tail indicates a wide dispersion in arrival times of the carriers at the back electrode and that the time t'_T defined in the figure characterizes the transit of the fastest carriers. Marshall and Owen [100] suggested that the initial spike results from the initial displacement of the injected carriers through extended states before the occurrence of the first "deep" trapping event, and that the long tail is determined by either a wide dispersion of "deep" trap release times and/or the (small) number of trapping events. Silver et al. [105] showed on the basis of a Monte Carlo calculation that current shapes of the form shown in Fig. 7.18(c) can be obtained if surface trapping is included in addition to bulk trapping. Fleming [106] argued that phonon-assisted tunnelling of the

References pp. 297—302

carriers in the mobility gap is the underlying mechanism for the wide dispersion. Scher and Montroll [104] proposed a stochastic transport model to explain the dispersion in arrival times. None but the last model, to be discussed in detail in Section 7.4.2, exactly predict the experimental current shape for a wide range of transit times.

Figure 7.18(d) shows that a log—log conversion of the signal shown in Fig. 7.18(c) produces a trace similar to the one shown in Fig. 7.18(b) but with different slopes. In this representation, the initial spike and plateau region are clearly de-emphasized.

(iii) Drift mobility

A drift mobility μ_d can be defined from a measurement of transit time t_T, applied field E and sample thickness L via

$$\mu_d = \frac{L}{t_T E}. \tag{7.28}$$

In order for the drift mobility to be a meaningful parameter characteristic of bulk transport, one has to ascertain that the transit time is well defined, as shown in Fig. 7.17 for example, and that it scales correctly with sample thickness L. The drift mobility μ_d obtained from eqn. (7.28) is often smaller than the microscopic or conductivity mobility μ_0. A widely proposed physical picture for this reduction is shallow trapping. In this case, the carriers move through extended states while rapidly communicating with a set of shallow traps of release time much shorter than the transit time. Experimentally, one determines the sum of the times the carrier is free and immobilized in shallow traps in the course of its transit through the sample. Since only during the free time does the carrier contribute to the current, the drift and microscopic mobilities are related by:

$$\mu_d = \mu_0 \frac{\tau_0}{\tau_0 + \tau_t} \tag{7.29}$$

where τ_0 is the mean free time in the extended state and τ_t is the average dwell time in a shallow trap. The evaluation of the reduction factor in eqn. (7.29) depends upon the details of the trap model. The most commonly used models are summarized below [40].

For a set of traps of density N_t at a discrete energy $\mathcal{E}_t$ below the extended state with mobility μ_0, eqn. (7.29) becomes:

$$\mu_d = \mu_0 \{1 + (N_t/N_c) \exp(\mathcal{E}_t/kT)\}^{-1} \tag{7.30}$$

which approximates to:

$$\mu_d = \mu_0 \frac{N_c}{N_t} \exp(-\mathcal{E}_t/kT) \tag{7.31}$$

for $\tau_0 \ll \tau_t$. N_c is the effective density of states at the band edge. For traps distributed uniformly from the band edge to a depth $\mathscr{E}_t$, one finds

$$\mu_d = \mu_0 \frac{N_c}{N_t} \frac{\mathscr{E}_t}{kT} \exp(-\mathscr{E}_t/kT) \tag{7.32}$$

If the density of states at the band edge falls off linearly with energy over a range $\Delta\mathscr{E}$, eqn. (7.29) yields

$$\mu_d = \mu_0 \frac{\Delta\mathscr{E}}{kT} \exp(-\Delta\mathscr{E}/kT) \tag{7.33}$$

It will be noted that for all trap models the drift mobility is thermally activated and, therefore, its temperature dependence alone is usually not sufficient to distinguish between the various models. The assignment of a particular trapping model will depend on the credibility of the values one derives for the various parameters introduced in the above equations. For an increasing number of amorphous solids, the drift mobility exhibits a field-dependent activation energy which originates in a field-dependent release of the carriers from the trapping centers.

(iv) Stochastic hopping transport

Initial studies of hole transport in a-As_2Se_3 brought to light some unusual features of transport properties in this amorphous solid [9, 96, 101]. It was noted that the width of the carrier sheet as inferred from the dispersion of the current signal did not exhibit the characteristic $t^{1/2}$ dependence, which is expected from Gaussian statistics [96]. In fact, it appeared that the dispersion relative to the transit time t_T remained constant: A plot of current signals recorded at a constant bias field for different sample thicknesses in relative units of current $I(t)/I(t_T)$ *versus* normalized time t/t_T yields a master curve independent of transit time [96]. In relative units, a Gaussian dispersion of the signal would vary as $t^{-1/2}$, hence, the tail of the current trace would become steeper as the transit time increases. The unusual dispersion of hole signals in a-As_2Se_3, which became known as "universality of current shape" [104], proved to be quite general. Similar universal behavior is observed for hole transport in the organic polymer poly(N—vinyl carbazole) (PVK) [107], electron and hole transport in the organic charge transfer complex of PVK with trinitrofluorenone [108, 109], charge transport in inert organic polymers doped with transport active molecules such as N—isopropyl carbazole in Lexan® polycarbonate [110], and hole transport in a-Se at lower temperatures [103]. A further puzzling observation was that drift mobility values determined from eqn. (7.28) are dependent upon sample thickness [101, 103]. Obviously, a conventional analysis of drift mobility measurements in systems with dispersive transport does not seem meaningful.

Most of the unusual aspects of charge transport in a-As_2Se_3 — and at the same time in the aforementioned inorganic and organic solids — were accounted for by the Scher and Montroll [104] theory. These authors propose as a transport mechanism a stochastic process which is characterized by an extremely wide distribution of individual event times. In fact, the distribution of event times may extend well into the time range of the experimental observation which is of the order of the transit time t_T. A prominent example of such a stochastic process, which is treated by Scher and Montroll, is hopping among localized states. For strong localization, the hopping time may be approximated by:

$$\tau_{\text{hop}} \simeq \gamma(T, \rho) \exp(2\rho/\rho_0) \exp(\mathcal{E}/kT) \tag{7.34}$$

where ρ is the hopping distance, ρ_0 the localization radius and $\mathcal{E}$ the activation energy. ρ_0 is defined by the exponentially decaying tail of the charge density at a distance r from the localization, namely $|\psi|^2 \simeq \exp(-r/\rho_0)$. $\gamma(\rho, T)$ is a mild function of both variables ρ and T compared to the exponential terms which have arguments of the order of 10. Therefore, fluctuations of the order of unity in the hopping distance ρ or activation energy $\mathcal{E}$ can readily produce fluctuations of the order of decades in τ_{hop}. Scher and Montroll assume that the intersite distance ρ, rather than the activation energy $\mathcal{E}$, is the dominant stochastic variable. This assumption seems to apply for a-As_2Se_3, in which case the current shape maintains its universality over a wide temperature range [97, 98].

With this background, we can now provide a qualitative description of the dispersive stochastic transport process and predict some features of the current signal. We use hopping transport as an illustrative example. For a detailed mathematical treatment, the reader is referred to Chapter 3 and the literature [104]. Suppose the carriers are injected at $t = 0$ at the surface of an infinitely-thick sample. As a result of the statistical variation in hopping distances, some carriers might immediately hop out of the generation region while others remain immobilized for some time. After each successful hop a carrier might encounter a long hopping distance, and as time goes on, an increasing number of carriers will experience such an event and become immobilized for an ever-increasing length of time. Hence, right from the start of the carrier drift, the current begins a continuous decrease and, for the infinitely-thick sample, eventually approaches zero. For a sample of finite thickness, the rate of the current decay begins to increase when the leading edge of the carrier sheet encounters the back electrode at which it is absorbed and ceases to contribute to the current. The time of this happening is operationally defined as transit time t_T.

It follows, then, that for sufficiently strong fluctuations in hopping distances the maximum of the carrier distribution remains close to the generation region, even for times $> t_T$, while, due to some carriers which initially encounter a succession of short hops, the leading edge of the carrier sheet may have penetrated far into the bulk of the sample (Fig. 3.11, Chapter 3,

p. 97). Therefore, the dispersion of the carrier sheet and the mean displacement of the charge from the generation region grow with time in the same manner, hence their ratio remains constant. This is the underlying physics for universality of the current shape. The difference from the Gaussian case is now obvious. There, the position of the maximum of the carrier sheet and the position of the mean are identical and the dispersion grows around this position (Fig. 3.10, Chapter 3, p. 96). Since the dispersion and mean position have different time dependences, namely $t^{1/2}$ and t, respectively, universality is not observed.

Scher and Montroll [104] have shown that the probability for a carrier to jump to the next site at time t after having arrived at $t = 0$ is a slowly decaying function which can be approximated by:

$$\psi(t) \simeq t^{-(1+\alpha)} \tag{7.35}$$

where α ($0 < \alpha < 1$) depends upon the microscopic parameters ρ_0, ρ and $\mathcal{E}$. α is not expected to be significantly field or thickness dependent but may decrease as the temperature is lowered. The slow variation of $\psi(t)$ ensures that over a wide range in time the carrier has substantial probability for a jump. This contrasts with the Gaussian case in which the probability for a jump is of the form:

$$\psi(t) \simeq \exp(-t/\tau) \tag{7.36}$$

and rapidly vanishes for $t > \tau$. On the basis of eqn. (7.35), Scher and Montroll show that the transient current decays algebraically as

$$i \simeq \begin{cases} t^{-(1-\alpha)} & t < t_T \\ t^{-(1+\alpha)} & t > t_T \end{cases} \tag{7.37}$$

The increase in the power exponent of the decay at $t \simeq t_T$ occurs when the leading edge of the carrier sheet encounters the absorbing substrate electrode. The time t_T, which has the meaning of the transient time, is experimentally easily determined. In logarithmic scales, the current trace should essentially appear as two straight lines intersecting at $t \simeq t_T$ with initial slope $-(1-\alpha)$ and final slope $-(1+\alpha)$. This is the type of current trace shown in Figs. 7.18(b) and 7.18(d). Universality is expected as long as the parameter α remains constant, i.e. traces recorded for different fields can be superposed by shifting along the logarithmic axes. From eqn. (7.37), it follows that the sum of power exponents, which are determined from the slopes of the current traces in the logarithmic representation, equals -2 and thus is independent of experimental parameters.

Scher and Montroll show that as a consequence of the algebraic distribution function, eqn. (7.35), the mean displacement of the carrier sheet depends on time as t^α, i.e. the mean velocity decreases with time. The nonlinear time dependence of the mean displacement leads to the relation

References pp. 297–302

$$t_T \propto L^{1/\alpha}. \tag{7.38}$$

Hence, the transit time depends superlinearly on the sample thickness L since $0 < \alpha < 1$.

The drift of the carriers is spatially biased by the applied field, which enhances the probability for a hop in the direction of the applied field compared to that against the field direction. Assuming that this asymmetry increases linearly with field strength, Scher and Montroll derive

$$t_T \propto E^{-1/\alpha}. \tag{7.39}$$

Thus, a superlinear relationship between transit time and field is predicted without invoking a field-dependent activation energy as a result of barrier lowering. It is clear that barrier effects can cause an additional field dependence. Both eqn. (7.38) and eqn. (7.39) are direct consequences of the probability function and, therefore, of the wide dispersion in hopping times. Combining eqns. (7.38) and (7.39), we find the scaling law:

$$t_T \propto \left(\frac{L}{E}\right)^{1/\alpha} \tag{7.40}$$

which is remarkably different from the conventional drift mobility relation, eqn. (7.28). For one thing, extracting a drift mobility from eqn. (7.40) in analogy to eqn. (7.28) would lead to $\mu_d \simeq (E/L)^{(1-\alpha)/\alpha}$. Field-dependent mobilities have become a familiar occurrence, but a thickness dependence makes such a definition doubtful. It is important to re-emphasize that the superlinear (E/L) dependence predicted by the stochastic hopping theory is a consequence of the distribution of event times, which extends into and beyond the time range t_T of the typical experimental observation. If the experimental conditions can be arranged such that the individual event times become short compared to the time of observation, the nonlinear effects disappear and, in particular, the mobility again becomes well defined. A significant such example is, of course, the d.c. limit. The transition between the two regimes is observed for hole transport in a-Se as a function of temperature [103].

A final important relation is contained in eqns. (7.37) and (7.40). Both the current shape and the thickness and field dependence of the transit time contain the parameter α. If there is no additional field dependence of the transport, the same α-values should be obtained from current shape and field dependence of the transit time. This correlation will be examined in detail in the next section.

7.4.3 Hole transport in amorphous As_2Se_3

(i) Introduction

Drift mobility experiments on a-As_2Se_3 films have been performed by a number of authors in different laboratories [9, 96—102]. In general agreement, only hole transport is observed, even in samples as thin as $\sim 1\,\mu m$ and fields as high as 7×10^5 V/cm. The transient signals are very dispersive, of the type shown in Fig. 7.18(b) or 7.18(d), and in conventional linear scales the transit time t_T is ill-defined. It has been verified that the observed transient signal is a manifestation of carrier drift across the sample rather than of range limitation. All authors agree that the transit time is thermally activated and decreases superlinearly with increasing bias field, i.e. the drift mobility is field dependent. The samples were either quenched from the melt or evaporated onto substrates held at various temperatures.

It has often been emphasized in the literature that the electrical properties of chalcogenides are relatively insensitive to preparatory conditions — at least in comparison to tetrahedrally bonded amorphous solids (Chapter 6). Nevertheless, some remarks of caution appear appropriate when data on differently prepared samples are compared. X-ray data indicate that the structure of films evaporated onto substrates held below 85°C differs from that of bulk glass [111]. Over a period of several days at room temperature, the evaporated film structure relaxes to a pseudo-equilibrium. With a step-wise increasing temperature, additional relaxation is observed until the film structure closely resembles that of the glass. No structural relaxation was observed after annealing at 175°C. Although this effect is not reflected in dark d.c. conductivity measurements [111], it might well influence drift mobility measurements [96]. The drift mobility increases by almost an order of magnitude, associated with a decrease in activation energy, when the temperature of the substrate onto which the film is evaporated is raised from ambient to above $\sim 175°C$. For substrates held at higher temperatures, no further increase is observed. Annealing samples evaporated onto room-temperature substrates, however, seems not to influence the mobility value.

Drift mobilities calculated from eqn. (7.28) are typically of the order of 10^{-5} cm^2/V.s at fields of $\sim 10^5$ V/cm. The low value was attributed to a trap-limited transport mechanism in which the holes, in addition to a possible rapid communication with shallow traps, are in correspondence with deep trapping states. Initially, the field dependence of the hole mobility in quenched glasses of a-As$_2$Se$_3$ was attributed to a barrier-lowering mechanism of the Poole—Frenkel type [20] which gave satisfactory agreement over a limited field and temperature range [100]. More recent experiments, however, seem to suggest that both drift mobility and dark d.c. conductivity in a-As$_2$Se$_3$ and a number of other amorphous chalcogenides, including a-Se, As—Se alloys and quaternary alloys, are more adequately represented by an expression of the type:

$$f(E) = f(0) \exp [a(T)E/kT] \qquad (7.41)$$

rather than by the Poole—Frenkel expression proposed earlier [112, 113]. f stands for the d.c. conductivity σ or the drift mobility μ_d. In the

References pp. 297—302

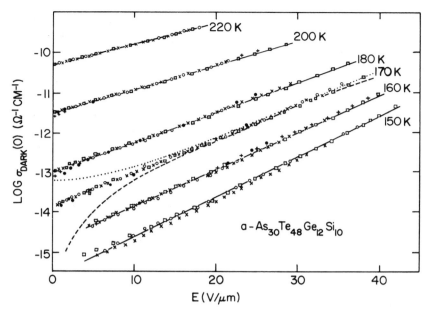

Fig. 7.19. Example of electric field and temperature dependence of d.c. dark conductivity observed in a number of amorphous solids. The 170K data have been fitted to two commonly used forms of field and temperature dependence: -- $\sigma = 2 \times 10^{-16} \exp(1.9 \times 10^{-2} E^{1/2})$ (Poole—Frenkel) ... $\sigma = 3.5 \times 10^{-9} \sinh(2.3 \times 10^{-5} E)$ (Bagley). (From Frenkel [20], Marshall and Miller [112], Bagley [114].)

Poole—Frenkel expression, the square root of the field appears in the exponent of eqn. (7.41) and the parameter a is temperature independent. Dark d.c. and drift mobility measurements give about the same value for the parameter $a(T)$ [112, 113]. $\mu_d(0)$ is the zero-field trap-limited drift mobility:

$$\mu_d(0) = A \exp(-\mathcal{E}_0/kT), \tag{7.42}$$

where $\mathcal{E}_0$ is the activation energy at zero field. Hence, the superlinear field dependence of the transport can be interpreted in terms of a field-dependent activation energy

$$\mathcal{E}(E) = \mathcal{E}_0 - ea(T)E. \tag{7.43}$$

The field dependence predicted by eqn. (7.41) is observed over a wide range of fields; in particular, to fields as low as 0.1 V/μm. An example of dark d.c. measurements at various temperatures on the quaternary glass $As_{30}Te_{48}Si_{10}Ge_{12}$ is shown in Fig. 7.19.

Several models have been proposed to explain the field-dependent transport mechanism [112, 113]. They include field enhancement of (i) thermally-activated release from traps into extended states (Poole—Frenkel)

[20], (ii) thermally-activated hopping along localized states (Bagley) [114], (iii) tunneling among "deep" localized states and (iv) tunneling from very shallow states directly into extended states, the former being located in the vicinity of the mobility edge separating extended from deeper localized states, *i.e.* in the region of Brownian motion (Mott) [115]. All but the last mechanism appear to predict a deviation from the superlinear to an ohmic field dependence as the value of the field decreases below ~ 10 V/μm. This would contradict the experimental observation at low fields, although the various models might provide a good fit to the data at high fields. This is illustrated in Fig. 7.19 for the two most commonly cited models, (i) and (ii). The field modulation of transport in the region of Brownian motion appears the most satisfactory model to account for the experimental data over the entire accessed field range [113]. In this model, no approach to ohmic behavior is expected since there will always exist sufficient shallow traps from which the tunneling emission rate can be field enhanced, even at the lowest fields. At present, no satisfactory analysis of such a model is available and much more theoretical and experimental study would be necessary to positively associate the field dependence of the drift mobility (and d.c. dark conductivity) with such a mechanism.

While transient and d.c. conductivity in a-As_2Se_3 exhibit the *same* field dependence, eqn. (7.41), in samples prepared by quenching from the melt [112, 113], their behavior at fields $\lesssim 10^5$ V/cm is distinctly *different* in samples prepared by evaporation [98]. In these samples, the field dependence of the transient conductivity is stronger at low fields than predicted by eqn. (7.41) and is of the form $\mu_d \propto E^n$ where $n \simeq 1$ is largely independent of temperature [98]. While the phenomenological expression eqn. (7.41) might be used to explain the d.c. conductivity, it is suggested that the algebraic field dependence of the transient conductivity in evaporated films be interpreted in terms of the stochastic hopping theory of Scher and Montroll [104], which has been introduced in Section 7.4.2. It is seen from eqn. (7.39) that this model provides the required field dependence, even at the lowest fields. It should be re-emphasized here that the Scher—Montroll model cannot be applied to explain the dark d.c. conductivity since in the d.c. experiment the observation time exceeds all event times.

The following paragraphs, which highlight an extensive experimental study of transient hole conductivity in a-As_2Se_3 [98] will show that the transient conductivity results can be interpreted in a consistent manner in terms of the Scher—Montroll model. Although Scher and Montroll in their theoretical analysis [104] use hopping as an example for a dispersive stochastic transport, this theory is not restricted to only this transport process. One can apply the same ideas to a multiple-trapping model but details of the theoretical predictions are expected to be different from the hopping model.

For hopping, with the hopping distance as a stochastic variable, the field, thickness and temperature dependence of the transit time and current shape are predicted in terms of one parameter (α). These predictions are in essential agreement with the observation for hole transport in a-As_2Se_3 and in molecularly-doped polymers for which hopping transport has been established from independent experiments [104, 108—110]. At present, no theory that treats multiple trapping within the framework of the Scher—Montroll theory is available, and it remains to be seen whether such a model can account for all the results, specifically the universality of the current shape with respect to field and temperature. On the other hand, the interpretation in terms of hopping has to rationalize the high value of 0.5—0.6 eV for the activation energy of the transit time, which is typically observed in a-As_2Se_3 and molecularly-doped polymers at low fields. This value certainly is too large to be purely polaronic and suggests that disorder effects might significantly contribute. It is also possible that the hopping process has to be visualized as a trap-controlled process in which the carrier displacement occurs via hopping with frequent interruption by trapping in a narrow band of deeper states. In this case, the energy difference between the hopping and trapping states would additively contribute to the observed activation energy. The hopping model of Scher and Montroll can readily be applied to this case. The following interpretation of the data has to be considered in this more general context.

(ii) Experimental analysis of the Scher—Montroll model

The samples were open-boat-evaporated onto aluminum substrates, which were held at least at 175°C during evaporation. The top electrode was mainly of semitransparent gold, but several samples with an aluminum top-electrode were also investigated. Some samples were overcoated with an insulating layer before the top electrode was applied. The shape of the transient current depends on the top-electrode material [9, 98]. The trace shown in Fig. 7.18(a) is more typical for aluminum and for samples overcoated with an insulator, whereas the trace shown in Fig. 7.18(c) is more characteristic for a gold top-electrode. The mechanism for the different current shapes is still debated and will not be discussed further at present.

(a) Current shape and universality

The insert of Fig. 7.20 shows three oscilloscope current traces A, B, C in linear I *versus* t scales recorded at the indicated sweep rates on a 62 μm a-As_2Se_3 film with an aluminum top-electrode. The main part of Fig. 7.20 shows the same current signals displayed on an x—y recorder in logarithmic current and time units. For clarification, the traces were shifted along the logarithmic current axis and the units on this scale are, therefore, relative.

The algebraic decay of the current, predicted by eqn. (7.37), is clearly manifest, and the transit time t_T can readily be identified in this representation, whereas in the linear scales no fiduciary time can be defined. To each

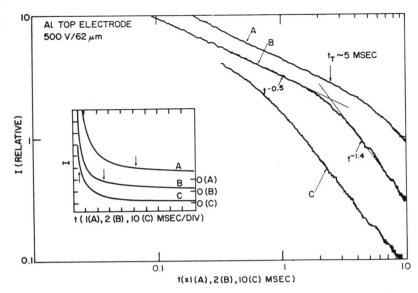

Fig. 7.20. Hole transient current signals in a-As_2Se_3 at room temperature recorded at the sweep rates indicated. The insert shows drawings from oscilloscope traces in linear scales; the main figure shows drawings from the x—y recorder traces of the same signals in logarithmic scales. The arrows in the insert indicate the transient times t_T determined from the break of the algebraic decay observed in the logarithmic representation. (From Pfister and Scher [98].)

side of t_T, the algebraic decay is observed over more than an order of magnitude in time, and the sum of the initial slope -0.5 and final slope -1.4 is close to the theoretical prediction of -2.

With increasing bias field, temperature, or hydrostatic pressure the log—log plot is shifted along the logarithmic scales to shorter times and larger currents, thus establishing that the current trace is characteristic of a carrier transport. Following the theory, the parameter α characterizing the current decay is not expected to vary significantly over the transit time range which can be accessed by changing experimental conditions. Hence, the slopes of the current signal in the logarithmic representation should be field, temperature, thickness and pressure independent (universality of current shape). To examine this feature, transient currents were recorded at different bias fields (or temperatures, or pressure) and then brought to coincidence by parallel shifting along the logarithmic axes. The resulting (universal) master plots obtained for a variation in field and temperature are shown in Fig. 7.21(a) and Fig. 7.21(b), respectively. For illustrative purposes, we schematically indicate in Fig. 7.21(a) for the two extreme transit times the relative dispersion of the tail portion of the current which is expected for a Gaussian spreading carrier sheet. We note from Figs. 7.21(a) and 7.21(b) that the

References pp. 297—302

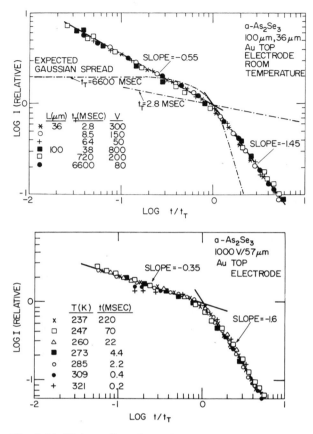

Fig. 7.21. Universality plots of current *versus* time in logarithmic scales for hole transients in a-As_2Se_3. The plots were obtained by parallel shifting of the current traces along the time and current axes. (a) Room temperature, variation of applied field and sample thickness. Expected spread of Gaussian dispersion is schematically indicated. (b) Fixed applied field and sample thickness, variation of temperature. (From Pfister and Scher [98].)

slopes observed for the different samples shown are not the same but both add up to a value close to -2. For thin samples ($\lesssim 30~\mu m$), it was generally found that the sum of the slopes is ~ -1.7 rather than -2 as predicted by theory. This deviation from the ideal value could be minimized by illuminating the sample with d.c. bulk-absorbed light prior to or during the transit of the carrier sheet.

It should be pointed out that for many samples the long tail was preceded by a cusp although the number of carriers in transit was less than CV. This effect was particularly noted when gold was the top electrode. Also, the initial part of the current shape could often be distorted by exposure to a flash sequence and could furthermore be sensitive to the length of time the sample was under d.c. bias in the dark. With aluminum as the top-electrode the current shape was less history-dependent. The details of these effects are not clear at present but are believed to be related to trapping in the region of the metal—insulator interface of thermal and photogenerated carriers and, associated with it, space-charge perturbation of the field and/or effects of surface release. It has recently been reported that at higher temperatures, gold diffuses into a-As_2Se_3 as positive ions [116]. Since during the electrode evaporation the temperature of the sample surface is expected to increase, in particular for the thin unsupported films, such an effect could strongly distort the initial part of the pulse shape. We restrict our discussion here to those current shapes that were reproducible, cusp free, and did not show effects of distortion. For more details on cases where transient signals contain cusps, we refer the reader to the literature [98, 105, 106, 117].

The temperature independence of the current shape is a central argument for the intersite distance ρ being the dominant stochastic variable rather than the activation energy $\mathcal{E}$. In the latter case, the carriers contributing to the tail of the current pulse are expected to sample hopping sites of which at least one has a larger activation energy than all hopping sites encountered by carriers contributing to the initial spike, and, in units relative to the transit time, the tail of the pulse would become sharper as the temperature is raised. Hence, universality with respect to temperature would not be observed, which is in obvious disagreement with the results shown in Fig. 7.21(b). The same argument can be put forward against a transport mechanism which involves multiple trapping [97, 98]. In this model, carrier transport is believed to occur in extended states with occasional "deep" trapping with a wide distribution of trap release times.

(b) Thickness and field dependence

The stochastic transport model predicts a superlinear increase of the transit time t_T with sample thickness L [eqn. (7.38)]. This behavior is independent of a model for the field dependence of the transport and thus can be examined separately. Transit times t_T measured at a constant field of 10 V/μm are plotted against the sample thickness L in Fig. 7.22. This graph clearly establishes a superlinear rather than the conventional linear t_T *versus* L relationship. The field dependence of the transit time is also superlinear and below $\sim$ 10—15 V/μm is of the form $t_T \propto E^{-n}$ where n scatters about the power exponent of the thickness dependence. Equality of the power exponents of the field and thickness dependence is theoretically expected for fields low enough that field enhancement of transport parameters by a mechanism such as barrier lowering is negligible, eqns. (7.38) and (7.39). For a-As_2Se_3, this appears to be the case for fields below $\sim$ 10—15 V/μm, which were typically applied at room temperature without risking electric breakdown. According to eqn. (7.40), one therefore expects that the transit times measured at various fields on samples with various thicknesses scale in units

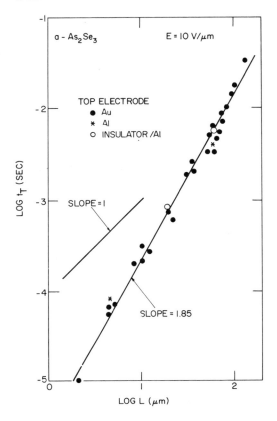

Fig. 7.22. Thickness dependence of transit time for hole transport in a-As$_2$Se$_3$ at room temperature. The indicated linear relationship is expected for well-defined mobility. (From Pfister and Scher [98].)

(E/L). The corresponding plot, Fig. 7.23, shows that over a wide thickness and transit-time range the predicted scaling is observed. The schematically indicated spread of the data which, in the units of Fig. 7.23, would be expected for a well-defined field-dependent drift mobility of the form $\mu_d \propto E^{0.85}$ by far exceeds the experimental scatter.

(c) *Correlation of current shape and field dependence of transit time*

Although the samples were prepared under identical conditions, the exponent n of the field dependence of the transit time varied between ~ 1.65 and 2.2 for different samples, and the spread of the transit times about the 1.85 power law in Fig. 7.23 is not due to data scatter of individual samples about that mean, but to systematic deviations of the power laws observed for the various samples. In fact, the scatter about the power law appropriate for individual samples is very small. According to eqns.

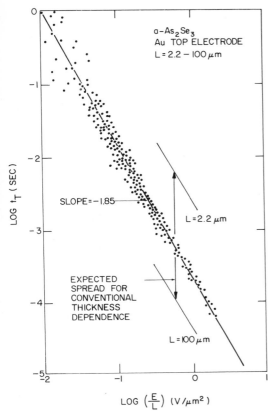

Fig. 7.23. Scaling of transit time with (E/L) for hole transport in a-As$_2$Se$_3$; 30 samples of thickness range 9.2—100 μm; ~ 12 points per sample; fields $\lesssim$ 10—15 V/μm; room temperature. Schematically indicated is the spread of the transit time expected for well-defined thickness-independent mobility $\mu_d \propto E^{0.85}$. (From Pfister and Scher [98].)

(7.38)—(7.40) the variation in field dependence should be paralleled by a corresponding variation in current shape since the α parameter in eqns. (7.38)—(7.40) is the same. Figure 7.24 shows that such a correlation is observed. Each point in this figure represents one sample. Plotted along the ordinate is the average α value obtained for each sample from master plots as shown in Fig. 7.21. The exponent n describing the field dependence of that particular sample is plotted along the abscissa.

(d) *Temperature and pressure dependence*

The temperature dependence of the transport constitutes an additional important test for the stochastic model. One expects that, at least for $E \lesssim 10$ V/μm, the field dependence of the transit time is temperature-independent since $t_T \propto E^{-1/\alpha}$ and α is approximately constant. The prediction of the

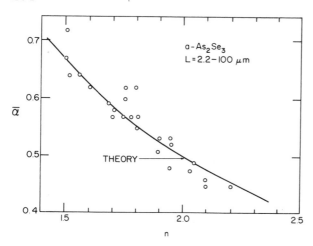

Fig. 7.24. Correlation between current shape and field dependence of transit time for hole transients in a-As_2Se_3 at room temperature. $\bar{\alpha}$ is the average of the α-values determined from the initial $(t < t_T)$ and final $(t > t_T)$ slopes of the transient signal in logarithmic units (see Fig. 7.21, eqn. (7.37)). n is the power exponent describing t_T versus E. Each point pertains to one sample. The theory predicts $\bar{\alpha} = 1/n$. (From Pfister and Scher [98].)

stochastic theory, therefore, is distinctly different from that of the phenomenological relationship, eqn. (7.41), in which case the field dependence is temperature dependent even at the lowest fields. Stated differently, in the latter case the activation energy is expected to increase continuously with decreasing applied field, whereas in the former case the activation energy should remain constant in the field range in which the entire superlinear field dependence of the transit time is dictated by the stochastic process without invoking barrier lowering. The experimental result shown in Fig. 7.25 provides strong evidence for the stochastic transport model. At fields below ~ 10 V/μm, $t_T \propto E^{-1/\alpha}$, $\mathcal{E}_0 \neq \mathcal{E}(E)$, $\alpha \neq \alpha(T)$; at fields above ~ 15 V/μm, $t_T \propto E^{-q}$, $\mathcal{E} = \mathcal{E}(E) < \mathcal{E}_0$, $q = q(T) > 1/\alpha$ where $\mathcal{E}$ decreases with field and q increases with temperature.

Pressure studies can provide an elegant measurement of important theoretical parameters. Using self-consistency arguments, Scher and Montroll [104] calculated $\rho/\rho_0 \simeq 5$. With this value, the pressure coefficient of the transit time due to the change of the overlap integral is estimated, using eqn. (7.34), as $(1/p)(d \ln t_T/dp) = -\frac{2}{3}(\rho/\rho_0)K \simeq -2.3 \times 10^{-3}$ kbar^{-1}, which is about a factor of 20 smaller than observed [102] ($K = 6.9 \times 10^{-3}$ kbar^{-1}). On this basis, it was suggested that the strong pressure dependence of the transit time results from a change in the activation energy rather than from a change in the hopping distance [102].

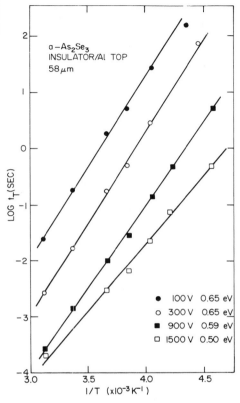

Fig. 7.25. Activation energy plot log t_T versus $1/T$ for hole transport in a-As_2Se_3 for various applied fields. At low fields ($\lesssim 10$ V/μm) the activation energy remains constant while at higher fields ($\gtrsim 15$ V/μm) it begins to decrease approximately in proportion with the field. (From Pfister and Scher [98].)

7.4.4 Amorphous selenium

(i) Introduction

The use of a-Se as photoreceptor in the xerographic electrophotographic process has generated widespread interest into studies related to transport [12, 13, 17, 32, 90—95, 118—123] and photogeneration [8, 11, 12, 18, 124]. Unlike a-As_2Se_3, a-Se is ambipolar and above ~ 200K both carriers have a well-defined thermally-activated drift mobility in accordance with the conventional definition, eqn. (7.28). In the following, we first briefly review the main results of drift mobility experiments for both carriers before we discuss the influence of impurities upon their transport properties.

References pp. 297—302

(ii) *Hole transport*

In Fig. 7.26, we show compiled data of the temperature dependence of the hole mobility in quenched and evaporated samples. For comparison, we

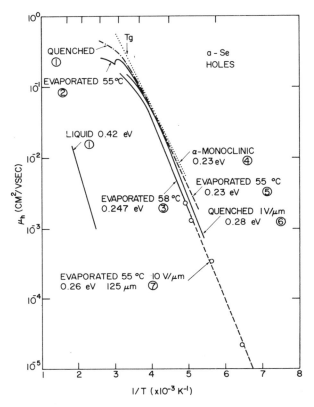

Fig. 7.26. Representative summary of hole drift mobility activation plots for hole transport in *a*-Se. Included are results on α-monoclinic and liquid Se. (From: ① Vengris *et al.* [121], ② Pai [12], ③ Grunwald and Blakney [93], ④ Caywood and Mead [123], ⑤ Tabak [94], ⑥ Marshall and Owen [95] and ⑦ Pfister [103])

include measurements on the liquid phase and on α-monoclinic single crystals. Although the samples were prepared under largely different experimental conditions, the results are in general agreement. Below ~ 250K the mobility falls off exponentially with an activation energy $\mathcal{E}_p \simeq 0.23-0.30$ eV. At higher temperatures, the mobility appears to approach saturation in the neighborhood of the glass transition at ~ 304K. The thermally-activated behavior has been attributed to a mechanism in which the charge flow occurs in extended states with frequent interruption by trapping in levels energetically close to the transport state (shallow-trap controlled transport)

[90—93]. This transport mechanism distinctly differs from the transport model discussed for a-As_2Se_3 since it involves transport in extended states and assumes that the trap release times are short compared to the transit time. The deviation from the exponential temperature dependence above ~ 250K has been attributed to phonon scattering of the holes in the valence band [93]. Using eqn. (7.30) and assuming a microscopic hole mobility $\mu_0^p \propto T^{-3/2}$, a good fit to the experimental data can be obtained when a value $\mu_0^p = 0.34 \pm 0.005$ cm^2/V.s is used at room temperature. There is a variety of experimental evidence that this value for the microscopic mobility is more reasonable than the earlier estimate [91] of $\mu_0^p = 60$ cm^2/V.s derived on the basis of a trap density of ~ 10^{20} cm^{-3} deduced from X-ray data. Photo Hall experiments suggest $\mu_0^p < 3$ cm^2/V.s, with ~ 0.15 cm^2/V.s a likely figure [125], while dark Hall mobility measurements at room temperature give $\mu_0^p = 0.37$ cm^2/V.s [32].

Representative hole transport parameters are summarized in Table 7.4. Of the parameters shown, the hole lifetime τ_p is the quantity most sensitive to sample purity and origin. Typically, τ_p can vary in the range $< 0.2\,\mu$s to 14 μs for samples prepared from different batches of 99.999% high purity selenium [126]. With increasing substrate (polymerization) temperature μ_d^p, $\mathscr{E}_p$ and N_v/N_t^p were found to increase [32, 93]. The lower values in Table 7.4 pertain to a substrate held at room temperature, the higher values to samples polymerized at 500K.

The drift mobility remains well defined with decreasing temperature down to 170K below which the dispersion rapidly increases [94, 95]. Typically, below ~ 200K μ_d^p becomes increasingly field dependent but for some samples a weak field dependence is observed at temperatures as high as room temperature [127]. Below ~ 30 V/μm, the mobility data can be characterized in terms of eqn. (7.41) with $a(T)$ ~ 10 Å at ~ 200K [95].

On decreasing the temperature below 170K, the pulse shape assumes the featureless form typically observed in a-As_2Se_3 and, finally, transit times can no longer be determined from the current traces when displayed in linear scales. Using the log—log representation of the signal described in Section 7.4.2, hole transport was explored down to temperatures as low as 100K [103]. In logarithmic units, the current traces assume a shape of the type expected for the stochastic hopping process [104]. Universality is observed with respect to field but not with respect to temperature, however [103]. At present, these studies are still in progress, but some preliminary results are of interest. Below ~ 200K, $t_T \propto L^m$ where $m > 1$ and increases with decreasing temperature. The open circles in Fig. 7.26 show that the low-temperature data join smoothly with the results obtained in a temperature range where the mobility is well defined in the conventional sense. Figure 7.27 shows the entire temperature dependence of the hole velocity for different fields with some representative current traces. One observes no change in the activation energy on lowering the temperature into the range where the dispersion rapidly increases and the t_T versus L relation becomes superlinear.

References pp. 297—302

TABLE 7.4

Representative hole (electron) transport parameters at room temperature for a-Se and, for comparison, α-monoclinic Se. μ_0 = microscopic mobility, μ_d = drift mobility, $\mathcal{E}$ = activation energy, τ = lifetime with respect to deep traps, N = density of states at the valence (conduction) band edge, N_t = density of shallow traps, S = capture cross-section of shallow traps. (Compiled from refs. 13, 17, 32, 90–95, 123, 125, 126, 131).

	μ_0 (cm²/V·s)	μ_d (cm²/V·s)	$\mathcal{E}$ (eV)	τ (µs)	N (cm⁻³)	$\dfrac{N}{N_t}$	S (cm²)
Amorphous Se holes	0.3–0.4	0.11–0.19	0.22–0.3 <250K	<50	10^{20}	10^3–10^5	3×10^{-16}
α-Monoclinic Se holes		0.2	0.23			10^3	
Amorphous Se electrons	0.32	$(5–8) \times 10^{-3}$	0.28–0.33	<50	10^{20}	$(1–9) \times 10^3$	
α-Monoclinic Se electrons	2	lattice controlled	0.25 <200K		10^{20}	10^6	10^{-13}

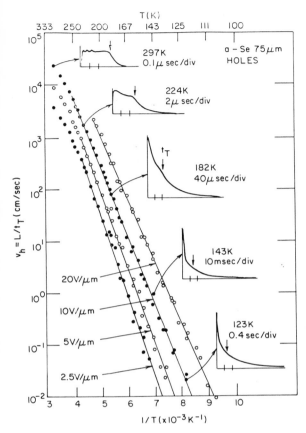

Fig. 7.27. Activation energy plot of the hole velocity in a-Se for various fields. Some representative current traces are shown. The arrows mark the transit time. (From Pfister [103].)

It is interesting to compare these findings with the results of similar measurements on a-Si in which case the activation energy for electron drift decreases discretely at ~ 250K from the room temperature value ~ 0.18 eV to ~ 0.09 eV [128, 129]. The lowering of the activation energy has been associated with a corresponding discrete change of the transport mechanism, namely from being shallow-trap controlled at temperature above ~ 250K to hopping among localized states at low temperatures. With this notion, the absence of a lowering of the hole activation energy in a-Se might indicate that in the temperature range from ~ 250K down to at least ~ 100K the type of hole transport mechanism does not change. It remains to be established by experiment whether the transport is a hopping or multiple-trapping process. In either case, a-Se offers the interesting possibility to follow a transport process from a high-temperature regime, in which thermal fluctuations largely overcome the randomness associated with the amorphous state and Gaussian statistics is applicable, to a low-temperature regime, in which the fluctuations of the microscopic transport parameters become dominant and lead to non-Gaussian behavior of the propagating carrier packet. The transition between the two temperature regimes is most drastically demonstrated by the dispersion of the pulse shape, which in relative time units exhibits the characteristic $t^{-1/2}$ dependence at high temperatures [32] and assumes universal behavior with respect to field at low temperatures [103].

The mechanism of transport in the high-temperature regime is still under discussion. The absence of a pressure effect upon the drift mobility in the temperature range 220—300K and pressure range up to 6 kbar has been used as an argument in favor of the shallow-trap-controlled model [118]. On the other hand, recent experiments question the generally accepted multiple-trap model [119, 122]. Thermopower, drift mobility and d.c. conductivity measurements in a temperature range encompassing the glass transition region appear to suggest that the exponential temperature dependence of the drift mobility is not due to multiple-trapping effects and that hole transfer is associated with Se_8 rings. It is interesting to note, in this context, that hole drift measurements in α-monoclinic Se (Se_8 puckered rings) give mobility values and activation energies essentially identical to those observed in a-Se except that above $\sim 250K$ no deviation from the thermally-activated behavior is indicated [123] (see Fig. 7.26). For comparison, we quote $\mu_d^p \sim 26\,cm^2/V.s$ in trigonal Se (helical chains) determined from acoustoelectric experiments [130].

(iii) Electron transport

The trap-controlled transport model has also been suggested to explain the electron drift mobility data. Again, the absence of a pressure effect on the electron transport has been considered as the most indicative evidence against a hopping process [118]. No electron drift data are presently available which extend down into the temperature range where the pulse shape becomes featureless. In the temperature range extending from $\sim 200K$ to $\sim 300K$ the mobility remains well defined and below $\sim 300K$ it is thermally activated with $\mathcal{E}_n \sim 0.28$—$0.33\,eV$. Typical electron transport parameters at room temperature are given in Table 7.4. The microscopic mobility μ_0^n has been determined from photo Hall effect measurements [125]. This value was used to determine the ratio of the densities of conduction band states to the density of shallow traps, eqn. (7.30). The deep-trapping lifetime is the quantity most sensitive to sample origin, a result also found for the hole lifetime. Electron lifetimes range from below $1\,\mu s$ to $50\,\mu s$ in samples prepared from batches of five to six 9's purity Se obtained from different suppliers [126].

A transition to a temperature-independent mobility is observed as the sample temperature is raised above the glass transition at $\sim 304K$ [93].

Unlike for holes, no temperature hysteresis is reported when the films are thermally cycled through the transition. Compared to the hole drift mobility, where the deviation from the exponential temperature dependence occurs gradually, the approach to saturation for the electron mobility is rather abrupt. This might perhaps be taken as evidence that the underlying mechanisms leading to saturation of the mobility at higher temperatures are different for electrons and holes. For electrons, a reversible increase of the trap density N_t^n above T_g associated with the increasing disorder as characterized by orientation, location and shape of constituent molecules has been suggested [93]. The electron transport parameters and substrate temperature appear to be correlated as found for holes; i.e., $\mathcal{E}_n$ and N_c/N_t^n increase with increasing substrate temperature [93].

In the earlier experiments, no field dependence of the electron drift mobility was reported. More recent experiments show an algebraic power dependence for μ_d^n with the exponent increasing from ~ 0.2 at room temperature to ~ 0.8 at 221K [127]. Again, the field and temperature dependence of μ_d^n can be described in terms of eqn. (7.41) with $a(T) \sim 10$ Å [112]. Hence, for both carriers in a-Se $a(T)$ has about the same value although their transport parameters are quite different (Table 7.4). This interesting result, together with the finding that eqn. (7.41) describes the field and temperature dependence of the electrical parameters μ_d and $\sigma_{d.c.}$ in a wide range of non-crystalline solids might indicate that the underlying principal mechanism of charge transport is determined by properties associated with the disorder in general rather than the specifics of the constituent molecules. Thus, a theoretical description of the overall transport mechanisms in disordered solids should provide answers which are largely insensitive to molecular parameters. The stochastic model discussed in Chapter 3 and summarized in Section 7.4.2 certainly is an important step towards such a unifying description. The universal features of transport in amorphous materials has interesting parallels related to other phenomena common to the amorphous state, namely, the WLF relaxation behavior of the glass transition [133], ω^s frequency dependence of the a.c. dark conductivity (Section 7.4.7) and the linear temperature dependence of the low-temperature specific heat [134].

Electron drift mobility experiments in α-monoclinic Se (Se$_8$ puckered rings) show a sharp transition from a lattice-controlled mobility ($\mu_0^n \propto T^{-3/2}$) at high temperatures to a trap-controlled mobility $\mu_d \propto \exp(-\mathcal{E}_n/kT)$ at low temperatures, where $\mathcal{E}_n \sim 0.25$ eV [131]. The close agreement between the activation energies in the trap-controlled region in α-monoclinic Se and a-Se might indicate that the type of trapping centers controlling the electron mobility might be common to both materials [131]. This interpretation seems to be supported by doping experiments which reveal a correlation between Se$_8$ ring population and electron drift mobility, thus suggesting that electron transport is related to the presence of Se$_8$ rings (Section 7.4.5). On

References pp. 297—302

the other hand, thermally cycling α-monoclinic Se to ~ 320 K leaves the electron mobility unchanged while the hole mobility is reduced, apparently due to an increase in the density of broken Se_8 rings [123]. More detailed information on the nature of the traps involved in the transport of charge in *a*-Se stems from detailed studies of impurity effects and will be discussed in the following section.

7.4.5 Effect of impurities on charge transport in *a*-Se

(i) Overview

Early studies related to the effect of impurities on charge transport in *a*-Se demonstrated that arsenic doping significantly decreased the electron drift mobility while the hole drift mobility was largely unaffected by arsenic concentrations less than ~ 2% [92]. Since that study, a number of papers on the same subject have been written.

The effects of the various impurities upon the electronic transport properties are conveniently grouped according to the valence of the impurity. These are (i) univalent additives (Cl, Tl), (ii) additives isoelectronic with Se (S, Te) and (iii) additives capable of introducing chain branching (P, As, Bi, Ge) [135]. The effects of additives within one group are very similar, so that we shall restrict ourselves to discuss representative members of each group, namely Cl, S, Te and As. We first provide the experimental evidence and defer its interpretation to the next section. In the following, all concentrations are given in units of atomic percent.

(a) Univalent additives; (for example, Cl)

The addition of small amounts of Cl (~ 20 p.p.m.) completely annihilates the electron response. In Cl-doped samples either the electron lifetime is $\leqslant 0.16\,\mu s$ or the Schubweg at 10^5 V/cm is $< 1\,\mu m$ [126]. Hole transport, on the other hand, is largely unaffected by Cl additives [126].

(b) Isoelectronic additives; (for example, S, Te)

The effect of isoelectronic additives is not as drastic as that of univalent additives and relatively large concentrations (~ 5% for S, 0.1% for Te) are necessary to influence the transport properties in a noticeable way. Although different workers report quantitatively somewhat different results, it is generally found that S and Te gradually decrease both electron and hole drift mobilities [135—137]. For both additives, the activation energy for electron transport remains essentially unchanged from its value observed in pure *a*-Se. No temperature measurements for hole transport in the Se—S system are reported, but studies on the Se—Te system indicate an increase of the hole activation energy from ~ 0.24 eV for pure Se to a final value ~ 0.33 eV reached at a loading of ~ 0.2% Te [127]. Carrier lifetimes appear most sensitive to S additives. The hole lifetime rapidly drops from the $10\text{—}15\,\mu s$ value

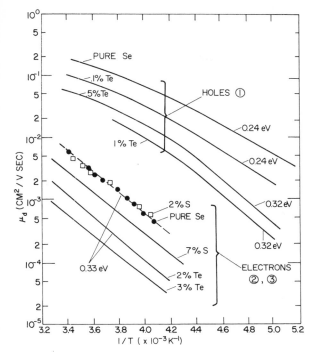

Fig. 7.28. Effect of isoelectronic additives on electron and hole mobility in a-Se. [From: ① Pai [127], ② Schottmiller et al. [135] (pure Se, 2%S, 2%Te), and ③ Kolomiets and Lebedev [136] (7%S, 3%Te).]

of pure Se to $<1\,\mu s$ at 2.4% S and then remains insensitive to a further increase in S content. The electron lifetime, on the other hand, slowly increases from $\sim 50\,\mu s$ (pure Se) to $\sim 150\,\mu s$ (12% S). Figure 7.28 summarizes representative mobility data.

(c) *Branching additives; (for example, As)*

The additive As is of significant importance since it allows one to continuously follow the transition between the two distinctly different (room temperature) transport characteristics observed in a-Se on the one hand and a-As_2Se_3 on the other hand. The effect on electron transport is summarized in Fig. 7.29, where room-temperature drift mobility and lifetime are shown in relative units normalized with respect to their values observed in pure Se. The solid lines in the figure are only for clarity. The reduction of the drift mobility appears larger in evaporated than in quenched films, which might indicate that in the latter specimen the assimilation of As is less destructive. In the quenched films, the activation energy increases continuously from ~ 0.33 eV for pure Se to ~ 0.44 for 8% As [132]. No such correlation was found in evaporated films in the reported 2% As concentration range [135].

References pp. 297—302

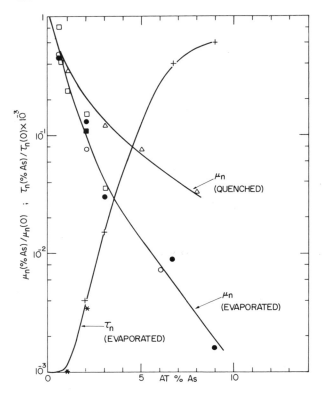

Fig. 7.29. Effect of As additives on electron mobility μ_n and lifetime τ_n in a-Se at room temperatures. The data were normalized with respect to their values of pure Se reported by the respective authors (From: ■ Hartke [92], ▲ Pai [127], △ Marshall et al. [132], +● Schottmiller et al. [135], *○ Kolomiets and Lebedev [136].)

The effect of As on the hole transport is rather spectacular [135, 137]. With small additions of As ($\sim 2\%$) the hole lifetime rapidly falls from 50 μs to below 0.03 μs while, in the same concentration range, the drift mobility remains well defined and retains its value for pure Se. Between 2—6% As only a lifetime-limited signal is observed. On raising the As concentration above 6%, the hole signal reappears, but now is comprised of an initial fast spike, characteristic of transport in pure Se, which is followed by a long featureless tail. Above $\sim 18\%$ As the initial fast response no longer can be resolved and the signal assumes the shape characteristic of a-As_2Se_3 (Fig. 7.18). In contrast to the electron drift mobility (Fig. 7.29) no gradual change of the hole drift mobility with increasing As content is observed and the transition from the value $\sim 10^{-1}$ cm^2/V.s below 2% As to $\sim 10^{-5}$ cm^2/V.s above 18% As reflects the destruction and buildup of two distinctly different transport mechanisms. In the intermediate concentration range, both mechanisms seem to operate.

(ii) *Interpretation of the experimental results*

An attempt to correlate molecular structure and electronic transport in pure and doped a-Se is rather speculative, particularly since the structure of a-Se itself is still very much debated [33]. Also, the interpretation of the drift mobility experiments is strongly influenced by the particular transport model in mind. Much of the experimental evidence available so far can be rationalized in terms of a trap-controlled mechanism as well as within the framework of phonon-assisted hopping, while, at the same time, objections can be raised against both lines of argument. In the following discussion, we describe the various interpretations of electronic transport in pure and doped a-Se without favoring any one of the proposed models.

The structural properties of a-Se have been summarized by Cooper and Westbury [33]. Different workers seem to agree that the structures of quenched and evaporated films do not differ significantly. This result seems to be manifest in the relative insensitivity of the electronic transport properties to sample preparation. Apart from this general observation, however, the question of the relative population of rings (Se_8) and chains that constitute a-Se remains far from settled.

Univalent additives (Cl, Tl) are believed to introduce deep traps of coulombic nature which are attractively charged when unoccupied [135]. Isoelectronic additives (S, Te) are expected to have relatively minor effects, because, within the framework of a shallow-trap-controlled carrier transport, these additives are likely to increase the density of shallow traps rather than to introduce deep-trapping states [136]. The increase of the carrier lifetime observed with some of these additives is associated with a longer residence time in shallow traps, due to an increase in their density, rather than to arise from a decrease in deep-trapping centers [136]. It has been suggested that isoelectronic additives decrease the Se_8 ring population while, at the same time, mixed ring species of the form $Se_{8-x}S_x$ are formed [135]. On this basis, electron transport can be correlated with the presence of Se_8 ring molecules. A similar, but less conclusive correlation seems to exist for hole transport [135]. Thus, according to this interpretation, localized or band states associated with the Se_8 structural unit may play an important role in electron and hole transport.

The correlation between electron transport and Se_8 ring population seems to be supported by the results observed for the branching additive As. Infrared and time-of-flight data on As—Se alloys suggest that the relative reduction of the Se_8 ring population and electron drift mobility with the addition of As are correlated [135, 138—140]. The Se_8 ring population extrapolates to zero at ~ 20% As addition. It has been suggested that As modifies the tail states of the conduction band by gradually shifting the mobility edge to higher energies [132]. In qualitative agreement with this interpretation the experimental activation energy increases continuously with As concentration [162].

The rapid decrease of the hole lifetime with increasing As concentration has been attributed to deep traps associated with the structural form $AsSe_{3/2}$ [135]. At low concentrations ($\sim 6\%$ As) these traps are isolated and the transport proceeds via states characteristic of pure Se. At higher As concentrations, a network structure of $(AsSe_{3/2})_n$ begins to evolve, which provides an additional but slow path for hole transport. The slow transport mechanism begins to dominate the fast transport through the Se-like states until, above 18%, only the response characteristic of a-As_2Se_3 is observed.

7.4.6 Hole transport in amorphous As_2S_3 and amorphous As_2Te_3

(i) Introduction

Information from time-of-flight measurements on the transport of excess carriers in the binary glasses As—S and As—Te unfortunately is very limited to date. Drift experiments on undoped a-As_2S_3 failed to time-resolve carrier transits because of range limitation [10], while similar studies on a-As_2Te_3 apparently were not undertaken because of the high dark-conductivity in this material. It would be highly desirable, however, to obtain information from drift mobility experiments on this class of chalcogenides since one has the unique possibility to nearly continuously tune the optical gap from ~ 2.32 eV (As_2S_3) [141] to ~ 1.7 eV (As_2Se_3) [141] to ~ 0.83 eV (As_2Te_3) [52, 142] which should provide insight into the much discussed density of states in the band gap. In the following sections, we first summarize transport data on pure [10] and CdI_2-doped [143] a-As_2S_3 obtained from time-of-flight and xerographic discharge measurements. We shall then describe recent transport-related experiments on a-As_2Te_3 [59, 60] which we feel are of interest to our discussion of hole transport in a-As_2Se_3 and a-Se presented in Sections 7.4.3 and 7.4.4, respectively.

(ii) Pure and CdI_2-doped amorphous As_2S_3

Time-of-flight experiments on evaporated a-As_2S_3 films were performed in the voltage mode in which case the time constant of the external circuit is large compared to the drift time [10]. Hence, for an ideal carrier sheet drifting with uniform constant velocity through the sample, the signal developed across the load resistor is a voltage ramp rising linearly with time for $t < t_T$ and remaining constant for $t \geqslant t_T$. Instead of the ideal ramp signal, for both carriers in a-As_2S_3 the voltage initially rises rapidly and then exhibits a dispersive tail also characteristic for a-As_2Se_3 [10]. For all fields applied, the electron signal is substantially smaller than the hole signal. In the latter case, the signal amplitude ΔV increases logarithmically with time over several decades following the light flash. For the electron signal, a logarithmic time dependence is observed for a short time period only and then ΔV saturates. For a carrier drift not severely hampered by trapping, $w = \mu_d \tau E > L$ and $\Delta V \propto L$, (eqn. (7.2)) whereas in the strong trapping limit $w < L$ and $\Delta V \propto w$

(eqn. (7.3)). L is the sample thickness, E the applied field and τ the lifetime with respect to deep traps. In no case, even for the thinnest samples (~ 1.9 μm) and the strongest applied fields (~ 3×10^5 V/cm), do sample thickness and signal amplitude correlate. This suggests than in a-As_2S_3, electronic transport is range-limited, i.e. the majority of the photogenerated charge does not traverse the sample within the time period of the experiment ($\lesssim 10$ ms). The much-reduced electron signal indicates an even more severe range limitation for this carrier.

The time-of-flight experiment is usually limited at the high-field end by excessive injection of charge from the metal—insulator interface or destructive breakdown. This restriction can often be overcome by using the xerographic discharge technique [146—148] in which method the unelectroded surface of the sample is charged by a corona and the decay of the surface potential is recorded while the sample is subjected to highly-absorbed light. With the xerographic discharge technique, fields strong enough to partially overcome the range limitation for hole transport can be applied to a-As_2S_3. Representative results are shown in Fig. 7.30 where the initial decay rate for

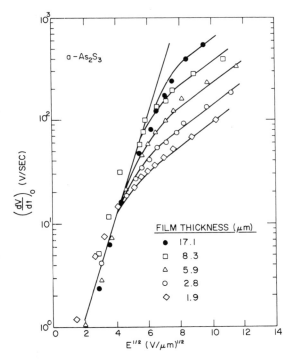

Fig. 7.30. Initial decay of the surface potential of a-As_2S_3 in a xerographic discharge experiment *versus* square root of initial field. (From Ing et al. [10].)

References pp. 297—302

constant light intensity of the surface potential dV/dt ($t = 0$) is plotted as a function of the square root of the initial field $\sqrt{E}$ for a number of samples with different thickness. It is noted that at low fields dV/dt ($t = 0$) rises sharply with field and that samples with different thickness scale in this plot. At higher fields, the field dependence of the initial discharge rate becomes weaker and, at the same time, exhibits an almost linear thickness dependence. The thickness dependence at high fields indicates that the majority of the holes transit the sample and the discharge is limited by field-dependent emission from the generation region. At low fields, in agreement with the time-of-flight result, the discharge is range limited and the field dependence of the discharge rate reflects the field dependences of Schubweg w and carrier generation efficiency. The intersection of the extrapolation to the high field and low field discharge rates determines the field for which $w \simeq L$.

Time-resolved hole transits under electron bombardment can be observed in a-As_2S_3 films doped with CdI_2 [143]. The shape of the hole transient-current pulse exhibits strong temperature and voltage dependence, which indicates that the shape is primarily determined by field-dependent release from the geminate recombination centers, i.e. from the generation region close to the surface, and that the stochastic process which is dominant in a-As_2Se_3 plays only a minor role. The hole mobility is thermally activated with $\sim 0.2\,eV$ and $\sim 0.1\,cm^2/V.s$ pre-exponential factor, thus $\mu_d^p \sim 10^{-4}$ $cm^2/V.s$ at room temperature. Compared to the activation energies for hole transport in a-As_2Se_3 ($\sim 0.55\,eV$) and a-As_2Te_3 ($\sim 0.14\,eV$) the activation energy for CdI_2-doped a-As_2S_3 is surprisingly low. One might speculate that CdI_2 reduces the activation energy from the value it would have in undoped a-As_2S_3 by a mechanism similar to that proposed for Ag-doped a-As_2S_3, in which case the reduction in activation energy for d.c. dark conductivity tentatively has been interpreted by a decrease of the range of localized valence states above the mobility edge [149]. In fact, one might speculate that Cd reduces the concentration of hole-trapping dangling bonds [143]. No electron transients were observed, even in the doped samples.

(iii) Amorphous As_2Te_3

In this section, we shall focus on aspects of hole transport in a-As_2Te_3 which were deduced from measurements of d.c. dark and photoconductivity [52, 56, 57, 59—61, 142, 150—153], recombination time [59, 60], thermopower [60, 151] and Hall mobility [60]. Extended-state conduction has been proposed as the likely transport mechanism around room temperature [57, 61, 153] and the treatment of the recombination kinetics in a framework similar to that applied to tetrahedrally bonded glasses as well as a large number of chalcogenide glasses appears to be in essential agreement with the photoconductivity data [50, 57, 58, 65—67, 72, 153, 154]. This treatment, which involves extended-state transport, has been described in great detail in Chapter 6 and has also been summarized in Section 7.3 of this chapter.

Alternatively, an internally consistent interpretation of a large body of experimental information can be provided if bandtail hopping is assumed to dominate the conduction mechanism at room temperature [59, 60]. This assumption, at first, appears surprising since the pre-exponential factor of the dark d.c. conductivity of $\sim 6 \times 10^2 \, \Omega^{-1} \, cm^{-1}$ is rather large for hopping conduction and usually is interpreted as indicative of extended-state transport [40]. Both models, extended-state and hopping transport, have their appeal and since the band model has been extensively described, we shall limit ourselves to a description of the reasoning that leads to the hopping interpretation.

One line of arguments stems from the interpretation of the thermopower, which is positive with $\mathcal{E}_s \simeq 0.3$ eV and the activation energies $\mathcal{E}_d$, $\mathcal{E}_p$, $\mathcal{E}_{\tau_0}$ and $\mathcal{E}_H$ for dark and photoconductivity σ_d and σ_p, recombination time τ_0, and Hall mobility, respectively. These activation energies are, if the sign is defined by $\exp(-\mathcal{E}/kT)$, $\mathcal{E}_d \simeq +0.44$ eV, $\mathcal{E}_p \simeq 0.14$ eV and $+0.17$ eV in the monomolecular and bimolecular recombination regimes, i.e. at temperatures above and below the photoconductivity maximum, respectively (Figs. 6.5, 7.13) $\mathcal{E}_{\tau_0} \simeq -0.26$ eV in the monomolecular recombination regime, and finally $\mathcal{E}_H \simeq 0.06$ eV. On the basis of phenomenological rate equations describing the two recombination regimes, one derives $\sigma_p \propto \mu \tau_0 F$ and $\sigma_p \propto \mu F^{1/2}$ where μ is the conductivity mobility and F the light intensity [59, 61, 79]. These relations independently predict that the mobility is thermally activated because the photoconductivity and the recombination time are activated with different energies. From the linear and square root regime one readily finds the respective activation energies for the mobility $\mathcal{E}_\mu \simeq +0.12$ eV and $+0.17$ eV, which are in reasonable mutual agreement. Evidence that the activated mobility represents hopping in tail states rather than thermal activation from shallow traps into extended states is deduced from the analysis of the thermopower results [60]. For hopping, $\mathcal{E}_\mu = \mathcal{E}_d - \mathcal{E}_s \simeq 0.14$ eV, which is in remarkable accord with the activation energies obtained from the d.c. photoconductivity experiment. Although the analysis of the Hall measurement is rather speculative at present [43], the observed thermally-activated Hall mobility with $\mathcal{E}_H \simeq (1/3)\mathcal{E}_\mu$ would tend to support hopping [155, 156] rather than extended-state [157] conduction, in which case the Hall mobility is expected to be temperature independent.

Starting from the Cutler—Mott [158] expression for the d.c. conductivity, the contributions to the total conductivity from extended-state and hopping transport can be derived assuming a density of states increasing linearly with energy from the bottom of the band tail and a linear energy dependence for both the hopping activation energy $\mathcal{E}_\mu$ and hopping distance ρ such that at the band edge $\mathcal{E}_\mu$ vanishes and ρ approaches an interatomic distance [60]. This assumption appears reasonable since close to the band edge the carriers execute non-thermal activated hops over interatomic distances [159, 160]. The field dependence of the conductivity observed at high fields is accounted

References pp. 297—302

for by letting $\mathcal{E}_\mu$ be a linearly-decreasing function of the applied field. With these assumptions, hopping conduction can be shown to prevail at room temperature while extended-state conduction becomes noticeable only at very high temperatures, close to melting, and, finally, in the liquid state two-band conduction begins to dominate [60, 145]. This model further accounts for the observed curvature of the log σ versus $1/T$ plot [152] and correctly predicts the higher activation energy (~ 0.55 eV) observed [94] in the temperature range 660—780K. The field dependence of the conductivity assumes the form of eqn. (7.41), where the approximately correct temperature dependence of the parameters $a(T)$ is predicted. The analysis yields ~ 50 Å for the average, and ~ 100 Å for the hopping distance at the bottom of the band tail. These numbers appear reasonable when compared to a value of ~ 120 Å obtained by Scher and Montroll [104] from the analysis of the time-of-flight results on a-As_2Se_3. Finally, the rather large value for the pre-exponential factor of the dark d.c. conductivity can be predicted which suggests that the pre-factor in itself is not necessarily a hallmark for the conduction process [60].

7.4.7 a.c. conduction

(i) Dark a.c. conduction

For a large number of amorphous solids, the dark a.c. conductivity $\sigma(\omega)$ exhibits over a wide range of frequencies an ω^s dependence where $s \lesssim 1$. The classical example for a frequency-dependent a.c. conductivity of this type is compensated crystalline silicon where it has been demonstrated that the a.c. conductivity observed at low temperatures is due to hopping of carriers among a random array of impurity sites [162]. A powerful argument for the interpretation of those experiments was that the impurity concentration could unambiguously be related to the a.c. conductivity. With this background, then, the observation of a similar frequency dependence of the dark conductivity in amorphous solids was very appealing since it appeared to reflect localized dissipation of electronic charge and thus a.c. conductivity was thought to provide a powerful method to probe the widely-discussed density of states in the band gap, in particular at the Fermi level and in the band tail. As a further attractive feature, the a.c. technique, in principle, enables one to examine purely bulk dissipative mechanisms since, as the frequency is increased, contact effects and internal barriers eventually should become capacitively short circuited. In the microwave frequency domain, where cavity perturbation and slotted line techniques are used, samples are not electroded and can be studied in the form of powders, suspensions or emulsions which do not require the preparation of thin films. This feature can be particularly advantageous in studies of active molecules dispersed in inert matrices [110].

Theoretical treatments of the ω^s dependence were provided by numerous authors [162—174] and have been reviewed extensively [40, 43, 174]. It is recognized that the exponent s is, in general, weakly temperature dependent and may decrease from $s \lesssim 1$ to $s \gtrsim 0.5$ as the temperature is raised. In principle, an almost-linear frequency dependence of the conductivity is obtained from any loss mechanism with sufficiently wide distribution in relaxation times [43, 167, 175, 176]. Hence, the generally observed ω^s behavior does not have to be a characteristic of a particular type of electronic conduction process but might, as has been argued, originate in the lack of long-range order. Recent experiments however, demonstrate that the absence of long-range order is not a necessary requirement for ω^s frequency dependence [177]. For instance, for both amorphous and crystalline As_2S_3, $\sigma_{dark} \propto \omega^{0.8}$ over a wide frequency range.

The interpretation of the a.c. data in terms of localized electronic dissipation is not without problems. For instance, assuming that electronic hopping at the Fermi level dominates the a.c. loss mechanism, which, based upon the weak temperature dependence of σ_{dark}, appears reasonable for most chalcogenides (see below), the density of states at the Fermi level $N(\mathscr{E}_F)$ can be determined from the Austin—Mott [169] formula. The derived densities which, in expectation with the model, scale with the band gap, *i.e.* decrease with increasing gap, typically range between 5×10^{17} cm^{-3} eV^{-1} (As_2S_3) and 5×10^{19} cm^{-3} eV^{-1} (Te—Si—Ge—As alloys) and thus are high enough to induce optical absorption and pin the Fermi level. Neither effect, however, has been substantiated satisfactorily by experiment. Photoinjection experiments demonstrate that in a-Se, for instance, the Fermi level is free to adjust to the Fermi level of the metal electrode [107], which is supported by the low value of $N(\mathscr{E}_F) \lesssim 2 \times 10^{14}$ cm^{-3} eV^{-1} derived from photoemission data, but is in striking disagreement with the conclusions from the a.c. conductivity measurements. Interestingly enough, however, more recent field-effect experiments placed a lower limit of $\sim 10^{20}$ cm^{-3} eV^{-1} on $N(\mathscr{E}_F)$, which tends to support the estimate from the a.c. conductivity [179].

Possible electronic dissipation mechanisms that lead to the observed frequency and temperature dependence of the a.c. conductivity have been reviewed by Mott and Davis [40]. In Fig. 7.31, we reproduce their informative schematic illustration of three conduction mechanisms (a—c). The frequency-independent mechanism shown by curve (a) represents the Drude loss due to carriers excited into extended states. The conductivity for this process is thermally activated with an energy corresponding to the difference between the mobility edge and the Fermi level ($\mathscr{E}_c - \mathscr{E}_F$). Curve (b) represents hopping transport of carriers in band tails. The conductivity in the d.c. limit is thermally activated with ($\mathscr{E}_A - \mathscr{E}_F$) + W_b where $\mathscr{E}_A$ is the energy at the band edge and W_b the hopping energy. The frequency-dependent part contains, as the dominant temperature term only, the number of carriers at $\mathscr{E}_A$ and thus is expected to be activated with ($\mathscr{E}_A - \mathscr{E}_F$). Curve (c) is the

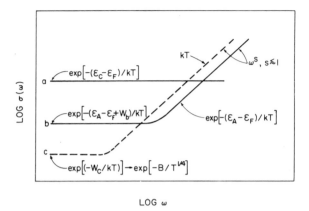

Fig. 7.31. Electronic conduction mechanisms and their frequency and temperature dependences; (a) Extended state (Drude), (b) Hopping in band tail state and (c) Hopping at the Fermi level. (From Mott and Davis [40].)

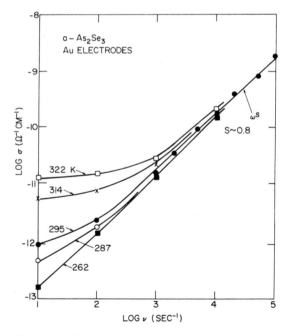

Fig. 7.32. Dark a.c. conductivity *versus* frequency of a-As_2Se_3 at various temperatures. (From Abkowitz *et al.* [177].)

contribution of the carriers hopping at the Fermi level. In this case, the d.c. current is activated with the hopping energy W_c if hopping occurs between nearest neighbors, but will exhibit the well-known $\exp(-BT^{-1/4})$ dependence [168] if, at still lower temperatures, hopping over longer distances becomes favorable. The frequency-dependent part contains no exponential temperature dependence and is expected to increase in proportion to T [169]. At any given temperature, the most lossy mechanism will dominate, and it is clear from Fig. 7.31 that only the temperature dependence of the entire $\sigma(\omega)$ curve can establish the dominant transport mechanism. An appreciated difficulty in analyzing the a.c. data is whether only one process is observed, process (b) for instance, or whether the process dominating the low frequency loss is different from the process observed at high frequencies.

In Fig. 7.32, we show as representative example for chalcogenide glasses the dark a.c. conductivity of $a\text{-}As_2Se_3$ at various temperatures. The sample was prepared by evaporation and was provided with gold electrodes. It is noted that, at low frequencies, the a.c. conductivity saturates. Referring to Fig. 7.31, the approach to d.c. conduction could be dominated by any one of the three conduction mechanisms (a—c). However, the low frequency loss is thermally activated with an energy about one half of the band gap. This tends to rule out process (c). A distinction between the mechanisms (a) and (b) most often is speculative because their activation energies should be similar and because, as has been discussed in Section 7.4.6, the magnitude of the pre-exponential factor of the d.c. conductivity does not seem to allow an unambiguous distinction between band-tail hopping and extended-state conduction. As a further complication, it is not clear at present whether the d.c. current is bulk-controlled or limited by emission from the contact region. In fact, it has been suggested [180] that the nature of the gold contact may be time dependent and could well change from blocking to injecting as time goes on. As indicated in Fig. 7.32, the frequency-dependent part of $\sigma_{a.c.}$ depends only weakly upon temperature, which suggests that process (c) dominates.

Often $\sigma_{a.c.}$ approaches a temperature-independent ω^2 dependence at high frequencies. The origin of this type of frequency dependence can be manifold. ω^2 dependence followed by a saturation can be characteristic of any limited sequence of hopping [174]. Also, correlation effects in the hopping mechanism would lead to a stronger than linear frequency dependence [165]. At the highest frequencies, the long low-energy tail due to phonon absorption processes may become important and give rise to the ω^2 dependence [181]. Further high-frequency loss mechanisms have been proposed to account for the ω^2 behavior in the 10^9-10^{13} Hz frequency range typical of a number of chalcogenide glasses [182]. At lower frequencies, the ω^2 behavior can also arise from sheet resistance of the electrode material [177, 183—185]. This is demonstrated in Fig. 7.33 by measurements on As_2S_3 single crystals provided with liquid electrodes of different conductivity

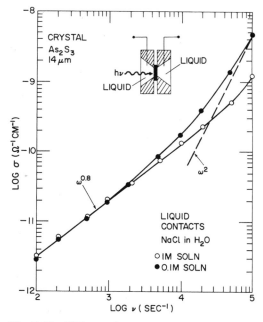

Fig. 7.33. Effect of contact resistance on frequency dependence of a.c. conductivity. (From Abkowitz et al. [177].)

[177]. The onset of a stronger than linear frequency dependence $\sigma_{a.c.}$ is clearly shifted towards lower frequencies as the salt solution of the electrode is diluted.

(ii) Photo a.c. conduction

No photoinduced dielectric loss at microwave frequencies is observed for chalcogenide glasses but a large effect is found, for instance, for Al-doped CdS [186]. The reason for the absence of a photo-induced effect in amorphous chalcogenides lies in the fact that, contrary to the ω^s dependence of $\sigma_{dark}(\omega)$, the photo a.c. conductivity in amorphous insulators typically is frequency independent and therefore the photoinduced a.c. loss at microwave frequencies can be swamped by the loss of the dark conductivity component [187]. In Al-doped CdS photo and dark a.c. conductivity exhibit the same (mild) frequency dependence [187].

A frequency-independent photo a.c. conductivity suggests either dissipation in extended states (process (a), Fig. 7.31) or an emission-limited process, in which case the dissipation is independent of transport parameters. Identification of the proper transport mechanism rests upon a knowledge of the relative magnitudes of carrier recombination and transit time and upon the nature of the electric contact. Loss mechanisms have been suggested for a number of amorphous insulators [187]. As an example, we mention

a-As_2Se_3 for which the photo a.c. loss is proposed to be extended-state dissipation. Thus, the path of electronic conduction under conditions of steady-state illumination with weakly bulk-absorbed light is different from that observed in the time-of-flight experiment. It is suggested that the carriers injected by a pulse of strongly-absorbed light rapidly fall into deep trapping centers and then transit the sample by hopping among those localized states whereas steady-state illumination maintains a carrier concentration in extended state, which is large enough to dominate the a.c. loss [187]. Using the photo a.c. technique, one therefore has the interesting prospective of selectively populating electronic states and studying the associated mechanisms of transport.

7.4.8 The mobility edge

The mobility edge, defined as to separate extended states from localized states, is a fundamental concept for electronic transport in noncrystalline solids [159, 160, 168, 189—192]. The localized states are believed to be a consequence of potential fluctuations associated with translational and compositional disorder and, thus, are an intrinsic property of the amorphous state and not the result of impurities. The localized states split off the respective bands and with increasing disorder extend and tail off into the energy gap. Charge transport above and below the mobility gap is characteristically different. Sufficiently above the mobility edge the carrier mean free path or, equivalently, the phase coherence length covers many interatomic distances; hence, the carriers suffer only occasional scattering and propagate without thermal activation. As the mobility edge is approached from the side of extended states, disorder increases and, correspondingly, the scattering length decreases. Eventually, the scattering length becomes of the order of an interatomic spacing and transport can no longer be described as being extended. The transition is reached when the mobility approaches a value of the order of $\sim 5 \text{ cm}^2/\text{V.s}$ [193, 194]. Deeper into the gap, the wave functions will become localized [192] and transport can proceed only via thermal activation, *viz.* phonon-assisted hopping among the localized states. An upper value for the mobility associated with phonon-assisted hopping is $\sim 4 \times 10^{-2} \text{ cm}^2/\text{V.s}$ which defines the position of the mobility edge in the band gap. Below the edge, the mobility decreases because the density of localized states tails off and the degree of localization becomes increasingly stronger. The transport mechanism bridging the mobility range 4×10^{-2}—$5 \text{ cm}^2/\text{V.s}$ is believed to involve Brownian motion with carriers jumping incoherently from site to site. Different from the hopping process occurring below the mobility edge, the process of Brownian motion is not thermally activated because the thermal energy is still large enough to overcome the fluctuation of the energies of the carriers at the individual localization sites.

Direct evidence in support of the mobility-edge concept stems from electron drift mobility experiments in a-Si prepared by glow discharge decomposition of silane [128]. These experiments were discussed extensively in Chapter 6 and have also been summarized in Section 7.4.4. In the following, we will summarize some experiments which appear relevant to a discussion on the existence of a mobility edge in the representative chalcogenides a-Se, a-As$_2$Se$_3$ and a-As$_2$Te$_3$.

(i) The absence of a pressure effect upon the drift mobility in a-Se is often cited as evidence for shallow-trap-controlled extended-state conduction since, as has been argued, pressure would increase the overlap if hopping were the transport mechanism [118]. Based upon the recent analysis of the transient current pulse in a-As$_2$Se$_3$, one would expect, however, that a pressure-induced change of the overlap might be too small to be observed. For a-As$_2$Se$_3$, Scher and Montroll [104] estimate $\rho/\rho_0 \simeq 5-10$ for the ratio of average hopping distance and localization radius which, in the experimentally accessed pressure range < 6 kbar [118] would amount to a pressure effect $\leqslant 10\%$. For a-Se ρ/ρ_0 is expected to be even smaller since the higher mobility and lower activation energy suggest a higher hopping site density and a more delocalized wave function.

(ii) Another argument in favor of shallow-trap-controlled transport in a-Se has been the good fit of eqn. (7.40) to the temperature dependence of the hole drift mobility when phonon scattering was assumed to control the microscopic mobility μ_0^p. The latter assumption, however, appears speculative since $\mu_0^p \simeq 0.3$ cm^2/V.s is rather small for extended-state transport. One also has to bear in mind that the deviation from thermally-activated behavior occurs on approaching the glass transition and the associated structural rearrangement might dominate the transport properties in this temperature range. In fact, for a number of organic systems for which there is strong evidence for hopping transport, the mobility deviates from the Arrhenius dependence below the glass transition in a fashion similar to that observed in a-Se [196] (Fig. 7.26).

(iii) The yield for internal photoemission from metal contacts into a-Se is essentially temperature independent [197]. This result suggests the existence of a weakly temperature-dependent mobility which would be evidence of a microscopic mobility. However, conclusions from the photoinjection experiments may not be relevant to time-of-flight experiments. The photoinjection yield is limited by emission of carriers across the image potential and, therefore, the derived mobility pertains to the properties of hot carriers close to the surface, which might be different from the bulk properties that control the transport [197].

(iv) For a-Se, no distinct lowering of the activation energy is indicated throughout the temperature range in which the hole pulse shape becomes progressively dispersive and, at constant temperature, assumes the universal shape expected from the hopping theory [103, 104].

(v) For a-As$_2$Se$_3$, extensive results from hole drift experiments favor the interpretation that the photoinjected holes are transported via hopping rather than via extended states with occasional deep trapping [97, 98, 102–104]. Extended-state transport in a-As_2Se_3 under conditions of steady-state illumination with weakly bulk-absorbed light is indicated in the frequency independence of the a.c. conductivity. This observation, however, has to be reconciled with the frequency-independent photo a.c. conductivity observed in hopping systems such as PVK: TNF. A possible explanation might be that, in the latter case, the a.c. conductivity is emission limited [187].

(vi) For a-As_2Te_3, the line of arguments proposed in favor of hopping conduction in band-tail states was presented in Section 7.4.6. The mobility remains activated over the accessed temperature range ($\sim$ 200K) with single activation energy. At higher temperatures ($\sim$ 650K) there seems to be evidence that extended-state conduction becomes important [60]. Recently, however, a decrease of the activation energies for the hole drift and d.c. conductivity in a-As_2Te_3 has been observed when temperature is lowered through $\sim$ 145K, which can be interpreted, in analogy to the results on electrons in a-Si, as due to the transition from shallow-trap-controlled to hopping transport [153]. Similar observations were also reported for the quaternary glass a-$As_{30}Te_{48}Si_{12}Ge_{40}$ [153].

REFERENCES

1 F.K. Dolezalek and W.E. Spear, J. Phys. Chem. Solids, 36 (1975) 819.
2 R.C. Chance and C.L. Braun, J. Chem. Phys., 59 (1973) 2269.
3 D.F. Blossey and R. Zallen, Phys. Rev. B, 9 (1974) 4306.
4 D.M. Pai, J. Chem. Phys., 52 (1970) 2285;
   G. Pfister and D.J. Williams, J. Chem. Phys., 61 (1974) 2416.
5 P.J. Melz, J. Chem. Phys., 57 (1972) 1694.
6 J.-P. Dodelet and G.R. Freeman, J. Chem. Phys., 60 (1974) 4657;
   G.R. Freeman and J.-P. Dodelet, Int. J. Radiat. Phys. Chem., 5 (1973) 371.
   J.-P. Dodelet, P.G. Fuochi and G.R. Freeman, Can. J. Chem., 50 (1972) 1617.
7 A. Mozumder, J. Chem. Phys., 60 (1974) 4300, 4305;
   W.F. Schmidt, Radiat. Res., 42 (1970) 73.
8 D.M. Pai and R.C. Enck, Phys. Rev. B, 11 (1975) 5163.
9 M.E. Scharfe, Phys. Rev. B 2 (1970) 5025.
10 S.W. Ing, Jr., J.H. Neyhart, and F. Schmidlin, J. Appl. Phys., 42 (1971) 696.
11 H.T. Li and P.J. Regensburger, J. Appl. Phys., 34 (1963) 1730.
12 D.M. Pai, in A.V. Patsis and D.A. Seanor (Eds.), Photoconductivity in Polymers, Technomic, in press.
13 M.E. Scharfe and M.D. Tabak, J. Appl. Phys., 40 (1969) 3230.
14 M. Tabak, Trans. AIME, 239 (1967) 330.
15 J.L. Hartke and P.J. Regensburger, Phys. Rev. A, 139 (1965) 970.
16 G. Lucovsky and M.D Tabak, in R.A. Zingaro and W.C. Cooper (Eds.), Selenium, Van Nostrand, New York, 1974, p. 788.
17 M.D. Tabak and P.J. Warter, Phys. Rev., 173 (1968) 899.
18 D.M. Pai and S.W. Ing, Phys. Rev., 173 (1968) 729.
19 J.E. Knights and E.A. Davis, J. Phys. Chem. Solids, 35 (1974) 543.

20 J. Frenkel, Phys. Rev., 54 (1938) 647.
21 L. Onsager, Phys. Rev., 54 (1938) 554; J. Chem. Phys., 2 (1934) 599.
22 P.J. Warter, Jr., in E.M. Pell (Ed.), Proc. Third Int. Conf. on Photoconductivity, Pergamon, Oxford, England, 1971, p. 311.
23 H. Seki, Phys. Rev. B, 2 (1970) 4877.
24 R.C. Enck, Phys. Rev. Lett., 31 (1973) 220.
25 H. Seki, J. Appl. Phys., 43 (1972) 1144.
26 R.C. Enck, Bull. Am. Phys. Soc., 211 (1974) 19.
27 J.L. Hartke, J. Appl. Phys., 39 (1968) 4871.
28 M. Silver and R.C. Jarnagin, Mol. Cryst. Liq. Cryst., 3 (1968) 461.
29 A. Mozumder and J.L. Magee, J. Chem. Phys., 47 (1967) 939.
30 J.-P. Dodelet, F.G. Fouchi and G.R. Freeman, Can. J. Chem., 50 (1972) 1617.
31 G. Juska, A. Matulionis and J. Viscakas, Phys. Status Solidi A, 4 (1971) 787.
32 G. Juska, A. Matulionis and J. Viscakas, Phys. Status Solidi, 33 (1969) 533.
33 W.C. Cooper and R.A. Westbury, in R.A. Zingaro and W.C. Cooper (Eds.), Selenium, Van Nostrand, New York, 1974, p. 87.
34 A. Eisenberg and A.V. Tobolsky, J. Polym. Sci., 46 (1960) 19.
35 M.E. Scharfe, unpublished.
36 F. Kosek and J. Tauc, Czech. J. Phys. B, 20 (1970) 94;
R. Zallen, R.E. Drews, R.L. Emerald and M.L. Slade, Phys. Rev. Lett., 26 (1971) 1564.
37 R.H. Bube, Photoconductivity of Solids, Wiley, New York, 1960.
38 A. Rose, Photoconductivity and Related Processes, Interscience, New York, 1963.
39 S.M. Ryvkin, Photoelectric Effects in Semiconductors, Consultants Bureau, New York, 1964.
40 N.F. Mott and E.A. Davis, Electronic Processes in Non-Crystalline Materials, Oxford Univ. Press, London, 1971.
41 B.T. Kolomiets and V.M. Lyubin, Phys. Status Solidi A, 17 (1973) 11.
42 J. Mort, in P.G. LeComber and J. Mort (Eds.), Electronic and Structural Properties of Amorphous Semiconductors, Academic Press, London and New York, 1973, p. 493.
43 H. Fritzsche, in J. Tauc (Ed.), Amorphous and Liquid Semiconductors, Plenum Press, London and New York, 1974, p. 221.
44 T.C. Arnoldussen, C.A. Menezes, Y. Nakagawa and R.H. Bube, Phys. Rev. B, 9 (1974) 3377.
45 E.A. Fagen and H. Fritzsche, J. Non-Cryst. Solids, 2 (1970) 180.
46 N.K. Kiseleva and B.T. Kolomiets, Sov. Phys.-Semicond., 7 (1973) 111.
47 B.T. Kolomiets, Yu.V. Rukhlyadev and V.P. Shilo, J. Non-Cryst. Solids, 5 (1971) 389.
48 B.T. Kolomiets, Yu.V. Rukhlyadev and V.P. Shilo, J. Non-Cryst. Solids, 5 (1971) 402.
49 See ref. 37, p. 390.
50 E.A. Fagen and H. Fritzsche, J. Non-Cryst. Solids, 4 (1970) 480.
51 R. Tsu, W.E. Howard and L. Esaki, J. Non-Cryst. Solids, 4 (1970) 322.
52 H.K. Rockstad, J. Non-Cryst. Solids, 2 (1970) 192.
53 R.F. Shaw, W.Y. Liang and A D. Yoffe, J. Non-Cryst. Solids, 4 (1970) 29.
54 J. Cornet and D Rossier, Philos. Mag., 27 (1973) 1335.
55 J.D. Dow and D. Redfield, Phys. Rev. B, 1 (1970) 3358; Phys. Rev. B, 5 (1971) 594.
56 J.M. Marshall, C. Main and A.E. Owen, J. Non-Cryst. Solids, 8—10 (1972) 760.
57 C. Main and A.E. Owen, Proc. 5th Int. Conf. on Amorphous and Liquid Semiconductors, Taylor and Francis, London, 1974, p. 783.

58 C. Main and A.E. Owen, in P.G. LeComber and J. Mort (Eds.), Electronic and Structural Properties of Amorphous Semiconductors, Academic Press, London and New York, 1973, p. 527.
59 T.D. Moustakas and K Weiser, Phys. Rev. B, 12 (1975) 2448.
60 A.J. Grant, T.D. Moustakas, T. Penney and K. Weiser, Proc. 5th Int. Conf. on Amorphous and Liquid Semiconductors, Taylor and Francis, London, 1974, p. 325.
61 K. Weiser, R. Fisher and M.H. Brodsky, Proc. 10th Int. Conf. on the Physics of Semiconductors, CONF—700801, NTIS, NBS, U.S. Dept. Commerce, Springfield, Va., 1970, p. 667.
62 K. Weiser, J. Non-Cryst. Solids, 8—10 (1972) 922.
63 W.E. Howard and R. Tsu, Phys. Rev. B, 1 (1970) 4709.
64 T.C. Arnoldussen, R.H. Bube, E.A. Fagen and S. Holmberg, J. Non-Cryst. Solids, 8—10 (1972) 933.
65 T.C. Arnoldussen, R.H. Bube, E.A. Fagen and S. Holmberg, J. Appl. Phys., 43 (1972) 1798.
66 K.P. Scharnhorst and H.R. Riedl, J. Appl. Phys., 43 (1972) 5142.
67 G.W. Taylor and J.G. Simmons, J. Phys. C, 7 (1974) 3067.
68 B. Cassanhiol, J. Cornet and D. Rossier, Proc. 5th Intern. Conf. on Amorphous and Liquid Semiconductors, Taylor and Francis, Ltd., London, 1974, p. 571.
69 J.G. Simmons and G.W. Taylor, J. Non-Cryst. Solids, 8—10 (1972) 947.
70 G.W. Taylor and J.G. Simmons, J. Non-Cryst. Solids, 8—10 (1972) 940.
71 J.G. Simmons and G.W. Taylor, J. Phys. C, 6 (1973) 3706.
72 J.G. Simmons and G.W. Taylor, J. Phys. C, 7 (1974) 3051.
73 W.E. Spear, R.J. Loveland and A. Al-Sharbaty, J. Non-Cryst. Solids, 15 (1974) 410.
74 M.H. Cohen, H. Fritzsche and S.R. Ovshinsky, Phys. Rev. Lett., 22 (1969) 1065.
75 W. Shockley and W.T. Read, Phys. Rev., 87 (1952) 835.
76 A. Madan, W.E. Spear and P.G. LeComber, J. Non-Cryst. Solids, in press.
77 W.E. Spear, in J. Stuke and W. Brenig (Eds.), Amorphous and Liquid Semiconductors, Taylor and Francis, London, 1974, p. 1.
78 H. Keller and J. Stuke, Phys. Status Solidi, 8 (1965) 831.
79 See ref. 39, p. 128.
80 See ref. 37, p. 279.
81 B.T. Kolomiets, T.N. Mamontova and V.V. Negreskul, Phys. Status Solidi, 27 (1968) K15.
82 S.G. Bishop and C.S. Guenzer, Phys. Rev. Lett., 30 (1973) 1309.
83 S.G. Bishop and D.L. Mitchell, Phys. Rev. B., 8 (1973) 5696.
84 F. Mollot, J. Cernogora and C. Benoit à La Guillaume, Phys. Status Solidi A, 21 (1974) 281.
85 B.T. Kolomiets, T.N. Mamontova and A.A. Babaev, J. Non-Cryst. Solids, 4 (1970) 289.
86 B.T. Kolomiets, T.N. Mamontova, E.A Smorgonskaya and A.A. Babaev, Phys. Status Solidi A, 11 (1972) 441.
87 J. Cernogora, F Mollot and C. Benoit à La Guillaume, Phys. Status Solidi A, 15 (1973) 401.
88 R.A. Street, T.M. Searle and I.G. Austin, J. Phys. C, 6 (1973) 1830.
89 R. Fisher, V. Heim, F. Stern and K. Weiser, Phys. Rev. Lett., 26 (1971) 1182.
90 W.E. Spear, Proc. Phys. Soc. London, Sect. B, (1957) 669; ibid., B76 (1960) 826.
91 W.E. Spear and H.P.D. Lanyon, Proc. Int. Conf. on Semiconductor Physics, Prague, 1960, Czech Acad. of Sciences, Prague, 1961, p. 987.
92 J.L. Hartke, Phys. Rev., 125 (1962) 1177.
93 H.P. Grunwald and R.M. Blakney, Phys. Rev., 165 (1968) 1006.
94 M.D. Tabak, Phys. Rev. B, 2 (1970) 2104.

95 J.M. Marshall and A.E. Owen, Phys. Status Solidi A, 12 (1972) 181.
96 M.E. Scharfe, Bull. Am. Phys. Soc., 18 (1973) 454.
97 H. Scher and G. Pfister, Bull. Am. Phys. Soc., 20 (1975) 322.
98 G. Pfister and H. Scher, Bull. Am. Phys. Soc., 20 (1975) 322.
99 B.T. Kolomiets and E.A. Lebedev, Sov. Phys.-Semicond., 1 (1967) 244.
100 J.M. Marshall and A.E. Owen, Philos. Mag., 24 (1971) 1281.
101 D.M. Pai and M.E. Scharfe, J. Non-Cryst. Solids, 8—10 (1972) 752.
102 G. Pfister, Phys. Rev. Lett., 33 (1974) 1474.
103 G. Pfister, Phys. Rev. Lett., 36 (1976) 271.
104 H. Scher and E.W. Montroll, Phys. Rev. B, 12 (1975) 2455.
105 M. Silver, K.S. Dy and I.L. Huang, Phys. Rev. Lett., 27 (1971) 21.
106 R.J. Fleming, J. Appl. Phys., 45 (1974) 4944.
107 J. Mort and A.I. Lakatos, J. Non-Cryst. Solids, 4 (1970) 117.
108 W. Gill, J. Appl. Phys., 43 (1972) 5033.
109 H. Seki, in J. Stuke and W. Brenig (Eds.), Proc. 5th Int. Conf. on Amorphous and Liquid Semiconductors, Taylor and Francis, London, 1974, p. 105.
110 J. Mort, G. Pfister and S. Grammatica, Solid State Commun., 18 (1976) 693.
111 A.J. Appling and A.J. Leadbetter, in J. Stuke and W. Brenig (Eds.), Proc. 5th Int. Conf. on Amorphous and Liquid Semiconductors, Taylor and Francis, London, 1974, p. 457.
112 J.M. Marshall and G.R. Miller, Philos. Mag., 27 (1973) 1151.
113 J.M. Marshall, F.D. Fisher and A.E. Owen, in J. Stuke and W. Brenig (Eds.), Proc. 5th Int. Conf. on Amorphous and Liquid Semiconductors, Taylor and Francis, London, 1974, p. 1305.
114 B.G. Bagley, Solid State Commun., 8 (1970) 354.
115 N.F. Mott, Philos. Mag., 24 (1971) 1281.
116 V.Kh. Biktimirova, B.I. Boltkas, Z.U. Borisova, T.D. Dzhafarov and A.A. Obraztsov, Sov. Phys.-Semicond., 8 (1975) 1412.
117 G. Pfister, to be published.
118 F.K. Dolezalek and W.E. Spear, J. Non-Cryst. Solids, 4 (1970) 97.
119 G. Juska, S. Vengris and J. Viscakas, in J. Stuke and W. Brenig (Eds.), Proc. 5th Int. Conf. on Amorphous and Liquid Semiconductors, Taylor and Francis, London, 1974, p. 363.
120 E.L. Rossiter and G. Warfield, J. Appl. Phys., 42 (1971) 2527.
121 S.A. Vengris, Yu.K. Viscakas, A.P. Sakalas and G.B. Juska, Sov. Phys. — Semicond., 6 (1972) 903.
122 G. Juska and S. Vengris, Phys. Status Solidi A, 16 (1973) K27.
123 J.M. Caywood and C.A. Mead, J. Phys. Chem. Solids, 31 (1971) 983.
124 P.J. Warter, Appl. Opt. Suppl., 3 (1969) 65.
125 J. Dresner, J. Phys. Chem. Solids, 25 (1964) 505.
126 M.D. Tabak and W.J. Hillegas, J. Vac. Sci. Technol, 9 (1971) 387.
127 D.M. Pai, in J. Stuke and W. Brenig (Eds.), Proc. 5th Int. Conf. on Amorphous and Liquid Semiconductors, Taylor and Francis, London, 1974, p. 355.
128 P.G. LeComber and W.E. Spear, Phys. Rev. Lett., 25 (1970) 509.
129 P.G. LeComber, A. Madan and W.E. Spear, J. Non-Cryst. Solids, 11 (1970) 219.
130 J. Mort, Phys. Rev. Lett., 18 (1967) 540.
131 W.E. Spear, J. Phys. Chem. Solids, 21 (1961) 110.
132 J.M. Marshall, F.D. Fisher and A.E. Owen, Phys. Status Solidi, 25 (1974) 419.
133 M.L. Williams, R.F. Landel and J.D. Ferry, J. Am. Chem. Soc., 77 (1955) 3701.
134 R.C. Zeller and R.O. Pohl, Phys. Rev. B, 4 (1971) 2029.
135 J. Schottmiller, M. Tabak, G. Lucovsky and A. Ward, J. Non-Cryst. Solids, 4 (1970) 161.
136 B.T. Kolomiets and E.A. Lebedev, Sov. Phys.-Solid State, 8 (1966) 905.

137 M.D. Tabak, in E.M. Pell (Ed.), Proc. Third Photoconductivity Conf., Stanford, 1969, Pergamon Press, Oxford, New York, 1971, p. 87.
138 G. Lucovsky, A. Mooradian, W. Taylor, G.B. Wright and R.C. Keezer, Solid State Commun., 5 (1967) 113.
139 G. Lucovsky, The Physics of Selenium and Tellurium, Montreal 1967, Pergamon Press, Oxford, 1969, p. 255.
140 G. Lucovsky, Mater. Res. Bull., 4 (1969) 505.
141 E. Felty and M. Myers, personal communication; see also ref. 40.
142 K. Weiser and M.H. Brodsky, Phys. Rev. B, 1 (1970) 791.
143 J. Banerji and J. Hirsch, Solid State Commun., 15 (1974) 925.
144 K. Weiser, A.J. Grant and T.D. Moustakas, in J. Stuke and W. Brenig (Eds.), Proc. 5th Int. Conf. on Amorphous and Liquid Semiconductors, Taylor and Francis, London, 1974, p. 355.
145 T.D. Moustakas, K. Weiser and A.J. Grant, Solid State Commun., 16 (1975) 575.
146 P.J. Regensburger, Photochem. Photobiol., 8 (1968) 429.
147 I. Chen, J. Appl. Phys., 43 (1972) 1137.
148 I. Chen and J. Mort, J. Appl. Phys., 43 (1972) 1164.
149 See ref. 40, p. 341.
150 N. Croituru, L. Vescar, C. Popesca and M. Lazarescu, J. Non-Cryst. Solids, 4 (1970) 493.
151 H.K. Rockstad, R. Flasck and S. Isawa, J. Non-Cryst. Solids, 8—10, (1972) 326.
152 L. Stowac, A. Abraham, A. Hruby and N. Zavetova, J. Non-Cryst. Solids, 8—10 (1972) 353.
153 J.M. Marshall and A.E. Owen, Philos. Mag., 31 (1975) 1341.
154 T. Minanni, A. Joshida and M. Tanaka, J. Non-Cryst. Solids, 7 (1972) 328.
155 D. Emin and T. Holstein, Am. J. Phys., 53 (1969) 439.
156 D. Emin, C.H. Seager and R.K. Quinn, Phys. Rev. Lett., 28 (1972) 814.
157 L. Friedman, J. Non-Cryst. Solids, 6 (1971) 329.
158 M. Cutler and N. Mott, Phys. Rev., 181 (1969) 1336.
159 N.F. Mott, J Non-Cryst. Solids, 8—10 (1972) 1.
160 M.H. Cohen, J. Non-Cryst. Solids, 2 (1970) 432.
161 J.T. Edmond, Br. J. Appl. Phys., 17 (1966) 979.
162 M. Pollak and T.H. Geballe, Phys. Rev., 122 (1961) 1745.
163 M. Pollak, Phys. Rev. A, 133 (1964) 564.
164 M. Pollak, Phys. Rev. A, 138 (1965) 1822.
165 M. Pollak, Philos. Mag., 23 (1971) 519.
166 A.K. Jonscher, J. Non-Cryst. Solids, 8—10 (1972) 293.
167 M. Pollak and G.E. Pike, Phys. Rev. Lett., 28 (1972) 1449.
168 N.F. Mott, Philos. Mag., 19 (1969) 835.
169 I.G. Austin and N.F. Mott, Adv. Phys., 18 (1969) 41.
170 H. Scher and M. Lax, J. Non-Cryst. Solids, 8—10 (1972) 497.
171 H. Scher and M. Lax, Phys. Rev. B, 7 (1973) 4491, 4502.
172 P.N. Butcher, J. Phys. C, 5 (1972) 1817.
173 A.K. Jonscher, in J. Stuke and W. Brenig (Eds.), Proc. 5th Int. Conf. on Amorphous and Liquid Semiconductors, Taylor and Francis, London, 1974, p. 1179.
174 A.K. Jonscher, in P.G. LeComber and J. Mort (Eds.), Electronic and Structural Properties of Amorphous Semiconductors, Academic Press, London, 1973, p. 329.
175 A.E. Owen, Prog. Ceram. Sci., 3 (1963) 77.
176 H. Fritzsche, J. Non-Cryst. Solids, 6 (1971) 49.
177 M. Abkowitz, D.F. Blossey and A.I. Lakatos, Phys. Rev. B, 12 (1975) 3400.
178 P. Nielsen, Phys. Rev., 6 (1972) 3789.
179 W.E. Spear, in J. Stuke and W. Brenig (Eds.), Proc. 5th Int. Conf. on Amorphous and Liquid Semiconductors, Taylor and Francis, London, 1974, p. 1.

180 S. Tutihasi, J. Appl. Phys., 47 (1976) 277.
181 I.G. Austin and E.S. Garbett, Philos. Mag., 23 (1971) 17.
182 K. Strom and P.C. Taylor, in J. Stuke and W. Brenig (Eds.), Proc. 5th Int. Conf. on Amorphous and Liquid Semiconductors, Taylor and Francis, London, 1974, p. 375.
183 C. Creveceour and H.J. deWitt, Solid State Commun., 9 (1971) 445.
184 A.I. Lakatos and M. Abkowitz, Phys. Rev., 3 (1971) 1791.
185 R.A. Street, G B. Davies and A.D. Yoffe, J. Non-Cryst. Solids, 5 (1971) 276.
186 J.L. Stone, C.R. Haden and S.A. Collins, J. Non-Cryst. Solids, 8—10 (1972) 614.
187 M. Abkowitz, A.I. Lakatos and H. Scher, Phys. Rev. B, 9 (1974) 1813.
188 N.F. Mott, Festkorperprobleme, 9 (1969) 22.
189 N.F. Mott, Adv. Phys., 16 (1967) 49.
190 N.F. Mott, Contemp. Phys., 10 (1969) 125.
191 For a recent review see N.F. Mott, in P.G. LeComber and J. Mort (Eds.), Electronic and Structural Properties of Amorphous Semiconductors, Academic Press, London, 1973, p. 1.
192 P.W. Anderson, Phys. Rev., 109 (1958) 1492.
193 N.F. Mott and W.D. Twose, Philos. Mag., 10 (1961) 107.
194 N.F. Mott, Philos. Mag., 17 (1968) 1259.
195 K. Weiser, Comments Solid State Phys., to be published.
196 D. Pai and G. Pfister, personal communication.
197 J. Mort, A.I. Lakatos and F. Schmidlin, J. Appl. Phys., 42 (1971) 5761.

CHAPTER 8

POLYMERIC PHOTOCONDUCTORS

W.D. GILL

8.1. Introduction
8.2. Charge transport
   8.2.1. General discussion
   8.2.2. Mobility
      (i) Mobility measurements
      (ii) Field-dependent mobility
      (iii) Concentration dependence of mobility
   8.2.3. AC conductivity
8.3. Charge generation and recombination
   8.3.1. Polymer films
   8.3.2. TNF:PVK charge transfer complex
8.4. Injection experiments
8.5. Conclusions

8.1 INTRODUCTION

In the last decade there have been significant advances in our understanding of the solid-state properties of organic materials. Most of this insight has been due to work on well-characterized organic crystals and has been described in Chapter 5 by H. Inokuchi and Y. Maruyama and references cited therein. In this same period, there has been intense interest in the electronic properties of organic polymers. For a number of reasons, the advance in our understanding of transport and photoconductivity in these materials has been slow. The two main obstacles to progress have been the complexity of disordered polymeric systems and the very low electrical conductivity of most of these materials. The whole subject of transport properties in disordered materials has itself evolved over this time span, but progress in this field has so far been limited to much simpler structural systems than organic polymers (see Chapters 6 and 7). The insulating nature of most polymer materials has made it impossible to use standard semiconductor measurement techniques with any success. Consequently, researchers have turned to photoconductivity measurements as a means of introducing excess charge carriers in attempts to measure transport properties.

The subject of the electrical properties of polymers is a vast one. In a

References pp. 332—334

recent review, Seanor [1] has considered the properties of major groups of polymers. It is important to recognize that the electrical properties of these major groups can be very different and because of fundamental differences in bonding, morphology or structure the dominant transport mechanisms are probably totally different. For example, the high conductivities of conjugated polymers are a consequence of extensive delocalization of conduction electrons along the polymer backbone in contrast to the situation in their saturated counterparts where this transport path is not available. In this chapter, we will restrict our discussion to that group of materials which can be best characterized as photoconducting polymers and monomer dispersions in polymers. These are the polymeric systems which have been of greatest technological interest as potential photoconductors. The photoconducting polymers can be roughly characterized by common structural features: a saturated polymer chain ensures low dark conductivity, and large, often pendant, planar structures with extended π-electron systems which can be efficiently photoexcited. The weak interaction of the π-electronic systems along saturated polymer chains suggests that very little difference should exist in the photoconducting properties of these polymers and concentrated dispersions of analogous monomer units in inert matrices. For this reason, we include some discussion of the transport properties of organic dispersions and glasses. As may be expected in a field such as organic photoconductivity, with its almost limitless variety of potential materials, there is an overwhelming literature dealing with various aspects of photoconductivity and transport. This chapter is not an attempt to review this extensive literature. The books by Gutmann and Lyons [2] and Katon [3] and several recent reviews [1, 4—8] provide more comprehensive surveys of this nature. Our purpose is rather to concentrate on those experiments on a much more limited number of materials where well-defined basic parameters of charge generation, injection and transport have been obtained providing a basis for understanding the dominant electronic processes of these complex systems. The first report of photoconductivity in an organic polymer was that of Hoegl [9] in poly-N-vinyl carbazole (PVK). This polymer has a high photoconductivity quantum efficiency, and consequently has been the subject of extensive research. A large part of the work described in this chapter has been on polyvinyl carbazole and charge transfer complexes of polyvinyl carbazole with 2,4,7-trinitro-9-fluorenone (TNF). This emphasis is a natural consequence of the efficiency of these materials resulting in a great amount of research and development work. The technological interest in these materials for photocopying systems has provided the broad, sustained effort which is necessary to make progress in understanding such new and complex materials. Photoconductivity studies in organic polymers are significantly harder than in inorganic semiconductor crystals where the generation and transport processes are relatively well-understood within the framework of band theory. In the disordered organic materials, photoconductivity has most often been a necessary preliminary step by which electronic carriers are introduced into the solid. The first goal of such injection has been to study the generation and transport processes themselves. Only after these basic mechanisms are understood will the complicated problems of trapping, recombination, sensitization, etc. become tractable. Consequently, the main part of the present discussion deals with electronic transport studies and free-carrier generation and injection experiments.

8.2 CHARGE TRANSPORT

8.2.1 General discussion

The study of electronic transport properties of highly-insulating, low-mobility materials presents many experimental problems which are not peculiar to organic polymers alone. The problems of defining the nature of contacts, avoiding space-charge effects and reaching true steady-state conditions have all become familiar obstacles to the reliability of d.c. conductivity results. Even with reliable conductivity data, further progress in understanding the transport mechanisms is blocked by an inability to separate the effects of charge carrier density from those of mobility. Hall effect techniques have proven ineffective for mobility determinations in these highly-insulating, low-mobility materials.

Fortunately, drift mobility techniques are ideally suited to low-mobility insulators and have been applied with great success in studies on organic crystals, inorganic molecular crystals, and a number of inorganic glasses. Here, the transient conductivity, due to excess free carriers injected by light, X-ray or electron-beam pulse excitation, is measured. Several transient conductivity experiments have now been applied to studies on polymer films with varying degrees of success. On a limited number of materials, notably PVK and TNF:PVK complexes, the experiments have been very successful for investigation of both charge transport and charge generation. In this section, we review the status of mobility studies on polymers. Heavy emphasis is placed on the work on PVK and the TNF:PVK complexes because of the almost unique success of these techniques on this class of materials. Successful mobility determinations in these materials have stimulated additional work using the TNF:PVK system as a prototype material. Thus, a quite extensive literature has appeared in the last few years from which the currently most complete picture of electronic processes in polymer films has emerged. The electric-field dependence of mobility and its concentration dependence are discussed in detail since they are the two most important aspects of transport in these materials. We conclude the section with a brief discussion of a.c. conductivity measurements which shed some light on transport processes.

References pp. 332—334

8.2.2 Mobility

(i) Mobility measurements

The mobility experiments on polymers and related organic systems can be divided into three techniques which we will call the charge-decay method, the transient-conductivity method, and the transit-time method. All three techniques are measures of transient response, a feature which in principle allows separation of carrier density and mobility as discussed in Chapter 2. The charge-decay experiments depend on the analysis of the initial decay rate of the surface potential. The transient-conductivity experiment analyzes the current transient in an inhomogeneously excited surface layer. The transit-time experiments are fundamentally different from the other two in measuring directly the drift time of injected excess carriers across the entire sample. No information on the magnitude of the carrier density is required for mobility determination in the transit-time or time-of-flight experiment.

The initial aim of all transient-conductivity experiments is to resolve the transit time of an injected carrier pulse. However, resolved transit times have only been observed in a few polymeric materials. The first reported direct drift mobility measurement in an organic polymer was by Vannikov [10] on high energy electron (5 Mev) irradiated polyvinyl acetate films. Measured hole mobilities were field dependent with a low field activation energy of 0.62 eV. At 18°C, the magnitude of the mobility was approximately 5×10^{-5} cm^2/V.s at 5×10^5 V/cm. The results were interpreted to represent transport dominated by carrier transitions between disconnected regions of continuous polyconjugation. Kryszewski *et al.* [11] reported mobilities on films of polyvinyl chloride and on a copolymer of acetonitrile and vinyl-pyridine. In these experiments, the transit time of carriers was measured after application of a voltage pulse. Mobilities of order 10^{-4} cm^2/V.s were obtained. Some of these results have been verified by recent results of Ranicar and Fleming [12] on polyvinyl chloride using photo-injected carriers. Since 1967, there has been a growing list of mobility measurements in polymers and polymer systems. Table 8.1 is a compilation of these mobility determinations. Several comments about this list are necessary to give a proper perspective on the status of mobility measurements in polymeric materials. First, Table 8.1 includes all polymeric materials including those whose dielectric properties have been of greater interest than their electronic transport properties [13—19]. Second, the interpretation of results for many of the polymers, especially those for which transit times were not resolvable, is open to question. In fact, for the non-transit-time techniques, reference to Table 8.1 shows that independent measurements on the same polymers are in major disagreement in every case. The problem lies with the complexity of interpreting the transients and the inevitable assumptions necessary to obtain carrier densities which are required to extract a mobility value with these experiments. The lack of well-defined, reproducible samples will also play a

Polyvinylcarbazole (PVK)   2, 4, 7-Trinitro-9-fluorenone (TNF)

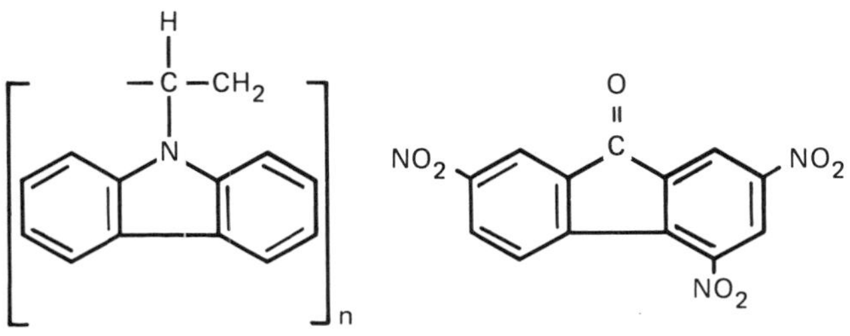

Fig. 8.1. Molecular structures of poly-N-vinyl carbazole (PVK) and 2,4,7-trinitro-9-fluorenone (TNF).

part in these experimental discrepancies. The third, and most crucial point for our present purposes of attempting to understand the mechanisms of transport is that very few studies have attempted or been capable of more than an order of magnitude determination of mobility and a rough temperature dependence. All the data indicate that the mobility in polymers is very low ($\leqslant 10^{-4}$ cm^2/V.s) and is generally an activated process. In order to understand the physical mechanisms dominating charge transport in disordered polymeric materials, we require detailed, reproducible measurements on well-defined samples showing the influence of structure, composition and purity.

For these reasons, the extensive work on mobility and charge generation in the related systems of PVK, TNF:PVK complexes and TNF dispersed in polyester assume a particular importance. The molecular structures of PVK and TNF are shown for reference in Fig. 8.1. Detailed and reproducible transport studies have been made on these materials using a number of different experimental and materials preparation techniques. Both hole and electron transport are observed and the effects of charge transfer complexing and varying composition are readily studied. These advantages have made of this group of materials a very successful prototype system for studies of generation and transport in extremely low mobility, disordered materials.

The first detailed mobility studies on a polymer were made by Regensburger [20] on PVK using a charge-decay technique. He reported detailed measurements as functions of electric field, film thickness and temperature. These results were interpreted in terms of an electric-field dependent hole mobility ($\mu \propto E^2$) with an activation energy of between 0.4 and 0.7 eV. Subsequently, Mort and Lakatos [21] measured the transient response due to photo-emission of holes from gold into PVK. The mobility was field dependent with a value of approximately 10^{-7} cm^2/V.s at 10^5 V/cm and an activation energy of 0.55 eV in good agreement with the results of Regensburger [20]. Long tails on the current transient pulses were observed indicating a dispersion in the effective transit times. Pai [22] also investigated

References pp. 332—334

TABLE 8.1

Compilation of mobility determinations in polymers

Polymer	Mobility (cm²/V.s) (−)	Mobility (?)	Mobility (+)	$\mathscr{E}_{act}$ (eV)	Method	Reference
Polyvinyl acetate	7×10^{-4}		10^{-5}–10^{-4}	0.62	Transit time (e-beam)	Vannikov [10]
Polyvinyl chloride	3×10^{-4}				Transit time (voltage)	Kryszewski et al. [11]
Acetonitrile vinyl-pyridine copolymer					Transit time (voltage)	Regensburger [20]
Polyvinyl carbazole			10^{-7}–10^{-6}	0.4–0.7	Transit time (photo)	Pai [22]
Polyvinyl carbazole			10^{-6}	0.36	Transit time (photo)	Szymanski and Labes [23]
Polyvinyl carbazole			10^{-3}	0.12	Transit time (photo)	
Polyvinyl carbazole			10^{-6}		Transit time (photo-emiss.)	Mort and Lakatos [21]
PVK:I$_2$ complex		0.5			Hall	Hermann and Rembaum [25]
PVK:I$_2$ complex	0.1		0.1		Transit time (photo)	Hermann [26]
PVK:TCNE complex	2×10^{-2}		2×10^{-2}		Transit time (photo)	Hermann
PVK:TNF 1:1 complex	10^{-7}–10^{-5}				Transit time (photo)	Seki and Gill [29]
PVK:TNF complexes	10^{-8}–10^{-6}		10^{-8}–10^{-5}	0.65–0.71	Transit time (photo)	Gill [24]
PVK:TNF complexes	10^{-10}–10^{-6}		10^{-8}–10^{-6}		Transit time (photo)	Mort and Emerald [34, 35]
Polyethylene	10^{-10}–10^{-7}				Charge decay (corona)	Davies [13]
Polythene		10^{-7}		0.7–1.2	Charge decay (corona)	Davies
Polystyrene			10^{-6}	0.2–0.75	Charge decay (contact)	Martin and Hirsch [14]
Polyethylene			10^{-10}		Transient cond. (e-beam)	Martin and Hirsch
Poly(ethylene terephthalate)	10^{-6}			0.2–0.35	Transient cond. (e-beam)	Martin and Hirsch

TABLE 8.1 (cont.)

Polymer	Mobility (cm^2/V.s)			$\mathscr{E}_{act}$(eV)	Method	Reference
	(−)	(?)	(+)			
Polymethyl methacrylate		2.5×10^{-11}		0.52	Charge decay (corona)	Reiser et al. [15]
Perspex		3.6×10^{-11}		0.48	Charge decay (corona)	Reiser et al.
Polybutyl methacrylate		2.5×10^{-10}		0.65	Charge decay (corona)	Reiser et al.
Lucite		3.5×10^{-9}		0.52	Charge decay (corona)	Reiser et al.
Polystyrene		1.4×10^{-11}		0.69	Charge decay (corona)	Reiser et al.
Butvar		4.5×10^{-11}		0.74	Charge decay (corona)	Reiser et al.
Vitel		4.0×10^{-11}		1.10	Charge decay (corona)	Reiser et al.
Polyisoprene		2.0×10^{-12}		1.10	Charge decay (corona)	Reiser et al.
Silicone		3.0×10^{-14}		1.70	Charge decay (corona)	Reiser et al.
Polyvinyl acetate (below T_g) (above T_g)		2.2×10^{-12}		0.48 1.20	Charge decay (corona) Charge decay (corona)	Reiser et al. Reiser et al.
Polyvinyl chloride	5×10^{-4}				Transit time (photo)	Ranicar and Fleming [12]
Polydiphenyl acetylene (sublimed)			2×10^{-4} 0.3	0.17	Transit time (e-beam) Transit time (e-beam)	Merkulov et al. [16] Merkulov et al.
Poly (ethylene terephthalate)	10^{-3}				Transient cond. (X-ray)	Hughes [17]
Polyethylene	10^{-3}			0.06	Transit time (voltage)	Matsumoto and Yahagi [18]
Poly (ethylene terephthalate)	1.2×10^{-4}		2.7×10^{-5}	0.3–0.32	Transit time (e-beam)	Hayashi et al. [19]

References pp. 332–334

the transient photoconductivity in PVK films but was unable to distinguish a transit time which could be associated with transit of a coherent carrier pulse. The observed signal which was proportional to the integrated current transient had an exponential nature defined by:

$$\Delta V(t) = \Delta V_0 [1 - \exp(-t/T_{\text{eff}})], \tag{8.1}$$

where T_{eff} was a function of applied field, film thickness and temperature. The parameter T_{eff} varied linearly with thickness at a constant field and could therefore be defined as an effective transit time. Measurements over a range of applied field from 10^4 to 10^6 V/cm showed $(T_{\text{eff}})^{-1}$ varied exponentially as the square root of the applied field. The temperature dependence of $(T_{\text{eff}})^{-1}$ was characterized by a field-independent activation energy of 0.36 eV. If T_{eff} is assumed to be an approximation to the transit time, Pai's results are in reasonable agreement with previous mobility determinations [20, 21]. Szymanski and Labes [23] reported measuring a field-independent drift mobility of order 10^{-3} cm^2/V.s in PVK with a thermal activation energy of 0.12 eV. However, the good agreement among independent studies of Regensburger [20], Mort and Lakatos [21], Pai [22] and Gill [24] suggest that this higher mobility determination was incorrect.

Hall mobility in PVK-iodine charge-transfer complexes has been reported by Hermann and Rembaum [25]. Hermann has also reported drift mobilities in PVK-iodine and PVK-tetracyanoethylene complexes [26]. For the iodine complexes (4 mole%I$_2$), the Hall mobility was 0.5 cm^2/V.s and the drift mobility ~ 0.02 cm^2/V.s for both holes and electrons. In the tetracyanoethylene complexes (4mole%TCNE), the drift mobilities were ~ 0.1 cm^2/V.s for holes and electrons. These values of mobility are anomalously high compared to all other determinations in PVK and PVK-based charge-transfer complexes.

The addition of 2,4,7-trinitro-9-fluorenone (TNF) to PVK resulting in charge-transfer (CT) complexing and introduction of CT bands in the visible part of the absorption spectrum was investigated by Lardon, Lell-Döller and Weigl [27]. At sufficiently high TNF concentrations, electrophotographic measurements reported by Schaffert [28] indicated that electron transport dominates the conductivity. This conjecture was verified by the electron drift mobility determination by Seki and Gill [29] on thin films of TNF:PVK in a 1:1 molar ratio with respect to the monomer units. The electron drift mobility was found to be strongly field-dependent with a value of $\sim 10^{-7}$ cm^2/V.s at 10^5 V/cm. The field dependence was an exponential of the same form as observed by Pai [22] for holes in PVK. The observation of well-defined electron transients in the TNK:PVK films was particularly significant, since the effects of material composition could be readily investigated in such a two-component system. Drift mobility studies over the entire composition range from pure PVK to pure TNF were carried out by Gill [24]. Measurements of the transient current under nearly

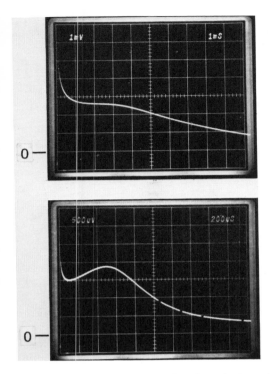

Fig. 8.2. Current transients showing electron transport in a 1:1 TNF:PVK film. Upper trace was obtained at 24°C and the lower trace at 65°C. (From Gill [24])

space-charge-limited conditions induced by 100 nanosecond duration pulses of high intensity u.v. light resulted in the well-resolved transit times shown in Fig. 8.2. Analysis of space-charge-limited transients for field-dependent mobilities by Scharfe [30] and by Gill and Kanazawa [31] show that the magnitude of the current transients will be decreased relative to the case for a field-independent mobility, but that the pulse shape is affected only a small amount and continues to show a sharp cusp at the transit time. The observed transients do not have the predicted pulse shapes, but rather show an initial rapid decay followed by a definite break with a long tail indicative of a strong spatial dispersion of the pulse. At higher temperatures, a definite cusp develops characteristic of the SCL conditions of the experiment and the pulse dispersion decreases. However, as Seki [32] has noted, the relative dispersion is unchanged when the transit time is varied by more than two orders of magnitude under small-signal conditions at constant temperature. Over most of the composition range of TNF in PVK both electron and hole mobilities were measurable in the same films. Both hole and electron mobilities increased exponentially as the square root of the field strength ($\mu \propto \exp[\gamma E^{1/2}]$) where γ was numerically the same for both carriers and for all film

References pp. 332—334

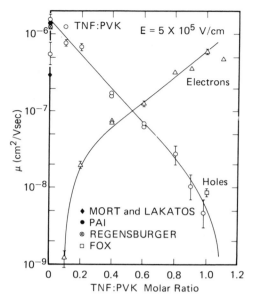

Fig. 8.3. Variation of hole and electron drift mobility with TNF:PVK molar ratio. (From Gill [24])

compositions. The composition dependence of the mobilities obtained by Gill is shown in Fig. 8.3. The hole mobility measured in pure PVK was in very good agreement with previous determinations [20—22]. This work showed that hole mobility is associated with PVK and electron mobility with TNF molecules.

Electron transport in TNF has been further investigated by Gill [33] by measuring drift mobility for a series of concentrations of TNF in a polyester matrix and for pure glassy TNF, liquid TNF and single crystals of TNF. For the TNF in polyester and the TNF glass, the mobility had the same field dependence as found in TNF:PVK films. However, in liquid TNF (*i.e.* above the crystal melting temperature) and in the TNF crystals, electron mobility was independent of field. Mobilities in all three forms of TNF were activated processes as shown in Fig. 8.4 with activation energies in the disordered solids where mobility was field dependent, being considerably higher than for the liquid and crystalline forms. Mobilities in the TNF:PVK system have also been measured by Mort and Emerald [34, 35] using an electrophotographic discharge technique. The derived mobilities are in good agreement with the results of Gill [24, 33]. This agreement extended to results on pure TNF where the method of sample fabrication was totally different. In the work by Gill, TNF films were prepared by quenching from the melt, while Emerald and Mort used vapor-deposited TNF films.

In the following sections, we will discuss in detail field-dependent mobilities

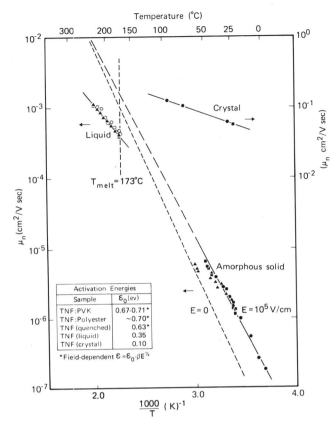

Fig. 8.4. Temperature dependence of electron mobility in liquid, amorphous solid and crystalline forms of pure TNF. (From Gill [33].)

and concentration dependences and the transport mechanisms which are appropriate to these low-mobility disordered materials. However, it is appropriate at this point to mention experiments by Hughes [17, 36] on TNF:PVK and poly(ethylene terephthalate) using nanosecond X-ray and ruby-laser pulse excitation. Initial high current magnitude is interpreted as a measure of intrinsic mobility. The free-carrier lifetime is very short so that subsequent transport proceeds by a trap-controlled mechanism characterized by the low effective mobilities measured by transit-time experiments. The intrinsic mobilities inferred by Hughes are $\sim 10^{-3}$ cm^2/V.s in TNF:PVK and $\sim 10^{-3}$ cm^2/V.s in poly(ethylene terephthalate). Hughes suggests that these low intrinsic mobilities may be due to small polaron formation.

(ii) Field-dependent mobility

Strong field dependence of the hole drift mobility in PVK was first reported

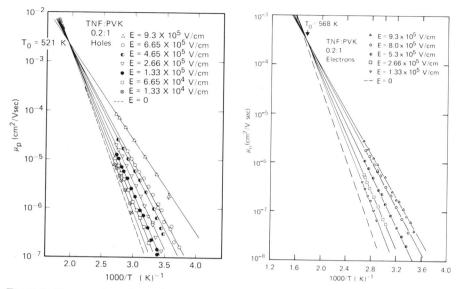

Fig. 8.5. Temperature dependences of hole and electron drift mobilities with applied field as a parameter. (From Gill [24].)

by Regensburger [20] and was subsequently observed by Gill [24] for both holes and electrons in TNF:PVK complexes. In the latter work, detailed temperature-dependence studies were reported showing that the observed field dependence was due to field dependence of the activation energies. The data shown in Fig. 8.5 illustrates this behavior for both carriers in a 0.2:1 TNF:PVK film. The field and temperature dependences of both hole and electron mobility fit an empirical relation of the form:

$$\mu = \mu_0 \exp\left[-(\mathcal{E}_0 - \beta E^{1/2})/kT_{\text{eff}}\right] \qquad (8.2)$$

with:

$$1/T_{\text{eff}} = 1/T - 1/T_0 \qquad (8.3)$$

where μ_0 is a function of film composition only, $\beta = 2.72 \times 10^{-5}$ eV $(V/m)^{1/2}$ is a constant coefficient for both carriers and all compositions, T is the temperature (K), and T_0 is the temperature at which extrapolated temperature dependences at different fields intersect. The values of μ_0, $\mathcal{E}_0$ and T_0 obtained for various TNF:PVK ratios are shown in Table 8.2. The observed field dependence of the mobilities can be most easily explained by applying the Poole–Frenkel [37, 38] mechanism to a trap-controlled mobility model. This model predicts the lowering of a Coulomb potential well in the direction of the applied field so that the effective trap depth, and hence the activation energy, decreases as the square root of the field. The model also predicts the same field dependence for holes and electrons since

TABLE 8.2

Parameters in eqns. (8.2) and (8.3) obtained for various TNF:PVK molar ratios. (From Gill [24].)

TNF:PVK	Holes			Electrons		
	$\mathscr{E}_0$(eV)	T_0(K)	μ_0(cm^2/V.s)	$\mathscr{E}_0$(eV)	T_0(K)	μ_0(cm^2/V.s)
0:1	0.65	660	2×10^{-2}	—	—	—
0.1:1	0.65	625	1×10^{-2}	0.67	595	1.6×10^{-5}
0.2:1	0.68	521	2.7×10^{-3}	0.71	568	3×10^{-4}
0.4:1	0.68	519	4.6×10^{-4}	0.68	545	1×10^{-4}
0.6:1	0.65	550	3.8×10^{-4}	—	—	—

the Poole–Frenkel coefficient $\beta_{\mathrm{PF}} = (e^3/\pi\kappa\kappa_0)^{1/2}$ depends only on the dielectric constant. The concept of an effective temperature can be easily introduced by assuming an energy distribution of the traps. Despite reasonable quantitative agreement with this model, there are serious objections to applying it to these polymeric systems. The most serious problem is that the Poole–Frenkel model requires a very high density of charged trapping centers of both polarities, an unrealistic situation in these materials.

Other models for field-dependent mobility have been suggested. Bagley [39] proposed that the kinetics of localized transport can lead to an activated field-dependent drift mobility proportional to $(1/E)\sinh(eE\lambda/2kT)$ where λ is the separation between localized states. This same field dependence was derived by Emtage [40] by including the effect of an electric field in Holstein's treatment of small polaron motion [40]. Tabak, Pai and Scharfe [41] have pointed out that this field dependence is too abrupt to explain the experimental results for field-dependent mobilities observed in amorphous Se, As$_2$Se$_3$ and PVK films. Seki [32] has proposed a model in which the energy levels of the localized hopping sites are assumed to vary spatially in a non-random fashion so that the spatial fluctuations can be characterized by an average amplitude U_0 over a distance λ_0. With this model, the mobility is given by an expression of the form:

$$\mu E = \lambda_0/\tau_0 \exp(-\rho/R_0)\exp(-U_0/kT)\,2\sinh(e\lambda_0 E/2kT) \qquad (8.4)$$

where τ_0 and R_0 are constants with dimensions of time and length respectively, and ρ is the average distance between hopping sites. Seki has used this model to fit Gill's data on TNF:PVK. An example of the fit for hole mobility data in 0.2:1 TNF:PVK is shown in Fig. 8.6. Similar fits are obtained for electron mobility and for different concentrations of TNF in PVK. However, the form of eqn. (8.4) differs considerably from that observed experimentally. Mobility data taken over a wide range of applied field by Fox [42] are shown in Fig. 8.7 together with a curve of the form expressed by eqn. (8.4). It is quite obvious that there is serious deviation from the experimental dependence at both high and low fields.

References pp. 332—334

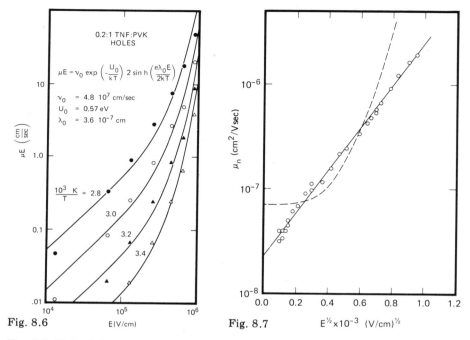

Fig. 8.6. Hole drift velocity in 0.2:1 TNF:PVK. The points are data from Gill [24]. The solid curves represent the fit to the data using Seki's model with the indicated parameters [32].

Fig. 8.7. Field dependence of mobility in TNF:PVK after Fox [42]. The dashed line shows dependence predicted by eqn. (8.4).

Two other aspects of the observed field-dependent phenomena in these materials are trap depths determined from thermally-stimulated conductivity (TSC) and pulse-shape analysis of the drift mobility transients. In measuring TSC in PVK films, Pai [22] observed that the temperature at which current peaks occur decreased as the applied field was increased, implying a decrease in the effective trap depth with applied field. Gill [43] has measured TSC in TNF:PVK over the concentration range from pure PVK to 1:1 molar ratios. Current peaks corresponding to emptying of both hole and electron traps were observed. The temperatures at the peak conductivities which were associated with the trap depths were found to be field dependent (see Fig. 8.8). The field dependence was found to be identical to that of the drift mobility activation energies. The field dependence of the TSC peaks reported by Pai [22] for PVK is in good agreement with that shown in Fig. 8.8. The pulse shapes observed in transient-conductivity experiments on PVK and TNF:PVK complexes exhibit several distinct characteristics which can be seen in Fig. 8.2. The initial current spike and the long tail following the transit time break

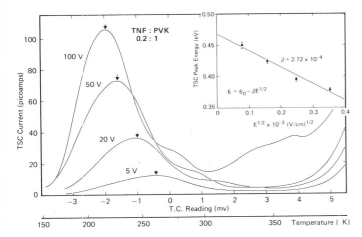

Fig. 8.8. Thermally-stimulated conductivity data showing the field dependence of conductivity peaks associated with traps. (From Gill unpublished data.)

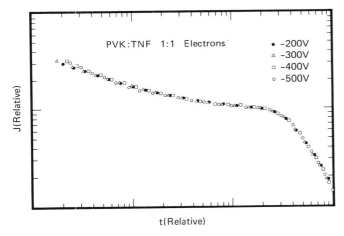

Fig. 8.9. Normalized current transients showing superposition of the pulse shapes for a range of applied fields. (From Šeki [32].)

are observed for both carriers and for all film compositions. Seki [32] has observed that the shape of the normalized photocurrent transients are identical over a wide range of applied field (Fig. 8.9). The shapes of non-space-charge perturbed current transients are found to superimpose even for transit times differing by more than two orders of magnitude. If the initial fast decay of the photocurrent and the long tails after the transit time are attributed to trapping effects, then the superposition of the pulse shapes implies that the carrier lifetimes and detrapping times scale with the transit

References pp. 332—334

time and are therefore strongly field dependent. Similar superposition characteristics have been reported by Scharfe [44] for current transients in amorphous As_2Se_3 films. Scher [45] has developed a stochastic transport (hopping) model of transient photoconductivity in amorphous solids which he has applied to both the As_2Se_3 and to TNF:PVK transients. The key feature of the stochastic model is the assumption of a distribution of waiting times between hopping events. The model predicts that as time increases, more carriers will encounter at least one larger waiting time at some site, leading quite naturally to the observed scaling of dispersion effects with transit time. A more detailed discussion of these points is given in Chapters 3 and 7 of this book.

Up to this point, we have dealt in considerable detail with the observed field-dependent behavior of mobility and trapping in PVK and PVK:TNF films. We might well ask whether field-dependent mobility is a usual feature of transport in polymers and other disordered organic materials. For polymers, the answer to this question is uncertain since very few reliable mobility determinations have been made. However, time-of-flight measurements on several polymers listed in Table 8.1 show field-independent mobilities with magnitudes $> 10^{-4}$ cm^2/V.s. In other disordered organic systems, results are mixed. In some liquid hydrocarbons, reported mobilities are considerably higher than for other organics and are field independent [46, 47]. In glassy 3-methylpentane, field-independent mobility in the range 10^{-2}—10^{-1} cm^2/V.s has been reported [48]. Studies of the mobility in pure 2,4,7-trinitro-9-fluorenone in the amorphous solid, liquid and crystalline states have shown field dependence of the mobility in the amorphous solid but no field dependence in the liquid or crystalline phases [33]. Mobility measurements in dispersions of the leucobase of malachite green were also reported to be field-independent with magnitudes in the range of 5×10^{-2} cm^2/V.s [49]. Field-dependent mobilities have also been observed in several disordered inorganic materials [48, 50, 51]. The phenomenon appears to be associated with very low mobilities ($< 10^{-5}$ cm^2/V.s) most often encountered in disordered materials. At this time, there is no really satisfactory model accounting for this behavior.

(iii) Concentration dependence of mobility

Probably the most important aspect of transport studies in TNF:PVK has been the compositional variability that is possible. Instead of merely measuring the mobility in another low-mobility solid, these experiments explored the effects of molecular concentration on carrier mobilities. The concentration dependence provided a very direct probe of the hopping transport of highly-localized carriers. The effects of TNF content on carrier mobilities were shown in Fig. 8.8. The strong compositional dependence was indicative of direct intermolecular hopping between carbazole units in the case of hole transport and hopping between TNF molecules for electron transport. Because

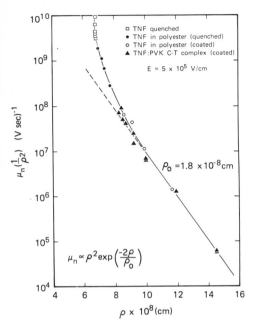

Fig. 8.10. Electron mobility as a function of the average separation of TNF molecules in TNF:PVK and TNF:polyester samples. (From Gill [33].)

TNF forms a weak charge-transfer complex with PVK, it was necessary to establish the effects of the complexing on the mobilities in order to properly understand the effects of concentration. The degree of complexing of TNF with PVK established by the optical absorption experiments of Weiser [52] are summarized in Table 8.3. Saturation of the degree of complexing at TNF:PVK ratios greater than about 0.6:1 and the observation of electron mobility in quenched films of pure TNF [24] eliminated the possibility that electron transport involved only complexed TNF. To distinguish whether the total TNF concentration or only the uncomplexed TNF was active in electron transport, mobility measurements were made as a function of TNF concentration in TNF:polyester films where no charge-transfer complexing occurs [33]. The results shown in Fig. 8.10 unequivocally demonstrate that the total TNF content in TNF:PVK is active. From these data, one can infer that electron transport is unaffected by complexing with PVK or by the nature of the polymer matrix in which it is dispersed. Photo-injection studies from amorphous selenium into TNF:PVK films [34] and into evaporated, amorphous films of pure TNF [35] lead to the conclusion that the electron transport states in TNF are the unperturbed lowest empty molecular orbitals of the TNF molecule. This conclusion is consistent with the observation that electron transport is unaffected by complexing of the TNF molecules with PVK [33] and with Weiser's conclusions from electro-absorption experiments

References pp. 332—334

TABLE 8.3

Degree of complexing in TNF:PVK films of different monomer ratio. (From Weiser [52].)

Ratio TNF:PVK	$N(10^{20}\,\text{cm}^{-3})$ Free PVK	$N(10^{20}\,\text{cm}^{-3})$ Free TNF	$N(10^{20}\,\text{cm}^{-3})$ Complexed TNF	$\dfrac{\text{Complexed TNF}}{\text{TNF}}\,\%$
0:1	36.2	0	0	0
0.1:1	30.6	0.93	2.37	71.8
0.2:1	25.9	1.96	4.00	67.3
0.4:1	19.0	3.93	6.10	60.7
0.6:1	14.9	6.30	6.60	51.2
0.8:1	12.0	8.20	6.95	45.9
0.9:1	10.8	9.05	7.01	43.7
1:1	9.9	9.9	7.00	41.5
1.1:1	9.1	10.7	6.90	39.2

[53] that charge transfer mainly takes place in the excited state. These experiments will be discussed in a later section.

For very low mobilities involving highly-localized carriers the mobility can be expressed in a form appropriate to diffusive motion:

$$\mu = \frac{e\rho^2}{6kT}\nu \qquad (8.5)$$

where ρ is the average jump distance and ν is an "attempt-to-jump" frequency. The hopping frequency ν can be expressed as:

$$\nu = \nu_0 \exp\left(-\frac{2\rho}{\rho_0} - \frac{\mathcal{E}}{kT}\right) \qquad (8.6)$$

where ν_0 is a constant characteristic of the electron in a particular localized state, ρ_0 is a constant parameter describing the spatial decay of the electronic wave functions, and $\mathcal{E}$ is an energy barrier between sites. If the concentration dependence of the mobility can be explained as a random hopping process, then a plot of $\ln(\mu/\rho^2)$ versus ρ should be linear, the slope giving a measure of the wave function overlap and hence a measure of the degree of localization of the excess carrier. At low TNF concentrations, the data of Fig. 8.10 fit the expected exponential relationship with the slope yielding a value of $\rho_0 = 1.8 \times 10^{-8}$ cm for the electron localization parameter. At higher TNF densities, where the average separation becomes comparable to the molecular dimensions, this simplified model is no longer adequate and the mobility deviates from the exponential behavior.

In dealing with hole transport, the question of whether the total carbazole concentration or only the uncomplexed carbazole is active has not been resolved. If the total carbazole concentration is active, dilution with TNF

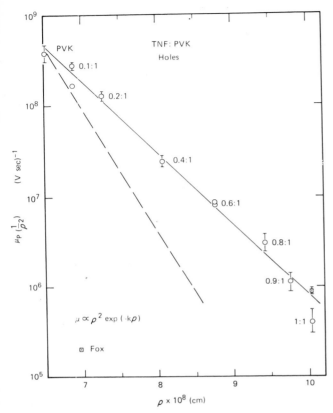

Fig. 8.11. Hole mobility as a function of the average separation of uncomplexed carbazole units. Dashed line shows concentration dependence if the total carbazole density is included. (From Gill [24].)

could not result in a simple increase in the mean carbazole separation since the carbazole spacing along the polymer would remain fixed. However, if only uncomplexed carbazole is active, then complexing with TNF will more nearly approximate a random dilution of the carbazole concentration. The latter assumption was made by Gill [24] in plotting the data shown in Fig. 8.11 for hole transport in TNF:PVK films. A value of $\rho_0 = 1.1 \times 10^{-8}$ cm was obtained for the hole localization parameter. The dashed line in Fig. 8.11 shows the mobility variation under the assumption that all carbazole units are active in transport. A value of $\rho_0 = 0.64 \times 10^{-8}$ cm results. Experiments analogous to those performed on TNF dispersed in polyester [33] have recently been reported for N-isopropyl carbazole doped polycarbonate [54].

Other than the studies discussed above, there are no published detailed studies of concentration dependence of mobility. Photoconductivity studies

References pp. 332—334

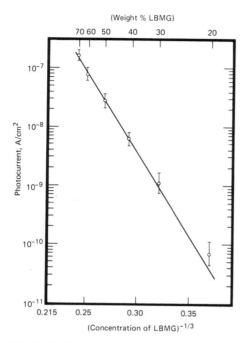

Fig. 8.12. Concentration dependence of photocurrent for molecules of the leucobase of malachite green. (From Mehl and Wolff [49].)

have focussed on the sensitizing properties of relatively low concentrations of dopants with the tacit assumption that mobility is a constant property independent of the matrix or host material. However, in a study of photoconductivity of solid solutions of the leucobase of malachite green (LBMG) in various organic media by Mehl and Wolff [49], concentration effects very similar to those observed in the TNF dispersions were reported. The concentration dependence of photocurrent shown in Fig. 8.12 was found to be independent of the matrix material which can be interpreted as an indication of direct interaction between leucobase molecules. An electron drift mobility $\mu \approx 5 \times 10^{-2}$ cm^2/V.s was determined for a single concentration which was given as evidence for a hopping model for transport. For hopping transport by direct interaction between leucobase molecules, the concentration dependence of the photocurrent must be mainly due to mobility effects. Comparison of this data with that for TNF dispersed in PVK or in polyester indicates a roughly comparable magnitude effect of concentration on the electron mobilities in TNF and LBMG dispersions.

8.2.3 AC conductivity

Extensive measurements have been made by Kanazawa and Schechtman [55] of the complex dielectric permittivity of TNF:PVK films as a function of frequency, temperature and d.c. bias. Measurements were made over the frequency range from 5×10^{-3} to 5×10^5 Hz. Complementary measurements were made of the transient-step response currents over the corresponding time domain. It was found that the observed transient currents for times up to one second could be accounted for in terms of dipolar relaxation processes with no need to invoke bulk generation or electrode injection of delocalized carriers. The a.c. dark conductivity was frequency-dependent, varying approximately from the 0.7 power of frequency at the lower frequencies to a nearly quadratic dependence above 10 kHz. In contrast, the a.c. photoconductivity was found to be essentially frequency-independent. This result was contrary to expectations since an 0.8 power dependence on frequency was expected for the localized hopping process [56] inferred from drift mobility results [24, 29]. In a recent study of a.c. conductivity and a.c. photoconductivity in amorphous and crystalline insulators, Abkowitz, Lakatos and Scher [57] report the frequency dependence in 1:1 TNF:PVK films over the frequency range from 20 Hz to 10^5 Hz. The dark conductivity varied linearly with frequency over the entire frequency range. The a.c. photoconductivity was independent of frequency in agreement with results of Kanazawa and Schechtman [55]. Abkowitz *et al.* interpret the a.c. dark conductivity as hopping among localized states. The frequency-independent a.c. photocurrent is explained in terms of carrier excitation to extended states. The hopping behavior observed in drift mobility experiments and the motion in extended states inferred from a.c. photoconductivity can be reconciled by assuming either very short carrier range (*i.e.* low-frequency limit for excited carriers moving in extended states) or by invoking emission-limited, and therefore mobility-independent, currents.

8.3 CHARGE GENERATION AND RECOMBINATION

A large part of the research on photoconductivity in polymers has been concentrated on measurements of efficiency and spectral response, and in attempts to improve these characteristics by the addition of impurities or dopants. Almost all our understanding of the basic processes of free-carrier generation in pure materials and of sensitizations by suitable dopants has been the result of work on molecular crystals. Because of the basically molecular nature of photoconducting polymer systems the same physical processes are expected to apply to generation and sensitization in these disordered molecular materials. The important processes by which excess charge is introduced by light can be roughly grouped as intrinsic, dye sensitization, charge-transfer absorption and photo-injection (internal photoemission) at some interface. The current status of research on photoconduction mechanisms in molecular crystals has been covered in an earlier chapter by Inokuchi and Maruyama. Studies of charge generation in polymer systems generally suffer from an inability to adequately separate-out charge-transport effects so that temperature and electric-field dependences of the generation process are difficult to assess. However, in recent years considerable detailed work has been possible on PVK films and on the TNF:PVK charge-transfer complexes where the use of transient techniques together with some knowledge of the transport behavior has allowed generation, injection and recombination processes to be studied. Work on these materials forms the basis for this section.

8.3.1 Polymer films

Hoegl [9] first investigated photoconductivity in polymers and also investigated sensitization by systematic doping experiments in an attempt to correlate chemical structure of the dopants with their effect on photoconductivity. Polymers investigated included PVK and other polymers containing aromatic and heterocyclic chain units. Low molecular weight organic compounds were also investigated by dispersing them in non-photoconducting polymer matrices. An electrophotographic discharge technique was used resulting in qualitative comparisons of photosensitivity. PVK was found to have the highest photosensitivity of the polymers investigated. Doping at the 0.1—2 mole% level with electron acceptors was found to increase the photosensitivity of the polymer films. The results of donor (D) and acceptor (A) sensitization were interpreted in terms of a charge-transfer (CT) interaction of the form

$$DA \underset{}{\overset{h\nu}{\rightleftharpoons}} D^+ A^-. \tag{8.7}$$

In PVK, the carbazole groups are electron donating. Doping with an acceptor molecule results in charge transfer from carbazole to form A^- at the acceptor site leaving a free hole to migrate through the polymer. Hoegl attributed all the photoconductivity effects he reported as impurity effects leading to CT sensitization. Regensburger [20] studied charge generation and transport in PVK and a layered system of PVK overlaid with a film of amorphous selenium. A transient photoresponse technique was used which made possible a direct measurement of the hole drift mobility. Both carrier generation and mobility in pure PVK were found to be strongly field dependent with both increasing in proportion to the square of the applied field. Electrophotographic gain in the range of 10^{-2}—10^{-1} was observed for fields of the order of 10^6 V/cm. The experiments on layered systems also established that efficient photoinjection of holes occurs at the Se/PVK interface. Dye sensitization of PVK has also been reported [19, 58]. Ikeda et al. [58] measured quantum yields of 10^{-2}—10^{-1} in sensitized PVK and yields about an order of magnitude lower in unsensitized films. These measurements were made at low fields ($\sim 10^4$ V/cm) where the photocurrent was approximately proportional to field. In his studies of transient photoconductivity in

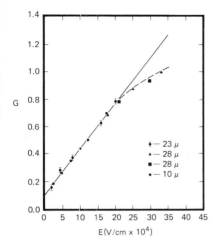

Fig. 8.13. The yield of carriers in PVK, G, in units of ion pairs per 100 eV energy deposited, *versus* the applied electric field, E. (From Hughes [59].)

PVK, Pai [22] measured the quantum efficiency or yield for highly-absorbed light as a function of field and temperature. The variation of yield with applied field was approximately $\exp(\gamma E^{1/2})$. At the highest fields ($\sim 10^6$ V/cm), the yield approached a value of 10^{-1}. This field dependence for surface photogeneration is much stronger than has been observed for bulk X-ray excitation experiments [59] discussed below. Hughes [59] suggests that the exponential dependence may be due to field-assisted thermal ionization of the ion pair, that is, the Poole—Frenkel effect. The field dependence of the carrier yield of X-ray excited PVK films was investigated by Hughes [59]. A pulse X-ray source with a pulse halfwidth of about 3 ns was used to generate carriers in the bulk PVK. The carrier yield, shown in Fig. 8.13, increased linearly with applied field from a zero field value of about 0.1 ion pairs/100 eV energy deposited. At about 2×10^5 V/cm the slope begins to drop off. These observations were explained using the Onsager theory for geminate (or initial) recombination of the excited electron—hole pairs [60]. This model treats the excited electron—hole pair at some initial separation r_0 diffusing in their mutual Coulomb field and the applied field. Increasing the applied field favors dissociation of the electron—hole pairs leading to increased free-carrier yields. In the low-field approximation, the yield G is given by:

$$G = \phi_0 [\exp(-e^2/4\pi\kappa\kappa_0 kTr_0)][1 + (e^3/8\pi\kappa\kappa_0 k^2 T^2)E] \tag{8.8}$$

where ϕ_0 is the limiting yield at high fields, e is the electronic charge, $\kappa\kappa_0$ is the dielectric constant of the medium, and E is the applied field. Besides a linear dependence of yield on applied field, eqn. (8.8) predicts that the slope/intercept ratio equals $(e^3/8\pi\kappa\kappa_0 k^2 T^2)$. For PVK, this ratio would be

References pp. 332—334

3.6×10^{-5} cm/V using $\kappa = 3.0$ for the high-frequency dielectric constant. From the data in Fig. 8.13, this ratio is 3.5×10^{-5} cm/V, in excellent agreement with theory. The value of the initial separation or thermalization length r_0 was estimated to be 60 Å.

8.3.2 TNF:PVK charge-transfer complex

As pointed out in the previous section, charge-transfer complex formation was recognized as the major sensitization mechanism in the earliest photoconductivity experiments on polymer films [9]. Charge-transfer complexes of PVK with the acceptor TNF have been of great importance to both the utilization and understanding of photoconductivity and transport in organic polymers. Because of interest in electrophotographic applications of PVK:TNF complexes, a considerable amount of research on these materials has been supported. The relatively good transport properties which have made commercial applications of these films possible have also made possible detailed studies of carrier generation processes. For these reasons, this discussion of CT complexes in organic polymers is limited to consideration of this prototype material. Following Hoegl's work [9], further investigation of charge-transfer sensitization of carbazole-based photoconductors was carried out by Lardon, Lell-Döller and Weigl [27]. Optical absorption studies together with electronic transport measurements using electrophotographic decay techniques clearly established the formation of CT complexes of carbazole with several acceptor molecules including TNF. Dramatic increases in photosensitivity, especially in the CT absorption bands, were observed with the addition of 1 to 2% acceptor molecules. Schaffert [28] reported on the development of an electrophotographic material using TNF:PVK complexes in which much higher TNF concentrations were used. The highest sensitivity system was a 1:1 molar ratio TNF:PVK complex formulated by Shattuck and Vahtra [61]. At these high TNF concentrations, transport is chiefly by electrons hopping through the acceptor sites in contrast to the situation in CT sensitized PVK where the predominant photocurrent is by holes hopping on donor sites. Other good CT sensitizers at high concentration levels were also reported including substituted anthraquinones and nitrated fluorenone-like structures.

Applying absorption and electro-absorption techniques, Weiser [52, 53] has investigated the CT absorption bands in TNF:PVK over a wide concentration range. The electro-absorption spectrum in Fig. 8.14 clearly distinguishes the two CT bands seen in absorption by Lardon et al. [27]. The electro-absorption was interpreted as a field-induced energy shift of the molecular states (Stark effect) since effects on the transition probability could be excluded for a disordered material with highly-localized states. Although the field dependence of the electro-absorption signal was quadratic, a consistent interpretation of the data could not be obtained by assuming a

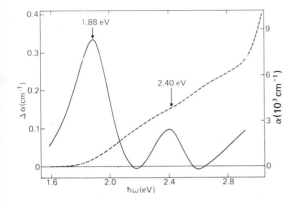

Fig. 8.14. Electro-absorption spectrum (solid curve) and absorption (dashed curve) of the charge transfer complex of 1:1 TNF:PVK. Sample thickness $L = 6.3\,\mu\text{m}$; $T = 295\text{K}$; $E = 8 \times 10^5$ V/cm. (From Weiser [53].)

simple quadratic Stark effect. The electro-absorption spectrum results from a shift of the molecular states due to interaction of the molecular dipoles with the external field. With excitation of the CT complex there is a large change in the molecular dipole moment leading to the observed quadratic dependence. Changes in the molecular dipole moments and the polarizability due to photo-excitation of the CT complex were derived and indicate that charge transfer mainly takes place in the excited state.

Photogeneration in TNF:PVK was measured by Melz [62] using a transient electrostatic-discharge technique. The strong field dependence of the carrier generation efficiency was attributed to competition with geminate recombination as discussed by Hughes (see above) in connection with PVK [58]. The photogeneration was viewed as a two-step process. In the first step, absorbed photons excite electrons to some bound state from which the electrons either decay to the ground state or else thermalize into a continuum state. The yield of electrons into the continuum states ϕ_0 was assumed to be field independent. In the second step, electron—hole pairs in the continuum states are treated as classical point charges diffusing under the influence of their mutual Coulomb attraction and the applied field. The probability that the electron—hole pair escapes recombination is determined by the Onsager theory [60]. Experiment and theory are compared in Fig. 8.15 for a 0.06:1 molar ratio of TNF:PVK. At fields greater than 7×10^4 V/cm, the Onsager theory fits the data for temperatures between 5°C and 50°C using values of $r_0 = 25$ Å for the thermalization length or initial electron—pair separation and $\phi_0 = 0.23$ for the auto-ionization probability. At lower fields, discrepancies between theory and experiment were not accounted for. The photogeneration efficiency increased by a factor of five in going from a 0.06:1 to 1:1 molar ratio of TNF to PVK monomer. The electric-field dependence was fitted to the Onsager theory using $r_0 = 35$ Å and $\phi_0 = 0.23$. Thus, the increased

References pp. 332—334

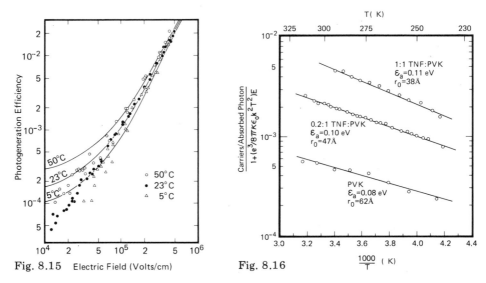

Fig. 8.15. Experimental results and theoretical predictions for photogeneration in 0.06 to 1.0 molar ratio TNF to PVK monomer films at three temperatures. Theoretical computations use $r_0 = 25$ Å and $\phi_0 = 0.23$ (From Melz [62].)

Fig. 8.16. Temperature dependence of the carrier quantum yield in PVK and TNF:PVK films. The thermalization lengths were obtained using $\mathcal{E}_a = (e^2/4\pi\kappa\kappa_0 r_0)$ from eqn. (8.8) (From Schechtman [63].)

photogeneration efficiency was attributed to increased initial electron–hole pair separation. Schechtman [63] has found that the photogeneration efficiency at low fields is extremely sensitive to space-charge effects due to previous illumination history. Using carefully dark-rested samples, linear photogeneration yield dependence on field was measured at low fields for all compositions from pure PVK to 1:1 TNF:PVK complexes. The experimental slope/intercept ratios were also in good agreement with the predicted ratio $(e^3/8\pi\kappa\kappa_0 k^2 T^2)$. The thermalization length r_0 was determined from the temperature dependence [see eqn. (8.8)] following the method used by Batt, Braun and Hornig [64] for anthracene. The temperature dependence is shown in Fig. 8.16 for several film compositions. These data clearly show the activated behavior predicted in eqn. (8.8) with $\mathcal{E}_a = (e^2/4\pi\kappa\kappa_0 r_0)$. The values of $r_0 = 62$ Å for pure PVK and 38 Å for the 1:1 TNF:PVK complex are in good agreement with the results of Hughes [59] and Melz [62] respectively. However, at low TNF concentrations, Schechtman's results give initial pair separations between the values for pure PVK and the 1:1 complex in contrast to Melz's result. The main differences in these experiments were the lower light fluxes and more controlled initial conditions of samples in Schechtman's work.

Bulk recombination of charge carriers in a 1:1 TNF:PVK molar ratio film

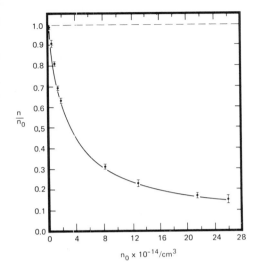

Fig. 8.17. The fraction of carrier pairs escaping recombination as a function of the initial concentration, n_0. The solid line is calculated from a simple theory of recombination due to Langevin [65]. The sweeping field was 1.1×10^5 V/cm in all cases. (From Hughes [36, 65].)

was observed by Hughes [36, 65] using a ruby laser pulse to generate very high carrier concentrations. Experimental results shown in Fig. 8.17 for the fraction of carrier pairs escaping recombination as a function of initial concentration n_0 were in good agreement with a diffusion-controlled recombination theory developed by Langevin [66]. The theory predicts that the recombination rate constant is proportional to the sum of the carrier mobilities $\gamma = \alpha(\mu_n + \mu_p)$ with $\alpha = e/\kappa\kappa_0$. Using this theory, Hughes showed that the fraction of initially created carriers which would escape bulk recombination is:

$$\frac{n}{n_0} = (E/n_0 \alpha L) \ln [1 + (n_0 \alpha L/E)], \qquad (8.9)$$

where E is the electric field and L is the film thickness. The result is independent of the carrier mobilities even if mobility is time dependent. The fit to the data of Fig. 8.17 was obtained with $\alpha = 5 \pm 1 \times 10^{-7}$ V/cm in good agreement with the predicted value of $\alpha = e/\kappa\kappa_0$ which for 1:1 TNF:PVK ($\kappa = 3.7$) was 4.9×10^{-7} V/cm. The recombination coefficient obtained using this value of α and $\mu = 5 \times 10^{-8}$ cm^2/V.s is $\gamma = 3 \times 10^{-14}$ cm^3/s. Hughes points out that the Onsager theory for geminate recombination and the Langevin theory of bulk recombination both depend on the premise that the carrier mean free paths are less than $r_0 = e^2/6\pi\kappa\kappa_0 kT$. Thus, materials showing geminate recombination should also display diffusion-controlled bulk recombination.

References pp. 332—334

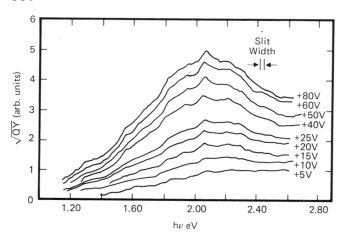

Fig. 8.18. Square root of the quantum yield $\sqrt{QY}$ (holes per absorbed electrons) *versus* photon energy $h\nu$ for photo-emission from gold into PVK. (From Lakatos and Mort [67].)

8.4 INJECTION EXPERIMENTS

In the preceding section, we have discussed charge generation and recombination experiments in essentially homogeneous polymer systems. In this section, we will review a number of experiments involving heterogeneous systems in which photogenerated carriers are injected into the polymer films from metals, inorganic photoconductors or other organic materials. Photo-injection offers many possibilities, both in probing the electronic structure of polymers and in photosensitizing polymeric transport materials using layered and fine particulate dispersion structures. Sensitization of PVK with a thin layer of amorphous selenium has been mentioned in our previous discussion of Regensburger's transport studies [20]. A thin (0.3 μm) layer of amorphous selenium was evaporated onto PVK films. Efficient hole injection into PVK was observed with a threshold for injection of about 6 V/μm which was interpreted as the field needed to overcome an injection barrier. The spectral response was essentially that of selenium and the gain was independent of Se layer thickness. Photo-emission of holes from gold, copper and aluminum into PVK were reported by Lakatos and Mort [21, 67]. Figure 8.18 shows the results for photo-emission from gold. The threshold energies for photo-emission were determined to be 1.28 eV, 1.48 eV and 1.53 eV for gold, copper and aluminum, respectively. From the sum of the metal work functions and the photo-emission threshold energy, an average value of 6.1 ± 0.4 eV was obtained for the position of the valence band edge in PVK relative to vacuum. Fine structure on the photo-emission curves of Fig. 8.18 was also observed for the other metals. This structure was interpreted as splitting of a narrow valence band in PVK by molecular vibrations. Since vibrational energies are known to be of the order of 0.1 eV [68], observation of this structure implies that the width of the valence band is significantly less than 0.1 eV.

Information on the energy level structure in TNF:PVK charge-transfer complexes was obtained by Mort and Emerald [34] from injection experiments from amorphous selenium into the organic films. An electrophotographic discharge technique was employed using 4300 Å light which was all absorbed in a 1-μm selenium layer deposited on the TNF:PVK films. TNF:PVK molar ratios from approximately 0.1:1 to 0.3:1 were measured. Injection from amorphous selenium into pure amorphous TNF layers was also reported [35]. The discharge rates for electron and hole injection could be correlated with known carrier mobilities [24, 33]. Since the discharge was space-charge-limited, mobility could be estimated for each composition. The observation that the discharge rate was mobility limited suggested that negligible barriers exist between transport states for electrons and holes in amorphous selenium and TNF:PVK. Based on the known mobility gap in amorphous selenium of ~ 2.1 eV [21, 69] the schematic energy level diagram shown in Fig. 8.19 was deduced. Electron transport states in the TNF:PVK complex are characteristic of the acceptor molecule TNF and hole transport states are characteristic of the carbazole monomer. This picture of the transport mechanism was also obtained by Gill [24, 33] based on the composition dependence of mobility.

8.5 CONCLUSIONS

Our present understanding of the electronic properties of polymeric photoconductors is at a very rudimentary level. Transport measurements, from which reliable basic parameters such as mobility are obtained, have been successful on very few materials. Of these materials, only the system consisting of poly-N-vinyl carbazole (PVK) and 2,4,7-trinitro-9-fluorenone (TNF) has been extensively investigated experimentally. Even in these materials where considerable experimental progress has been made, many of the observed phenomena such as field dependence of mobilities are not understood. Investigation of the transport properties of a much broader range of polymeric materials will probably be necessary before a detailed understanding of the physical mechanisms is obtained. Experience with PVK and TNF:PVK has shown that reasonably good transport properties, manifested by the ability to transport charge across a sample without extensive deep trapping, is the key to successful application of many experimental techniques. This ability has made it possible to measure mobilities, carrier lifetimes, recombination coefficients, quantum yields, and photo-injection properties with a good deal of confidence and accuracy. The materials criteria necessary to realize these transport properties in polymeric systems have not yet been defined. The nature of deep trapping centers is unknown even to

References pp. 332—334

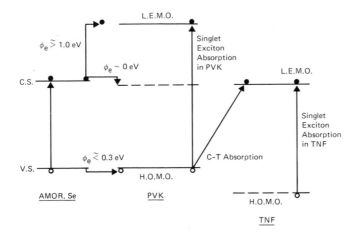

Fig. 8.19. Schematic diagram of relative energy-level structure of amorphous selenium and the charge-transfer complex PVK:TNF deduced from injection experiments. (From Mort and Emerald [34].)

whether they are intrinsic or extrinsic. Certainly, higher standards of purification and more extensive structural characterization of polymeric photoconductors will be required to broaden the range of materials in which these types of experiments will be successful. A number of commercial applications of polymeric photoconductors have now been made. This interest in the application of the electronic properties of these materials has already resulted in a rapid growth of research in this field. There is also a rapidly growing interest in the electronic properties of a much broader range of organic solids including insulating polymers, high-conductivity organics, organic crystals and biological systems. We can expect continually more rapid development of our knowledge of polymeric photoconductors as research in each of these related fields progresses.

REFERENCES

1 D.A. Seanor, in A.D. Jenkins (Ed.), Polymer Science, North-Holland, London—Amsterdam, 1972, p. 1233.
2 F. Gutmann and L.E. Lyons, Organic Semiconductors, Wiley, New York, 1967.
3 J.E. Katon, Organic Semiconducting Polymers, Marcel Dekker, New York, 1968.
4 L.I. Grossweiner, Organic Photoconductors in Electrophotography, Most Associates, Inc., Marblehead, Mass., 1970.
5 A. Rembaum, Encyclopedia of Polymer Science and Technology, Vol. 11, Interscience, 1969, p. 318.
6 P. Erlich and M.M. Labes, Encyclopedia of Polymer Science and Technology, Vol. 11, Interscience, New York, 1969, p. 338.
7 P.J. Reucroft, in K.C. Frisch and A. Patsis (Eds.), Electrical Properties of Polymers, Technomic, Westport, Conn., 1972, p. 59.

8   R.G. Kepler, to be published.
9   H. Hoegl, J. Phys. Chem., 69 (1965) 755.
10 A.V. Vannikov, Sov. Phys.-Solid State, 9 (1967) 1068.
11 M. Kryszewski, A. Szymanski and J. Swiatek, J. Polym. Sci. C, 16 (1968) 3915.
12 J.H. Ranicar and R.J. Fleming, J. Polym. Sci, A—2, 10 (1972) 1321.
13 D.K. Davies, J. Phys. D, 5 (1972) 162.
14 E.H. Martin and J. Hirsch, Solid State Commun., 7 (1969) 783; J. Non-Cryst. Solids, 4 (1970) 133; J. Appl. Phys., 43 (1972) 1001.
15 A. Reiser, M.W.B. Lock and J. Knight, Trans. Faraday Soc., 65 (1969) 2168.
16 E.I. Merkulov, A.V. Vannikov and I.D. Mikhailov, Sov. Phys.-Solid State, 13 (1972) 2243.
17 R.C. Hughes, Photogr. Sci. Eng., to be published.
18 S. Matsumoto and K. Yahagi, Jpn. J. Appl. Phys., 12 (1973) 930.
19 K. Hayashi, K. Yoshino and Y. Inuishi, Jpn. J. Appl. Phys., 12 (1973) 754.
20 P.J. Regensburger, Photochem. Photobiol., 8 (1968) 429.
21 J. Mort and A.I. Lakatos, J. Non-Cryst. Solids, 4 (1970) 117.
22 D.M. Pai, J. Chem. Phys., 50 (1969) 3568.
23 A. Szymanski and M.M. Labes, J. Chem. Phys., 50 (1969) 3568.
24 W.D. Gill, J. Appl. Phys., 43 (1972) 5033.
25 A.M. Hermann and A. Rembaum, J. Appl. Phys., 37 (1966) 3642.
26 A.M. Hermann, in K.C. Frisch and A. Patsis (Eds.), Electrical Properties of Polymers, Technomic, Westport, Conn., 1972, p. 103.
27 M. Lardon, E. Lell-Döller and J.W. Weigl, Mol. Cryst. Liq. Cryst., 2 (1967) 241.
28 R.M. Schaffert, IBM J. Res. Dev., 15 (1971) 75.
29 H. Seki and W.D. Gill, Proc. Second Int. Conf. on Conductivity in Low-Mobility Materials, Eilat, 1971, Taylor and Francis, London, 1971, p. 409.
30 M.E. Scharfe, Phys. Rev. B, 2 (1970) 5025.
31 W.D. Gill and K.K. Kanazawa, J. Appl. Phys., 43 (1972) 529.
32 H. Seki, Proc. 5th Int. Conf. on Amorphous and Liquid Semiconductors, Garmisch-Partenkirchen, 1973, Taylor and Francis, London, 1974, p. 1015.
33 W.D. Gill, Proc. 5th Int. Conf. on Amorphous and Liquid Semiconductors, Garmisch-Partenkirchen, 1973, Taylor and Francis, London, 1974, p. 901.
34 J. Mort and R.L. Emerald, J. Appl. Phys., 45 (1974) 175.
35 R.L. Emerald and J. Mort, J. Appl. Phys., 45 (1974) 3943.
36 R.C. Hughes, J. Chem. Phys., 58 (1973) 2212.
37 J. Frenkel, Phys. Rev., 54 (1938) 647.
38 R.M. Hill, Philos. Mag., 23 (1971) 59.
39 B.G. Bagley, Solid State Commun., 8 (1970) 345.
40 T. Holstein, Ann. Phys. (New York), 8 (1959) 343.
41 M.D. Tabak, D.M. Pai and M.E. Scharfe, J. Non-Cryst. Solids, 6 (1971) 357.
42 S.J. Fox, Electrophotography, Second Int. Conf., SPSE Publication, 1974, p. 170.
43 W.D. Gill, unpublished data.
44 M.E. Scharfe, Phys. Rev. B, 2 (1970) 5025.
45 H. Scher, Proc. Fifth Int. Conf. on Amorphous and Liquid Semiconductors, Garmisch-Partenkirchen, 1973, Taylor and Francis, London, 1974, p. 135.
46 R.M. Minday, L.D. Schmidt and H.J. Davis, J. Chem. Phys., 50 (1969) 1473; ibid., 45 (1971) 3112.
47 W.F. Schmidt and A.O. Allen, J. Chem. Phys., 52 (1970) 4788.
48 Y. Maruyama and K. Funabashi, J. Chem. Phys., 56 (1972) 2342.
49 W. Mehl and N.E. Wolff, J. Phys. Chem. Solids, 25 (1964) 1221.
50 M.D. Tabak, Phys. Rev. B, 2 (1970) 2107.
51 J.M. Marshall and A.E. Owen, Philos. Mag., 24 (1971) 1281; Phys. Status Solidi A, 12 (1972) 181.

52 G. Weiser, J. Appl. Phys., 43 (1972) 5028.
53 G. Weiser, Phys. Status Solidi A, 18 (1973) 347.
54 J. Mort, G. Pfister and S. Grammatica, Solid State Commun., 18 (1976) 693.
55 K.K. Kanazawa and B.H. Schechtman, in M.M. Perlman (Ed.), Electrets, Charge Storage and Transport in Dielectrics, Dielectric and Insulation Division, The Electrochemical Society, 1973, p. 405.
56 M. Pollack, Philos. Mag., 23 (1971) 519.
57 M. Abkowitz, A. Lakatos and H. Scher, Phys. Rev. B, 9 (1974) 1813.
58 M. Ikeda, K. Morimoto, Y. Murakami and H. Sato, Jpn. J. Appl. Phys., 8 (1969) 759; ibid., 8 (1969) 931.
59 R.C. Hughes, Chem. Phys. Lett., 8 (1971) 403; J. Chem. Phys., 55 (1971) 5442.
60 L. Onsager, Phys. Rev., 54 (1938) 554.
61 M.D. Shattuck and U. Vahtra, U.S. Patent 3,484,237, June 13, (1966); December 16, (1969).
62 P.J. Melz, J. Chem. Phys., 57 (1972) 1694.
63 B.H. Schechtman, to be published.
64 R.H. Batt, C.L. Braun and J.F. Hornig, J. Chem. Phys., 49 (1968) 1967; Appl. Opt. Suppl., 3 (1969) 20.
65 R.C. Hughes, Appl. Phys. Lett., 21 (1972) 196.
66 P. Langevin, Ann. Chim. Phys., 28 (1903) 289, 443.
67 A.I. Lakatos and J. Mort, Phys. Rev. Lett., 21 (1968) 1444.
68 J.H. Sharp, J. Phys. Chem., 71 (1967) 2587.
69 J. Mort and H. Scher, Phys. Rev. B, 3 (1971) 334.

CHAPTER 9

NON-POLAR LIQUIDS

WERNER F. SCHMIDT

9.1. Introduction
9.2. Radiation-induced conductivity
    9.2.1. Low field, low LET radiation
    9.2.2. High field, low LET radiation
    9.2.3. Corpuscular radiation
    9.2.4. Ion yields
9.3. Photoionization and photoeffect
    9.3.1. Non-polar liquids
    9.3.2. Solutes in non-polar liquids
    9.3.3. Photoeffect
9.4. Transport properties of charge carriers
    9.4.1. Mobility
    9.4.2. Recombination coefficient
    9.4.3. Diffusion constant
    9.4.4. Electron attachment
9.5. High energy radiation detectors with non-polar liquids
    9.5.1. Liquid-filled ionization chambers
    9.5.2. Detectors with electron multiplication
9.6. Purification techniques and measurement cells
    9.6.1. Preparation of liquid samples
    9.6.2. Measurement cells

9.1 INTRODUCTION

The electrical conductivity of metals and crystalline semiconductors has received extensive attention for many decades and great advances have been made towards the understanding of charge transport in these materials. Recently, the electronic properties of disordered systems have received increasing interest and many experimental and theoretical investigations have been carried out for non-crystalline solids. Charge transport in non-polar liquids and glasses has also been studied for many decades from different points of view. Physicists and electrical engineers were interested in the insulating properties of non-polar liquids and sought the intrinsic conductivity of non-polar liquids. Measurements of the self-conductivity of liquid hydrocarbons showed that the specific conductivity decreased the more the

References pp. 384—388

liquids were subjected to various purification procedures and no value was found which could be attributed to the natural conductance. The discovery of the ionization of gases by high energy radiation, such as X-rays or nuclear radiation, led rather early also to the investigation of radiation-induced ionization in insulating liquids. J.J. Thomson [1] reported in 1897 that vaseline oil showed an increased electric conductivity during irradiation with X-rays. A similar effect was found by P. Curie [2] in 1902 who studied the influence of radium rays on several liquids. From 1908 to 1913, a series of very beautiful investigations were carried out by Jaffé [3—7] and his coworker Van der Bijl [8] which led to the formulation of the theory of columnar ionization [6] in which the spatially inhomogeneous formation and recombination of charge carriers along the track of an ionizing particle were treated. Although gross simplifications had to be assumed for the solution of the mathematical problem, Jaffé's approach and an extension by Lea [9] for weakly-ionizing radiations is still of interest. The transport properties of radiation-induced charge carriers have been investigated since the days of Jaffé and many reports have been published by Adamczewski and his coworkers [10, 11]. Radiation chemists in their research on the interaction of ionizing radiation with liquids became interested in the yield and the transport properties of charge carriers in non-polar liquids while investigating the primary processes which follow the absorption of energy. Measurements of charge-carrier yields were performed and, with the advancement of the understanding of the primary process and improvements in purification procedures and sample preparation, observation of excess electrons and holes in liquid hydrocarbons became possible.

Charge-carrier transport in liquefied rare gases has been studied for several decades. Large ionization pulses due to single α-particles and γ-quanta were observed in liquid argon indicating the possible application in liquid-filled ionization chambers. Recently, electron avalanches have been observed in liquid xenon, and multi-wire proportional chambers filled with liquid xenon, for track measurements of elementary particles, have been described (cf. Section 9.5). On the other hand, investigations of excess electrons in liquid helium have received much interest since the excess charge can be considered as a probe for the structure of the liquid. Several reviews have appeared [12—14] and, thus, this subject will be treated in a rather general manner.

In this chapter, the radiation-induced conductivity of non-polar liquids (dielectric constant $\kappa \approx 2$) is reviewed. The ionization process leads to the formation of charge carriers with an excess energy. This energy is transferred to the molecules of the liquid by numerous collisions and, eventually, charge carriers of both signs will have obtained thermal energies. We shall use the terms charge carrier, ion or excess electron, for charge carriers in thermal equilibrium with their surroundings unless specified otherwise. The energy absorbed by a medium when high energy radiation is traversing it is given by the radiation dose measured in rad (1 rad = 100 erg g^{-1}). The yield of ions is usually expressed as the G-value which represents the number of ion pairs produced by absorption of 100 eV. Therefore, the dose is frequently also given in eV g^{-1} or for a specific liquid volume in eV cm^{-3}. The quality of the radiation is measured by the linear energy transfer (LET), a quantity which describes the energy transferred to the medium on a certain length of the path of the high energy particle or quant. It is usually given in keV/μm and densely-ionizing radiation, such as α-particles, exhibit a high LET, while high energy X-rays and γ-rays produce few spatially distributed ionization events and have a small LET.

In Sections 9.2.1 and 9.2.2 radiation-induced conductivity by low LET radiation is discussed. Ionization by α-particles is reviewed in Section 9.2.3. The measurement of transport properties of slow charge carriers which are usually observed in normal purity liquids is described in Section 9.2.1, while in Section 9.4.1 the motion of excess electrons in non-polar liquids is discussed. Only very few reports on the direct photoionization of dielectric liquids by ultraviolet light have been published and the results are reviewed in Section 9.3.1. Ionization of solutes (Section 9.3.2) is experimentally easier to achieve especially by the use of high intensity lasers with which biphotonic ionization is possible. Injection of electrons by photoeffect is discussed in Section 9.3.3. These experiments are important for the measurement of the energy level of the conduction electron state in non-polar liquids. In Section 9.4, the transport properties of excess electrons are mainly discussed. In Section 9.5, applications of non-polar liquids in nuclear radiation detectors are reviewed. In Section 9.6, the various purification procedures are compiled. Each experiment requires purification with respect to a different effect. For electron mobility measurements, a very low level of electron scavengers is necessary while the self-conductivity might not be extremely low. At present, the measured effect is the only criterion for the purity of the liquid since the impurity concentrations involved are well below the sensitivity of conventional analysis techniques. All equations in this chapter are in the International System of units if not specifically mentioned otherwise.

9.2 RADIATION-INDUCED CONDUCTIVITY

9.2.1 Low field, low LET radiation

(i) Steady-state conductivity

When a dielectric liquid is irradiated with ionizing radiation of constant intensity, an increase of the electrical conductivity results. In a parallel plate measurement with a cell of electrode area A and electrode separation L the specific conductivity is given by:

$$\sigma = \frac{i}{V}\frac{L}{A} \tag{9.1}$$

where i is the ionization current with an applied voltage V. In the steady state, ion losses due to recombination are compensated by the rate of charge-carrier production g and at low electric field strengths

$$\frac{dn}{dt} = 0 = g - \alpha n^2 \tag{9.2}$$

α is the volume recombination coefficient and n the steady-state charge-carrier concentration. The number of charge carriers of one sign which is produced per $cm^3 s^{-1}$ is proportional to the absorbed dose rate D_R in eV $cm^{-3} s^{-1}$ and is given by

$$g = D_R G_{fi} \times 10^{-2} \tag{9.3}$$

G_{fi} is the free-ion yield, the number of charge carriers measured at zero field strength when 100 eV have been absorbed. If the penetration depth of the radiation is large compared to the cell dimensions, a homogeneous distribution of positive and negative charge carriers will be obtained and the observed conductivity is related to the charge-carrier density n and the mobility μ by:

$$\sigma = en(\mu_+ + \mu_-) \tag{9.4}$$

with e the electronic charge.

Combining eqn. (9.2) and eqn. (9.3) gives:

$$n = \left(\frac{G_{fi} D_R}{10^2 \alpha}\right)^{1/2} \tag{9.5}$$

and from eqn. (9.4) and eqn. (9.5) the specific conductivity as a function of dose rate is obtained:

$$\sigma = \frac{e(\mu_+ + \mu_-)}{10} \left(\frac{G_{fi} D_R}{\alpha}\right)^{1/2} \tag{9.6}$$

and hence

$$\sigma \propto D_R^{0.5} \tag{9.7}$$

The presumption for this dependence is that negative and positive carriers with only one mobility μ_- and μ_+, respectively, are produced. Rewriting eqn. (9.6) yields an expression for the free ion yield

$$G_{fi} = 10^2 \frac{\sigma^2/e^2}{D_R} \cdot \frac{\alpha}{(\mu_+ + \mu_-)^2} \tag{9.8}$$

For the determination of this value, the following quantities have to be determined:
(a) the steady-state conductivity σ during irradiation with a dose rate D_R;
(b) the volume recombination coefficient α;

(c) the mobilities μ_+ and μ_- of the charge carriers; and
(d) the absorbed dose rate D_R.

Measurement of the steady-state ionization current in liquid hydrocarbons have been carried out by several authors [15—21] and it was found that at low voltages i increased linearly with V and that in this region $\Delta i/\Delta V$ varied with the square root of the dose rate. Figure 9.1 shows the ionization current as a function of the applied voltage for n-hexane irradiated with 1.5 MeV X-rays. The current at zero voltage is not zero. It is partly due to currents induced in the cables of the circuit and partly due to contact potentials. For the conductivity $\Delta i/\Delta V$ is taken and in Fig. 9.2 these values are plotted *versus* dose rate. The square root dependence was observed over several orders of magnitude of σ and D_R. The ratio σ^2/D_R should be a constant quantity for each liquid. Since σ depends on the mobilities μ_+ and μ_- of the charge carriers σ^2/D_R may vary from one study to another since different impurities present in the liquid may have influenced the mobilities. Typical values of σ^2/D_R and the mobilities are given in Table 9.1. Values of σ^2/D_R reported by other authors for other liquids exhibit the same order of magnitude [16, 18—21].

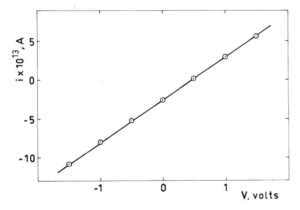

Fig. 9.1. Ionization current *versus* cell voltage in n-hexane, irradiated with 1.5 MeV X-rays. $L = 0.62$ cm, $A = 2.78$ cm^2, $D_R = 3.2 \times 10^{11}$ eV cm^{-3} sec^{-1}. (From Hummel and Allen [15])

(ii) Volume recombination coefficient

If the liquid is irradiated with a short pulse of radiation, or a continuous irradiation is suddenly interrupted, then a decay of the conductivity in time is observed which, at low voltages, is due to recombination. The change in charge-carrier concentration n is given by

$$\frac{dn}{dt} = -\alpha n^2 \tag{9.9}$$

References pp. 384—388

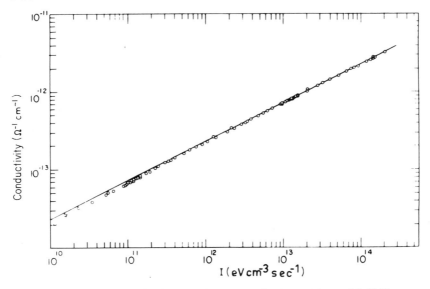

Fig. 9.2. Conductivity of n-hexane during irradiation with 1.5 MeV X-rays as a function of dose rate. (From Hummel and Allen [15])

TABLE 9.1

Typical values of σ^2/D_R and the charge-carrier mobilities at $T = 24°C$. (From Hummel and Allen [15])

Liquid	σ^2/D_R Ω^{-2} cm eV^{-1} s^{-1}	$\mu_+ \times 10^3$ cm^2 V^{-1} s^{-1}	$\mu_- \times 10^3$ cm^2 V^{-1} s^{-1}
n-Hexane	5.4×10^{-38}	0.68	1.3
1,4-Dioxane	1.5×10^{-38}	0.80	0.48
Carbon tetrachloride	1.6×10^{-38}	0.40	0.33

Integration yields:

$$\frac{1}{n(t)} = \frac{1}{n(0)} + \alpha t \tag{9.10}$$

where $n(0)$ is the concentration of the charge carriers at the end of the pulse or at the moment of interruption. Direct measurements of α are possible with the clearing-field method first introduced into radiation research by Rutherford [22]. An ionization chamber is irradiated with a pulse of X-rays. At time t after the end of the pulse, a high voltage V is applied. The field produced by this voltage has to be high enough to sweep out of the irradiated volume all remaining charge carriers $n(t)$. The collected charge is measured and by varying the time t after the end of the irradiation $n(t)$ is obtained and from eqn. (9.10) α can be determined. This method was applied to the investigation of ion recombination in dielectric liquids by Jaffé [5], van der Bijl [8], Adamczewski [10] and Gazda [23]. Since the carrier concentration decreases after interruption of irradiation due to volume recombination, the conductivity and the ionization current also decay in time. From eqn. (9.10), the ionization current as a function of time is:

$$\frac{1}{i(t)} = \frac{1}{i(0)} + \left[\frac{\alpha}{e(\mu_+ + \mu_-)} \frac{L}{AV} \right] t \tag{9.11}$$

and from the measurement of this dependence the ratio $\alpha/(\mu_+ + \mu_-)$ can be determined. For eqn. (9.9) to hold, it is necessary that carrier losses due to diffusion out of the volume and neutralization on the electrodes are negligible. Several authors have carried out measurements of $\alpha/(\mu_+ + \mu_-)$. However, agreement is not always very good [15, 16, 24—27].

A different method for the determination of the recombination coefficient was described by Careri and Gaeta [28]. Two beams of charge carriers of opposite sign were generated in a parallel plate cell with polonium alpha-sources on each electrode. Under the influence of the applied electric field, the ions generated at each electrode moved across the cell. By shifting one source with respect to the other the two ion beams could be superimposed. The current decreased due to recombination and α could be determined. The measurements were carried out in liquid helium.

(iii) Mobility of ionic charge carriers

A number of different methods for the determination of mobilities of charge carriers in dielectric liquids have been developed. The most prominent method is the "time-of-flight" where the movement of some spatial discontinuity in the density of charge carriers is measured. The time it takes a group of ions to traverse a certain distance in an electric field is measured. In a uniform field, the transit time t_T to cross a gap L is given by

$$t_T = \frac{L^2}{\mu_\pm V} \tag{9.12}$$

Charge carriers are usually produced either by ionization by low energy X-rays or α-particles. Low energy X-rays interact predominantly by photoeffect and generate secondary electrons which have a small range. Thus, well-defined layers of charge carriers can be generated. In the method introduced by Gzowski and Terlecki [29], a thin layer of the liquid adjacent to the voltage electrode of a conductivity cell is irradiated with a pulse of X-rays. At the end of the pulse, a voltage is applied which drives ions of one sign across the gap to the collector electrode [Fig. 9.3(a)]. The current due to the motion of the charge carriers can either be integrated or recorded directly. Ideally, a square wave current signal should be observed with the current ceasing the moment the charge carriers reach the collector electrode.

References pp. 384—388

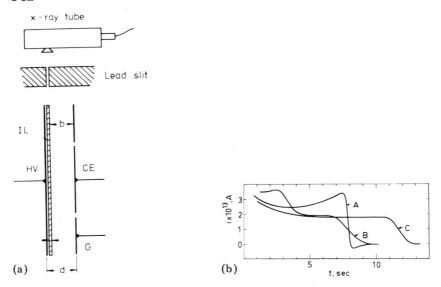

Fig. 9.3. Thin-layer method (Gzowski method) for mobility measurements. (a) Experimental set-up; IL ion layer; HV, high voltage electrode; CE collector; G guard ring. (b) Current recordings. (From Hummel et al. [24])

Diffusion broadening of the layer, space-charge effects and other limiting conditions lead to a deviation from this ideal behavior as can be seen in the examples of Fig. 9.3(b). The thin-layer method has been extensively used by the Polish group in Gdansk and their results have been reviewed by Adamczewski [10]. Hummel et al. [24] injected the ion layer at various positions between the electrodes and determined the transit time from the slope of t_T versus position x between the electrodes. The application of the thin-layer method requires the charge of the layer to be small enough so that the change of the field strength across the layer due to space charge is negligible in comparison with the externally applied field, i.e. $Q/\kappa\kappa_0 \ll V/L$. With $\kappa \approx 2$, $Q/\kappa\kappa_0 = 5.7 \times 10^6$ V cm^{-2} is obtained when Q is given in C cm^{-2}. For an external field of $V/L = 10^2$ V cm^{-1}, the condition $Q \ll 1.8 \times 10^{-11}$ C cm^{-2} follows. With $Q \approx 1.8 \times 10^{-13}$ C cm^{-2}, $E = 10^2$ V cm^{-1} and $(\mu_- + \mu_+) = 10^{-3}$ cm^2 V^{-1} s^{-1}, a current density of $j = 1.8 \times 10^{-14}$ A cm^{-2} is obtained. Very small currents are involved and extreme restrictions on the self-conductivity of the liquid are imposed. A variation of the thin-layer method is the thick-layer method. Here ions of both signs are generated near the voltage electrode. The electric field is applied all the time, so that the liquid volume is filled with ions of one sign. If the radiation is interrupted, a linear decay of the ionization current results from which the transit time t_T of the particular ions can be measured (Fig. 9.4).

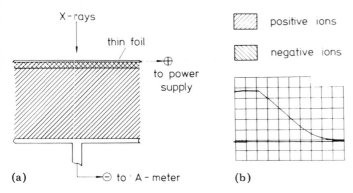

Fig. 9.4. Thick-layer method for mobility measurements. (a) spatial distribution of ions; (b) oscilloscopic recording of the decay of the ionization current. (From Schmidt [16].)

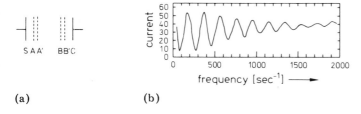

Fig. 9.5. Shutter method for mobility measurements. (a) S, Electrode with radioactive α-source; AA$'$ and BB$'$, two pairs of grids; C, collector electrode; (b) Recording of ion current *versus* frequency of the grid voltage. Positive charge carriers in liquid helium, $E = 106$ V cm^{-1}. (From Meyer and Reif [34].)

Negative charge carriers can be generated by photoeffect on one electrode. LeBlanc [30] reported measurements on n-hexane and n-heptane. Electrons were liberated from an aluminium electrode by a 1 ms u.v. light flash. Terlecki [31] and Chong and Inishi [32] also used this method and were able to measure mobilities in n-hexane up to field strengths of 0.5 MV cm^{-1}. Other variations of these principles can be found in the literature and a comprehensive review has been given by Hummel and Schmidt [33]. It is also possible to generate charge carriers continuously in a certain part of the liquid and create layers of ions by switching potentials on electrodes and grids. Meyer and Reif [34, 35] used two pairs of grids placed between two plane electrodes [fig. 9.5(a)]. At the electrode S, charge carriers are generated by an alpha-source. A voltage is applied across the cell and the grids are maintained at the potential corresponding to their position between the plates. Between each pair of grids, an a.c. or pulsed d.c. voltage is applied with the direction of the field between the grids opposite to the direction of the external field for a certain time τ. Charge carriers passing through the first pair of grids (shutter) will only pass through the second shutter if their

References pp. 384—388

drift time is a multiple of the period of the shutter voltage. Changing the shutter voltage continuously leads to oscillations in the current reaching the collector electrode. From the distance of consecutive maxima or minima the drift time between the two shutters can be determined [Fig. 9.5(b)]. Several variations of this basic idea have been described [36—42].

(iv) Mobility of excess electrons

Most of the methods described above are also applicable for the measurement of excess electron mobility. Excess electrons in liquid hydrocarbons usually have a greater mobility than negative ions so that their motion in an electric field leads to a separation in time from the motion of the ions. Schmidt and Allen [43, 44] applied a method originally developed by Hudson [45] for the measurement of the electron drift velocity v in gaseous argon. A parallel-plate ionization chamber is subjected to a step function of X-rays. The initial rise of the ionization current is due to the fast carrier only. The time dependence of the current is given by

$$i(t) = \begin{cases} 0 & \text{for } t \leqslant 0 \\ 2i_{max} \left(\frac{v}{L}\right)\left[t - \frac{1}{2}\left(\frac{V}{L}\right)t^2\right] & \text{for } 0 \leqslant t \leqslant t_T \\ i_{max} & \text{for } t \geqslant t_T \end{cases} \tag{9.13}$$

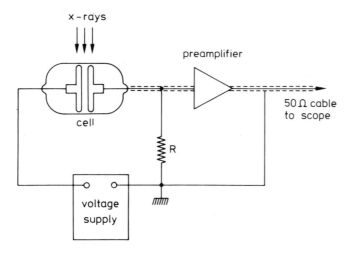

Fig. 9.6. Circuit for drift velocity measurements.

The current was measured with a circuit shown in Fig. 9.6. The current was converted into a voltage which was amplified and recorded oscillographically. Figure 9.7 shows a typical oscilosope trace obtained with the Hudson method. The slower continuing rise of the ionization is due to the

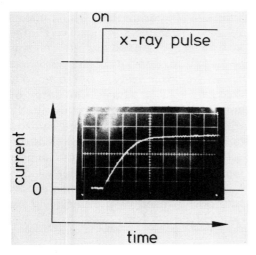

Fig. 9.7. Trace of the ionization current obtained by the Hudson method in n-pentane, 5 μsec/div., 43.3 kV cm^{-1}. (From Schmidt and Allen [44].)

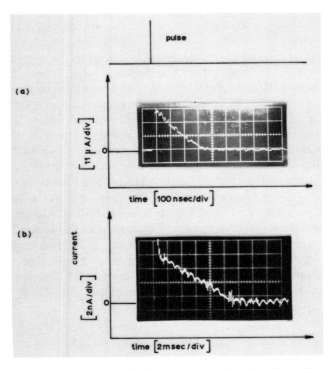

Fig. 9.8. (a) Decay of electron current in liquid methane after irradiation with a 5 ns pulse of 15 MeV X-rays. (b) Decay of the ion current in liquid methane. (From Schmidt and Bakale [46].)

References pp. 384—388

positive ions for which eqn. (9.13) also holds. A similar method which makes use of the large difference in the mobilities of negative and positive charge carriers was used by Schmidt and Bakale [46, 47]. A parallel-plate cell was irradiated with a pulse of high energy X-rays. The duration of the pulse was short compared to the electron drift time. Since, effectively, a homogeneous distribution of positive and negative charge carriers is produced, the decay of the ionization current after the pulse consists of two linear parts distinctly separated in time. The measuring circuit was the same as for the Hudson method. Typical oscilloscope traces obtained are shown in Fig. 9.8. Measurement of the pulse conductivity during a pulse of high energy X-rays can also give the electron mobility according to eqn. (9.4) if the charge-carrier yield is known. Since $\mu_- \gg \mu_+$, the conductivity at the end of a short pulse is proportional to the electron mobility. This method has been used extensively by Freeman et al. [48–50] and by Conrad and Silverman [25].

(v) *Ion diffusion coefficient*

Mobility and diffusion coefficient D of ordinary ions are connected by the Nernst–Einstein equation

$$\frac{\mu}{D} = \frac{e}{kT} \tag{9.14}$$

To test the validity of this relation for charge carriers produced in dielectric liquids by radiation, a separate measurement of the diffusion coefficient is required. A very ingenious application of the clearing-field technique to this problem has been made by Gazda [51, 52]. Ions of one sign are produced in a given volume and the decay of the ion concentration with time is observed. It is assumed that ions coming into contact with the electrodes or the wall of the measurement cell are neutralized. The ions are introduced into the liquid volume by irradiating the liquid near the voltage electrode with a collimated beam of low-energy X-rays. An applied electric field separates the ions of the thin layer and after a time $\tau > t_T = L^2/\mu_\pm V$ the liquid volume is filled with ions of one sign. Then the electric field (V/L) is turned off and ion losses occur due to diffusion to the wall. After a time t, a clearing field is applied and all ions which have remained in the liquid volume are collected and measured. The use of a parallel-plate cell with a guard-ring electrode assembly allowed the problem to be treated as one-dimensional and the diffusion equation was solved with the following boundary conditions

$n(x, t) = 0 \quad \text{for } x \leqslant 0, x \geqslant L$

$n(x, 0) = \text{const} \quad \text{for } 0 < x < L$

$$\frac{\partial n}{\partial t} = D \frac{\partial^2 n}{\partial x^2} \tag{9.15}$$

An approximate solution for times $t \geqslant 4.5 \times 10^{-2} L^2/D$ is

$$n(x, t) = \frac{4}{\pi} n(x, 0) \exp\left(-\frac{\pi^2}{L^2} Dt\right) \sin\left(\pi \frac{x}{L}\right) \qquad (9.16)$$

The total charge remaining at time t is given by:

$$Q(t) = \frac{8}{\pi^2} A\, n(x, 0)\, eL \exp\left(-\frac{\pi^2}{L^2} Dt\right) \qquad (9.17)$$

with $n(x, 0)\, eLA = Q_0$ the initially produced charge. D can be obtained from the measurement of $Q(t)$ as

$$D = \frac{[\ln Q(t_1) - \ln Q(t_2)]\, L^2}{\pi^2 (t_2 - t_1)} \qquad (9.18)$$

(vi) Measurement of the free ion yield

The yield of free ions G_{fi} which escape geminate recombination can be determined from the low-field conductivity [*cf.* Section 9.2.1(i)] or by a clearing-field method. In the technique described by Schmidt and Allen [53—55], the liquid is ionized by a short pulse of radiation. Immediately following the pulse, a clearing field is applied so that all free ions are collected on the electrodes before any volume recombination has time to occur. The collected charge is measured as a function of the pulse dose and Fig. 9.9 shows an example of the data obtained. The dose per pulse has to be small enough so that volume recombination during the pulse can be neglected, *i.e.*:

$$\alpha D_R\, G_{fi}\, t_p \ll 100 \qquad (9.19)$$

which follows from eqn. (9.2), t_p is the pulse length.

The self-conductivity of the liquid has to be extremely low so that the charge collected in the absence of radiation is sufficiently small. If this condition is not fulfilled (as in the case of diethyl ether) then the charge produced by the radiation adds to the intrinsic charge and a linear dependence is observed.

9.2.2 High field, low LET radiation

(i) Extrapolated free ion yield

At higher field strengths, all free ions are collected and no volume recombination occurs, *i.e.*

$$\alpha n^2 = 0 \qquad (9.20)$$

In a gas ionization chamber this would be represented by the saturation current. In dielectric liquids we observe, however, a further increase of the ionization current due to the effect of the electric field on the geminate recombination process. In the stationary state, the change of the ion concentration is zero:

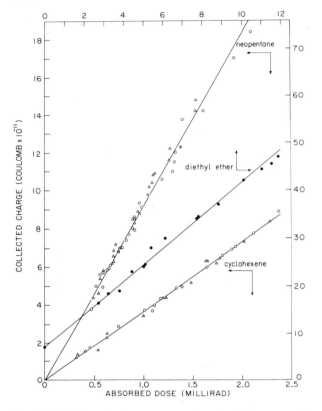

Fig. 9.9. Collected charge *versus* dose in the clearing-field method for the determination of the free ion yield. (From Schmidt and Allen [54].) Cell volume: cyclohexene 3.02 cm³; diethyl ether 1.4 cm³; neopentane 1.8 cm³.

$$-\frac{dn}{dt} = n(\mu_+ + \mu_-)\frac{V}{L^2} - G(E)D_R \times 10^{-2} = 0 \qquad (9.21)$$

where $G(E)$ is the yield of ions collected with the applied field E. Since the ionization current is given by:

$$i = ne(\mu_+ + \mu_-)\frac{A}{L}V \qquad (9.22)$$

the field strength dependence of $i(E)$ is given by

$$i(E) = eAL\,G(E)\,10^{-2}D_R \qquad (9.23)$$

The ionization current is proportional to D_R. This proportionality is an important prerequisite for the use of liquid-filled ionization chambers for dosimetry (*cf.* Section 9.5). Experimentally, it was found that in many liquids $G(E)$ is a linear function of E at moderate high fields (up to some tens kV cm^{-1}) [10, 56—61]. The ionization current can be written as:

$$i = f(E) + cE \qquad (9.24)$$

where $f(E)$ represents the saturation function for the free ions. The constant c describes the influence of the electric field strength on the geminate recombination. Since $f(E)$ is constant at higher fields the extrapolation of the linear part of the current *versus* field strength dependence to zero field strength gives the free ion yield $G_{fi} = G(0)$. All three methods — low-field conductivity [Section 9.2.1(i)] clearing-field technique [Section 9.2.1(v)] and current extrapolation — gave the same G_{fi} values for the liquids where all three methods have been used.

(ii) Geminate recombination

At higher field strengths, the constant c of eqn. (9.24) becomes a function of E and essentially two different theories have been developed to describe the influence of an external electric field on the process of geminate recombination. Already in 1913 Jaffé formulated his famous theory of columnar ionization [6]. The ionization is assumed to occur along the track of high-energy charged particles (ionization columns). Diffusion and recombination change the initial distribution of the ions and, by neglecting the influence of the Coulomb forces on the diffusion process, Jaffé derived a differential equation describing the change of the charge carrier concentration n in the column with time in the absence of an external electric field

$$\frac{\partial n}{\partial t} = D \left(\frac{\partial^2 n}{\partial r^2} + \frac{1}{r} \frac{\partial n}{\partial r} \right) - \alpha n^2 \qquad (9.25)$$

Cylindrical geometry was applied for the column. Positive and negative charge carriers were assumed to exhibit the same diffusion coefficient D. Jaffé solved eqn. (9.25) by neglecting the recombination term first and then taking into account the influence of recombination. One solution of eqn. (9.25) was a Gaussian distribution (parameter b) of the charge-carrier density around the axis of the column

$$n(t) = \frac{N_0}{1 + \frac{\alpha N_0}{8\pi D} \ln\left(\frac{4Dt + b^2}{b^2}\right)} \frac{\exp\left(-\frac{r^2}{4Dt + b^2}\right)}{\pi(4Dt + b^2)} \qquad (9.26)$$

The diffusion leads to a broadening of the distribution while the recombination leads to a decrease of the total number of charge carriers N_0 in the column. The presence of an external electric field influences the recombination and different orientations of the direction of the ionization columns and the electric field were considered. The overall effect of the field on the

References pp. 384—388

ionization current was found to be given by:

$$\frac{1}{i(E)} = \frac{1}{i_s}\left(1 + \frac{\alpha N_0}{8\pi D} f(z)\right) \tag{9.27}$$

where i_s represents the saturation current for the case that no geminate recombination occurs. The function $f(z)$ varies as E^{-1} for high field strengths. A plot of i^{-1} versus E^{-1} should give a straight line with the intercept i_s^{-1}. With the absorbed dose rate, the total ionization yield G_{ti} can be calculated. Jaffé pointed out that the application of eqn. (9.27) in the case of low LET radiation was questionable since most of the ionization events are spatially distributed in spurs, clusters of 2 to 3 ion pairs, rather than ionization columns. Later, however, several authors applied eqn. (9.27) for the determination of G_{ti} ($\overline{W}$ values) from the X- or γ-ray-induced ionization currents in dielectric liquids [58, 59, 61–67]. Some justification comes from Lea's treatment who extended the Jaffé theory to the case of low LET radiation [9]. As the density distribution, a three-dimensional Gaussian was assumed. Kramers [68] investigated the problem of columnar ionization and solved eqn. (9.25) by neglecting the diffusion term first. He found that at high fields saturation of the ionization current is approached in a way given by

$$i(E) = i_s\left(1 - \frac{\text{const}}{E}\right) \tag{9.28}$$

While in the Jaffé theory and its extensions multi ion-pair distributions were discussed, Onsager [69] treated the recombination behaviour of a single ion pair separated by a distance r embedded in a dielectric medium (κ) with and without the influence of an external electric field. With no external field, the escape probability is given by:

$$\varphi(r) = \exp\left(-\frac{r_c}{r}\right) \tag{9.29}$$

where r_c is the distance between the two charge carriers where the energy of the coulomb attraction equals kT

$$r_c = \frac{e^2}{kT 4\pi\kappa\kappa_0} \tag{9.30}$$

The influence of an external electric field was expressed by an infinite double series in r, E, and θ the angle between the field and the line connecting the ions

$$P(r, \theta, E) = \exp\left[-\frac{r_c}{r} - \frac{eEr}{2kT}(1 + \cos\theta)\right] \sum_{m,n=0}^{\infty} \frac{\left[\frac{eE}{2kT}(1 + \cos\theta)\right]^{n+m} r_c^m r^n}{m!(m+n)!}$$

$$\tag{9.31}$$

Integration over θ for an isotropic distribution of ion yields $P(r, E)$. It is interesting that this formalism has wide applicability since it has also been shown to explain (see Chapters 7 and 8) charge photogeneration in amorphous chalcogenides and organic solids.

For high-energy X-rays or γ-rays over 80% of the energy is deposited in spurs, consisting of an average of 2 to 3 ion pairs. Recombination takes place in each spur until only one ion pair separated by a distance r is left. For not too large a field strength E (usually up to 20 kV cm^{-1}) higher powers in E may be neglected and the yield of separated ions $G(E)$ is given by

$$G(E) = \left(1 + \frac{e^3}{8\pi\kappa\kappa_0 k^2 T^2} E\right) G_{fi} \tag{9.32}$$

A plot of the ionization current *versus* field strength should be linear and the ratio of the slope to the intercept is obtained to give

$$P = \frac{e^3}{8\pi\kappa\kappa_0 k^2 T^2} \tag{9.33}$$

This value depends on T and κ only. Many measurements in this range of field strength have been published and values for P obtained are compiled in Table 9.2. The value P^{-1} represents the field strength necessary to double

TABLE 9.2

Slope/intercept values P obtained from the dependence of the ionization current on the field strength

Liquid	T[K]	P_{obs} cm V$^{-1} \times 10^4$	P_{th}	Reference
n-Hexane	298	0.60	0.58	56
	285	0.65	0.63	
	268	0.75	0.70	
	246	0.88	0.81	
	228	0.97	0.94	
	219	1.12	1.01	
	298	0.56	0.58	66
	293	0.68 ± 1		57
Iso-octane	300	0.40		58
	210	0.93	0.95	
	260	0.65	0.75	
	315	0.54	0.50	
Carbon disulfide	293	0.43	0.46	a
Liquid oxygen	90	8.0	7.9	58
Liquid nitrogen	77	11 ± 2	11.2	b

[a] H. Wurst, Diplomarbeit, HMI Berlin (1971)
[b] W.F. Schmidt, unpublished results (1969)

References pp. 384—388

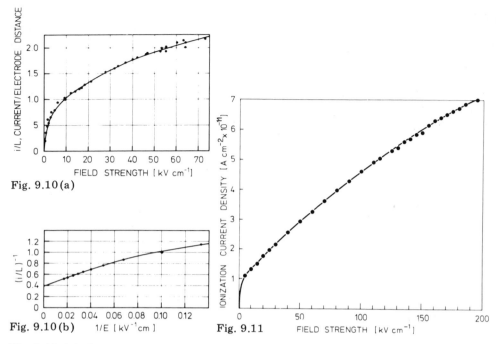

Fig. 9.10. (a) Ionization current *versus* electric field strength for carbon disulfide irradiated with 120 keV X-rays. Electrode distance 0.3 and 1 mm (room temperature) (b) Plot of data from (a) to obtain saturation current by extrapolation to infinite field strength. (From Mohler and Taylor [62].)

Fig. 9.11. Ionization current *versus* electric field strength for n-hexane irradiated with ^{60}Co-γ-rays. $T = 25°$C. (From Mathieu [70], the figure was kindly made available to us by Dr. J. Mathieu, Toulouse.)

the free ion yield. The good agreement between the observed values and the values obtained by the Onsager formula, especially the temperature dependence, gives some support to the picture of the last ion pair in a spur whose recombination is influenced by the field.

(iii) Ionization currents at high electric fields

Measurements of the radiation-induced conductivity of dielectric liquids have been carried out for over 70 years. Most measurements were done at electric field strengths below 100 kV cm^{-1}. In many cases, no data on the radiation intensity are given and sometimes no volume proportionality was observed. One of the first quantitative investigations of the X-ray induced ionization was made by Mohler and Taylor with carbon disulfide [62]. Measurements were carried out at two electrode distances and in Fig. 9.10 the dependence of the ionization current per unit volume on the electric field at constant radiation intensity is shown. At 75 kV cm^{-1}, the ionization current is about 0.75 of the saturation current. By applying eqn. (9.27) Mohler and Taylor determined the energy required for the generation of one ion pair to be $\overline{W}_l = 24$ eV which is slightly smaller than the gas phase value of $\overline{W}_g = 26$ eV. Pao [58] investigated the radiation-induced conductivity of iso-octane up to field strengths of 40 kV cm^{-1}. The ionization current was proportional to cell volume and radiation intensity. The ion yield at the highest field strength was twice the extrapolated free ion yield at zero field. A plot of the data according to eqn. (9.27) gave approximately straight lines for different temperatures with the same i_s. This seems to indicate that the total number of ions produced is independent of temperature. Similar results were obtained by Gibaud [61] and Ullmaier [59]. Lenkeit and Ebert [64] and Ebert et al. [65] used Jaffé's theory for the estimation of $\overline{W}$ values in various alkanes. Values much higher than the corresponding values for the hydrocarbon vapours were obtained. Branched hydrocarbons gave lower $\overline{W}$-values than the straight chain alkanes. The maximum field strengths applied in these experiments were between 50 and 80 kV cm^{-1}. At these field strengths, only 10—20% of the total ion yield is collected by the field and the extrapolation is rather inaccurate. Mathieu [66, 70] tried to avoid this ambiguity by adjusting the parameters of Jaffé's theory and the function $f(z)$ to fit his experimental results. He measured the ionization current in n-hexane up to 180 kV cm^{-1} (Fig. 9.11) and obtained a $\overline{W} = (37 \pm 3)$ eV which is still 50% higher than the gas phase value ($\overline{W} = 24$ eV). Much more favorable experimental conditions for extrapolation to the total ion yield were found in neopentane. Measurements of the γ-ray-induced ionization current up to 140 kV cm^{-1} by Schmidt [67] gave ion yields of $\sim$80% of the expected total yield (Fig. 9.12). The extrapolated yield of $\overline{W} = 24.8 \pm 10\%$ agrees with the gas phase value. The addition of a small amount of SF$_6$ led to a drastic reduction of the ion yield.

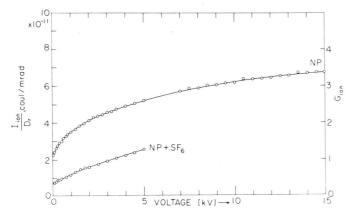

Fig. 9.12. Ionization current versus applied voltage for neopentane irradiated with ^{60}Co-γ-rays. Electrode distance 1.08 mm. (From Schmidt [67].)

References pp. 384—388

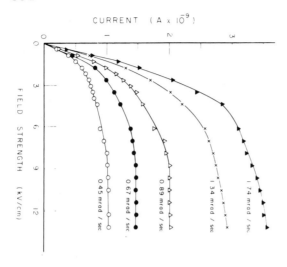

Fig. 9.13. Ionization current *versus* electric field strength for liquid argon irradiated with 1.5 MeV bremsstrahlung at several dose rates. (From Klassen and Schmidt [72].)

The only liquids where saturation currents have been measured are liquid argon and liquid xenon which exhibit also relatively high electron mobilities [*cf.* Section 9.4.1]. Ullmaier [71] irradiated liquid argon with low energy X-rays and obtained saturation currents above 50 kV cm^{-1}. He determined $\overline{W} = (25.7 \pm 3)$ eV. Klassen and Schmidt [72] investigated the ionization currents in liquid argon induced by 1.5 MeV bremsstrahlung. Their results are shown in Fig. 9.13. At low dose rates, saturation currents were obtained and $\overline{W} = (22.5 \pm 3)$ eV was estimated. Gas phase values range from 23.7 to 27.9 eV [73, 74]. Ion yields in liquid xenon were investigated by Robinson and Freeman [50]. They measured the conductivity produced by a 100 ns pulse of X-rays as a function of the field strength. Above 15 kV cm^{-1} only a very slight increase in yield with field strength occurred and the total ion yield was estimated to be $G_{ti} = 16$ which corresponds to $\overline{W} = 6.25$ eV. Electron avalanches at still higher field strengths were observed in liquid xenon by Derenzo *et al.* [75]. The test cell was a coaxial arrangement with a wire anode (diameter of several μm) and a cylindrical cathode. Irradiation with γ-quanta from a radioactive source produced ionization pulses the amplitude of which increased drastically above a certain voltage with increasing voltage. Electron multiplication near the anode takes place and the first Townsend coefficient as a function of the field was estimated. The values vary from $\alpha(400 \text{ kVcm}^{-1}) = (470^{+600}_{-470})$cm^{-1} to $\alpha(2 \text{ MV cm}^{-1}) = (4.47 \pm 0.26) \times 10^4$ cm^{-1}.

9.2.3 Corpuscular radiation

Ionization experiments with corpuscular radiation in dielectric liquids were carried out in order to test Jaffé's theory of columnar ionization. Most investigations were carried out with α-particles but some experiments have been reported where ionization currents generated by fast neutrons and protons were measured. As in the case of X- and γ-irradiation, the current–field strength dependence is characterized by two parts: a rapidly rising portion and a linear increase at higher field strengths. The linear part can be described by

$$i(E) = i_0 + cE \qquad (9.34)$$

Chybicki [76—78] investigated the ionization currents induced by α-particles in n-hexane and n-heptane. He found that only 0.2% of the total ionization yield appears as free ions which corresponds to a $G_{fi} = 0.008$. Recently, an extensive study of the ionization of dielectric liquids by α-particles was carried out by Ramy [79]. Free ion yields were obtained and the application of Jaffé's theory yielded $\overline{W}$-values comparable to the gas phase values.

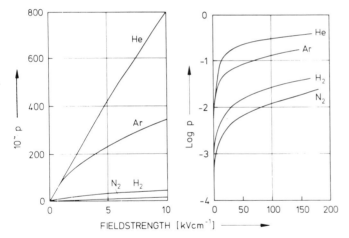

Fig. 9.14. Relative ionization current *versus* electric field strength in liquid nitrogen (77 K), hydrogen (20 K), argon (88 K), and helium (4 K); $p = i/k_s$, i ionization current, i_s saturation current in the gas phase. (From Gerritsen [82].)

Richards [80] used a collimated beam of α-particles to study the ionization in n-hexane. The experimental set-up allowed irradiation perpendicular and parallel to the field. The ionization current was the same for both directions in contradiction to the Jaffé theory. Richards and Chybicki concluded from their experiments that the charge carriers collected at low and moderate field strength stem from ionization events generated by δ-rays at some distance

References pp. 384—388

TABLE 9.3

$\overline{W}$-values for various liquids

Compound	T[K]	$\overline{W}_{liq}$ [eV/ion pair]	$\overline{W}_{gas}$	Radiation	Method	Reference
Argon	85	25.7 ± 3		LE X-rays	SC	[71]
		22.5 ± 3	26.2 ± 0.2 [73]	HE X-rays	SC	[72]
		26	26.3 ± 0.1 [73]	^{239}Pu-α	JEC	[83]
		23.6 $^{+0.5}_{-0.3}$		0.976 MeV electrons	SC	[84]
Xenon	170—200	6.25	21.9 ± 0.3 [73]	HE X-rays	EY	[50]
		17.3	21.9 [73]	^{210}Po-α	JEC	[85]
Neopentane	296	24.8 ± 2.5	24[a]	HE X-rays	JEC	[67]
Carbon disulfide	RT	24	26[b]	LE X-rays	JEC	[62]

[a] Value is an estimate from ref. 67.
[b] Value from J.J. Thomson, cited in ref. 62.

SC: Saturation current; JEC: Extrapolated Saturation Current from Jaffé theory; EY: Extrapolated Yield; RT: Room Temperature
HE: High Energy; LE: Low Energy.

from the α-track. Gerritsen [81, 82] investigated the ionization produced by α-particles in liquid nitrogen, hydrogen, helium and argon. The field dependence of the ionization current is shown in Fig. 9.14. From the data, he concluded that agreement with Kramer's modification of Jaffé's theory was obtained. The measurements in liquid helium and liquid argon gave much higher ionization currents than in liquid nitrogen and hydrogen. This indicates that the interaction between non-thermal electrons and inert gas atoms is much smaller than between electrons and molecules. Current pulses produced by single α-particles can be observed in liquid argon. Swan [83] measured the pulse amplitude as a function of the field strength up to 60 kV cm^{-1} and extrapolated with Jaffé's theory, $\overline{W} = 26$ eV. Current pulses generated by the passage of α-particles in liquid argon were also observed by Takahashi et al. [84]. In their experiments, however, the pulse amplitude remained constant above 15 kV cm^{-1} and $\overline{W} = (23.6^{+0.5}_{-0.3})$ eV was determined, which is lower than the gas phase value. Ionization of liquid xenon by α-particles was studied by Konno and Kobayashi [85]. They estimated $\overline{W} = 17.3$ eV which is much lower than the value in the gas phase. All $\overline{W}$-data are compiled in Table 9.3 and are compared with gas phase values. At least in the case of the liquefied rare gases $\overline{W}_{liq}$ seems to be smaller than $\overline{W}_{gas}$ which is correlated to a similar decrease in ionization energy (cf. Section 9.3).

9.2.4 Ion yields

One of the interesting values in the ionization of matter by high energy radiation is the number of ions produced by the absorption of a given amount of energy. This value is usually denoted by G and it is the number of ion pairs generated per 100 eV. Sometimes it is more convenient to use the energy necessary to produce one ion pair $\overline{W}$. Both values are connected by

$$G = \frac{100}{\overline{W}} \qquad (9.35)$$

(i) Free ion yields

In non-polar liquids, most of the charge carriers produced by high energy radiation recombine in a very short time ($<10^{-10}$ s) and only a small fraction escapes the geminate recombination process. This yield G_{fi} of so called "free ions" is important for the understanding of radiation chemical reactions in liquids and it is influenced by the interaction of electrons with the molecules of the liquid. G_{fi} values are determined in the absence of an electric field so that any influence of the electric field on the escape probability [cf. Section 9.2.2(ii)] is eliminated. First values were obtained from the measurement of the radiation-induced low-field conductivity by Hummel and Allen [15] and by Freeman and Fayadh [86]. Later, many G_{fi} values were obtained with the clearing-field technique [53—55] and also from extrapolation of the

TABLE 9.4

G_{fi}-values in various liquids for low LET-radiation

Liquid	T[K]	κ	G_{fi}	Method	Reference
Methane	110	1.67	0.8 ± 0.1	CF, EY	[87]
Ethane	183	1.8	0.13	CF	[87]
Propane	183	1.9	0.076	CF	[87]
n-Butane	296	1.76	0.19	CF	[55]
n-Pentane	296	1.84	0.145	CF	[55]
Isopentane	296	1.84	0.17	CF	[55]
Neopentane	296	1.78	0.86	CF	[55]
n-Hexane	296	1.89	0.13	CF	[55]
			0.1	LFC	[15]
			0.13	EY	[56]
n-Heptane	296	1.93	0.13	CF	[55]
n-Octane	296	1.94	0.12	CF	[55]
Iso-octane	296	1.94	0.33	CF	[55]
n-Nonane	296	1.97	0.12	CF	[55]
n-Decane	296	1.99	0.12	CF	[55]
Cyclohexane	296	2.02	0.15	CF	[55]
Benzene	296	2.28	0.05	CF	[55]
Carbon tetrachloride	296	2.23	0.096	CF	[55]
			0.07	LFC	[24]
Carbon disulfide	296	2.63	0.31	CF	[55]
			0.35	EY	[a]
Tetramethylsilane	296	1.84	0.74	CF	[55]

[a] H. Wurst, Diplomarbeit, HMI Berlin, (1971).

CF: Clearing-Field method; LFC: Low-Field Conductivity; EY: Extrapolated Yield.

TABLE 9.5

Influence of LET on the free ion yield at $T = 296$ K

Liquid	G_{fi}	Radiation	References
n-Hexane	0.13	^{60}Co-γ	[57]
	0.05	^{37}Ar-β	[24]
	0.04	^{3}T-β	[163]
	(0.008)	^{239}Pu-α	[76]
	0.0058	^{210}Po-α	[79]
2-Methylpentane	0.15	HE X-rays	[55]
	0.006	^{210}Po-α	[79]
2,2-Dimethylbutane	0.30	HE X-rays	[55]
	0.006	^{210}Po-α	[79]
n-Heptane	0.13	HE X-rays	[55]
	0.0056	^{210}Po-α	[79]
Iso-octane	0.33	HE X-rays	[55]
	0.0062	^{210}Po-α	[79]

HE: 1.5 MeV bremsstrahlung.

steady-state ionization current at higher field strength to zero [56, 57, 87]. While the low-field conductivity method requires the separate measurement of the mobility of positive and negative charge carriers, in the clearing-field technique only the charge is measured. This method is insensitive to impurities which alter the mobility by charge transfer. Temperature dependence of G_{fi} and free ion yields in mixtures were studied [53—56, 87, 88]. Table 9.4 summarizes some of the G_{fi} values obtained in various dielectric liquids under low LET irradiation, while in Table 9.5 the influence of the LET on the free ion yield is demonstrated. Many more values can be found in a recently published review by Hummel and Schmidt [33]. The yield of free ions is determined by two processes: (a) the thermalization of the initially energetic electron from the ionization process leading to an average separation r between electron and parent positive ion and (b) the recombination (escape) process once thermal energies for electron and positive ion are reached. Onsager [69] calculated the escape probability for a single ion pair separated by a distance r. Even in this approximation, there will be a distribution of separation distances and the free ion yield will be given by

$$G_{fi} = G_{ti} \int_0^\infty f(r) \exp(-r_c/r) dr \qquad (9.36)$$

Several papers concerning $f(r)$ in various liquids have been published [89, 90]. With the proper distribution function, it should be possible to obtain the temperature and field-strength dependence.

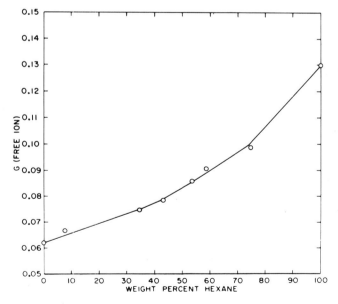

Fig. 9.15. Dependence of G_{fi} values in mixtures of n-hexane and n-hexene-1 on the composition at $T = 296\,\text{K}$. (From Schmidt and Allen [55].)

References pp. 384—388

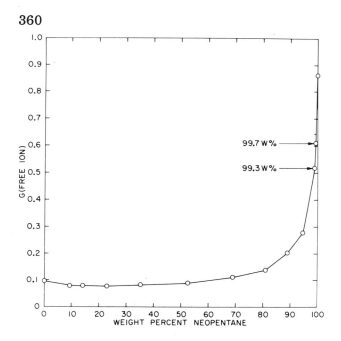

Fig. 9.16. Free ion yield in mixtures of neopentane and carbon tetrachloride at $T = 296$K. (From Schmidt and Allen [55].)

Studies of the free ion yield in mixtures of non-polar liquids showed that, in some cases, the influence of the components was additive (*cf.* Fig. 9.15) on the slowing down of the electron, while, in mixtures of a hydrocarbon and an electron scavenger, already very small amounts of the scavenger led to a drastic reduction of the free ion yield (*cf.* Fig. 9.16). In these cases it must be concluded that non-thermal electrons on their way outward from their positive parent ions are captured at a smaller average separation distance r so that the escape probability is decreased. Recently a detailed analysis of these effects has been given by Mozumder [91, 92]. This effect was only pronounced in liquids which exhibited a large free ion yield indicating a large average separation. In *n*-hexane such a reduction was not observed at low concentrations and at higher solute concentrations the direct effect of the radiation on the solute cannot be neglected.

(ii) *Total ion yields*

In gases under irradiation with high energy radiation, saturation currents are observed and the energy $\overline{W}$ for the generation of one charge carrier pair can be determined. This energy was found to be two to three times the ionization energy of the gas atoms or molecules. The excess energy is distributed in electronic and vibronic excitations of the molecules or atoms. Due to the strong influence of geminate recombination in liquids, it is very difficult to collect all charge carriers produced. Saturation currents have been observed in liquid argon and xenon and $\overline{W}$-values were obtained. In the liquid hydrocarbons, no saturation currents were obtained at the highest electric field strengths applied. Attempts have been made to estimate $\overline{W}$-values by extrapolation of the current *versus* electric field strength dependence to infinite fields or adjusting the total ionization yield to give the correct temperature dependence of the free ion yield. Generally, one would expect a reduction in the $\overline{W}$-value since the ionization energy also decreases from the gaseous to the liquid state (*cf.* Section 9.3). The use of the Jaffé theory for the extrapolation may be justified in cases where the ion yield at the highest field strength represents well over 50% of the total expected yield. In some cases, extrapolations have been carried out, however, from yields as low as 20% so that the extrapolated value is unreasonably out of scale. Table 9.3 summarizes $\overline{W}$-values in liquids and the corresponding vapors.

9.3 PHOTOIONIZATION AND PHOTOEFFECT

9.3.1 Non-polar liquids

Very few investigations have been carried out on the direct photoionization of dielectric liquids. The gas phase ionization potentials of the rare gases range from 24.6 eV for helium to 12.08 eV for xenon, while the saturated hydrocarbons exhibit values around 10 eV. A reduction of this quantity in the condensed phase can be expected. Halpern and Gomer [93] estimated from field-ionization experiments in liquid argon that the ionization energy in the liquid phase is 13 eV as compared to 15.68 eV in the gas phase. Still, these energies correspond to wavelengths of the far ultraviolet. For photoionization experiments, cells with suitable windows such as LiF or sapphire are necessary in order to admit the light into the liquid. Since it is strongly absorbed in the photoionization region, the penetration is usually very small and charge carriers are generated in a thin layer behind the window only. Various other effects can lead to charge-carrier production under illumination with light and the determination of the photoionization threshold requires a careful consideration of other possible sources of charge carriers. A very thorough investigation of the intrinsic photoconductivity of liquid xenon has been published by Roberts and Wilson [94]. The liquid xenon was contained in a cell with a lithium fluoride (LiF) window of 1-mm thickness. Their experimental set-up is schematically shown in Fig. 9.17. One electrode was in contact with the liquid, while the other was outside the cell attached to the LiF window. An aperture of about 40 mm^2 admitted the light beam. The liquid layer was 2-mm thick. Currents induced in the LiF or due to electron emission at the window—liquid interface did not present a major problem. The cell was placed into a vacuum ultraviolet spectrograph

References pp. 384—388

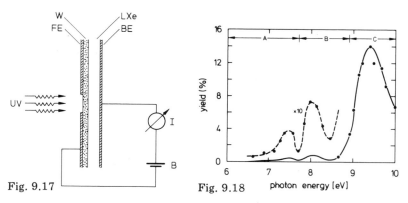

Fig. 9.17. Experimental arrangement for the investigation of photoionization of liquid xenon. (From Roberts and Wilson [94].) FE: front electrode; BE: back electrode; W: window; LXe: Liquid xenon.

Fig. 9.18. Photo-yield in liquid xenon as a function of the quantum energy. (From Roberts and Wilson [94].)

and the incident light intensity was monitored with a sodium salicylate-coated photomultiplier. This light detector has a constant quantum efficiency independent of energy in the range up to 10 eV, so that the photomultiplier current could serve as a relative measure of the photon flux. Absolute quantum yields were determined with a gold film deposited on the inner side of the LiF window in an empty cell. The absolute quantum flux could thus be determined with an accuracy of approximately 20%. Typically, 10^6 photons/s were admitted into the cell. The photocurrents were measured with a vibrating reed electrometer either in an integrating mode or by the voltage drop across a large resistor. Currents as low as 10^{-15} A could be measured which corresponds to a carrier generation of 6×10^3 s^{-1}. The absolute photo-yield of charge carriers was determined with an accuracy of about 30%. Carriers are generated behind the window in a thin layer and drift towards their respective electrodes. Since the front electrode is blocked by the LiF window, accumulation of space charge can occur. It was found, however, that this effect did not represent a serious problem. Space charge, which accumulated after longer times, could be removed by illumination with white light for a short time.

The dependence of photo-yield on the light energy was characterized by three regions which are shown in Fig. 9.18. Region A was attributed to ionization of impurities in the bulk of the xenon. In region B, excitons are created which can directly ionize impurities and, in region C, electrons and holes are generated. The xenon was not pure enough to characterize the charge carriers of region C and it was assumed that the current was carried by O_2^- and X_2^+ ions which were produced by electron scavenging and hole conversion. The photo-yields in regions A and B saturated with the applied

TABLE 9.6

Ionization threshold for argon and xenon in the various aggregate states

Substance	Phase	T[K]	I[eV]	Reference
Argon	gas		15.68	a
	liquid	85	13.0	[93]
	solid	20	14.3	b
Xenon	gas		12.08	a
	liquid	163	8.9	[94]
	solid	135	9.28	[165]

[a] Handbook of Chemistry and Physics, Chem. Rubber Publishing Co., Cleveland.
[b] Estimated value from ref. 164.

electric field, while, in region C, it was not possible to obtain saturation currents with field strengths up to 1.2 kV cm^{-1}. The highest intrinsic photoionization yield observed was 25%. Lower yields were observed in impure samples. The photoionization yield decreases above 9.5 eV. This effect could be attributed to the generation of excitons of higher energy. An exciton at 10.3 eV has been observed in solid and liquid xenon and its decay could lead to non-conducting states. The ionization threshold in liquid xenon at 163K was found to be $I_{liq} = 8.9$ eV. As in the case of argon [93], condensation leads to a marked reduction of the ionization threshold by several eV. Table 9.6 summarizes the data obtained for the rare gases. Almost no information is available on the ionization threshold in liquid hydrocarbons. Vermeil et al. [95] investigated the intrinsic photoconductivity of propane, isopentane, and methylpentene at four different wave lengths (123.6 nm; 147 nm; 160 nm; 184.9 nm) and estimated that the ionization energy is reduced by approximately 1 eV in all these liquids as compared to the gas phase.

9.3.2 Solutes in non-polar liquids

The study of the photoionization of molecular substances in non-polar liquids began as early as 1942 with the work of Lewis and Lipkin [96]. They observed photoionization of rigid organic solutions at 77K at quantum energies much smaller than the gas phase ionization potential. Since that time, interest in photoconductance of organic solutions has increased steadily and a wealth of information has been accumulated which is impossible to treat in this short section. We will, therefore, concentrate on some recent experiments which are interrelated to the properties of excess electrons in liquid hydrocarbons. A prominent solute with which many investigations have been carried out is N, N, N', N'-tetramethyl-p-phenylenediamine (TMPD).

References pp. 384—388

```
        CH₃      CH₃
          \     /
           N
           |
           C
          ⁄ ⁀
        HC    CH
        |     ||
        HC    CH
          ⁀  ⁄
           C
           |
           N
          ⁄ \
        CH₃   CH₃
```

It can be dissolved in hydrocarbons and exhibits an absorption spectrum shown in Fig. 9.19. Biphotonic and monophotonic ionization has been observed in 3-methylpentane (3MP) by Albrecht and coworkers [97] and by Jarnagin and coworkers [98, 99]. The biphotonic ionization proceeds via the triplet state:

$S_0 + h\nu \to S_1$

$S_1 \to T_1$

$T_1 + h\nu \to M^+ + e^-$

where S_0 represents the ground state of TMPD, S_1 the lowest excited singlet state and T_1 the lowest excited triplet state. The energy of T_1 is 2.9 eV above S_0. The first excited singlet band begins at approximately 3.1 eV above S_0. The gas phase ionization potential has been reported as 6.35 eV [100], 6.2 eV [101], or 6.7 eV [102]. Houser and Jarnagin [99] illuminated a TMPD-3MP solution with a light flash from a xenon lamp (320—410 nm) or a frequency doubled ruby laser (347.1 nm) and observed the pulse conductivity. The peak current was proportional to the number of charge carriers generated (no volume recombination during the pulse) and it increased with the square of the light intensity. The decay of the pulse photocurrent is determined by the properties of the charge carriers. No photocurrents were observed unless the spectral range of the light flash overlapped the lowest singlet band. One-photon ionization was observed in the wavelength interval from 240—280 nm [98]. The photoionization threshold was estimated to be 4.5 eV. This decrease in ionization potential is mainly due to the polarization energy P_+ of the positive ion which can be estimated from Born's equation (in the c.g.s. system)

$$P_+ = \frac{1}{2}\left(\frac{e^2}{R}\right)\left(\kappa - \frac{1}{\kappa}\right) \tag{9.37}$$

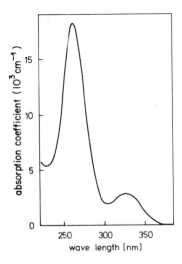
Fig 9.19

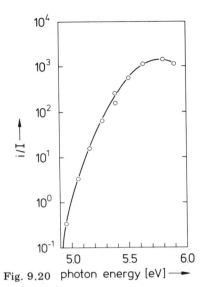
Fig. 9.20

Fig. 9.19. Absorption spectrum of TMPD in liquid 3-methylpentane. (From Piloff and Albrecht [97].)

Fig. 9.20. Normalized photocurrents *versus* photon energy for photoionization of TMPD, solvent: 85 Mol% n-hexane, 15 Mol% tetramethylsilane. (From Holroyd and Tauchert [110].)

to equal 1.7 eV assuming a value of 2 Å for R. Smaller shifts of the ionization threshold are due to the variation of the electron affinity from one liquid to another. Investigation of this effect by Holroyd [103, 104] yielded relative values for the energy V_0 of the electron conduction state in several liquid hydrocarbons (*cf.* Section 9.3.3). The ionization threshold I_s of TMPD in solution is given by:

$$I_s = I_g + P_+ + V_0 \tag{9.38}$$

where I_g is the gas phase ionization potential, P_+ the polarization energy of the TMPD$^+$ and V_0 the electron affinity of the liquid. P_+ can be estimated either from eqn. (9.37) or by direct measurement in a liquid with a known V_0 value. For n-pentane, Holroyd determined experimentally a value of $P_+ = -1.6$ eV. Values for other hydrocarbons were adjusted with the respective dielectric constant. Figure 9.20 shows a typical example for the photocurrents in TMPD solutions. The observed photocurrent i increased continuously with decreasing wavelength. It was converted to a relative quantum yield RQY by:

$$\text{RQY} = [(i-i_0)/(I-I_{sc})]\left(\frac{\text{PMR}}{\%\text{T}}\right)\frac{1}{F} \tag{9.39}$$

References pp. 384—388

with PMR the response of the light monitor, %T the percentage of light transmitted through the windows of thermostat and cell, i_0 the dark current, I the photomultiplier current, I_{sc} the current for scattered light and F the fraction of absorbed light by the TMPD.

F is given by:

$$F = \frac{OD_{sol} - OD_{liq}}{OD_{sol}} A_{sol} \qquad (9.40)$$

where OD_{sol} and OD_{liq} represent the optical density of the solution and of the pure liquid, respectively, and A_{sol} the fraction of the incident light absorbed by the solution. Plots of RQY *versus* quantum energy were used for the determination of V_0 by observing the shift of the ionization onset $I(s)$, which was defined as the photon energy at which RQY equals a certain minimal value. V_0 was then obtained from

$$V_0 = I(s) - I_g - P_+ \qquad (9.41)$$

This shift in ionization onset was also observed with perylene where the photoionization occurs at shorter wave lengths. The relative V_0 values obtained agreed with those of the TMPD measurements. Values for V_0 are found in Table 9.7. Besides the investigation of the energy levels, photoionization of solutes can serve as a means of generating excess charge carriers in dielectric liquids. Takeda, Houser and Jarnagin [98] observed the transient photocurrents in TMPD solutions and estimated electron mobilities in some silanes. Devins and Wei [105] investigated the motion of excess electrons in *n*-hexane. Charge carriers were generated by biphotonic ionization of TMPD with a pulsed nitrogen laser. Biphotonic ionization of anthracene by a laser pulse in various liquids was used by Beck and Thomas [106] for the generation of excess electrons and the study of transport properties and electron attachment reactions. Since the excess electron mobility in many hydrocarbons is several orders of magnitude greater than the mobility of the positive ion, the current immediately after the pulse is carried mainly by electrons. The decay is determined either by attachment to impurities or recombination.

9.3.3 Photoeffect

Emission of electrons from metal electrodes in non-polar liquids can be achieved by photoeffect. The work function of the metal in the liquid ϕ_{liq} is reduced or increased by the energy V_0 of the conducting state of the electron in the liquid, *i.e.*

$$\phi_{liq} = \phi_{vac} \pm V_0 \qquad (9.42)$$

If $\phi_{liq} > \phi_{vac}$, then V_0 is the energy necessary to introduce the electron from the vacuum into the conducting state in the liquid. $\phi_{liq} < \phi_{vac}$ means that

TABLE 9.7

V_0 values for non-polar liquids obtained by photoelectric measurements and by photo-ionization of TMPD

Liquid	T[K]	V_0 [eV]	Method	Reference
Argon	85	−0.33	PE	[114]
Helium	4.2	+1.05	PE	[115]
Tetramethylsilane	295	−0.62	PE	[108]
		−0.59	PI	[104]
Neopentane	295	−0.43	PE	[108]
		−0.38	PI	[104]
Isooctane	295	−0.18	PE	[108]
		−0.33	PI	[104]
Neohexane	295	−0.15 ± 0.03	PE	[111]
		−0.26	PI	[104]
n-Pentane	295	−0.01	PE	[108]
n-Hexane	295	+0.04	PE	[108]
		+0.10	PI	[104]
		+0.16 ± 0.05	PE	[111]
n-Tetradecane	295	+0.21	PI	[104]
Cyclopentane	295	−0.28	PE	[108]
		−0.21	PI	[104]
Cyclohexane	295	+0.01	PI	[104]
Benzene	295	−0.14 ± 0.05	PE	[111]
Toluene	295	−0.22 ± 0.03	PE	[111]
Methane	109	0.0	PE	[112]
	95	0.0, −0.12	PE	[113], [116]
Ethane	182	+0.02	PE	[112]

PE: Photoeffect; PI: Photoionization.

the energy of the conducting state in the liquid is lower by V_0 as compared to the vacuum level. In vacuum, the photocurrent density emitted from a metal electrode as a function of the light frequency ν is described by the Fowler equation [107]:

$$j = \alpha A T^2 F(x) \tag{9.43}$$

where $F(x)$ is the Fowler function:

$$F(x) = \exp(x) - \frac{\exp(2x)}{2^2} + \frac{\exp(3x)}{3^2} - \frac{\exp(4x)}{4^2} + \frac{\exp(5x)}{5^2} - \ldots \tag{9.44}$$

with $x = (h\nu - \phi_{vac})/kT$, A a constant and α the quantum yield. Holroyd [108] found that eqn. (9.43) also described the current density for the emission in liquids. Investigations on hydrocarbons were carried out by Holroyd et al. [108, 109, 110], Schiller et al. [111], Kevan and Noda [112] and Tauchert and Schmidt [113]. The experiments consist in measuring $j(\nu)/I(\nu)$ over a range of frequencies ($I(\nu)$ is a relative measure of the light

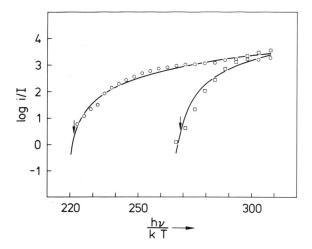

Fig. 9.21. Fowler plot of photocurrent from a zinc electrode in liquid xenon (W. Tauchert, [116], ○ xenon (161K); □ vacuum.

intensity). The data are plotted as a function of $h\nu/kT$ and the work function ϕ is obtained by shifting the experimental curve in x-direction by Δx and in y-direction by Δy until it coincides with the Fowler function $F(x)$. The work function is obtained from $\Delta x = \phi/kT$. Shifts Δy represent the efficiency of electron escape from the image potential barrier near the cathode [109]. Measurements in liquid argon and helium were reported by Lekner et al. [114] and W.T. Sommer [115], respectively. No Fowler-type emission currents were observed although the photocurrents were shifted by -0.33 eV in the case of argon and $+1.05$ eV in the case of liquid helium. Recent measurements by Tauchert [116] on a zinc electrode in liquid xenon gave $V_0 \approx -(0.6 \pm 0.1)$ eV. Figure 9.21 shows the photocurrent as a function of the light frequency in liquid xenon and in vacuum. Drift mobility measurements on electrons injected in hydrocarbons by photo-effect were performed by Minday, Schmidt, and Davis [117] (cf. Section 9.4).

9.4 TRANSPORT PROPERTIES OF CHARGE CARRIERS

9.4.1 Mobility

One of the basic experiments in the investigation of the physical properties of charge carriers in any medium consists in the measurement of their drift velocity under the influence of an applied electric field. Due to the interaction with the atoms and molecules of the medium, a constant drift velocity will be obtained at a particular field strength and the drift velocity increases proportionally to the electric field E

$$v_d = \mu E \qquad (9.45)$$

μ is called the mobility and is measured in $cm^2 V^{-1} s^{-1}$ if v_d is given in $cm\, s^{-1}$ and E in $V\, cm^{-1}$. Various different methods for mobility measurements have been described in the literature [cf. Section 9.2.1(iii)] and many reports have been published on radiation-induced charge carriers in dielectric liquids. While there is general agreement about the character and the mobilities of charge carriers in liquid rare gases, the picture is less clear in the field of hydrocarbons. Although measurements have been performed since the beginning of this century, well-defined charge carriers (as excess electrons) have only recently been observed in liquid hydrocarbons and their properties investigated. The study of excess electrons and positive charge carriers in liquefied rare gases has been the subject of research for over 20 years and a wealth of information has been gathered. Several reviews and summarizing articles have appeared [12—14] so that this field can be treated here in a rather general way.

TABLE 9.8

Electron mobilities in liquefied rare gases

Liquid	$T[K]$	μ_{el} $cm^2\, V^{-1}\, s^{-1}$	Reference
^{4}Helium	4.2	2×10^{-2}	[121]
^{3}Helium	3.0	3.7×10^{-2}	[121]
Neon	25	1.6×10^{-3}	[118]
Argon	85	475	[166]
Krypton	115	1800	[166]
Xenon	161	1900	[166]

(i) *Liquefied rare gases*

Measurements of the electron mobility in liquefied rare gases gave values which fall into two groups: (a) low mobilities $\mu_{el} \ll 1\, cm^2\, V^{-1}\, s^{-1}$ and (b) high mobilities $\mu_{el} \gg 1\, cm^2\, V^{-1}\, s^{-1}$. In Table 9.8, some data are compiled. In liquid helium and neon, the electron mobility is low which is indicative of an ionic-type motion. In liquid argon, xenon and krypton, the electron mobility reaches high values indicating a weak interaction between the electron and the atoms of the liquid. The physical state of the excess electron in liquid helium has been described as a localized state where the electron resides in a bubble of 28-Å diameter. This bubble formation occurs mainly due to the strong repulsive electron—helium-atom interaction. The confinement of an electron in a bubble is energetically more favorable than the state of a quasi-free electron. The low mobility of the excess electron in neon has also been explained by the formation of a bubble state [118]. The high mobilities in liquid argon, xenon and krypton can be explained by a quasi-free particle model where the magnitude of the mobility is determined by scattering processes. At higher field strength, the mobility decreases and the drift velocity increases with $E^{1/2}$. At still higher field strengths, a constant drift velocity is obtained which does not change with increasing field strength. Lekner [119] developed a theory for the motion of electrons in liquid argon which described the dependence of the drift velocity on the field strength up to 10^4 V cm^{-1}. In liquid xenon and krypton, deviations from proportionality of v_d and E occur already at small field strengths (~ 50 V cm^{-1}) and the drift velocity remains constant above 10 kV cm^{-1}. Electrons in liquid argon, xenon and krypton behave similarly to electrons in germanium crystals where the field dependence of the drift velocity is explained on the basis of electron—phonon interactions [120]. Positive charge carriers in liquefied rare gases were investigated by Davis, Rice, and Meyer [121—123]. Low mobilities of the order of 10^{-3}—10^{-4} cm^2 V^{-1} s^{-1} were observed in liquid Ar, Kr, and Xe, and it could be shown that the species observed are most likely the di-cations Ar_2^+, Kr_2^+, and Xe_2^+. In liquid helium, a mobility of $\mu_+ \approx 2 \times 10^{-2}$ cm^2 V^{-2} s^{-1} was observed.

(ii) Other liquefied gases

Very few data have been published on the mobility of excess charge carriers in liquid nitrogen, hydrogen and other liquefied gases. Only low mobility values were measured and no unambiguous information on the kind of charge carriers was obtained. Stepwise changes of ionic mobilities with increasing field strength were found in liquid helium II [124], liquid argon and nitrogen [125] and in carbon tetrachloride [125]. However, the results on liquid helium were not verified by Steingart and Glaberson [126]. Table 9.9 summarizes some data.

(iii) Hydrocarbons and related compounds

(a) *Electrons*

Electron drift mobilities have been measured for a variety of liquids and the dependence on temperature and electric field strength yields important information about the transport mechanism. Drift mobilities were obtained either from time-of-flight measurements [44, 46, 47, 117, 127—129] or from the measurement of the radiation-induced conductance by taking into account the yield of charge carriers [25, 48—50, 87]. At low field strengths, the drift velocity increased proportionally to the field strength and the mobility remained independent of the field strength. In Table 9.10, some data on the electron mobility in various liquids are compiled. The influence of the molecular structure on the magnitude of the mobility is quite surprising. For instance, in methane at $T = 111$K the mobility is 400 cm^2 V^{-1} s^{-1}, while in ethane at the same temperature a value of 10^{-3} cm^2 V^{-1} s^{-1} was

TABLE 9.9

Mobilities of positive and negative charge carriers in liquefied gases

Liquid	T[K]	μ_- [cm^2 V^{-1} s^{-1}]	μ_+	Reference
^{4}Helium	4.2		5.3×10^{-2}	[167]
Hydrogen	21	8.6×10^{-3}	8.3×10^{-3}	[168]
		3×10^{-2}		[169]
			4.5×10^{-2}	[93]
Deuterium	21	9×10^{-3}		[169]
Oxygen	77	8×10^{-3}	8×10^{-3}	[93]
		1×10^{-3}	1×10^{-3}	[170]
Nitrogen	77	8×10^{-3}		[169]
			1×10^{-2}	[93]
			2×10^{-3}	[37]
			1×10^{-2}	[93]
		1.5×10^{-3}	1.2×10^{-3}	[170]
Carbon monoxide	68	4×10^{-4}	8×10^{-4}	[170]

TABLE 9.10

Electron mobilities in liquid hydrocarbons and tetramethylsilane

Liquid	T[K]	μ [cm^2 V^{-1} s^{-1}]	Method	Reference
Methane	111	400 ± 50	TL	[47]
Ethane	200	0.8	TL	[128]
	111	10^{-3}	TL	[128]
Propane	200	0.12	TL	[171]
n-Butane	296	0.4 ± 0.1	Hudson	[44]
n-Pentane	296	0.16 ± 0.01	Hudson	[44]
		0.07	PE	[117]
Neopentane	296	55 ± 5	Hudson	[44]
		70	PE	[117]
		50	LFC	[49]
n-Hexane	296	0.09 ± 0.01	Hudson	[44]
		0.07	PE	[117]
		0.22 ± 0.01	LFC	[25]
Cyclohexane	296	0.35 ± 0.03	Hudson	[44]
		0.45	LFC	[49]
2,2-Dimethyl-butane	296	10 ± 1	Hudson	[44]
Iso-octane	296	7 ± 2	Hudson	[44]
Benzene	300	0.6	PE	[127]
	296	0.1	LFC	[50]
Toluene	300	0.54 ± 0.1	PE	[127]
Tetramethylsilane	296	90 ± 5	Hudson	[44]

TL: Thick Layer; PE: Photoelectric injection; LFC: Low-Field Conductivity; Hudson: Hudson method.

References pp. 384—388

found. In neopentane at $T = 296$K the electron mobility is 50–70 cm^2 V^{-1} s^{-1} while in n-pentane 0.16 cm^2 V^{-1} s^{-1} was measured. In most hydrocarbons, a positive temperature coefficient for the low field mobility was found with the influence of the temperature being stronger for the smaller mobilities. The electron mobility remained constant in tetramethylsilane and decreased with increasing temperature in liquid methane. Since the temperature dependence has been studied in most cases over a rather limited temperature range only, Arrhenius plots usually give straight lines and the activation energies are the higher the lower the mobility at a specific temperature. Extrapolation of the Arrhenius plots to $T^{-1} = 0$ yielded values between 10^2 and 10^3 cm^2 V^{-1} s^{-1} which is usually identified with the mobility in a conduction band [130, 131]. Since this mobility should depend on the temperature also, no straight line is expected and the extrapolation is ambiguous.

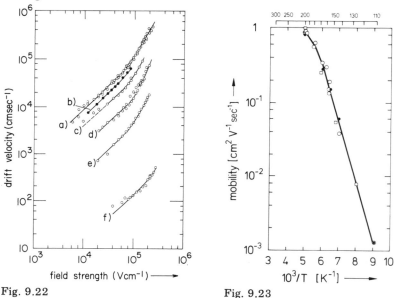

Fig. 9.22.   Fig. 9.23

Fig. 9.22. Electron drift velocity in liquid ethane as a function of field strength and temperature. (From Schmidt et al. [128].)

Fig. 9.23. Arrhenius plot of electron mobility in liquid ethane. (From Schmidt et al. [128].)

A detailed investigation of the excess electron transport in ethane has been carried out recently [128]. The dependence of the electron velocity on the field strength and of the mobility on the temperature were studied over a wide interval. In Fig. 9.22, the field strength dependence of the drift velocity is shown. At higher field strengths, the drift velocity increases more

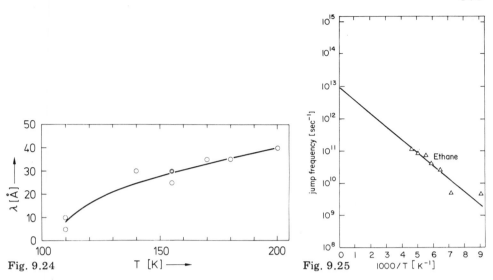

Fig. 9.24. Temperature dependence of jumping distance in liquid ethane. (From Schmidt et al. [128].)

Fig. 9.25. Temperature dependence of jump frequency in liquid ethane. (From Schmidt et al. [128].)

than proportional and the field strength at which this increase occurs is shifted towards higher values the lower the temperature. The temperature dependence of the low-field mobility is shown in Fig. 9.23. The low values of the electron mobility and the positive temperature coefficient seem to indicate that the electron is localized most of the time and motion between localized states occurs by thermal activation. A model described by Bagley [132] was used to analyze the data and to obtain information on the jumping distance and jumping frequency. In Figs. 9.24 and 9.25, the temperature dependence of jumping distance Λ and jumping frequency ν are shown. At the present, two types of electron transport can be found in liquid hydrocarbons which differ with respect to the influence of the electric field strength. In liquids which exhibit a relatively large low-field mobility $\mu(0)$, a decrease of the mobility is observed at higher field strengths (*e.g.* methane, neopentane), while in liquids with a relatively small low-field mobility an increase of the electron mobility is observed above a certain field strength (*e.g.* ethane, propane). In Fig. 9.26, these two groups of hydrocarbons are contrasted by displaying the electron drift velocity as a function of field strength.

(b) *Anions*

Excess electrons usually have a short life time in non-polar liquids due to rapid reaction with impurities. In most drift experiments reported in the

References pp. 384—388

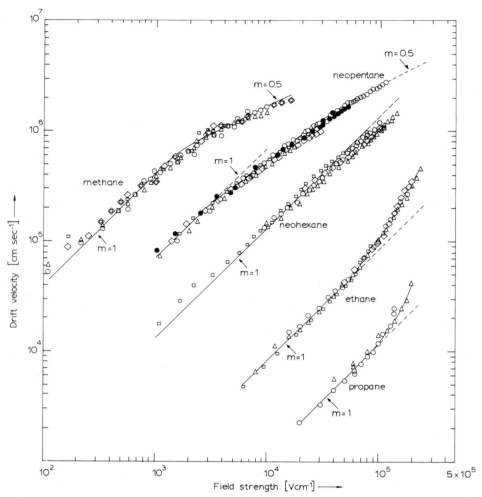

Fig. 9.26. Electron drift velocity as a function of electric field strength in various hydrocarbons: methane 111K; neopentane 296K; neohexane 296K; ethane 200K; propane 195K.

past, which yielded mobilities of the order of 10^{-3}—10^{-4} cm^2 V^{-1} s^{-1}, ionic species rather than excess electrons were observed. Some authors found Walden's rule ($\mu_- \eta = $ const., where η is the viscosity of the liquid) to be obeyed [10] while others did not [24]. The nature of the negative charge carrier remained obscure, sometimes O_2^- was suspected. But other more efficient scavengers might have converted the electron into a heavy negative ion by the capture process

$$e^- + S \to S^-$$

So far, no values for the mobility of defined anions in non-polar liquids have been reported.

(c) *Positive charge carriers*

Many measurements of positive charge carriers produced by high-energy radiation in dielectric liquids have been reported over the decades [10]. Usually, values slightly smaller than those for anions were found. The identity of the species observed remained unknown. Charge-transfer reactions to impurities can lead to an energetically more stable ion than the parent positive ion. The investigation of electron-scavenging reactions in irradiated hydrocarbons led, however, to the indirect conclusion that in cyclohexane and n-hexane the positive parent ion from the ionization process should have a mobility of about $0.05\mu_{el}$ (μ_{el} electron mobility). Recently, the direct observation of what might be called hole conduction has been observed in cyclohexane and the measured mobility confirmed the estimates [133].

9.4.2 Recombination coefficient

Volume recombination coefficients were measured either by the clearing-field technique or by observing the decay of the ionization current after interruption of the irradiation. The values obtained are characteristic for the motion of slow ionic species and the Debye formula [134] which links the volume recombination coefficient and the mobility was found to be valid, *i.e.*:

$$\frac{\alpha}{(\mu_- + \mu_+)} = \frac{4\pi e}{\kappa \kappa_0} \qquad (9.46)$$

where e represents the electronic charge, κ the relative dielectric constant and $\kappa_0 = 8.86 \times 10^{-14}$ F cm^{-1}. Recently the direct observation of recombination of excess electrons and positive ions in hydrocarbons has been reported. Baxendale et al. [135] observed the decay of the optical absorption spectrum of excess electrons in n-hexane and methylcyclohexane. Allen and Holroyd [129] determined recombination rate constants by observing the decay of the ionization current after a pulse of ionizing radiation. The data reported so far are summarized in Table 9.11.

9.4.3 Diffusion constant

The diffusion constant and mobility of ordinary ions are connected by the Nernst—Einstein equation

$$\frac{\mu_\pm}{D_\pm} = \frac{e}{kT} \qquad (9.47)$$

References pp. 384—388

TABLE 9.11

Volume recombination coefficients for electrons and positive ions in non-polar liquids

Liquid	T[K]	α [l mole^{-1} s^{-1}]	α [cm^3 s^{-1}]	Reference
n-hexane	293	2×10^{14}	3.3×10^{-7}	[135]
	295	4.7×10^{13}	7.8×10^{-8}	[129]
Methyl cyclohexane	293	8×10^{13}	1.33×10^{-7}	[135]
n-pentane	295	7.9×10^{13}	1.32×10^{-7}	[129]
Tetramethylsilane	295	4.8—5.5×10^{16}	8—9×10^{-5}	[129]

TABLE 9.12

Rate constants for electron attachment to solutes in non-polar liquids

Liquid	Solute	T[K]	k [l mole^{-1} s^{-1}]	Reference
Methane	SF_6	110	4×10^{14}	[172]
	N_2O	110	8.5×10^{11}	[172]
	O_2	110	8.4×10^{11}	[172]
Ethane	SF_6	195	1×10^{13}	[173]
n-Hexane	SF_6	295	2×10^{12}	[138]
	N_2O	295	1×10^{12}	[138]
	CCl_4	295	1.3×10^{12}	[138]
	O_2	293	1.5×10^{11}	[136]
	biphenyl	293	1.2×10^{12}	[136]
	C_2H_5Br	293	1.5×10^{13}	[129]
Neopentane	SF_6	294	2×10^{14}	[138]
	N_2O	294	2.3×10^{12}	[138]
	CCl_4	296	3×10^{13}	[129]
Tetramethylsilane	SF_6	294	2×10^{14}	[138]
	N_2O	294	7.5×10^{11}	[138]

At 298K, this ratio is 39 V^{-1}. For slow ionic-type charge carriers in n-hexane, diffusion coefficients and mobilities were measured in separate experiments by Gazda [51, 52] and Gzowski and Terlecki [29] and values of $\mu_-/D_- = 34.5$ V^{-1} and $\mu_+/D_+ = 42.4$ V^{-1} at 298K were obtained in reasonable agreement with the predicted value.

9.4.4 Electron attachment

Bimolecular rate constants k_s for the attachment of electrons to solutes S:

$$e^- + S \underset{k_s}{\rightarrow} S^-$$

to form negative ions S^- have been measured for a variety of solutes in many liquid hydrocarbons. Good electron scavengers in the gas phase such as SF_6, N_2O, CCl_4, O_2 etc. also react rapidly with electrons in non-polar liquids.

Table 9.12 gives some characteristic values. Several hypotheses have been put forward for the variation of the rate constant with the mobility of the electron [106, 137] or the energy of the conducting state [138]. Recently, the equilibrium of electron attachment and detachment has been observed for biphenyl and carbon dioxide in tetramethylsilane, isooctane and neopentane [139].

9.5 HIGH ENERGY RADIATION DETECTORS WITH NON-POLAR LIQUIDS

Since the ionization current of a dielectric liquid is a function of the dose rate, liquid-filled measurement cells can be used as radiation detectors and dosimeters. As early as 1928, Stahel [140] described a small ionization chamber filled with n-hexane for the relative dosimetry of X- and γ-rays. Later, extensive investigations on the use of dielectric liquids in ionization chambers were carried out by Adamczewki and coworkers [10, 141, 142] and by Blanc, Mathieu and coworkers [143–147]. Recently, the discovery of electron multiplication in liquid xenon has led to the construction of xenon-filled multi-wire proportional counters for track measurement of elementary particles [148, 149].

9.5.1 Liquid-filled ionization chambers

Liquid-filled ionization chambers with hydrocarbons are operated at an electric field strength where the ionization current is a linear function of the field and the current is proportional to the dose rate (*cf.* Section 9.2.2). The lowest detectable dose rate depends on the self-conductivity of the liquid. In n-hexane, a dose rate of 1 mrad/s produces a specific conductivity of 1.4×10^{-14} Ω^{-1} cm^{-1} at 10^3 V cm^{-1}. Since the self-conductivity can be decreased by proper purification, a lower limit of 10^{-8} rad/s is possible. A requirement for the use of liquid-filled ionization chambers in dosimetry is tissue equivalence, *i.e.* the radiation dose in the liquid should have the same dependence on the energy of the radiation as in the tissue. For γ-rays, the ratio of the mass energy absorption coefficient has to be independent of energy. The tissue dose D_t is then given by:

$$D_t = D_{liq} \frac{(\mu/\rho)_t}{(\mu/\rho)_{liq}} \qquad (9.48)$$

where D_{liq} represents the liquid dose and $(\mu/\rho)_{t,\, liq}$ the mass energy absorption coefficient for the tissue or liquid, respectively. For compounds, (μ/ρ) can be calculated from the (μ/ρ) values of the atoms and in Fig. 9.27 the variation of D_{liq}/D_t with radiation energy is shown for the case of n-hexane and carbon tetrachloride. Both liquids are not tissue equivalent for γ-rays below 10^5 eV. A mixture (4 Vol.% CCl$_4$ in n-hexane) of both liquids,

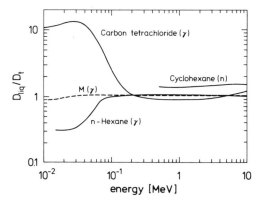

Fig. 9.27. The ratio of liquid dose to tissue dose as a function of energy for γ-rays and neutrons. M: mixture of 4 Vol.% CCl$_4$ and n-hexane. (From Mathieu [70] and Schmidt [151].)

however, is tissue equivalent within 10% for γ-rays from 10^4 eV to 10^7 eV. Liquid hydrocarbons are especially suited for the detection of fast neutrons which collide mainly with the H-atoms and produce recoil protons. The dose transmitted by fast neutrons to a medium composed of i different nuclei is given by:

$$D = n_\mathscr{E} \mathscr{E} \sum_i \sigma_i f_i N_i \qquad (9.49)$$

where $n_\mathscr{E}$ is the number of neutrons of energy $\mathscr{E}$, σ_i the cross-section of the ith kind of nuclei, f_i the fractional energy loss $f_i = 2mM/(m+M)^2$, N_i nuclei per gram [150]. The energy dependence of the neutron dose for cyclohexane and tissue is the same and between 0.5 MeV and 20 MeV $D_{\text{C-Hex}}/D_t = 1.45$ [151]. Hydrocarbons with a brutto formula of $(CH_2)_n$ are tissue equivalent for fast neutrons. Besides tissue equivalence, radiation equilibrium has to be maintained in the sensitive volume. For γ- and X-rays, aluminium electrodes are sufficient. In the case of fast neutrons, the electrodes should be either grids or thin conducting layers on polyethylene discs. The sensitive volume must be enclosed by a layer of $(CH_2)_n$ material the thickness of which should be larger than the maximum range of the recoil protons. Liquid ionization chambers for fast neutrons have been described by Schütze [152], Schmidt [151], Blanc et al. [144, 153] and Ladu et al. [154, 155]. One practical disadvantage of liquid ionization chambers compared to gas or solid devices is the sensitivity to mechanical shocks and perturbations.

A very important finding of Blanc et al. [144, 153] was the fact that the extrapolation of the linear part of the ionization-current—field-strength dependence to zero current intersected the abscissa at different points for

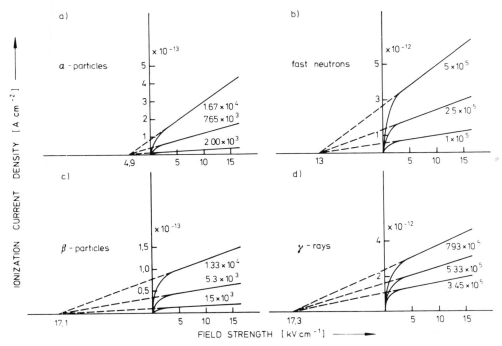

Fig. 9.28. Ionization current density *versus* electric field strength for different types of radiation (a) 5.3 MeV; (b) 14.7 MeV; (c) $E_{max} = 65$ keV; (d) $E_\gamma = 1.17$ and 1.33 MeV. (From Mathieu [70], reproduced with kind permission of Dr. Mathieu, Toulouse).

TABLE 9.13

Correlation between the intersection E_c and the Linear Energy Transfer (LET) in *n*-hexane (From Mathieu [70])

Radiation	Energy [MeV] E_c	[kV cm^{-1}]	LET [keV μm^{-1}]	RBE
γ-rays	1.25	$-(17 \pm 0.5)$	0.52	1
β-particles	0.065	$-(17 \pm 0.5)$	0.5	1
Fast neutrons	14.7	$-(13 \pm 0.5)$	11 ± 3	4
α-particles	5.3	$-(5 \pm 1)$	97 ± 20	20

different types of radiation qualities. In Fig. 9.28, current—field strength curves are shown for the irradiation of *n*-hexane with α-particles, β-particles, γ-rays and fast neutrons. As can be seen, the intersection E_c is correlated with the LET of the radiation. In Table 9.13, E_c and LET values and r.b.e. values are compiled (r.b.e. — relative biological efficiency, a measure of the necessary increase of the dose if a particular biological effect is to be produced by X- or γ-rays compared to the particular type of radiation). The intersection E_c in mixed radiation fields depends on the contribution of each

References pp. 384—388

radiation to the total dose rate and, after calibration with fields of known composition, this method can be used for dosimetry in mixed fields and for the determination of partial doses. Liquid argon ionization chambers have found interesting application in high-energy elementary particle physics. Multi-plate arrangements have been constructed for the detection and calorimetry of electromagnetic and hadronic showers [156, 157].

9.5.2 Detectors with electron multiplication

The high mobility of electrons in liquid argon and xenon and the possibility of measuring saturation currents made these liquids especially promising for use in radiation detectors. In 1968, Alvarez [148] speculated that, indeed, electron multiplication might be observable in these liquids and, in subsequent work at the Lawrence Radiation Laboratory, it was demonstrated that amplification of primary charge-carrier yield from elementary particles can be obtained by electron collisional ionization. The detectors are constructed as counter tubes with a very thin center wire (diameter $\sim 10\,\mu$m) surrounded by a cylinder (diameter ~ 1 cm). In the vicinity of the positive wire, excess electrons are accelerated to energies sufficient for further ionization. Depending on the voltage applied, a proportional region and a Geiger region can be observed in liquid xenon. Multi-wire proportional counters have been constructed and applied to track measurements of elementary particles [148, 149]. Similar experiments have been reported by Russian and French groups [158—160].

9.6 PURIFICATION TECHNIQUES AND MEASUREMENT CELLS

9.6.1 Preparation of liquid samples

In the experiments described in the preceding sections, liquid samples of different degrees of purity were required. Measurements of ionization currents require samples with a low self-conductivity. Although the quality of commercially-available liquids has been improved considerably since the early days, extensive drying procedures are still necessary in order to obtain liquid samples suitable for ionization measurements at high-field strengths and low radiation dose rates. The intrinsic conductivity of non-polar liquids has been studied for decades and no unambiguous answer has been given yet to the question of the "natural" self-conductivity of a dielectric liquid. Many investigations were carried out on n-hexane and, in 1931, Nikuradse [161] undertook an extensive purification procedure which decreased the self-conductivity of n-hexane from $10^{-13}\,\Omega^{-1}\,\text{cm}^{-1}$ to less than $10^{-19}\,\Omega^{-1}\,\text{cm}^{-1}$. The apparatus consisted of a series of stills and electrolysis cells, and the liquid was continuously circulated through the measurement cell. A self-conductivity of 10^{-18}—10^{-19} Ω^{-1} cm^{-1} can be obtained by a purification technique that combines chemical and physical methods. Commercially-available liquids are usually supplied either as chemically pure, approximately 99 mole per cent, or as research-grade liquids with a stated purity of usually 99.9 mole per cent. A first step in the purification consists in the removal of major impurities either by chemical reactions or chromatography through activated columns of silica gel, molecular sieve or other materials. The next step is a fractionated distillation from a drying agent, preferably one which reacts with water, such as lithium aluminium hydride or sodium potassium alloy. The measurement cells may be heated to several hundred degrees centigrade under vacuum prior to filling. Before the measurements are carried out, electrolysis in an auxiliary cell at a field strength of several kV cm^{-1} can lead to a further decrease of the self-conductivity of one order of magnitude. For measurements at fields above 100 kV cm^{-1}, extremely dust-free samples are necessary and filtering of the liquid through glass frits with pores of 1 μm diameter is mandatory. Also, the cell has to be flushed with dust-free liquid prior to the filling of the test sample. The presence of oxygen or other electron-attaching impurities at low concentrations does not seem to influence the steady-state ionization currents. This indicates that the originally produced charge carriers are converted into the most stable species by attachment or charge-transfer processes to other impurities. Excess electron mobility measurements require extreme purification with respect to electron-attaching impurities. All operations are carried out under vacuum. Since the rate constants for electron reactions with good scavengers are of the order of 10^{12}—10^{14} $l\,mol^{-1}\,s^{-1}$, the concentration of those impurities has to be less than 1 $\mu mol/l$ in order to obtain electron lifetimes of microseconds. For liquid hydrocarbons, passage of the vapors over barium mirrors has produced samples suitable for drift velocity measurements with photoelectric generation of electrons [117]. In other experiments, the vapors were passed over activated silica gel, molecular sieve and charcoal, and condensed into a vacuum. Degassing was achieved by trap-to-trap distillation and pumping. The liquid was then condensed into a bottle, sealed off and pre-irradiated with ^{60}Co-γ-rays to a dose of several tens kilorads. The vapors were again passed through activated columns of silica gel etc. and, finally, the cell was filled by condensation. If the electron lifetime was still too short, pre-irradiation of the filled cell sometimes improved the situation [46, 47, 48, 135].

Rare gases and hydrogen are usually supplied in high purity and only nitrogen, oxygen, carbon dioxide, and water vapor must be removed. Activated charcoal, molecular sieves, and silica gel, or metal getters (barium, calcium, tantal) are applied. The activation of the material requires heating to 500—600°C under high vacuum. During the passage of the gas, the column should be maintained at a lower temperature (*e.g.* dry ice temperature) in order to improve the absorption capacity. Oxygen can also be removed by

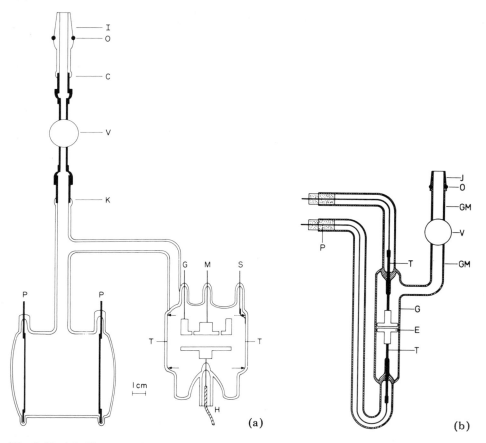

Fig. 9.29. (a) Glass cell for clearing-field measurements; I, standard taper joint; O, viton O-ring; C, K, Covar-to-Pyrex seals; V, valve; P, Pt electrodes of auxiliary cell; G, guard ring; S, T, tin-oxide screen; M, measuring electrode; H, high voltage electrode. (From Schmidt and Allen [54].) (b) Cell for mobility measurements; (From Bakale and Schmidt [47]) I, standard taper joint; O, viton O-ring, GM, glass-metal seal; V, valve; P, rubber plugs; T, tungsten wire; G, glass; E, electrodes.

a catalyst (BTU from BASF). Nitrogen is removed either by pumping or by passage over hot calcium chips where calcium carbide is formed. Other physico-chemical methods such as gas chromatography or zone melting should be considered but have hardly been applied in the purification of non-polar liquids for these conductivity experiments.

9.6.2 Measurement cells

For ionization measurements with X-rays or γ-rays, measurement cells of different geometry and construction have been described. Most often parallel-plate chambers are used consisting of a voltage electrode and a collector electrode sometimes surrounded by a guard ring. Coaxial geometry has been employed in dosimeters. The body of the cell is frequently made of glass, metal or plastics. Glass cells require additional screening to prevent leakage currents across the glass surface or through the bulk. Tin-oxide coatings on the inside or silver paint on the outside give the required properties. The glass is also the insulator between the collector and the voltage electrode. Since glass exhibits an ionic conductivity at higher temperatures, care has to be taken if measurements above 100°C are performed.

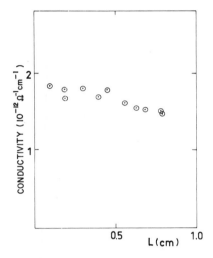

Fig. 9.30. Radiation-induced conductivity as a function of electrode separation in a test cell with Pt-electrodes. Liquid: n-hexane. (From Hummel [163].)

Figure 9.29(a) shows a cell used for clearing-field measurements, while, in Fig. 9.29(b), a cell is shown which has been used for electron mobility measurements. Cells with a metal body are especially suited for steady-state work. Frequently, ceramic, glass or teflon are used as insulators for the electrodes. Electrodes can be made of aluminium, stainless steel, platinum etc. Brass electrodes tend to corrode and gold plating is advisable. Quantitative measurements of the ionization by X-rays or γ-rays require accurate dosimetry. Bragg's principle [162] is fulfilled only if the material of the electrodes and the wall has the same average atomic number as the liquid. Aluminium is a good approximation for hydrocarbons. Platinum electrodes yield a higher secondary-electron emission from the electrodes and give a higher dose rate which is not easily taken into account. Figure 9.30 shows the variation of the specific conductivity with electrode distance in n-hexane in a cell with platinum electrodes. For the investigation of electron

References pp. 384—388

avalanches in liquid xenon, a coaxial cell with a thin center wire and an outer cylinder has been used [75]. In all experimental set-ups, irradiation of the cables and of liquid volume other than the space between the electrodes can lead to spurious pulses and currents which sometimes are greater than the original effect. Cells for photoeffect or photoionization require suitable windows, usually quartz. For the investigation of the photoeffect, parallel electrode plates were used where the anode was formed by a grid allowing the passage of light to the cathode [108].

ACKNOWLEDGEMENTS

The author wishes to thank his colleagues, Drs. Bakale, U. Sowada, W. Tauchert, and Dr. Yoshino for stimulating discussions. Support of the author's research projects on electron transport in dielectric liquids by the Deutsche Forschungsgemeinschaft is gratefully acknowledged.

REFERENCES

1 J.J. Thomson, Nature, 55 (1897) 606.
2 P. Curie, C.R. Acad. Sci., 134 (1902) 420.
3 G. Jaffé, Ann. Phys. (Leipzig), 25 (1908) 257.
4 G. Jaffé, Ann. Phys. (Leipzig), 28 (1909) 326.
5 G. Jaffé, Ann. Phys. (Leipzig), 32 (1910) 148.
6 G. Jaffé, Ann. Phys. (Leipzig), 42 (1913) 303.
7 G. Jaffé, Radium (Paris), 10 (1913) 126.
8 H.J. van der Bijl, Ann. Phys. (Leipzig), 39 (1912) 170.
9 D.E. Lea, Proc. Cambridge Philos. Soc., 30 (1934) 80.
10 I. Adamczewski, Ionization, Conductivity and Breakdown in Dielectric Liquids, Taylor and Francis, London, 1969.
11 I. Adamczewski and B. Jachym, Acta Phys. Pol., 30 (1966) 767.
12 S.A. Rice, Acc. Chem. Res., 1 (1968) 81.
13 J. Jortner, Excess electron states in liquids, Actions Chim. Biol. Radiat., 14 (1970) 1.
14 J. Jortner, S.A. Rice and N.R. Kestner, Electrons in liquids. In O. Sinanoglu (Ed.), Quantum Chemistry, Istanbul Lectures, Academic Press, 1965.
15 A. Hummel and A.O. Allen, J. Chem. Phys., 44 (1966) 3426.
16 W.F. Schmidt, Z. Naturforsch., 23b (1968) 126.
17 G. Bakale, E.C. Gregg and R.D. McCreary, J. Chem. Phys., 57 (1972) 4246.
18 K. Hayashi, Y. Yamazawa, T. Takagaki, F. Williams, K. Hayashi and S. Okamura, Trans. Faraday Soc., 63 (1967) 1489.
19 T. Takagaki, K. Hayashi, K. Hayashi and S. Okamura, Ann. Rep. Jpn. Assoc. Radiat. Res. Polym., 6 (1964—1965) 13, AEC-tr-6806.
20 G.R. Freeman, J. Chem. Phys., 39 (1963) 988.
21 K. Takada, T. Takagaki, K. Hayashi, K. Hayashi and S. Okamura, Ann. Rep. Jpn. Assoc. Radiat. Res. Polym., 8 (1966—1967) 84, AEC-tr-6937.
22 E. Rutherford, Philos. Mag., 44 (1897) 422.
23 E. Gazda, Acta. Phys. Pol., 27 (1965) 881.

24 A. Hummel, A.O. Allen and F.H. Watson, J. Chem. Phys., 44 (1966) 3431.
25 E.E. Conrad and J. Silverman, J. Chem. Phys., 51 (1969) 450.
26 P.H. Tewari and G.R. Freeman, J. Chem. Phys., 49 (1968) 4394.
27 E.C. Gregg and G. Bakale, Radiat. Res., 42 (1970) 13.
28 G. Careri and F.S. Gaeta, Proc. 7th Int. Conf. Low Temp. Phys., Toronto, 1960, Univ. of Toronto Press, 1961, p. 505.
29 O. Gzowski and J. Terlecki, Acta Phys. Pol., 18 (1959) 191.
30 O. LeBlanc, J. Chem. Phys., 30 (1969) 1443.
31 J. Terlecki, Nature, 194 (1962) 172.
32 P. Chong and Y. Inuishi, Technol. Rep. Osaka Univ., 10 (1960) 545.
33 A. Hummel and W.F. Schmidt, Radiat. Res. Rev., 5 (1974) 199.
34 L. Meyer and F. Reif, Phys. Rev., 110 (1958) 279.
35 F. Reif and L. Meyer, Phys. Rev., 119 (1960) 1164.
36 B. Halpern, J. Lekner, S.A. Rice and R. Gomer, Phys. Rev., 156 (1967) 351.
37 B.L. Henson, Phys. Rev. A, 135 (1964) 1002.
38 V. Essex and P.E. Secker, Brit. J. Appl. Phys. Ser. 2, 2 (1969) 1107.
39 V. Essex and P.E. Secker, Brit. J. Appl. Phys. Ser. 2, 1 (1968) 63.
40 H.T. Davis, S.A. Rice and L. Meyer, J. Chem. Phys., 37 (1962) 947.
41 S. Cunsolo, Nuovo Cimento, 21 (1961) 76.
42 L. Bruschi and M. Santini, Rev. Sci. Instrum., 41 (1970) 102.
43 W.F. Schmidt and A.O. Allen, J. Chem. Phys., 50 (1969) 5037.
44 W.F. Schmidt and A.O. Allen, J. Chem. Phys., 52 (1970) 4788.
45 D.D. Hudson, U.S. A.E.C., Rep. No. MDDC—524, 1946.
46 W.F. Schmidt and G. Bakale, Chem. Phys. Lett., 17 (1972) 617.
47 G. Bakale and W.F. Schmidt, Z. Naturforsch., 28a (1973) 511.
48 P.G. Fuochi and G.R. Freeman, J. Chem. Phys., 56 (1972) 2333.
49 J.P. Dodelet and G.R. Freeman, Can. J. Chem., 50 (1972) 2667.
50 M. Robinson and G.R. Freeman, Can. J. Chem., 52 (1974) 440.
51 E. Gazda, Acta Phys. Pol., 24 (1963) 209.
52 E. Gazda, Nature, 200 (1963) 767.
53 W.F. Schmidt and A.O. Allen, Science, 160 (1968) 301.
54 W.F. Schmidt and A.O. Allen, J. Phys. Chem., 72 (1968) 3730.
55 W.F. Schmidt and A.O. Allen, J. Chem. Phys., 52 (1970) 2345.
56 A. Hummel and A.O. Allen, J. Chem. Phys., 46 (1967) 1602.
57 A. Jahns and W. Jacobi, Z. Naturforsch., 21a (1966) 1400.
58 C.S. Pao, Phys. Rev., 64 (1943) 60.
59 H. Ullmaier, Z. Phys., 178 (1964) 44.
60 A. Jachym and B. Jachym, Acta Phys. Pol., 34 (1968) 879.
61 R. Gibaud, J. Chim. Phys. Phys. Chim. Biol., 64 (1967) 521.
62 F.L. Mohler and L.S. Taylor, J. Res. Nat. Bur. Stand., 13 (1934) 659.
63 L.S. Taylor, J. Res. Nat. Bur. Stand., 17 (1936) 557.
64 S. Lenkeit and H.G. Ebert, Naturwissenschaften, 51 (1964) 237.
65 H.G. Ebert, J. Booz and R. Koepp, Z. Phys. Chem. (Frankfurt am Main), 43 (1964) 304.
66 J. Mathieu, D. Blanc, P. Caminade and J.P. Patau, J. Chim. Phys. Phys. Chim. Biol., 64 (1967) 1679.
67 W.F. Schmidt, Radiat. Res., 42 (1970) 73.
68 H.A. Kramers, Physica (Utrecht), 18 (1952) 665.
69 L. Onsager, Phys. Rev., 54 (1938) 554.
70 J. Mathieu, Etude physico-chimique de la conduction induite par les rayonnements nucléaires dans les liquides diélectriques. Application à la dosimétrie, Thesis, University of Toulouse, Toulouse, 1968.
71 H.A. Ullmaier, Phys. Med. Biol., 11 (1966) 95.

72 N.V. Klassen and W.F. Schmidt, Can. J. Chem., 47 (1969) 4286.
73 Int. Committee Radiol. Units, Rep. 10b, NBS Handbook 85, U.S. Department of Commerce, Washington, D.C., 1964.
74 J. Booz and H.G. Ebert, Strahlentherapie, 120 (1963) 7.
75 S.E. Derenzo, T.S. Mast, Z. Zaklad and R.A. Muller, Phys. Rev. A, 9 (1974) 2582.
76 M. Chybicki, Acta Phys. Pol., 30 (1966) 927.
77 M. Chybicki, Acta Phys. Pol., 34 (1968) 285.
78 M. Chybicki, Acta Phys. Pol., 34 (1968) 445.
79 J.P. Ramy, Thesis, Toulouse, 1971.
80 E.W.T. Richards, Proc. Phys. Soc. London, Sect. A, 66 (1953) 631.
81 A.N. Gerritsen, Physica (Utrecht), 14 (1948) 381.
82 A.N. Gerritsen, Physica (Utrecht), 14 (1948) 407.
83 D.W. Swan, Proc. Phys. Soc. London, 85 (1965) 1297.
84 T. Takahashi, M. Miyajima, S. Konno, T. Hamada, S. Kubota, H. Shibamura and T. Doke, Phys. Lett., 44A (1973) 123.
85 S. Konno and S. Kobayashi, Sci. Papers Inst. Phys. Chem. Res. (Tokyo), 67 (2) (1973) 57.
86 G.R. Freeman and J.M. Fayadh, J. Chem. Phys., 43 (1965) 86.
87 M.G. Robinson, P.G. Fuochi and G.R. Freeman, Can. J. Chem., 49 (1971) 3657.
88 J.P. Dodelet, K. Shinshaka, U. Kortsch and G.R. Freeman, J. Chem. Phys., 59 (1973) 2376.
89 G.C. Abell and K. Funabashi, J. Chem. Phys., 58 (1973) 1079.
90 J.P. Dodelet, P.G. Fuochi and G.R. Freeman, Can. J. Chem., 50 (1972) 1617.
91 A. Mozumder, J. Chem. Phys., 60 (1974) 4300.
92 A. Mozumder and M. Tachiya, J. Chem. Phys., 62 (1975) 979.
93 B. Halpern and R. Gomer, J. Chem. Phys., 51 (1969) 1048.
94 I. Roberts and E.G. Wilson, J. Phys. C., Solid State Phys., 6 (1973) 2169.
95 C. Vermeil, M. Matheson, S. Leach and F. Muller, J. Chim. Phys., 61 (1964) 596.
96 G.N. Lewis and D. Lipkin, J. Am. Chem. Soc., 64 (1942) 2801.
97 H.S. Pilloff and A.C. Albrecht, J. Chem. Phys., 49 (1968) 4891.
98 S.S. Takeda, N.E. Houser and R.C. Jarnagin, J. Chem. Phys., 54 (1971) 3195.
99 N. Houser and R.C. Jarnagin, J. Chem. Phys., 52 (1970) 1069.
100 G. Briegleb and J. Czekulla, Z. Elektrochem., 63 (1959) 6.
101 Y. Nakato, M. Ozaki, A. Egawa and H. Tsubomura, Chem. Phys. Lett., 9 (1971) 615.
102 R. Forster, Nature, 183 (1959) 1253.
103 R.A. Holroyd, J. Chem. Phys., 57 (1972) 3007.
104 R.A. Holroyd and R.L. Russell, J. Phys. Chem., 78 (1974) 2128.
105 J.C. Devins and J.C. Wei, 4th Int. Conf. on Conduction and Breakdown in Dielectric Liquids, Dublin, July 1972, p. 13.
106 G. Beck and J.K. Thomas, J. Chem. Phys., 57 (1972) 3649.
107 R.H. Fowler, Phys. Rev., 38 (1931) 45.
108 R.A. Holroyd and M. Allen, J. Chem. Phys., 54 (1971) 5014.
109 R.A. Holroyd, B.K. Dietrich and H.A. Schwarz, J. Phys. Chem., 76 (1972) 3794.
110 R.A. Holroyd and W. Tauchert, J. Chem. Phys., 60 (1974) 3715.
111 R. Schiller, Sz. Vass and J. Mandics, Int. J. Radiat. Phys. Chem., 5 (1973) 491.
112 S. Noda and L. Kevan, J. Chem. Phys., 61 (1974) 2467.
113 W. Tauchert and W.F. Schmidt, Z. Naturforsch., 29a (1974) 1526.
114 J. Lekner, B. Halpern, S.A. Rice and R. Gomer, Phys. Rev., 156 (1967) 351.
115 W.T. Sommer, Phys. Rev. Lett., 12 (1964) 271.
116 W. Tauchert, Thesis, Free University of Berlin, 1975.
117 R.M. Minday, L.D. Schmidt and H.T. Davis, J. Chem. Phys., 54 (1971) 3112.
118 R.J. Loveland, P.G. LeComber and W.E. Spear, Phys. Lett. A, 39 (1972) 225.

119 J. Lekner, Phys. Rev., 158 (1967) 130.
120 W. Shockley, Bell Syst. Tech. J., 30 (1951) 990.
121 H.T. Davis, S.A. Rice and L. Meyer, Phys. Rev. Lett., 9 (1962) 81.
122 H.T. Davis, S.A. Rice and L. Meyer, J. Chem. Phys., 37 (1962) 1521.
123 H.T. Davis, S.A. Rice and L. Meyer, J. Chem. Phys., 37 (1962) 947.
124 G. Careri, S. Consolo, and P. Mazzoldi, Phys. Rev. A, 136 (1964) 303.
125 L. Bruschi, G. Mazzi and M. Santini, Phys. Rev. Lett., 25 (1970) 33.
126 M. Steingart and W.I. Glaberson, Phys. Rev. A, 2 (1970) 1480.
127 R.M. Minday, L.D. Schmidt and H.T. Davis, J. Phys. Chem., 76 (1972) 442.
128 W.F. Schmidt, G. Bakale and U. Sowada, J. Chem. Phys., 61 (1974) 5275.
129 A.O. Allen and R.A. Holroyd, J. Phys. Chem., 78 (1974) 796.
130 K. Fueki, D.-F. Feng, and L. Kevan, Chem. Phys. Lett., 13 (1972) 616.
131 H.T. Davis, L.D. Schmidt and R.M. Minday, Chem. Phys. Lett., 13 (1972) 413.
132 B.G. Bagley, Solid State Commun., 8 (1970) 345.
133 M.P. de Haas, J.M. Warman, P. Infelta and A. Hummel, Chem. Phys. Lett., 31 (1975) 382.
134 P. Debye, Trans. Electrochem. Soc., 82 (1942) 265.
135 J.H. Baxendale, C. Bell and P. Wardman, Chem. Phys. Lett. 12 (1971) 347.
136 J.H. Baxendale and E.J. Rasburn, J. Chem. Soc. Faraday Trans. 1, 70 (1974) 705.
137 B.S. Yakovlev, I.A. Boriev and A.A. Balakin, Int. J. Radiat. Phys. Chem., 6 (1974) 23.
138 A.O. Allen, T.E. Gangwer and R.A. Holroyd, J. Phys. Chem., 79 (1975) 25.
139 R.A. Holroyd. T.E. Gangwer and A.O. Allen, Chem. Phys. Lett., 31 (1975) 520.
140 E. Stahel, Strahlentherapie, 31 (1929) 582.
141 I. Adamczewski, Ionization chambers with liquid and their practical application, Proc. Symp. Selected Topics Radiation Dosimetry, IAEA, Vienna, 1961.
142 I. Adamczewski, Liquid-filled ionization chambers as dosimeters, Phénomènes de conduction dans les liquides isolants, Colloq. Int. Centre Nat., Rech. Sci., No. 179, Grenoble, 17—19 September, 1968, p. 21.
143 D. Blanc, J. Mathieu and J. Boyer, Fonctionnement, à la temperature ambiante, de chambres d'ionisation remplies d'un dielectrique liquide, Proc. Conf. Nuclear Electronics, Vol. 1, IAEA, Vienna, 1962, p. 285.
144 D. Blanc, J. Mathieu and P. Vermande, Nucl. Instrum. Methods, 21 (1963) 349.
145 D. Blanc, J. Mathieu and L. Torres, Nucl. Instrum. Methods, 27 (1964) 353.
146 D. Blanc, J. Mathieu and P. Vermande, Health Phys., 11 (1965) 63.
147 D. Blanc, J. Mathieu, J.P. Patan, H. François and G. Soudain, Health Phys., 2 (1966) 1589.
148 R.A. Muller, S.E. Derenzo, R.G. Smits, H. Zaklad and L.W. Alvarez, UCRL Rep., 20 (1970) 135.
149 S.E. Derenzo, R. Flagg, S.G. Louie, F.G. Mariam, T.S. Mast, A.J. Schwemin, R.G. Smits, H. Zaklad and L.W. Alvarez, CONF—72 (1972) 0923—9.
150 G.S. Hurst, Brit. J. Radiol., 27 (1954) 353.
151 W.F. Schmidt, Atompraxis, 10 (1964) 157.
152 W. Schütze, Patentschrift Nr. 924226, Kl.21g, Gr.1801, BRD 1955.
153 D. Blanc, J. Mathieu and P. Vermande, J. Phys. (Paris), 24 (1963) 31A.
154 M. Ladu, M. Pelliccioni and M. Roccella, Nucl. Instrum. Methods, 34 (1965) 178.
155 M. Ladu and M. Pelliccioni, Nucl. Instrum. Methods., 39 (1966) 339.
156 J. Engler, B. Friend, W. Hofmann, H. Keim, R. Nickson, W. Schmidt-Parcefall, A. Segar, M. Tyrrell, D. Wegener, T. Willard and K. Winter, Nucl. Instrum. Methods, 120 (1974) 157.
157 G. Knies and D. Neuffer, Nucl. Instrum. Methods, 120 (1974) 1.
158 A.F. Pisarev, V.F. Pisarev and G.S. Revenko, Rep. Joint Inst. for Nuclear Research, P 13—6450.

159 B.A. Dolgoshein, V.N. Lebedenko and B.U. Rodionov, JETP Lett., 11 (1970) 351.
160 J. Prunier, R. Allemand, M. Laval and G. Thomas, Nucl. Instrum. Methods, 109 (1973) 257.
161 A. Nikuradse, Z. Phys. Chem. A, 155 (1931) 59.
162 G.J. Hine and G.L. Brownell (Eds.), Radiation Dosimetry, Academic Press, New York, 1956.
163 A. Hummel, Thesis, Free University, Amsterdam, 1967.
164 G. Baldini, Phys. Rev., 128 (1962) 1562.
165 U. Asaf and I.T. Steinberger, Phys. Lett., 41A (1972) 19.
166 L.S. Miller, S. Hower and W.E. Spear, Phys. Rev., 166 (1968) 871.
167 L. Meyer, H.T. Davis, S.A. Rice and R.J. Donelly, Phys. Rev., 126 (1962) 1927.
168 N. Zessoules, J. Brinkerhoff and A. Thomas, J. Appl. Phys., 34 (1963) 2010.
169 B. Halpern and R. Gomer, J. Chem. Phys., 51 (1969) 1031.
170 R.J. Loveland, P.G. LeComber and W.E. Spear, Phys. Rev. B, 6 (1972) 3121.
171 G. Bakale, U. Sowada and W.F. Schmidt, 1974 Ann. Rep., Conf. Electr. Insul. Dielectr. Phenom., Downington, Pa., Oct. 1974, Natl. Acad. Sci., Washington, D.C.
172 W.F. Schmidt, G. Bakale and W. Tauchert, 1973 Ann. Rep., 42nd Conf. Elect. Insul. Dielectr. Phenom., Varennes, Canada, Oct. 1973, Natl. Acad. Sci., Washington, D.C., 1974.
173 G. Bakale, V. Sowada and W.F. Schmidt, J. Phys. Chem., 79 (1975) 3041.

CHAPTER 10

PHOTOELECTRONIC SEMICONDUCTOR DEVICES

DIETER BONNET

10.1. Introduction
10.2. Light sensors and detectors
   10.2.1. Performance criteria
   10.2.2. Polycrystalline films and layers
   10.2.3. Single crystal devices for visible and near infrared radiation
   10.2.4. Single crystalline materials for medium and far infrared light
   10.2.5. Heterojunction photodiodes
10.3. Imaging systems
   10.3.1. Imaging tubes with electron beam readout
   10.3.2. Self-scanned solid state systems
10.4 Optical memories and image storage
10.5 Image intensifiers and converters
10.6. Solar cells

10.1 INTRODUCTION

The sense of vision is generally considered to be the most important sense in humans without which our culture as it is would not have been possible. Therefore, it is not surprising that a huge amount of technical and scientific effort has been invested in enhancing, extending and diversifying the photoreceptor in its function as a light detector and, later, as an image pick-up system which allows for the transmission and regeneration or processing of the received optical information. Since photoconductivity was first observed in selenium in 1873, research into photosensitive systems has steadily increased, leading to an almost exponential growth during the past 30 years. In the first half of this century, systems based on the external photoelectric effect, such as the vacuum photocells and the still-used orthicon television camera, rose to prominence, whereas systems based on the internal photoelectric effect, such as the selenium cell, were in a state of relative stagnation. However, starting in the 1940's, more and more materials became available in very pure form and development of solid-state systems grew quickly. Consequently, they soon clearly dominated in many applications. Today, vacuum electronic systems are still widely used in television cameras, but development effort is concentrated on all-solid photosensitive systems

References pp. 418—420

culminating today in the charge-transfer image-sensing systems. The success of this work can be estimated from the fact that the receptors of these solid-state imaging systems exhibit quantum efficiencies close to 100% compared to 1—2% of the human eye [1]. Due to lack of space, we can only give the reader a concentrated account of photosensitive systems utilizing the internal photoelectric effect in solids in this chapter. This includes pure photoconductive, but also photodiode, systems. These two types of solid photosensors cannot be treated separately where applications are concerned because they can be substituted for one another in many cases. Emphasis is being put on systems which are presently used or which, in the view of experts involved, will be used in the near future. Therefore, in some cases, only the more important or more advanced examples of a class of materials are treated in some detail (*e.g.* InSb of the II—V semiconductor group). Electrophotographic systems utilizing high-resistivity low-mobility photoconductors such as selenium, or organic materials in the form of thin films, are treated in Chapter 11 of this book and are therefore excluded here.

10.2 LIGHT SENSORS AND DETECTORS

10.2.1 Performance criteria

In order to compare and evaluate photodetectors, it is necessary to have detailed information on three properties:
 (a) region of spectral sensitivity;
 (b) absolute sensitivity or minimum detectable power; and
 (c) speed of response [2—4].
The first property is the most unambiguous one. The long-wavelength edge of a semiconductor detector is usually determined by the electronic transition which leads to the excited free carrier, *i.e.* the band gap for an intrinsic detector. The response for shorter wavelengths usually decreases as the absorption coefficient increases, as surface recombination plays a lifetime-limiting role for carriers generated near the surface. This decrease can be modified, *e.g.* by the position of the $p-n$ junction in a diode detector relative to the position of light absorption or by special antireflective coatings. In practice, it is therefore useful to have a curve of relative spectral sensitivity of the detector considered at one's disposal. The measurement of absolute sensitivity, *e.g.* at the point of maximum relative sensitivity, is more problematic. Whereas for the application of photoconductors in the visible region of the spectrum it is often sufficient to know values such as light-to-dark current ratio or gain number, it is usually important to consider noise sources because noise of one kind or the other limits the detectable signal. Therefore, it is necessary at least to give the noise equivalent power (NEP). It is the r.m.s. radiant power incident on the detector which produces an output

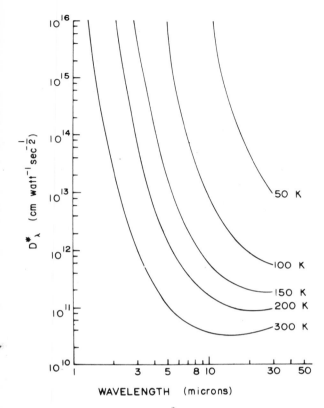

Fig. 10.1. Dependence of D_λ^* at spectral peak on long-wavelength threshold for several background temperatures. Calculations assume ideal photoconductive detectors having a 2π-steradian field of view. (From Levinstein [2].)

r.m.s. signal S equal in magnitude to the r.m.s. detector noise N

$$\text{NEP} = PN/S \tag{10.1}$$

Here P is the incident r.m.s. power. The dimensions of NEP are therefore watts. The quality of a detector increases with decreasing NEP under defined conditions. It is more convenient to have a value which becomes larger with increasing sensitivity of the detector. Such a value has been formed which makes the comparison between different detectors easier by normalizing it to an amplifier bandwidth of 1 Hz and a detector area of 1 cm² :

$$D^* = \frac{S/N}{P}\left(\frac{\Delta f}{A}\right)^{1/2} \tag{10.2}$$

where D^* is called the detectivity and has units of cm Hz$^{1/2}$ W^{-1}.

In order to avoid misunderstandings, it is necessary to state the conditions

References pp. 418—420

under which a D^*-value has been measured. For a black body radiation detector in the infrared spectrum, a black body of 500K is usually employed, the radiation of which is chopped at 900 Hz and an amplifier bandwidth of 5 Hz is employed. These numbers are given in parentheses behind the D^* sign. Although there is still some ambiguity in the definition of NEP (sometimes the detector area, sometimes the amplifier bandwidth is included), the D^* values can really be compared if the conditions of measurement are given. Figure 10.1 gives some calculated D^* curves, for an ideal photoconductive detector, as a function of long-wavelength threshold with the background temperature as a parameter.

The third characteristic value for a radiation detector is its response time. Often the response time corresponds to the lifetime of free photoexcited carriers. Nevertheless, the behaviour of rise and decay times may be quite different and not at all of an exponential nature. Furthermore, rise and decay times are often dependent on light intensity which further complicates the picture. This may be the case for photoconductive CdS layers, where some scepticism on single numbers for response time is indicated. For fast detectors with response times of less than 10^{-6}s, one value of response time or limiting frequency is usually sufficient as a measuring value.

10.2.2 Polycrystalline films and layers

(i) CdS films

One of the first large-scale applications of highly sensitive photoconductors was the use of CdS in light meters for photographic cameras, replacing the less sensitive selenium photoelements. This development owes much to the fact that CdS polycrystalline materials may have the major photoelectric properties of single crystals. A rather simple method can be employed to produce sensitive sintered layers of CdS and CdSe and mixed systems [5, 6]: the II—VI material in powder form, mixed with $CdCl_2$ and $CuCl_2$, is spread onto a suitable substrate, *e.g.* in the form of an aqueous slurry and heated above the melting point of $CdCl_2$ for a predetermined time. The $CdCl_2$ acts mainly as a flux for the recrystallization process and the $CuCl_2$ provides the activator Cu and the coactivator Cl. The resulting impurity concentration in the recrystallized layers is rather high, so that impurity bands are formed. The spectral response differs considerably from that of the pure material extending, in the case of CdS, into the red spectral region. This is desirable for a light meter replacing the human eye. These layers have to be used with current flow in the plane of the film because of their porosity after the recrystallization process. They have a relatively large free surface area which is exposed to ambient atmosphere. This may lead to strong effects by adsorption and photodesorption of oxygen and water, influencing the photoelectric properties. Moreover, the recrystallization process suffers from problems regarding reproducibility. Application for light meters in photographic cameras does not demand high standards of reproducibility and stability, thus making the use of these layers possible. Nevertheless, considerable effort has been spent on the development of better layers. Firstly, binder layers have been studied consisting of CdS powder dispersed in a resin [6], which gives surface protection to the individual grains. As the resin adds series resistances, the sensitivity was lower and the improvement in overall properties was not too high.

The next improvement was to use evaporated layers. However, when CdS is evaporated, the resulting films do not have a sufficiently high resistivity to permit their use as photoconductors because of a non-perfect stoichiometry: sulfur vacancies which act as donors are created under these conditions. Therefore a doping and sensitizing process has to be added [7—9], resulting in similar disadvantages as in the case of sintered powder layers. A considerable improvement of the system should result if dense, pinhole-free thin films are used which need no thermal after-treatment. These are sensitive immediately after the deposition process and thus lead to "through-thickness" photoconductor films. Such films should be of satisfactory stoichiometry from the start because they avoid sulfur vacancy donors. Films like this are produced by offering an excess of sulfur during deposition. Two methods have been studied: sputter deposition and co-evaporation of sulfur and cadmium in different ratios. Both methods have recently given promising results. In the sputtering experiments [10], H_2S has been added to increase the S content of the films, in the co-evaporation process, an excess of S over Cd has been maintained during film growth for the same reason [11]. In both cases, films with considerable photosensitivity resulted without further sensitizing treatment. Dark resistivities of more than $10^{11}\,\Omega\,cm$ and up to $10^8\,\Omega\,cm$ have been obtained by the co-evaporation method and the sputtering method, respectively, without after-treatment, indicating good stoichiometry. Gain values of several thousand have been measured for green light in CdS films of these types. One special dilemma which has been observed, even in the latest work, has always been encountered with CdS photoconductors: highly sensitive systems show long response times at low illumination levels. It seems to be quite certain that the sensitizing center formerly attributed to Cu doping is always coupled with shallow traps below the conduction band which lead to slow response times. No method has yet been suggested to avoid this. Major problems do not arise when sufficient light is available to saturate the traps; this leaves open the way to applications needing large-area photoconductors or photoconductor arrays at sufficient illumination intensity.

(ii) PbS films

For a considerable period of time, PbS photoconducting films and layers have been of great practical interest because they were the only high-sensitivity solid-state detectors operating in the infrared region between 1 and

References pp. 418—420

4 μm [5, 6, 12]. PbS films have been prepared by two methods: chemical reaction and vacuum evaporation. In the former method, the starting materials are lead acetate and thiourea, which react in a basic aqueous solution and form PbS films on a suitable substrate. The latter method uses PbS powder as a starting material and high vacuum evaporation for film formation. In both cases, n-conducting films are obtained which have to be activated in order to become sufficiently photo-sensitive. The activation process mainly involves the incorporation of oxygen to compensate for the donors (sulfur vacancies similar to CdS). The sensitivity is best at the compensation point when the conductivity type changes from n to p and the resistivity is at its maximum. Doping is effected by heating the films in the presence of oxygen at temperatures up to 350 to 400°C. The doping effect is reversible: heating in vacuum removes the oxygen. A great variety of individual processes and formulations have been worked out in the course of time. Details may be found in several review articles. In the case of vacuum evaporation, the films can conveniently be deposited onto the surface of special Dewar flasks which are used to cool the detector. With cooling, the long-wavelength edge of sensitivity shifts from 2.5 μm at 300K to longer wavelengths due to an energy gap variation of 4×10^{-4} eV/degree. As expected, the detectivity of the devices increases on operation at low temperatures (typically 77K). Detectivities greater than 1×10^{12} cm Hz$^{1/2}$ W^{-1} can be obtained at the wavelength corresponding to peak sensitivity and optimum chopping frequency. D^*, as a function of chopping frequency, is constant between 1 and 10 kHz for good detectors. The D^* value indicated approaches the value for an ideal background-limited detector. Because of the good quality and homogeneity of these films, especially of the chemically-deposited layers, detector arrays can be fabricated quite easily for thermal-imaging systems. Elements 10 μm × 10 μm in size are feasible. No size dependence of the detectivity has been observed.

Although these devices have been in use for a long time, there is still some dispute about the mechanism of their photosensitivity. Two theories are being discussed:

(a) The usual mechanism of photoconductivity, suggesting that light absorption creates additional charge carriers which contribute to the current.

(b) A barrier modulation concept, according to which the intercrystalline barrier produced by depleted regions at the surfaces of the individual crystallites is decreased by light, enabling carriers to flow more easily from grain to grain. Because of substantial evidence, the general opinion is more inclined towards the first classical photoconductive theory.

10.2.3 Single crystal devices for visible and near infrared radiation

(i) Silicon diodes

Silicon has an energy gap of 1.1 eV and, therefore, as an intrinsic photosensitive material should be sensitive to radiation of wavelength below

1.1 μm. Because silicon is an "indirect" semiconductor, its spectral absorption coefficient increases rather slowly at decreasing wavelength and reaches a value of 10^5 cm^{-1} at wavelengths below 500 nm. Thus, photons of this range are mainly absorbed within a considerable distance from the surface of the single crystal, where their lifetime is only slightly affected by surface recombination states. This results in a range of high photosensitivity in silicon photodiodes, which practically covers the whole visible and the near infrared spectrum. Of particular interest are photodiodes in various forms rather than photoconductors. This is because it is difficult to make nearly intrinsic materials of sufficient dark resistance, while it is easy to produce high-quality diodes which also have the advantage of a potential high speed of response [2, 13, 14]. To achieve optimum efficiency, especially in the 1-μm region where the absorption coefficient is as low as 10^{-3} cm^{-1}, the active region in which charge-carrier generation takes place has to be rather thick, i.e. 100—200 μm. If a diode is produced in a silicon chip by the usual planar technique, the $p-n$ junction itself does not extend over more than a few μm. This means that most of the charge carriers are generated in a field-free region 100—200-μm thick and have to diffuse to the junction where they accelerate, separate and contribute to the photocurrent of the diode. This is one of the reasons why silicon photodiodes have to be made from single crystals. Polycrystalline materials would never allow for such long diffusion lengths. Nevertheless, even the most simple silicon diodes have response times around 1 μs, as can be observed in typical solar cells. A reverse bias under which the diodes are operated does not significantly increase the response time because the field is concentrated in the small-junction area and the thick diffusion region remains field free. Nevertheless, response times in the nanosecond region are desired for many applications. The solution to this problem lies in the $p-i-n$ diode, in which a broad intrinsic region of 100—200-μm thickness is generated between the n and p regions by suitable doping procedures. In this case, the application of a reverse bias generates a field in this wide region and charge carriers excited therein are accelerated immediately and traverse the whole diode structure much faster than in the simple diode. Today, diodes of this configuration are commercially available which have response times of 0.1 ns for light between 550 and 1000 nm. Due to the progress made in materials technology, the diodes have very low reverse dark currents and quantum efficiencies of nearly 100%. Thus detectivities of around 2×10^{12} cm Hz$^{1/2}$ W^{-1} are achieved. This means that $p-n$ and $p-i-n$ silicon diodes almost reach the performance of ideal detectors in any background-limited application and are consequently suitable for most practical applications. Naturally they do not provide the gain of II—VI photoconductors and, thus, need an amplifier to furnish a sufficient signal current. There are, however, applications which are not background limited; for example, in the field of laser communications. The narrow bandwidth of the laser allows for the use of narrow-band optical filters which exclude

References pp. 418—420

much of the normal background radiation. Furthermore, bandwidths of 1 GHz or more are of interest. In these cases, the noise-limiting detector performance can be the dark current in the detector or the amplifier noise or the quantum noise of the signal carrier. In many cases, the preamplifier noise is the limiting factor with normal $p-i-n$ diodes. Here, a gain mechanism is desirable which amplifies signal and detector noise until the latter equals the preamplifier noise. The avalanche effect in a suitable diode is such a gain mechanism. A photogenerated carrier is multiplied by exciting secondary carriers with the energy gained from the applied field. Such a diode has to be made of a very good semiconductor material in order to avoid local avalanches (microplasmas) and achieve a predictable low-noise performance. In recent years, materials problems of this type have been solved and avalanche diodes are available which are linear over many orders of magnitude of incident light level [15]. The response time is mainly limited by the transit time of the photoexcited and multiplied carriers to the junction. This problem is similar to that encountered in normal $p-n$ photodiodes. A solution comparable to that in $p-i-n$ junctions has been found here: in the "reach-through" avalanche diode, the narrow multiplying junction is followed by a broad quasi-intrinsic region (π-region) followed again by a p^+ region which forms the back contact. The majority of the charge carriers are generated in the π-region and are accelerated towards the junction where they multiply. The multiplication time is short compared to this time. Response times around 1 ns can thus be achieved.

The maximum of the spectral sensitivity of these diodes can be shifted by an antireflective coating. It is usually positioned between 900 and 1000 nm. In this range, the quantum efficiency lies very close to 100%. Thus, these diodes are very well suited to the detection of laser pulses at 900 nm, corresponding to the emission of GaAs diodes. This fit is one of the many reasons why silicon diodes have been developed so extensively. Present technology makes it possible to manufacture single- and multi-element avalanche diodes with uniform gain over the diode area of up to 20 mm^2. Useful gains of several hundred are feasible.

(ii) Germanium diodes

Compared to silicon, germanium has a smaller energy gap (0.7 eV) and a longer wavelength edge of photosensitivity (1.6 μm), extending the sensitivity from 1.1 μm to 1.6 μm. Germanium diodes have had a development similar to that of silicon diodes [2]. $p-i-n$ diodes have been produced which show response times of less than 1 ns. Avalanche diodes have also been investigated. Large-area diodes such as in the case of silicon have not been fabricated. The performance of germanium diodes is inferior to that of silicon diodes for the basic reason that the energy gap is lower, so that the reverse saturation current in these diodes is increased. As the simple planar technology used for silicon cannot be applied to germanium, the fabrication technology is much more complicated. As the laser lines which are important for fast detection, *i.e.* between 0.9 and 1.06 μm, lie in the range of silicon diodes, interest in an additional photodiode was not so great as to justify a similar intense technological development for germanium.

10.2.4 Single crystalline materials for medium and far infrared light

(i) InSb detectors

Indium antimonide is a semiconductor with an energy gap of 0.16 eV, corresponding to a wavelength of 7.6 μm at room temperature. Thus, it is a potential candidate for an infrared detector [2]. In addition, its charge carriers, especially the electrons, possess extremely high mobilities. As it is not very difficult to grow good crystals, the potential of InSb as an infrared detector has been recognized at a relatively early stage [16]. Photoconductive, photovoltaic and photoelectromagnetic detectors have been produced commercially. These detectors operate at low or normal temperatures, depending on the application. As the energy gap is highly temperature dependent, the long-wavelength edge of sensitivity moves to shorter wavelengths at lower temperatures. At 77 K it lies at 5.9 μm. The sensitivities lie near the theoretical value and are limited by the background radiation fluctuation. At 77 K and a field of view of 90°, the detectivity is 1.8×10^{11} cm Hz$^{1/2}$ W^{-1}. At present, interest in these systems is concentrated on the manufacture of detector arrays, the trend being towards planar technology in order to produce the whole array on a single crystal chip.

(ii) Extrinsic photoconductors

It is known from experience with CdS that impurity centers in the energy gap of semiconductors give rise to optical absorption and photoconductivity beyond the intrinsic absorption edge of the base material. This indicates the way to infrared detectors which make use of germanium, one of the better-known and well-developed semiconductors; the first photodetectors for radiation around 10 μm were based on photoexcitation from impurity centers in germanium [17, 18]. In contrast to small-gap semiconductors, germanium had been available in a very pure form and the energetic position of many impurity levels had been established. Group I or II elements may be substituted for germanium atoms and give rise to two or three impurity levels whose depth increases with rising atomic weight. Of major importance are the elements gold, mercury, copper and zinc, which form impurities having ionization energies small enough to make them useful for the detection of long-wavelength radiation between 5 and 40 μm. The long wavelength cut-offs lie at 9, 14, 30 and 40 μm. These impurities can be introduced only in a relatively low concentration of about 10^{16} cm^{-3}. Therefore, their absorption coefficients are very low, typically 1 cm^{-1}, compared to 10^5 cm^{-1} for intrinsic absorption processes. Thus, large-volume detectors have to be used in

References pp. 418—420

conjunction with integrating spheres recollecting the transmitted radiation. Linear dimensions for germanium detectors of 1mm are quite common. At the same temperature, the sensitivity of these bulky detectors is considerably lower than that of comparable intrinsic detectors like the small-gap detectors developed later. Consequently, these detectors have to be operated at rather low temperatures, in the range from 4 to 30K. This limitation has seriously restricted their industrial use. Nevertheless, for about ten years they have been, and, in some cases, still are the most sensitive detectors of their kind and have found wide application in science. Their detectivities are usually greater than 10^{10} cm $Hz^{1/2}$ W^{-1}. Their response times are 1 μs and lower. Figure 10.2 gives some sensitivity data for the most important detectors.

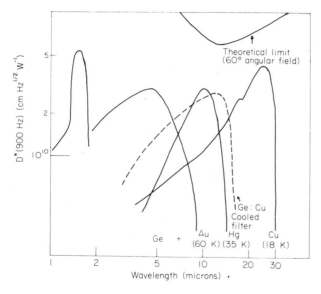

Fig. 10.2. Spectral response curves of various extrinsic germanium detectors and their operating temperature (field of view $\simeq 60°$). (From Levinstein [18].)

(*iii*) *Small-band-gap materials: $Hg_{1-x}Cd_xTe$ and $Pb_{1-x}Sn_xTe$*

At higher temperatures, intrinsic materials can show the same photosensitivity as the extrinsic photodetectors on the basis of germanium. Operation with high sensitivity is possible at liquid nitrogen temperature [19]. Therefore, extensive research has been devoted to the technological development of the two candidates offering sensitivities in the atmospheric transmission window between 8 and 12 μm. These are the mixed semiconductor systems $Hg_{1-x}Cd_xTe$ [20] and $Pb_{1-x}Sn_xTe$ [21]. By adjusting the relative composition (x) of the two constituent semiconductors, the long-wavelength sensitivity limit for $Hg_{1-x}Cd_xTe$ ranges between 4 and 30 μm and that for $Pb_{1-x}Sn_xTe$ between 0.8 and 30 μm.

Both materials presented considerable technological difficulties. $Pb_{1-x}Sn_xTe$ is less problematic for the following reasons: the liquidus—solidus curves in the phase diagram lie closer together than in the HgTe/CdTe system. The further these curves are separated, the more will small local temperature variations lead to variations in composition. The vapor pressures of the constituting elements lead, tin, and tellurium are closer together, fabrication is simpler and stability better. In the case of HgCdTe, the devices deteriorate at temperatures above 100°C because mercury diffuses out. Furthermore, thin epitaxial films of PbSnTe of good quality have recently been made, giving the promise of cheaper materials. In both cases, good crystals have been made by various crystal-growing techniques. Both photoconductive and photovoltaic detectors have been studied. Photovoltaic diodes are technically more important because they are faster and lend themselves to the application of planar fabrication technology. They also show higher sensitivities than photoconductors. Photoconductive gain has not been observed in these materials owing to short carrier lifetimes. The dark current in these materials is considerable compared with that of large-gap materials. The detectivity value for photoconductivity in $Pb_{1-x}Sn_xTe$ is 30 times lower than for diodes because it is difficult to achieve doping levels much lower than 10^{18} cm^{-3} in this class of materials. It is not such a problem in $Hg_{1-x}Cd_xTe$, but, even here, photovoltaic detectors in which an intrinsic region is created in the form of a depleted region in the p—n junction are more sensitive and desirable. Therefore, the question of doping these materials has been thoroughly investigated and a number of doping techniques has been studied. p—n junctions can be produced in both materials by the diffusion of lattice constituents (mercury in the case of $Hg_{1-x}Cd_xTe$) which produces a change in type from p to n by a change in stoichiometry. In the case of PbSnTe, diffusion of the impurity antimony also changes p-type to n-type. These techniques lead to good diodes with high photosensitivity, but their reproducibility does not seem to be satisfactory.

Consequently, alternative methods have been studied: proton bombardment of p-type material converts the surface region of both materials to n-type by the production of donor-like lattice defects. This resulted in diodes of low leakage currents and great reproducibility. Subsequently, a new method of ion implantation was employed which furnished even better results, namely doping with antimony which led to even lower leakage currents in the case of PbSnTe after annealing. This is ascribed to the decreased lattice damage caused by chemical doping as compared to proton bombardment. The production of Schottky barriers, which are the simplest diode structures, has likewise been investigated. Excellent results have been obtained in the case of $Pb_{1-x}Sn_xTe$. The metal—semiconductor contact seems to exhibit very few interfacial states to give an almost ideal Schottky barrier. Very high D^* values have been measured. The simplicity of the method makes it very promising for the manufacture of thin-film diodes and photodiode arrays.

References pp. 418—420

While the speed of response of these detectors does not play a significant role in black body radiation detection, it is of paramount importance to the demodulation of CO_2 laser light. The response times of good diodes of both materials are not limited by carrier lifetimes or transit times to or across the junction but by the dielectric relaxation time: they are RC-limited. This makes $Hg_{1-x}Cd_xTe$ more advantageous because its dielectric constant is 16 compared to a value of 400 for PbSnTe. At the present time, the cut-off frequency lies in the region of 1 GHz for both materials. Good diodes of both materials show detectivity values of 10^{10} cm Hz$^{1/2}$ W^{-1} for 10-μm radiation at 77K. Under these conditions, calculations of the theoretical limit give values of $\sim 10^{12}$ cm Hz$^{1/2}$ W^{-1}. This leaves room for further improvement. At the same sensitivity, operation at higher temperatures might be possible by using better quality materials. The $Hg_{1-x}Cd_xTe$ system might have better potential as a high speed detector because of the shorter response time. $Pb_{1-x}Sn_xTe$ has the more stable materials properties, leading to more sturdy "all-round" detectors. Because of the excellent homogeneity achieved, especially in the $Pb_{1-x}Sn_xTe$ crystals, the production of two-dimensional arrays by planar doping technique looks very promising so that, in the future, tubes may be operating in the spectral region of the atmospheric window between 8 and 12 μm.

10.2.5 Heterojunction photodiodes

The term heterojunction means a semiconductor system which contains a more or less continuous region where two different semiconductors meet and generally the conduction type changes [22, 23]. This transition region leads to favorable and less favorable properties. It is an advantage, especially where photodetection is concerned, that both semiconductors mutually compensate the surface recombination states at the interface to some extent. When two semiconductors with different energy gaps meet, and light passes through the material with the larger gap, the sensitivity of the system compared with a small-gap semiconductor is usually extended into the short-wavelength region until the large-gap semiconductor acts as an absorber. Thus, heterojunction photodiodes usually have broader spectral response than the pure small-gap material. Furthermore, p—n heterojunctions can be made from materials which individually cannot be doped p and n as can many II—VI semiconductors. Up till now, except for a very few cases, heterojunction photodiodes could not compete with the "classical" homojunction photodiodes, mainly because of problems related to materials technology. Nevertheless, realistic calculations show that if these difficulties can be overcome, heterojunctions could exhibit high detectivity values [24].

10.3 IMAGING SYSTEMS

Photoelectronic imaging is concerned with the pick-up and transmission of pictures to more-or-less distant places [25]. As these pictures are to be viewed with the human eye, they must consist of a minimum of picture points and the picture sequence for moving or changing objects also has to be at a minimum level. The picture information has to be transmitted via electrical channels, mostly single channels, so that the information in the individual image points cannot be transmitted at the same time in parallel, but must be ordered in time and transmitted successively. Typically, there are around 500 × 500 image points per picture in commercial television systems, and around 30 pictures per second have to be picked up and transmitted. Thus, the imaging system must permit the picture to be dissected thus enabling sequential read-out of the individual points, *i.e.* scanning the image. As mechanical scanning systems are of interest only in very rare cases, non-mechanical methods had to be found. Imaging systems in general consist of an array of photosensitive point elements, which are interrogated serially and periodically. There are two essentially different methods of interrogation:

(a) electron-beam scanning where an electron beam is steered over the photodetector array and thus connects them to a current source; or

(b) solid-state switching and connecting the individual detectors to a terminal point from which the video signal issues. The first principle is realized in all present television pick-up tubes and, especially where photoconductivity is concerned, in the vidicon and the vidicon-type cameras. The second principle, which is technologically much more sophisticated, has led, after intensive research to some very promising solid-state imaging systems. These two systems in their most important embodiments are discussed in the following sections.

10.3.1 Imaging tubes with electron-beam readout

(*i*) *Vidicons*

The electron beam is the most simple scanning device which has been considered for imaging systems because it can be scanned virtually lag-less, continuously or by random access across a flat or curved surface which is to be interrogated. Beams with diameters of less than 40 μm can be produced. One of the few principal disadvantages is its limitation in current density: values of more than $0.1 \, A/cm^2$ are difficult to achieve, as the beam is defocussed by its own electrostatic forces.

The first practical imaging device using the combination of electron beam and photoconducting target has been called the vidicon [26]. Figure 10.3 illustrates its operation: the faceplate of a suitable vacuum tube containing the electron-beam generating system carries a photoconducting film on top of a transparent conductor. The transparent conductor serves as the signal electrode leading to the video amplifier. The electron beam scans the

References pp. 418—420

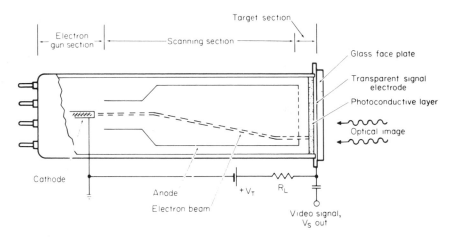

Fig. 10.3. Schematic drawing of a vidicon imaging tube, showing the most important elements of the device. (From Sehr and Zuleeg [25].)

photoconductor, the "target", linewise and charges the surface of the photoconductor up to a certain potential, e.g. 10 V. When the frame time, i.e. the time until the beam returns to an individual image point, is 1/30 s, as is typical of television systems, the photoconductor has to keep the charge without illumination practically undiminished for this time. When light falls onto this image point, the ensuing photocurrent reduces the deposited charge and the electron beam has to recharge this point. This recharge leads to a charging current in the signal line which is the video signal belonging to the point being charged. Suitable synchronization and synchronization signals allow for the coordination between the information in the line and the image point and thus the reconstitution of the image. First of all the photoconductor is required to retain the charge in the dark during a frame period. A dark resistivity of more than 10^{12} Ω cm follows from the condition $\rho\kappa\kappa_0 \leqslant 1/30$ s. It proved to be very difficult to find a photoconductor satisfying this condition with the additional requirement that the response time of photoconductivity should be less than 1/30 s. This second requirement is important, if the reconstituted picture is not to exhibit the effect of a time lag. Here the region of low illumination is most critical, as, generally, the response time of photoconductors rises with decreasing illumination. Naturally, the photoconductor should be as sensitive as possible to provide a maximum signal. In 1950, a photoconductor was found which met these conditions reasonably well. It consisted of a polycrystalline Sb_2S_3 film which had a sufficiently high dark resistance. The dark current is low in these films because of space-charge limitations due to traps. As this condition breaks down at about 15 V, the target cannot be charged to higher potentials. The most undesirable property of the vidicon is the long response time at low illumination. This limits its use as a camera tube under low illumination conditions. Thus, the vidicon did not entirely replace the bulky image orthicon.

An intensive search has been made to find more suitable photoconductors including the II—VI compounds and especially CdSe. Pure CdSe films with high resistivity have been made by improving the stoichiometry of the film, and the resulting vidicons showed improved sensitivity [27]. However, their instability apparently gave rise to problems; therefore the CdSe layer was reinforced by an As_2S_3 layer as an electron landing site [28]. But new complications arose which again appear to be detrimental to the absolute sensitivity. A final evaluation of these systems is not yet possible.

(ii) Plumbicons

Similar to the improvements observed in photosensors when changing from pure photoconductive systems to photodiode systems, the plumbicon shows improved performance over the photoconductive vidicon [29]. The plumbicon is a p—i—n structure of large area having an n-type layer formed adjacent to the transparent SnO_2 signal electrode, followed by an intrinsic layer which represents the major part of the layer. The electron beam impinges on the very thin p-type layer. This layer has to be very thin in order to avoid lateral conduction and smearing of the picture. As the central intrinsic region is free from shallow traps, the response time of the layer is short so that the human eye does not detect any image lag. It is actually so short that the plumbicon is also used for high-speed motion analysis. The plumbicon exhibits a spectral sensitivity similar to that of the human eye. It can be varied to a certain degree by changing the thickness of the layer. The red-sensitivity decreases with thinner layers. By addition of PbS the red-sensitivity can be increased. The vacuum evaporation process, by which the PbO layer is deposited on the front glass of the tube, leads to very homogeneous and stable layers. The photocurrent I_{pc} follows the relation $I_{pc} = CF^\gamma$, where C is a constant and F the luminous flux with a fixed γ over a large interval of F, not only for white light but also for the three wavelength bands used in color TV. This makes the plumbicon very well suited for the triple-tube color television cameras.

(iii) Silicon diode vidicon

The next logical step in vidicon development was an attempt to use silicon as the photosensitive material. As silicon photodiodes have very high resistance in the reverse direction, the basic idea was to use an array of individual photodiodes integrated in a single silicon wafer [30]. Similar to the previously-described targets, the silicon diodes are charged by the electron beam and discharge during the frame time by photoconduction and thereby integrate the photocurrent over this time. The starting material is n-type and the diodes are produced by photolithographic techniques and p-diffusion. As the manufacture of satisfactory silicon diodes does not present any difficulties,

References pp. 418—420

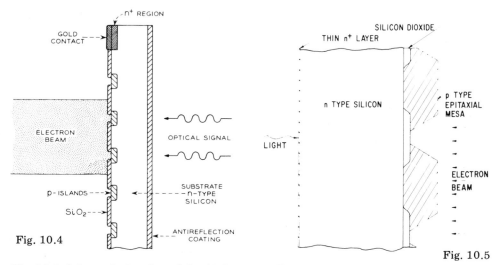

Fig. 10.4. Schematic drawing of the diode array. (From Crowell et al. [30].)

Fig. 10.5. Cross-section of an epitaxial diode array target. The diagram is drawn approximately to scale for 15-μm centers. The n^+ layer is normally 0.1—0.2-μm thick. (From Blumenfeld et al. [31].)

the problems which arose during the development of these devices were mainly connected with the very large number of diodes which have to be produced: half a million diodes with center-to-center spacing of 20 μm have to be produced on an area of 12.5 mm × 12.5 mm. Not a single defective diode is tolerated by the TV public, as this shows as a white or black point on the screen. The silicon target is fabricated by diffusion-doping the diodes of the array and subsequently depositing a silicon dioxide layer between the diodes. This insulates the undiffused n-type base material from the electron beam and reduces surface leakage. A schematic view of the target is illustrated in Fig. 10.4. However, the charge deposited onto the SiO_2 layer leaks away slowly, thus distorting the charge pattern and field and adversely affecting the diode recharge. To avoid this, a thin resistive layer 1000 Å in thickness and of 5×10^{13} ohm/square surface resistivity is deposited onto the entire array. This film dissipates the charges deposited onto the SiO_2 and improves the properties. The deposition of this film nevertheless is rendered difficult because of the necessary close tolerances in film properties. Somewhat later, a different remedy against charge-up of the SiO_2 was found: the n-type wafer is covered with an SiO_2 film, and holes for the diodes are etched through the oxide layer. Following this step, p-type silicon is grown epitaxially onto the n-Si openings in the oxide. Growth is continued until the epitaxial silicon grows onto the SiO_2 and nearly fills the SiO_2 portions between diodes, leaving only very narrow strips of uncovered oxide which do not disturb operation [31]. A schematic view of this target is shown in Fig. 10.5.

The silicon vidicon has a number of desirable properties: the response is extremely fast, the spectral sensitivity is so broad that it covers the visible range and the near infrared region up to 1.1 μm, its sensitivity is high and — what is important compared with earlier vidicons — it does not show burn-in under intense illumination, e.g. when the camera is accidentally directed towards the sun or another intense light source.

(iv) Infrared-sensitive vidicons

In principle, a vidicon-like imaging tube, which is sensitive in the infrared region, can be made using a small-gap semiconductor as the target material. However, we have to observe the general condition for the dark conductivity in vidicons which requires that the dark resistivity must exceed $10^{11}\,\Omega\,\text{cm}$. Since resistivities of this magnitude cannot be obtained in semiconductors with band gaps below 1.0 eV at room temperature, it follows that a wavelength response beyond 1.24 μm is not possible without cooling the photoconductor. A number of developments have been reported for the near-infrared region up to 4-μm wavelength: starting from the know-how concerning the plumbicon PbO layers, mixed lead oxysulfide layers have been produced which, due to the smaller band gap of PbS, are sensitive to 2.5 μm at 77K [32]. A slight lag, in the form of an after-image following high illumination, has been observed. Analogous to the silicon-diode array target a germanium target has been produced [32]. Since the band gap of germanium is 0.7 eV, the maximum sensitivity is obtained at 1.5 μm. Furthermore, the developing technology for InAs allowed for the construction of an InAs-diode array target which can be operated at the normal TV frame time of 1/30 s at 77K. The substrate is *n*-type: *p*-diodes are produced by suitable masking and diffusion techniques [33]. The maximum sensitivity lies at 3.25 μm.

For the detection of light of longer wavelength, representing the thermal radiation of the bodies to be viewed, a fundamental problem arises; the background radiation, characterized by the number of incident quanta, increases considerably, leading to a high discharge current. On the other hand, because of the electron gun design, the signal current is limited to 1 μA. Thus, the higher the background radiation, the nearer is the tube to background saturation, and contrast decreases [34]. With the aid of a flood electron gun, which floods the target with an unfocused beam and compensates for part of the background-induced current, this limit can be pushed further into the infrared region. Nevertheless, non-quantum detectors such as bolometric or pyrometric targets will finally supersede the photoconductive films or diode arrays in tubes scanned by electron beams. Also, infrared detectors over which the image is scanned by mechanical means and which are cooled to much lower temperatures are becoming of interest for long-wavelength radiation, despite the fact that the charge integration property cannot be utilized.

References pp. 418—420

10.3.2 Self-scanned solid-state systems

Whereas, today, imaging tubes are still the only image pick-up systems used commercially, there is a clear trend towards pure solid-state imaging devices. Means of replacing the easily steerable electron beam contacting the individual sensor elements — the silicon diodes in the diode vidicon — by an essentially two-dimensional solid contacting system had to be found. There are two different lines of development: in the first, a system of crossed conductors is employed, where contact is achieved by simultaneously connecting two lines of the array to the output and thus making contact and interrogating the element at the crossing point of the two conductor lines. Here, the sensor array is not self-scanned, but scanned by an auxiliary system which can be situated at the periphery on the same chip. The second system actually scans itself, insofar as the image information is not interrogated for each element individually from the outside, but transferred from image element to image element to the periphery of the array. These principles, called x—y-addressed systems and charge-transfer systems are discussed in the following two sections.

(i) x—y-addressed solid-state imaging systems

The most simple photosensor which has been investigated for solid-state imaging is the photoconductor. The contacting lines are a system of crossed metallic conductors. A photoconductor is positioned at each intersection and connected in series with a rectifying diode, as is shown schematically in Fig. 10.6. The diodes are required in order to avoid "cross-talk". This means that a non-illuminated photoconductor between two conductors n and m may be short circuited by a series of illuminated photoconductors which also connect these two lines by a detour. If each photoconductor has a diode in series, at least one reverse biased diode is found at this detour, preventing cross-talk. Interrogation of the array is effected by peripheral shift registers or similar switches.

As can easily be seen, in general there is no light integrating mechanism such as in the electron-beam addressed tubes. At the moment of interrogation, only the photoconductivity due to the light incident at this moment is measured. In the most well-known system of this type, CdS is used as a photoconductor [35]. In this case, by correct sensitization, the lifetime of photoexcited carriers may reach quite high values, as indicated in Section 10.2.2. Therefore, at the moment of interrogation, photoexcited charge carriers which have been generated one lifetime ago are still present. This is an effective mode of light integration called "excitation integration". This system has been implemented using thin-film technology with up to 256×256 elements [35]. Moreover, the interrogating peripheral system has been improved by means of thin-film transistors. Operation as an imaging system was demonstrated in 1966. As in the case of photosensors and electron tube

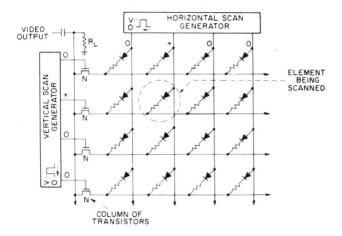

Fig. 10.6. Equivalent circuit for the completely integrated 180 × 180 element array, showing method for attaching scan generators and coupling out the video signal. (From Weimer et al. [35].)

imaging, when the first successes with these new principles became known, the technology of silicon diodes was set into motion. A good silicon $p-n$ diode has a resistivity under reverse bias which allows for charge integration if (and only if) it can be charged and then insulated during the frame time. This is the same principle as that used in the silicon-diode vidicon. Solid-state switches giving very high resistances in the switched-off state became available with the emergence of MOS technology. By this method, it is possible to charge a silicon diode with a gate pulse over the source-drain channel of the MOS transistor, and by removing the gate potential, or reversing it, actually insulate it. After the frame time, the diode can be recharged, the recharge current giving the picture-point information.

Linear switched silicon-diode arrays were realized first [36]. A schematic view illustrating the principle of operation is given in Fig. 10.7. The silicon diode is shown as an ideal diode in parallel with the capacitance of the diode. One switching MOS transistor is needed for each photodiode. The switching electronics may be integrated in the same silicon chip, leading to a very compact system. A two-dimensional array is more complex [37]: two switching MOS transistors are needed for each image point, one as an x-addressing switch, the other as a y-addressing switch (Fig. 10.8). Technical complications arise because of the numerous crossovers of the x- and y-line conductors. Linear diode arrays with up to 1000 picture elements and two-dimensional picture arrays with up to 100 × 100 diodes are commercially available [38]. Scanning frequencies of 1 MHz have been achieved for linear arrays, whereas two-dimensional arrays only furnished 100 kHz. With further development, these arrays can be expected to reach the resolution of commercial television within a few years.

References pp. 418—420

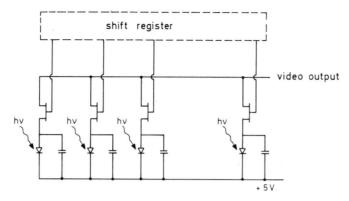

Fig. 10.7. Equivalent circuit for a self-scanned linear photodiode array.

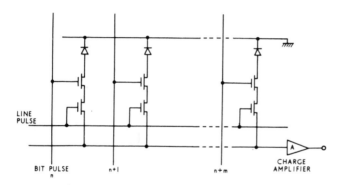

Fig. 10.8. Equivalent circuit of a section of a two-dimensional photodiode array. (From Noble [37].)

Because of their inherent current gain, phototransistors are more light-sensitive than diodes. Therefore, research has been conducted on both transistor arrays and diode arrays [39]. Arrays comprising about 100 × 100 phototransistors have been fabricated. This line of development has been abandoned because MOS technology combined with the charge-storage mode in diodes permits easier fabrication and higher production yield.

(ii) *Charge-transfer imaging systems*

The principle of operation of this class of devices is very simple [40—44]. By proper biasing relative to the semiconductor substrate, a linear array of insulated electrodes on the free surface of a semiconductor chip, *e.g.* silicon, can create inversion regions or potential wells in the semiconductor. Charge carriers present in such a well, *e.g.* a photogenerated carrier, may be spilled from one well into the next by properly biasing the different wells relative

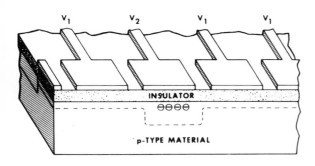

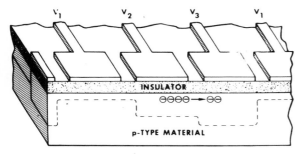

Fig. 10.9. A section of the basic three-phase CCD element: (a) storing charge; (b) transferring charge. The indicated potentials applied to the electrodes are $V_3 \geqslant V_2 > V_1$. (From Tompsett [41].)

to each other. This is illustrated in Fig. 10.9. The electrodes are coupled electrically in three sets which can be biased externally. The upper part of the figure shows electrons present in the well below the electrode with the potential V_2. This charge may be moved to the right by differently biasing the three sets of electrodes and creating an asymmetry, thus preventing the charge carriers from moving to the left, and inducing them to move to the lower energy level to the right. After this step, the well, which the electrons are now in, is again raised into the position of the well in the upper picture, with the difference that the charge is now one period further to the right. When a well of this type is created, it is depleted of electrons. In the dark, it fills in a time which is long compared to the times considered here for imaging (*i.e.* the frame time), by thermally-generated carriers. If the semiconductor is illuminated, minority carriers are created which flow into the nearest well. They accumulate until they are moved on, thus facilitating a charge integration

References pp. 418—420

mechanism essential for imaging devices. A linear imaging array has accurately spaced wells, existing during the period of the frame time, which collect photogenerated carriers. By activating the charge-transfer mechanism, the carriers are moved quickly to the end of the array, where they are transferred into the output amplifier. This simplest three-line surface-charge-coupled array has some disadvantages: as the charges move directly below the insulating oxide in a relatively perturbed lattice, their mobility is reduced and they tend to be caught by interface traps. This leads to slow operation and relatively high transfer inefficiency. This latter criterion is very important. As many transfers are necessary for arrays of reasonable resolution, the inefficiency has to be as low as possible.

This has led to the development of the buried channel device where the substrate (p-Si) is first doped in a shallow n-type region and then covered with oxide and the metal electrodes. In this way, the potential wells are generated with their minimum in the less disordered n-Si away from the silicon oxide interface so that efficiency and speed are improved considerably. The gaps between the metal electrodes have to be as small as possible, typically around 3 μm, to ensure good overlapping of the wells during transfer. These small gaps can nevertheless be electrostatically charged and thus deform the potential wells underneath and transfer irregularities may result. This can be avoided by another structure which actually simplifies the mechanism: the two-phase system. If the electrodes generating the wells are asymmetrical, the three-phase-system operation can be reduced to a two-phase system. This can be achieved by a stepped oxide structure in which the electrodes overlap, separated by an insulating layer avoiding the influence of static charges on the profile of the potential wells. An alternative method is to produce an asymmetric profile in the n-Si layer. These approaches have furnished excellent results; transfer inefficiencies lie well below 10^{-4}. When these arrays are used as photosensor arrays, two questions arise; that of illumination and that of image smearing during transfer. The surface of the arrays is covered with the electrode material, usually aluminum layers, which are non-transparent to light. These can, at least in part, be replaced by deposited doped polycrystalline silicon which is 80% transparent to light. The other alternative is illumination from below. To this end, the silicon chip has to be reduced to a thickness of 25—100 μm. This requires some skill, considering that the diameter of a large array is about 2 cm. In the first case, the response time is increased due to higher resistance of the polycrystalline silicon, whereas the second solution involves technological difficulties. The second question, that of smearing, may be explained as follows: if, for example, one charge-transfer line is interrogated, the information on the illumination at one particular position is carried past different positions and is disturbed there by new inflowing charge due to the illumination at this point. This leads to severe smearing of the picture. This undesired effect can be alleviated by adding a second charge-transfer line which is not illuminated. The first line is

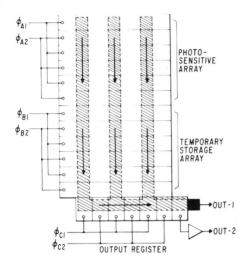

Fig. 10.10. Frame transfer charge-coupled imaging array with a temporary storage array. (From Séquin et al. [45].)

illuminated during the frame time allowing for charge accumulation in the individual wells. Subsequently, this charge is transferred very quickly into the second array of wells, where it is stored and transferred quite slowly toward the output. Meanwhile, the next integration step takes place in the first line. When the transfer from line to line is fast enough, smearing is negligible. This principle has been successfully applied in two-dimensional arrays [45]: this is shown in Fig. 10.10. Two arrays can be seen: the photosensitive integration array and the storage array. The transfer between arrays and within the individual arrays is effected vertically. Imaging arrays of this type with 106×124 positions are commercially available and provide good pictures with frame times of 1/30 s.

One problem still has to be mentioned here. In the case of excessive illumination of one image point, the potential well may literally flow over, spilling charge into the neighboring wells. This "blooming" effect may be avoided by the incorporation of overflow drains in the form of strongly-doped lines in the arrays transporting the excess charge to ground [46].

The fabrication requirements are very stringent for this type of imaging system, as one defect element may lead to the dropout of a whole line. These charge-transfer imaging systems should be an order of magnitude more sensitive than the silicon-diode vidicon because the capacitance of its output stage can be two orders of magnitude lower. Compared to diode arrays, there is considerably less synchronous noise generated on the chip as the read-out is fundamentally different. Synchronous noise on the output of the diode array is caused by capacitive feed-through from clock and address pulses.

References pp. 418—420

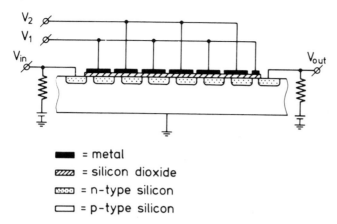

Fig. 10.11. Cross-section of an integrated MOS bucket brigade. (From Collet and Esser [42].)

Another different embodiment of the charge-transfer principle is the "bucket brigade" system. The wells are generated by diffusion and the transfer is effected by a two-phase conductor system (Fig. 10.11). These systems have been of less interest with regard to imaging because the transfer efficiency has usually been lower. Reference is made to the literature for a more detailed discussion [41—44]. It is generally expected that highly-sensitive imaging devices with the full resolution of commercial television will be available in the course of a few years. This will mean that very small, sturdy image pick-up cameras will be available, combining long life with low power consumption.

10.4 OPTICAL MEMORIES AND IMAGE STORAGE

Recently, the gap between the memory systems needed and existing memories has increased considerably [42]. This is the reason for an enhanced development effort into optical storage systems which are expected to provide for increased capacity at decreased access time. Information capacities of more than 10^{10} bits with access times of much less than a second, preferably in the millisecond range, are being discussed. An optical memory of this type is a system which stores two-dimensional optical information, like a picture which is erasable, allowing for more-or-less complete regeneration or change of the information content. This last requirement illuminates the difference between optical memories and imaging systems such as photographic processes which produce permanent images. Photographic devices and systems are excluded in the following discussion.

Most optical memories are written-in optically and, after write-in, may be

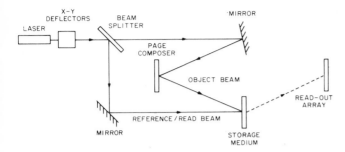

Fig. 10.12. Generalized arrangement for a read/write-*in-situ*/erase holographic optical memory. (From Anderson [50].)

interrogated non-destructively by optical means [48, 49]. The most simple concept of writing-in the time-ordered information comprises a deflected modulated laser beam which writes the information bit-wise into the storage medium. However, the vast amount of information to be written would require a light deflector of a complexity not feasible at present. Therefore, the writing step is preceded by the composition of a bundle of information in the form of a page composer (*cf.* Fig. 10.12). In this page composer a non-permanent transparency generated, for example by matrix addressing, is projected into the actual memory by a light deflector in one step in the form of a hologram. If, for example, the page composer has 1000×1000 positions which can be addressed by shift registers of present design and if a light deflector with 100×100 positions is available (which will probably be the case in the near future), 10^{10} bits can be written. The advantage of the holographic storage lies in the fact that minor defects in the storage medium do not affect the stored information. The hologram is reconstructed on the sensitive surface of a photodiode-, or charge-coupled, imaging array. Most of the mass-storage mediums discussed at present have photoconductor films as the light sensitive element [47–49]. In most cases, the photoconductor is connected in series with a recording system which is sensitive to electric fields above a critical value. The field which is induced in the field-sensitive medium is too low to have an effect when the photoconductor is not illuminated, but high enough when the photoconductor is illuminated. This effect should be irreversible and permanent without an electric field under read-out illumination, but reversible in an electric field with some sort of flood-illumination.

The principle of operation of a promising system of this type is the ferroelectric photoconductor device (FEPC) illustrated in Fig. 10.13. A thin ferroelectric layer, together with a photoconductor layer is sandwiched between two transparent conducting electrodes (*e.g.* lead—tin-oxide films). The crystallites of the FE are pre-poled by high field and flood illumination. By the application of a reverse bias, the crystallites are switched under the action of an increasing field at the illuminated sites to a less-polarized state.

References pp. 418—420

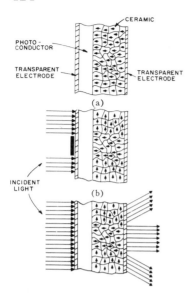

Fig. 10.13. Photoconductor controlled ferroelectric ceramic light valve (CERAMPIC). States illustrated are (a) the pre-poled ceramic; (b) the image storage process in which some domains are switched to a lower polarization remanence; and (c) viewing the image with light transmitted through the CERAMPIC. Light is scattered by the switched domains. (From Flannery [49].)

After removing the field, the depolarized parts of the layer scatter light and appear dark under suitable illumination. Contrast ratios of 100:1 with resolution capabilities of 40 lines/mm in a 0.25-mm-thick FE-plate have been reported [49]. Other systems use liquid crystals or thermoplastic layers instead of the FE electro-optical layers. The former need additional heat for writing-in [47–49]. The electro-optical system has much less contrast, the liquid-crystal system is rather slow. This also applies to the thermoplastic system.

The ease of preparation of organic films on different substrates has led to their use as photoconductors on an experimental scale in order to show the feasibility of some systems [52]. However, they suffer from low photo-conductivity because of their poor carrier mobility. The inorganic photo-conductors discussed in Section 10.2.2 are more promising here [53]. If the powerful argon laser with its main emission line at 514 nm for write-in is used, CdS would be the most suitable material because its maximum sensitivity lies in this region. Because of the required speed, some mixed CdS–CdSe alloy system could be of greater interest. Contrary to earlier assumptions, considerable gain can be achieved in II–VI films under conditions of non-ohmic contacts as those prevailing here [51]. Thus, the full power of II–VI films can be used to obtain fast and sensitive systems. Experimentally, gains of nearly 30 have been measured in pure CdSe polycrystalline films with response times in the region of 10 μs.

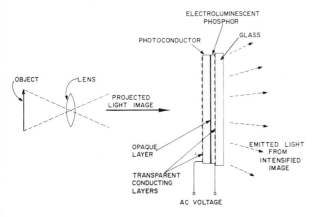

Fig. 10.14. Sandwich-type two-layer intensifier. (From Nicoll [54].)

10.5 IMAGE INTENSIFIERS AND CONVERTERS

The structure of such systems closely resembles that of the image-storage systems discussed in the preceding section, *cf.* Fig. 10.14 [54, 55]. A two-dimensional photoconductor is connected in series with a two-dimensional electroluminescent layer. Light which is incident on the photoconductor induces photoconductivity and a photocurrent perpendicular to the layer in turn produces luminescence in the second layer. If both layers are sufficiently thin, the resolution is not impaired by lateral conductivity and a flat system results. If the photoconductor is sensitive to visible light, the system is a light intensifier, in the case of infrared-sensitivity it is called an image converter. In the first case, a net amplification is expected. As in good photoconducting films, gain-values of around 100 may be expected, a luminescence efficiency of more than 1% is necessary for an amplification effect. As efficiencies of much more than 1% have not yet been attained in thin-film electroluminescence [56], these devices have lost much of their previous attraction. This is even more the case as their vacuum-tube rivals, especially the channel image intensifiers, have reached luminance gains of up to 10^5 [57]. The situation is quite similar for the image converters. They need infrared-sensitive photoconductors with gains which have not been attained for wavelengths above 1μm. Therefore, here too, the hope of developing systems of satisfactory characteristics has decreased in favor of vacuum-tube systems.

10.6 SOLAR CELLS

A solar cell is essentially a large-area photodiode which, in contrast to the detector diodes, is not operated with reverse bias voltage, but in the photovoltaic mode, where the diode may feed electrical energy into an external

References pp. 418—420

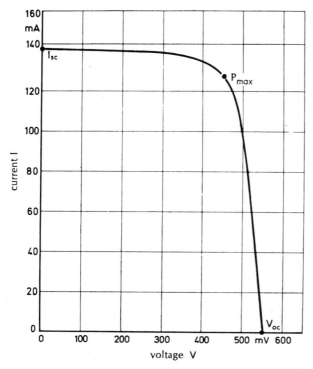

Fig. 10.15. Current—voltage curve of a typical 2 × 2 cm² silicon solar cell. I_{sc} and V_{oc} refer to the short-circuit current and open-circuit voltage, respectively.

circuit under illumination. They further differ from photodiodes in that speed is not important and adaptation is not for monochromatic or narrow-band radiation, but to the given broad spectrum of solar radiation [58]. Figure 10.15 gives the most important part of the current—voltage curve of a typical silicon solar cell. For a given irradiation, there is a point where the extractable power is at a maximum. This maximum power divided by the radiation power incident on the cell represents the efficiency, the most important characteristic of a solar cell. Usually, efficiencies are given for light of the intensity and spectral distribution of the solar irradiation above the atmosphere. As in photodiodes, the spectral range of sensitivity is determined by the energy gap of the semiconductor used. Light quanta of an energy lower than the energy gap are not absorbed. The higher the energy gap of the semiconductor, the higher is the open-circuit voltage, but the lower is the number of quanta utilized and, thus, the short-circuit current. The smaller the energy gap, the more quanta of the spectrum are absorbed and, hence, the higher is the short-circuit current, but the smaller the

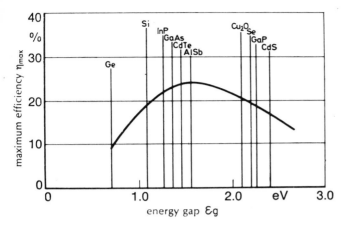

Fig. 10.16. Theoretical values of the maximum efficiency of a solar cell as a function of the energy gap of the semiconductor for an ideal p—n junction. (From Loferski [59]).

open-circuit voltage. From this argument, which can be made quantitatively [59], it appears that for a given spectrum there is a value of the energy gap for which the theoretical efficiency has a maximum. This function is shown in Fig. 10.16. The historical solar cell development did not start from this theoretical basis, but followed the more practical approach using available semiconductors. Thus, silicon was investigated first as a material for solar cells. Silicon of high purity was available in the form of single crystals soon after 1950, and in 1954 it was possible to make solar cells with efficiencies of more than 5%. Silicon cells with an efficiency of 10% were produced in 1962. The efficiency has not increased much since that time, but the resistance of the cells to particle radiation in space has been greatly improved. Standard cells have surface areas of $4\,cm^2$, a thickness of about 300 μm, and generate 60 mW of electrical power in space. Today, scientists are confident that, after further technological development it will be possible to produce industrial cells with efficiencies of around 15%.

Parallel to this development, a considerable amount of work was invested in the development of solar cells using different semiconductors. The more important systems were the GaAs system for high-efficiency single crystal cells and the II—VI compounds for large-area thin-film cells (mainly CdS, CdTe and mixtures or heterojunctions using, say, Cu_2S). Nevertheless, the silicon cell has remained the only cell which has been applied on a large scale in space. These other systems should have some consequences where energy conversion for terrestrial application is concerned. Here, new ways for the manufacture of new cheap systems have to be found.

References pp. 418—420

REFERENCES

1 A. Rose, The relative performance of human vision, electronic vision and photographic film, Photogr. Sci. Eng., 18 (1974) 589.
2 H. Levinstein, Characterization of infrared detectors, in E.K. Willardson and A.C. Beer (Eds.), Semiconductors and Semimetals, Vol. 5, Academic Press, New York, London, 1970, p. 3.
3 T.S. Moss, G.I. Burrell and B. Ellis, Semiconductor Optoelectronics, Butterworths, London, 1973.
4 P.W. Kruse, L.D. McGlauchlin and R.B. McQuistan, Elements of Infrared Technology, Wiley, New York, 1963.
5 R.H. Bube, Comparison of solid-state photoelectronic radiation detectors, Trans. Metall. Soc. AIME, 239 (1967) 291.
6 R.H. Bube, Photoconductivity of Solids, Wiley, New York, 1960.
7 F.V. Shallcross, Methods of deposition and physical properties of polycrystalline II—VI films, RCA Rev., 28 (1967) 569.
8 A. Vecht, Methods of activating and recrystallizing thin films of II—VI compounds, in G. Hass and R.E. Thun (Eds.), Physics of Thin Films, Vol. 3, Academic Press, New York, London, 1966, p. 165.
9 K.W. Boer, A.S. Esbitt and W.M. Kaufmann, Evaporated and recrystallized CdS-layers, J. Appl. Phys., 37 (1960) 2664.
10 D.B. Fraser and H. Melchior, Sputter-deposited films with high photoconductivity through film thickness, J. Appl. Phys., 43 (1972) 3120.
11 D. Beecham, Cadmium/sulfur isothermal source for CdS deposition, Rev. Sci. Instrum., 41 (1970) 1654.
12 D.E. Bode, Lead salt detectors, in G. Hass and R.E. Thun (Eds.), Physics of Thin Films, Vol. 3, Academic Press, New York, London, 1966, p. 275.
13 P.H. Wendland, Silicon photodiodes come into their own, Opt. Spectra, 7 (Oct. 1973) p. 33.
14 H. Melchior, Sensitive high-speed photodetectors for the demodulation of visible and near infrared light, J. Lumin., 7 (1973) 390.
15 P.P. Webb, R.J. McIntyre and J. Conradi, Properties of avalanche photodiodes, RCA Rev., 35 (1974) 234.
16 E.H. Putley, Solid state devices for infrared detection, J. Sci. Instrum., 43 (1966) 857.
17 E.H. Putley, Modern infrared detectors, Phys. Technol., 4 (1973) 202.
18 H. Levinstein, Extrinsic detectors, Appl. Opt., 4 (1965) 639.
19 I. Melngailis, Small bandgap semiconductor infrared detectors, J. Lumin., 7 (1973) 501.
20 D. Long and J.L. Schmit, Mercury—cadmium telluride and closely related alloys, in E.K. Willardson and A.C. Beer (Eds.), Semiconductors and Semimetals, Vol. 5, Academic Press, New York, London, 1970.
21 I. Melngailis and T.C. Harmann, Single-crystal lead—tin chalcogenides, in E.K. Willardson and A.C. Beer (Eds.), Semiconductors and Semimetals, Vol. 5, Academic Press, New York, London, 1970.
22 B.L. Sharma and R.K. Purohit, Semiconductor Heterojunctions, Pergamon, Oxford, 1974, p. 135.
23 A.G. Milnes and D.L. Feucht, Heterojunctions and Metal-Semiconductor Junctions, Academic Press, New York, London, 1972, p. 118.
24 B.L. Sharma, S.N. Mukerjee, J.K. Modi, Detectivity calculations for n—p heterojunction detectors, Infrared Phys., 11 (1971) 207.
25 R. Sehr and R. Zuleeg, Imaging and display, in E.K. Willardson and A.C. Beer (Eds.), Semiconductors and Semimetals, Vol. 5, Academic Press, New York, London, 1970, p. 467.

26 P.K. Weimer, S.V. Forgue and R. Goodrich, The vidicon photoconductive camera tube, Electronics, 23 (May 1950) 70.
27 K. Shimizu and Y. Kiuchi, Characteristics of the new vidicon-type camera tube using CdSe as a target, Jpn. J. Appl. Phys., 6 (1967) 1089.
28 K. Shimizu, O. Yoshida, S. Aihara and Y. Kiuchi, Characteristics of experimental CdSe vidicons, IEEE Trans. Electron Devices, ED—18 (1971) 1058.
29 E.F. de Haan, A.v.d. Drift and P.D.M. Schampers, The "Plumbicon", Philips Tech. Rev., 25 (1964) 133.
30 M.H. Crowell, T.M. Buch, E.F. Labuda, J.V. Dalton and E.J. Walsh, A camera tube with a silicon diode array target, Bell. Syst. Tech. J., 46 (1967) 491.
31 S.M. Blumenfeld, G.W. Ellis, R.W. Redington and R.H. Wilson, The epicon camera tube: An epitaxial diode array vidicon, IEEE Trans. Electron Devices, ED—18 (1971) 1036.
32 M. Berth and J.J. Brissot, Targets for infrared television camera tubes, Philips Tech. Rev., 30 (1969) 270.
33 C.W. Kim and W.E. Davern, InAs charge-storage, photodiode infrared vidicon targets, IEEE Trans. Electron Devices, ED—18 (1971) 1062.
34 J.A. Hall, Problem of infrared television camera tubes *vs.* infrared scanners, Appl. Opt., 10 (1971) 838.
35 P.K. Weimer, G. Sadasiv, J.E. Meyer, Jr., L. Meray-Horvath and W.S. Pike, A self scanned solid state image sensor, Proc. IEEE, 55 (1967) 159.
36 O.P. Weckler, Operation of p—n junction photodetectors in a photon flux integrating mode, IEEE J. Solid-State Circuits, SC—2 (1967) 65.
37 P.J.W. Noble, Self scanned silicon detector arrays, IEEE Trans. Electron Devices, ED—15 (1968) 202.
38 R.H. Dyck and G.P. Weckler, A new self-scanned photodiode array, Solid State Technol., 14 (July 1971) 37.
39 E. Arnold, M.H. Crowell, R.D. Geyer and D.P. Mather, Video signals and switching transients in capacitor phototransistor image sensors, IEEE Trans. Electron Devices, ED—18 (1971) 1003.
40 W.S. Boyle and G.E. Smith, Charge coupled semiconductor devices, Bell Syst. Tech. J., 49 (1970) 587.
41 M.F. Tompsett, Charge transfer devices, J. Vac. Sci. Technol., 9 (1972) 1166.
42 M.G. Collet and L.J.M. Esser, Charge transfer devices. In Advances in Solid State Physics (Festkörperprobleme), Vol. XIII, Vieweg, Braunschweig, 1973, p. 337.
43 J.E. Carnes, Charge-coupled imaging state of the art. In H. Weiss (Ed.), Solid State Devices, 1973, Institute of Physics, London and Bristol, 1974, p. 83.
44 J.E. Carnes and W.F. Kosonocky, Charge-coupled devices and applications, Solid State Technol., 17 (April 1974) 67.
45 C.H. Séquin, D.A. Sealer, W.J. Bertram, M.F. Tompsett, R.R. Buckley, T.A. Shankof and W.J. McNamara, A charge coupled area image sensor and frame storage, IEEE Trans. Electron Devices, ED—20 (1973) 244.
46 W.F. Kosonocky, J.E. Carnes, M.G. Kovac, P. Levine, F.V. Shallcross and R.L. Rodgers, Control of blooming in charge coupled imagers, RCA Rev., 35 (1974) 3.
47 Th.K. Gaylord, Optical memories, Opt. Spectra, June 1974, 29.
48 J. Bordogna, S.A. Keneman and J.J. Amode, Recyclable holographic storage media, RCA Rev., 33 (1972) 227.
49 J.B. Flannery, Jr., Light-controlled light valves, IEEE Trans. Electron Devices, ED—20 (1973) 941.
50 L.K. Anderson, Ferroelectrics in optical memories and displays: A critical appraisal, Ferroelectrics, 3 (1972) 69.
51 B.S. Sharma and R.R. Mehta, Photoconductor memory device, Ferroelectrics, 3 (1972) 225.

52 W.C. Stewart, R.S. Mezrich, L.S. Cosentino, E.M. Nagle, F.S. Wendt and R.D. Lohman, An experimental read-write holographic memory, RCA Rev., 34 (1973) 31.
53 D.B. Frazer, Sputtered films for display devices, Proc. IEEE, 61 (1973) 1013.
54 F.H. Nicoll, Solid state image intensifiers, in S. Larach (Ed.), Photoelectronic Materials and Devices, van Nostrand, Princeton, 1965, p. 313.
55 Z. Szepesi, Thin film PC—EL sandwich type image intensifiers, Thin Solid Films, 13 (1972) 397.
56 E.E. Loebner, The futures of electroluminescent solids in display applications, Proc. IEEE, 61 (1973) 838.
57 D.L. Emberson and R.T. Holmsharo, Some aspects of the design and performance of a small high-contrast channel image intensifier, in L. Maron (Ed.), Advances in Electronics and Electron Physics, Vol. 33 A, Academic Press, 1972, p. 133.
58 H. Fischer, Physics and technology of photovoltaic solar energy conversion. In Advances in Solid State Physics (Festkörperprobleme), Vol. XIV, Vieweg, Braunschweig, 1974, p. 153.
59 J.J. Loferski, Theoretical considerations governing the choice of the optimum semiconductor for photovoltaic solar energy conversion, J. Appl. Phys., 27 (1956) 777.

CHAPTER 11

ELECTROPHOTOGRAPHY

F.W. SCHMIDLIN

11.1. Introduction
11.2. Physical basis of charged pigment electrophotography
    11.2.1. The process steps in two practical systems
    11.2.2. Idealized charged pigment xerography (CPX) and photoactive pigment electrophotography (PAPE)
11.3. Statistical physics of single toner development
11.4. The electric field in xerography
11.5. Macroscopic image transformations in CPX
    11.5.1. The input image
    11.5.2. Latent image formation
    11.5.3. Development
    11.5.4. Connection between physical and conventional image transformation descriptors
11.6. The fundamental processes of latent image formation
    11.6.1. Energetics of the latent image forming process
    11.6.2. The three fundamental processes
    11.6.3. The quantum efficiencies of CPX (η_x) and PAPE (η_p)
    11.6.4. Connection between η_x and the photoinduced discharge characteristic
11.7. Photoreceptor material requirements
    11.7.1. Photo-requirements
    11.7.2. Dark requirements
11.8. Discussion of practical xerographic photoreceptors
11.9. Conclusions

11.1 INTRODUCTION

The practical use of electrostatic forces to print images on paper began with the invention of Chester F. Carlson [1]. Since Carlson's invention, a wide variety of different imaging systems have been developed which similarly exploit electrostatic phenomena, and, by now, the list of inventions and general literature on the subject has grown enormously. Several books [2, 3] and review articles [4—6] are now available which discuss the practical aspects of electrophotography and provide extensive bibliographies [2, 4]. In this chapter, we primarily focus on a particular theoretical view [7] which was specifically formulated to help elucidate the fundamental differences between the various electrostatic imaging methods, identify their

References pp. 476—478

limitations and establish the material properties required for optimal performance. Although the task is still far from complete, enough details have been added to the original formulation of the theory that it is now possible to understand quantitatively the distinguishing characteristics of xerography (Carlson's invention), its inherent limitations, and the material properties required for the photoreceptor.

The theory of charged pigment electrophotography (CPE) was mainly prompted by the combination of Carlson's invention of xerography and Tulagin and Carreira's invention of photoelectrophoresis [8]. The process steps in both of these are reviewed in Section 11.2.1 (and are recommended for preliminary study by the reader if they are not familiar). The unifying concept that led to the theory of CPE is that development in both cases is driven by an imagewise variation in the electrostatic monopole force QE, where Q is the charge on a charged pigment particle (toner) and E is the local electric field. Traditionally, the term "toner" refers to the dry pigment particles in a xerographic system. However, we shall extend its meaning to any "charged pigment particle", whether it is dry or wet (as when suspended, for example in an electrophoretic ink). It was also recognized that this provided the key to a quantitative formalism which then unfolded by taking a two-dimensional variation in QE over the focal plane where development takes place; *i.e.*

$$\delta(QE) = Q\delta E + E\delta Q \qquad (11.1)$$

It is evident from this that the total force variation tending to insert or withdraw toner from an image plane may arise from either a field variation δE or a charge variation on the toner δQ. If either of these is simply a natural fluctuation, it represents noise; but if either is generated by a real optical image (during the latent image forming step), it represents a latent image. The physical basis of imaging in both systems is then completed by formulating the specific interaction between a real image and a photoreceptor (or toner) which generates the δE or δQ latent images.

The main concerns of this chapter are the physical processes and materials required to form the latent images. However, the required strength of the latent images and system geometry (including photoreceptor thickness in xerography) are determined by development. Therefore, it is appropriate (if not essential) to first discuss the development step and establish a suitable classification scheme in order to clarify the domain in which the formalism applies.

Many different methods have been invented for developing xerographic latent images (thought of more broadly as an electrostatic charge pattern) which do not even involve toner. In particular, methods have been invented in which an electrostatic stress acting on the surface of a fluid can be identified as the dominant force driving development (*e.g.* Frost [9] and a non-electrophoretic liquid-ink process [10]). These systems are fundamentally different from those based on variations in QE. In fact, an interesting and far reaching consequence of this difference is that development in the surface-stress-driven cases can occur continuously, whereas development in QE-driven systems must occur in discrete units of the optical opacity produced by the pigment in a single toner. This is necessary because the QE force is only meaningful in the sense that it acts on the toner as a whole. (It is assumed that a toner particle is not pulled apart.)

Because of the above, it is apparent that both $Q\delta E$ and $E\delta Q$ are meaningful as development driving forces only in particulate imaging systems; yet, they are clearly generic of broad classes. For this reason, it was decided to call $Q\delta E$-driven systems, charged pigment xerography (CPX), and $E\delta Q$-driven systems, photoactive pigment electrophotography (PAPE). The two together then define charged pigment electrophotography (CPE). Unfortunately, PAPE is somewhat of a misnomer because δQ can be produced by either direct interaction of light and the charged pigment particles or by association with a neighboring photoactive material (such as a xerographic plate). But the intended distinguishing characteristic of PAPE is that δQ represents the latent image whether the toner itself is actually photoactive or not.

Important subdivisions of CPX and PAPE arise from consideration of possible non-imagewise forces that may exist during development to oppose the electrostatic driving force. This was first realized via an order of magnitude analysis [7] which showed that fluctuations in the adhesion of a toner to an electrode could account for the limited sensitivity of CPX systems. An order of magnitude evaluation of the non-imagewise forces envisioned to exist in practical CPX development systems [6] then enabled their separation into three classes. The first class consists of aerosol [11, 12] and electrophoretic development [13, 14]. They are designated "viscosity-controlled" because Stoke's drag is the only force acting on the toner to oppose the QE force. The second class consists of touchdown [15—17] and magnetic brush [18—20]. These are designated "adhesion-controlled" because a static adhesive force holds the toner to a carrier vehicle that transports the toner to the latent image. The third and final class contains cascade development [21, 22] alone. In this system, toner is made available for development by tumbling carrier beads over the photoreceptor. The toner adheres to the carrier beads in the same way as in a magnetic brush. However, in cascade, the beads acquire free kinetic energy in the tumbling process, and the inertial force acting on the toner during collisions helps the electrostatic force overpower adhesion. Since the inertial force is large enough to play an important role in the release of toner from the carrier beads, cascade is referred to as a "mixed inertia—adhesion controlled system".

It may be realized that viscous drag is merely a rate limiting factor; so given sufficient time for development to take place, it turns out to have no effect on the system input—output characteristic. Adhesion, on the other

References pp. 476—478

hand, can have a profound effect on the input—output characteristic, and to explain how this arises, we shall examine this case in detail. We shall not discuss cascade development any further; but from the standpoint of the photoreceptor requirements, it can be said that they are essentially the same as for the adhesion-controlled case.

As already indicated, development in CPE is necessarily quantized much like it is in silver halide photography — with the physical insertion (or extraction) of a whole toner into (or from) an image plane being the equivalent of chemical conversion (or not) of a whole silver halide grain into an opaque silver particle. This suggests that one should be able to describe development in CPE via a statistical description of how a single toner develops. It only requires the neglect of direct toner—toner interactions. Of course, neglect of toner—toner interactions may be hard to justify in many cases, particularly if the toner particles are stacked on top of each other at the moment the development force is applied. On the other hand, if the toner is delivered by beads covered with a fractional monolayer of toner, as in a magnetic brush, then it is reasonable to expect that a single toner description should be a good approximation. In this case, the only way toner particles interact is through their charge, which can be taken into account self-consistently via the local electric field.

It turns out that many of the important limitations of practical systems can be understood via a statistical treatment of single toner development. Therefore, we describe herein a reformulation of the theory of CPE from such a point of view. The physical processes which convert a real image into a δE or δQ latent image are first described in Section 11.2.2 using idealized single toner models. By neglecting all non-imagewise forces and noise, it is possible to show that these systems can respond to a single incident photon. This serves as an instructive reference for comparison with the performance of practical systems. In Section 11.3, a statistical description of development is carried through under the realistic conditions of noisy adhesion between a toner and carrier. It may be realized that the mere existence of any static force (like adhesion) does not, in itself, lead to any fundamental system limitations. This is because any uniform force can be biased out: the fluctuations cannot. It is shown that this noisy adhesion results in lower bounds for the workable range of size and charge of toner, and a minimum latent image strength (δE). The latter is 10^5 times stronger than in the idealized case, and is close to the minimum field required to drive development in practical systems.

The physical nature of the latent image formation processes are described in Section 11.2.2 using a small-amplitude image. This is enough to describe the processes and formulate quantum efficiencies, designated η_p and η_x for PAPE and CPX respectively. However, it turns out that both quantum efficiencies are strongly field dependent and η_x is even dependent on the structure of the in-going image. To explain this, and prepare for extension of the analysis to large-amplitude images, the solution of Poisson's equation for a periodic charge pattern is reviewed in Section 11.4.

The actual extension of the formalism to large-amplitude images is then carried through for CPX in Section 11.5. Here a periodic image is transformed through a complete system from an input density variation to an output density variation. The objective of this section is to show how the input—output characteristic for a complete system is dependent on the underlying toner and photoreceptor properties.

In Section 11.6, the quantum efficiencies of CPX and PAPE are formulated in terms of the fundamental processes of charge generation, displacement of the photogenerated charges through a material, and finally the transfer *versus* trapping of the displaced charges at a material boundary. The fundamental difference between η_p and η_x is shown to lie in the transfer *versus* trapping event. In PAPE, one and only one of the photogenerated charges must transfer off a toner, and in CPX the charge which approaches the surface of the photoreceptor must be trapped. This requirement, plus the need for a photoreceptor to accept corona charge, turn out to be sufficient conditions for showing that the surface material of a xerographic photoreceptor must be fundamentally incompatible with the bulk.

In Section 11.7, a set of criteria for the selection of materials for building efficient xerographic photoreceptors is formulated. The results in this section may prove interesting to readers interested in photoconductivity in general. In particular, it is shown that the traditional high-resistivity (or long dielectric relaxation time) criteria for the selection of a photoreceptor material is meaningless, if not misleading. What is needed for maximum sensitivity instead are blocking contacts, and this leads to a generation-limited dark current instead of a space-charge-limited (SCLC) dark current. Thus, xerography arises as an exception to the general argument by Rose [23] that SCLC is a desirable state of dark conduction for maximum photoresponse. The apparent dilemma here is resolved by showing that the assumption of an ohmic contact (which leads to SCLC in the dark) is not necessary in general for maximum photoresponse.

In Section 11.8, the properties of several practical xerographic photoreceptors are discussed in terms of the criteria derived in Section 11.7. It is shown that the maximum field variation per unit input density is comparable in magnitude in all cases, indicating a similar capability for reproducing low input density images with the same or equivalent developer. This assumes an optimally adjusted exposure and development bias voltage, and a similar adhesive interaction between the photoreceptor and toner. The main differences between the photoreceptors discussed lie in their required exposure energy and operating speed (as limited by the transit time for a photogenerated charge to cross the photoreceptor). Cyclic ability of the photoreceptors is not compared because of too little data. However, available data on their dark discharge characteristics are summarized.

References pp. 476—478

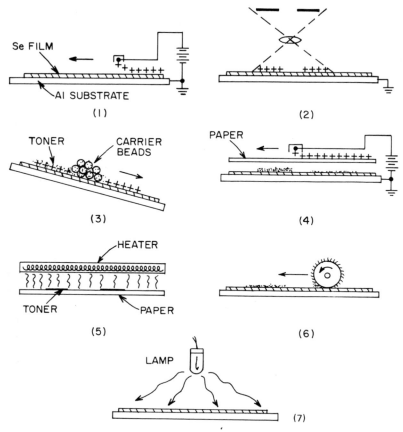

Fig. 11.1. The seven basic steps in the xerographic process: (1) Sensitization, (2) Exposure, (3) Development, (4) Transfer, (5) Fixing, (6) Cleaning, (7) Erase. (From Tabak et al. [24].)

11.2 PHYSICAL BASIS OF CHARGED PIGMENT ELECTROPHOTOGRAPHY

11.2.1 The process steps in two practical systems

In a typical xerographic system, there may be as many as seven distinct process steps as illustrated in Fig. 11.1. Step 1 consists of depositing a uniform electrostatic charge over the surface of a thin layer of photoconductive material (50 microns of selenium, say) coated over a conductive substrate. The charging device is typically a corona unit as illustrated. The second step consists of focusing a real optical image of the object to be recorded (or reproduced) onto the photoreceptor. This selectively discharges the photoreceptor and creates a latent image in the form of a surface charge pattern (or dipole array) over the surface of the photoreceptor. Step 3 consists of delivering electrostatically charged pigment particles (toner) to the latent image via some kind of vehicle — usually small beads of the order of 100-μm diameter. The toner and bead materials are usually chosen to charge each other triboelectrically in such a polarity as to develop "black" where the original object is "black". The next two process steps (4 and 5) consist of transferring the developed toner to paper and fusing it in place to avoid subsequent smear or smudge. This completes the basic processing of the in-going image. The final two steps (6 and 7) prepare the photoreceptor for repeated use. They consist of cleaning off any residual toner remaining on the photoreceptor after transfer, and erasure of any residual charge left on the surface of the photoreceptor (or trapped inside). These last two steps are necessary only because of the imperfect nature of the development, transfer and photoreceptor discharge processes. Otherwise there would be no toner left on the photoreceptor after transfer and the photoreceptor charging process would automatically erase any previous latent image. In xerographic systems wherein the paper bearing the final image is also photoresponsive (such as ZnO), it is evident that Step 4 as well as Steps 6 and 7 are simply not present.

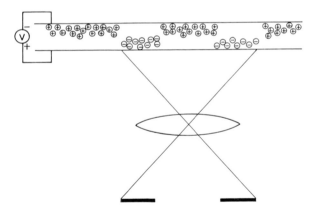

Fig. 11.2. Image formation by charge reversal of pigment in photoelectrophoresis.

Now, photoelectrophoresis can be described as a one (or two) step imaging process as illustrated in Fig. 11.2. The physical structure consists of a layer of pigment particles suspended in a hydrocarbon fluid and sandwiched between a pair of electrodes. This capacitive-like structure is placed in the focal plane of a lens system and a voltage is applied simultaneously with exposure. (To do this, of course, one electrode must be transparent.) The

References pp. 476—478

pigment and fluid materials are preselected so that the pigment particles acquire one polarity in the dark (say negative), and the opposite polarity in the light. As a result, the exposed and unexposed pigment particles are attracted to opposite electrodes. If the particles have sufficient room to separate, the image may become immediately visible. But to enhance visibility of the images, or produce a permanent record, the electrodes may be subsequently peeled apart.

Although photoelectrophoresis may be viewed as a single-step process, it is useful to separate the total process into the following two distinct steps: the reversal (or change) in the charge on the toner may be identified as a latent image forming step, and the physical separation of the oppositely charged toner particles may be identified as development. Both of these are essential (and measurable) events regardless of whether the two actually overlap in time during exposure or whether the toner remains immobile until the electrodes are physically peeled apart. For any particular toner, the two steps are always distinct and must occur in order. The charge must be reversed in polarity before the toner can be attracted to the opposite electrode.

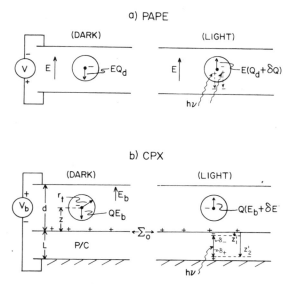

Fig. 11.3. Microscopic view of single particle imaging in (a) PAPE and (b) CPX.

11.2.2 Idealized charged pigment xerography (CPX) and photoactive pigment electrophotography (PAPE)

The physical processes creating an imagewise driving force in PAPE and CPX are illustrated in Fig. 11.3. To formulate these processes in a manner compatible with eqn. (11.1), we must express the in-going image as a two-dimensional variation in the local number of photons per unit area, designated δn_p. In general, there is also an average exposure, but for simplicity we can presently choose the minimum exposure as zero in the local "dark" areas. Then δn_p also becomes the absolute photon flux in the "light" areas. To provide a further simplification (as needed, in particular, for the CPX case) we also consider a small-amplitude in-going image of the form:

$$\delta n_p(x, y) = \frac{\delta n_{pk}}{2} (1 + \cos kx), \qquad (11.2)$$

where $k = 2\pi/\lambda$ and λ is wavelength. Thus the "dark" and "light" areas correspond to points where $kx = \pi$ and 0 respectively. In the "light", $\delta n_p = \delta n_{pk}$, the peak-to-peak amplitude. The case of large-amplitude exposures and the shift in the average field due to the average exposure is discussed in Section 11.5.

Referring now to Fig. 11.3(a), we suppose a photoactive toner possesses a negative charge (say 1 electron) in the dark. In the light, we indicate the ejection of two electrons by two incident photons. Designating the probability that each ejection event is successful by η_p, the net difference in charge between a toner particle in the light and dark is given by:

$$\delta Q = e \eta_p a \delta n_p \qquad (11.3)$$

where e is the charge of a single electron and a is the projected area of the toner. This equation holds at any point where the local exposure is δn_p, and may be regarded as the definition of a quantum efficiency for PAPE, η_p. The underlying microscopic processes comprising η_p are discussed in Section 11.6.

It is evident that if δQ is larger (and opposite in sign) to the net charge on a toner particle in the dark (Q_d), then a reversal in the electrostatic force is achieved the moment $\delta Q > Q_d$. The toner may then be driven toward the opposite electrode. It is of interest to note that δQ may be as small as one electron, in principle, providing a uniform non-image force (such as gravity) is introduced as a bias to just marginally push the unexposed toner toward the positive electrode. Q_d could then be zero.

Turning now to the CPX system illustrated in Fig. 11.3(b), we first note a change in the physical structure. It consists of a precharged (sensitized) photoreceptor, with a counter-electrode placed above the toner. The purpose of this counter-electrode is to provide a means for adjustment of the bias field in the dark (E_b).

The latent image forming interaction in this system is the conversion of an in-going photon into an electric dipole; and the microscopic physical processes involved in this conversion are discussed in depth in Section 11.6. Here, we shall simply indicate the result of the interaction in terms of the electric field acting on the toner due to the electric dipoles (or image dipoles).

References pp. 476—478

To calculate the electric field produced by the image dipoles, it is necessary to solve Poisson's equation for the actual position of the dipole ends, which are designated z_1' and z_2' [cf. Fig. 11.3(b)], and the particular solution required is reviewed in Section 11.4. From this solution, it can be shown that the periodic part of the electric field can be written in the form:

$$\delta E = \frac{f_k}{\kappa_p \kappa_0} \delta \Sigma_I ; \qquad (11.4)$$

where:

$$\delta \Sigma_I = -e \eta_x(k) \delta n_p ; \qquad (11.5)$$

$\kappa_p \kappa_0$ is the permittivity of the photoreceptor; f_k is a dimensionless quantity which depends upon the system geometry, the spatial frequency k and the coupling distance z between the toner and photoreceptor surface; $\delta \Sigma_I$ is an *effective* surface charge density which would produce the same electric field acting on the toner as the actual image dipoles; and η_x is a xerographic quantum efficiency. The latter is formulated in terms of fundamental charge generation, displacement and trapping processes in Section 11.6. The minus sign in eqn. (11.5) simply indicates that the image dipoles are of such a polarity that their associated field pushes the toner away from the photoreceptor.

Since eqns. (11.4) and (11.5) are formulated from first principles they require no other interpretation. However, it is instructive and useful to note that they can be alternately viewed as macroscopic definitions of $\delta \Sigma_I$ and η_x. To do this, one begins by assuming *a priori* that the actual image dipoles produce a specific field variation δE. Such an assumption is obviously valid on physical grounds, and the actual connection is exhibited by the microscopic formulation. This field is then viewed as arising alternatively from equivalent surface charges, or full-length dipoles. *A priori*, this is non-obvious but again it is verifiable from Poisson's equation, as indicated in Section 11.4. The connection between δE and the equivalent full-length dipoles is f_k, by definition (and the appropriate expression for f_k is described in detail in Section 11.4). Finally, with δE and f_k known, eqn. (11.4) defines $\delta \Sigma_I$ uniquely, and eqn. (11.5), in turn, defines $\eta_x(k)$ uniquely.

Now with either approach to the description or definition of η_x, it can be seen that it depends on k, the photoreceptor thickness L and the distribution of dipole-end positions z_1' and z_2'. As a result, it clearly differs from the familiar photoconductivity definition of quantum efficiency, namely the number of charges crossing the photoconductor per incident photon. On the other hand, the two definitions of quantum efficiencies are closely related; for, as shown in Section 11.4, η_x reduces to the conventional definition for either a uniform image ($k \to 0$) or actual full-length dipoles ($z_1' = 0$ and $z_2' = L$). The final theoretical expression for η_x is given in Section 11.6, where the actual distribution of z_1' and z_2', as determined by the dynamics of carrier motion within the photoreceptor, is taken into account.

Because of the key role f_k plays in CPX, its physical significance deserves emphasis. As eqn. (11.4) shows, f_k quantifies the conversion of a surface charge variation into an electric field. Thus it can be thought of as a "coupling coefficient", coupling (at a distance) the toner to the latent-image dipoles. Though more rigorously, it couples toner to the equivalent full-length dipoles.

After examining (in Section 11.4) the specific dependence of f_k on k and the system geometry, it can be seen that its maximum value is typically of the order of unity. Assuming η_x is also a maximum (~ 1), it can be readily computed from eqns. (11.4) and (11.5) that one photon in an area covered by a 10-μm-diameter toner would produce δE of 30 V/m. By adjusting the bias field E_b to half this value, toner would be attracted to the photoreceptor in the "dark" and repelled in the "light". Such would be the nature of the ideal one-photon xerographic system.

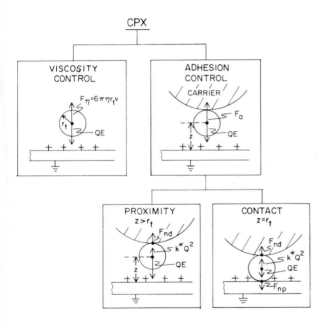

Fig. 11.4. Subclassification of CPX imaging systems according to the dominant non-image force that may control (or interfere) with single toner development.

11.3 STATISTICAL PHYSICS OF SINGLE TONER DEVELOPMENT

The first step toward the description of real systems is to include other forces that may act on the toner during development. The principal ones are viscosity and adhesion, and their essential difference is illustrated in Fig. 11.4. In the viscosity-controlled case it is evident that a toner particle will settle onto the photoreceptor with certainty if the QE force is so directed. In the adhesion-controlled case, on the other hand, where a toner particle is delivered to a latent image by a carrier, the toner will transfer from the carrier to the photoreceptor only when the *net* static force including adhesion is directed toward the photoreceptor. This event, like the viscosity-controlled case, will again occur (or not) with certainty for any particular toner, providing the forces are appropriately unbalanced. But when a toner particle is selected at random from a realistic developer, the forces acting on it can only be known statistically. To take this into consideration we shall next formulate the probability of transfer or development in terms of the underlying parameters and forces that control the process. We shall first write the force balance criterion for the development of some specific toner, and then formulate the relevant physics in statistical terms.

In general, the net force acting on a toner particle can be written as a quadratic function in Q and E. And to a good approximation, a criterion for release from the carrier can be simply written:

$$QE > F_n + k^* Q^2 \tag{11.6}$$

where

$$F_n \simeq F_{nd} - F_{np}, \tag{11.7}$$

$$k^* \simeq \frac{1}{16\pi\kappa_0 r_t^2} \left[\frac{\kappa_d - 1}{\kappa_d + 1} - \frac{\kappa_p - 1}{\kappa_p + 1} \frac{r_t^2}{z^2} \right], \tag{11.8}$$

$$E = E_c + E_I - E_b - E_t. \tag{11.9}$$

F_{nd} and F_{np} are very short-range non-Q dependent forces between the toner—delivery vehicle and toner—photoreceptor respectively. We approximate them, when in contact, as step functions proportional to the local contact areas, and zero otherwise. Thus, $F_{np} = 0$ for what we designate as a proximity toner ($z > r_t$) and finite for what we designate as a contact toner ($z = r_t$), as illustrated in Fig. 11.4. κ_d and κ_p are the dielectric constants of the carrier and photoreceptor, respectively. Hence $k^* Q^2$ is the electrostatic image force toward the carrier, reduced by the electrostatic image force toward the photoreceptor. E is the net local field and is comprised of many components: E_c is the field due to the corona charge; E_I is the field due to the latent image dipoles; E_b is an applied bias; and E_t is due to other toner that may be present in the system. It is evident that E in eqn. (11.6) must be normal to the carrier. To avoid unnecessary complexity, we shall assume the carrier surface is parallel to the photoreceptor surface (which is perpendicular to the z-direction).

A more rigorous force balance criterion would reveal a dependence of both F_n and k^* on the dielectric constant of the toner. Also, F_n would contain an added term proportional to E^2. For typical toner dielectric constants, however, these effects are small and may be neglected. In fact, their neglect is a necessary condition to justify the hypothesis that QE is the dominant development force.

Now E, Q, r_t, z, F_{nd} and F_{np}, all vary independently of each other for different toner and different delivery events. Therefore, in a complete description of development they must all be treated as independent stochastic variables. The general problem then is to calculate the probability that eqn. (11.6) is satisfied given the distribution functions for all of the above variables.

For this chapter, however, it is enough to examine the conditional probability of development for given values of z and r_t. Also, since F_{nd} and F_{np} appear only as a difference in F_n, their distributions may be folded together to find a single distribution of F_n. This reduces the number of variables in the immediate problem to three (E, Q, F_n), with k^* constant.

For later interpretation of the results, it proves effective to formulate the development probability in terms of the moments of the variables involved. In general, all the moments of each variable would be required, but for our purpose here it is enough to examine the dependence of the development probability on the first two moments of E, F_n and Q alone. We shall designate the mean of a quantity, Q say, by $\bar{Q}$ and the variance by $\sigma^2(Q)$, or simply σ_Q^2.

Description of the results is now facilitated by introducing a new stochastic variable, defined by the relation

$$E^* \equiv E - \left(\frac{F_n}{Q} + k^*Q\right) \tag{11.10}$$

In terms of this quantity, the development criterion expressed by eqn. (11.6) simply becomes $E^* > 0$. A density function for E^*, denoted $h(E^*)$, can then be generated from assumed (or actual) density functions for E, F_n and Q. Finally, the development probability we seek can be written

$$P\{E^* > 0 | r_t, z, \text{all moments of } (E, F_n, Q)\} = \int_0^\infty h(E^*) dE^* \tag{11.11}$$

Representative $h(E^*)$ curves for a proximity toner (h_p) and a contact toner (h_c) are sketched in Fig. 11.5(a). These two cases differ from each other significantly because F_{np} and F_{nd} are non-negative and their distributions fall off slowly in the positive direction. Since $F_{np} \equiv 0$ for a proximity toner, h_p tends to be one-sided, with a slow tail-off as $E^* \to -\infty$. For a contact toner, on the other hand, the distribution tends to be more symmetrical, with h_c tailing off slowly in both directions. The curves shown here are constructed for the special case of $\bar{E} = 0$ and $\sigma^2(E) \sim 0$; i.e. $\sigma^2(E)$ is considered vanishingly small compared to other factors causing the spread

References pp. 476—478

a)

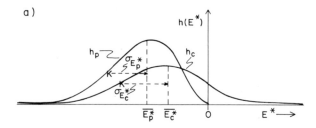

b)

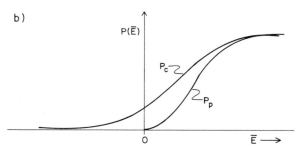

Fig. 11.5. (a) Representative density functions for the development of a single proximity or contact toner designated h_p and h_c, respectively. (b) The development probability for a proximity (P_p) or contact (P_c) toner as a function of the average local electric field acting on the toner.

in E^*. We also neglect any "wrong sign" toner, $Q < 0$. Under these circumstances, adhesion to the carrier clearly prevents any development of proximity toner, and implies $h_p(E^*) = 0$ for $E^* > 0$. For a contact toner, on the other hand, adhesion to the photoreceptor may overpower adhesion to the carrier, meaning $h_c(E^*)$ is finite for $E^* > 0$.

The remaining distinctive features of h_p and h_c can be understood by simply examining the mean and variance of E^*. Assuming $\sigma^2(Q) < 2.5\bar{Q}^2$ (which is only marginal in practice),

$$\bar{E}_* = \bar{E} - \left(\frac{\bar{F}_n}{\bar{Q}} + k^*\bar{Q}\right) \tag{11.12}$$

$$\sigma^2(E^*) \simeq \sigma^2(E) + \frac{\sigma^2(F_n)}{\bar{Q}^2} + \left[\frac{\bar{F}_n}{\bar{Q}^2} - k^*\right]^2 \sigma^2(Q) \tag{11.13}$$

The variance in E^* is obtained here by simply differentiating eqn. (11.10). The difference between a contact and proximity toner can now be explained in terms of differences in the distributions of F_n. We recall first that F_n is the total adhesive force as defined by eqn. (11.7). Thus $\bar{F}_n = \bar{F}_{nd} - \bar{F}_{np}$ and $\sigma^2(F_n) = \sigma^2(F_{nd}) + \sigma^2(F_{np})$. These expressions follow from the statistical independence of F_{nd} and F_{np}. For the proximity case, we put $\bar{F}_{np} = 0$ and $\sigma^2(F_{np}) = 0$. The effect of this is illustrated in Fig. 11.5(a) by a shift of $\bar{E}_p^*$ in the negative direction relative to $\bar{E}_c^*$. It also results in a smaller $\sigma^2(E^*)$ for a proximity toner. Note that appropriate subscripts are now attached to the moments of E^* to emphasize that they are different in the two cases.

The effect of a local increase in $\bar{E}$ may now be envisioned graphically as a uniform shift of h_p and h_c in the positive E^* direction. This follows from eqn. (11.12). The final development probability, according to eqn. (11.11) can thus be envisioned as the area under that part of an $h(E^*)$ curve that extends beyond $E^* = 0$. Representative results for the proximity and contact cases are illustrated in Fig. 11.5(b).

It may be realized that $\bar{E}$ may change with either the bias field E_b or an exposure. In the case of an exposure, the imagewise variations in $\bar{E}^*$ are given by:

$$\delta\bar{E}^* = \delta\bar{E} = \delta\bar{E}_I, \tag{11.14}$$

as follows from eqns. (11.12) and (11.9). Note that from the definition of E_I, an imagewise variation in $\bar{E}$ is now $\delta\bar{E}_I$. The average part of an exposure would also change the second and higher moments of E; but it can be shown, for visible light, that $\sigma^2(E)$ due to an exposure is of the order of 10^6 times smaller than the other terms appearing in eqn. (11.13). Hence, from the standpoint of image transformations it is enough to know the dependence of P on the first moment of E alone. This is designated $P(\bar{E})$, as an abbreviated way of writing $P\{E^* > 0 | \bar{E}, \ldots\}$.

Now the variations in $\bar{E}$ induce variations in P_c and P_p. Thus, the slope of $P(\bar{E})$ is a key quantity in CPX. In fact, it may be thought of as a "single-particle development gamma"; and under certain limiting conditions (emphasized in Section 11.5.3) it also represents the development gamma for a system.

From the previous discussion of how $P(\bar{E})$ is constructed, it follows that:

$$P' \equiv \frac{dP}{d\bar{E}} = h(E^* = 0); \tag{11.15}$$

and a corollary to this is that the maximum P' is

$$P'_{max} = h_{max} \simeq \frac{1}{2\sigma(E^*)} \tag{11.16}$$

This shows that $\sigma(E^*)$, which is physically the total noise in the forces controlling the development of a single toner particle, completely characterizes the maximum gamma. Thus, the scale of the latent image signal, $\delta\bar{E}_I$, required to produce a given output signal (δP) is entirely set by "noise". This noise is neither the output noise [25] nor a manifestation of the input noise, but a combination of factors which contribute to the total noise in the forces which control the development of a single toner particle;

References pp. 476–478

in most practical cases, it is almost entirely adhesion noise. The fact that latent-image and development-noise factors must be mixed to quantify the development of a single toner particle is a significant and general result which evidently applies to any particulate imaging system.

The dependence of $\sigma(E^*)$, and hence P', on the physical control variables is expressed by eqn. (11.13). For typical $\sigma(Q)$, $\sim \bar{Q}/3$ say, it can be shown that $\sigma(E^*)$ is a minimum for $\bar{Q} \simeq (\bar{F}_n/k^*)^{1/2}$. This is the optimal toner charge, for which the minimum $\sigma(E^*)$ is $\sigma_{min}(E^*) \simeq \sigma(F_n)(k^*/\bar{F}_n)^{1/2}$. If $\sigma(F_n) \simeq \bar{F}_{nd} \simeq 10^{-8}$ N, as deduced from centrifuge measurements [26] on typical 10-μm-diameter toner, we find $\sigma_{min}(E^*) \simeq 10^6$ V/m. As indicated above, it takes $\delta\bar{E} \simeq 2\sigma_{min} \simeq 2 \times 10^6$ V/m, to produce good output images. We thus find, that a realistic imaging system, with adhesion included, requires at least 10^5 times more field contrast to drive development than was estimated earlier for the ideal one-photon system. But what is more remarkable, is that 2×10^6 V/m is less than a factor of 10 from the maximum field contrast physically achievable with the best photoreceptor materials.

If we further assume now that $\bar{F}_n$ is proportional to contact area, which is proportional to toner size (or local radius of curvature), then $\sigma_{min}(E^*) \propto r_t^{-1/2}$. Thus, larger toner particles could relax the required driving field, but it is generally known that toner particles larger than 10 μm would result in excessive granularity in the output image. Smaller toner particles, of course, would require an even larger field contrast than was calculated above.

To sum up the above, we have identified that $\bar{F}_n$ not only determines the size and charge of the toner, but it also determines the field contrast that must be created by the photoreceptor. And what is most remarkable, is that the field contrast required to operate typical systems is so close to being physically unattainable. Thus, if $\bar{F}_n$ had proven to be much larger than it evidently is, in two-component developers, xerography probably would not yet exist — at least in many of its more familiar forms known today.

11.4 THE ELECTRIC FIELD IN XEROGRAPHY

An essential preliminary to the description of image formation and development for large-amplitude input images is a full description of the electric field throughout the system. We consider the system geometry shown in Fig. 11.3(b), and a surface charge density on the photoreceptor of the form

$$\Sigma(x, y) = \Sigma_0 + \Sigma_k \cos kx \qquad (11.17)$$

This represents a general case in which Σ is a superposition of corona charge (Σ_c) and an effective surface charge due to image dipoles.

By solving Poisson's equation for the system defined above, it can be shown [27] that the electric field in the development region above the photoreceptor can be written in the form:

$$E_z = f_0 \left(\frac{\Sigma_0}{\kappa_p \kappa_0} - \frac{V_b}{L} \right) + f_k \frac{\Sigma_k}{\kappa_p \kappa_0} \cos kx \qquad (11.18)$$

$$E_x = b_k \frac{\Sigma_k}{\kappa_p \kappa_0} \sin kx \qquad (11.19)$$

where $\kappa_p \kappa_0$ and L are the permittivity and thickness of the photoreceptor respectively; V_b is the bias voltage applied to the development electrode; and

$$f_k = \frac{1}{\frac{1}{\kappa_p} + \frac{\tanh kd}{\tanh kL}} \frac{\cosh k(d-z)}{\cosh kd} \qquad (11.20)$$

$$f_0 = \left(\frac{1}{\kappa_p} + \frac{d}{L} \right)^{-1} = \text{(limit of } f_k \text{ as } k \to 0) \qquad (11.21)$$

$$b_k = f_k \tanh kd \qquad (11.22)$$

κ_0 is the permittivity of free space, and d is the photoreceptor to counter-electrode spacing [cf. Fig. 11.3(b)].

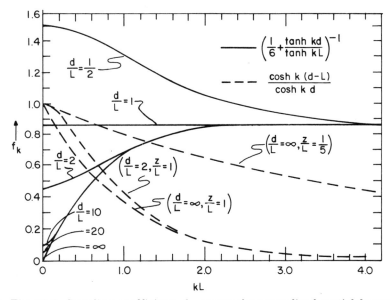

Fig. 11.6. Coupling coefficient, f_k, versus the normalized spatial frequency (kL). f_k is the product of a solid curve and a dashed curve, and d, z and L are illustrated in Fig. 11.3(b).

References pp. 476—478

The key quantity here which plays a large role in shaping many of the characteristic features of xerography is the "coupling coefficient" f_k. It directly relates the electric field to the charge amplitude Σ_k and its physical significance was discussed earlier. To illustrate its specific dependence on spatial frequency (k), and system geometry (d, L, z), the two separate factors comprising f_k are plotted separately in Fig. 11.6. The first factor, which contains all the dependence on photoreceptor thickness (L), is shown as a family of solid curves for different electrode spacings (d). The range of d of practical interest is for $d \gg L$. In this case, all the curves closely follow the case for $d = \infty$, except for $kd < 1$, where $f_k \to {\sim} L/d$ and the k-dependence disappears. The second factor, shown by the dashed curves, illustrates the dependence of f_k on the coupling distance z. For $z < d$, as it is physically required for the toner to be between the electrode and the photoreceptor, the second factor is approximated extremely well by $\exp(-kz)$. To show the weak dependence of this factor on d, the extreme choices of $d = \infty$ and $d = 2L$ are shown for the case of $z = L$.

The net effect of both factors is to produce a low-frequency fall-off which is dependent on the photoreceptor thickness and a high-frequency fall-off due to the coupling distance z. Physically, the low-frequency fall-off occurs because the field of remote image dipoles is reversed in direction with respect to the field of local dipoles. The counter-electrode screens this reverse field, but it is difficult to get the electrode close enough ($d \simeq L$) to be effective. The high-frequency fall-off occurs because of the cancellation of the field due to alternate half periods in dipole orientation. This cancellation becomes extremely effective for $z > \lambda/2$, and constitutes the physical basis of Gundlach's rule [29] for the limited resolution of xerography.

A useful approximation to f_k, for $kz < 1$, is simply

$$f_k \simeq \begin{cases} L/d, \text{ for } k < 1/d \\ kL, \text{ for } 1/d < k < 1/L \\ (1 + 1/\kappa_p)^{-1}, \text{ for } k > 1/L \end{cases} \tag{11.23}$$

This shows that $f_k \propto L$ for $kL < 1$, which encompasses the predominant frequency range for most images of practical interest. Thus, the field available to drive development for a fixed surface charge contrast is essentially always proportional to the photoreceptor thickness L. An example which illustrates the preponderance of cases in which this conclusion applies is ordinary type for which the dominant wave number is $k \sim 10^{-2}$ μm^{-1}. Thus it would take photoreceptors thicker than $100\,\mu m$ to render the above rule inapplicable. We shall make use of this result in Section 11.7 as an essential input to the establishment of photoreceptor material requirements.

Another solution of Poisson's equation that is needed to formulate the xerographic quantum efficiency (in Section 11.6) is that of a periodic dipole array whose ends are at fixed distances z'_1 and z'_2 from the surface of the photoreceptor [see Fig. 11.3(b)]. This solution can again be obtained from Schaffert [27]. However, to obtain this, the regions inside the photoreceptor above and below the charge sheets at z'_1 (or z'_2) should be treated as separate layers. The solutions for a negative charge sheet at z'_1 and a positive charge sheet at z'_2 are then superposed. By comparing the result with eqns. (11.18) and (11.19), it can be shown that the electric field everywhere above the surface of the photoreceptor is reduced below what it would be if the same dipoles were full-length by the factor

$$f_\delta(k, z'_1, z'_2) = f_{\delta_1}(k, z'_1) - f_{\delta_2}(k, z'_2) \equiv \frac{\sinh k(L - z'_1)}{\sinh kL} - \frac{\sinh k(L - z'_2)}{\sinh kL}$$
(11.24)

This is a complex function, and its meaning can be more readily seen by the approximate form

$$f_\delta(k, z'_1, z'_2) \simeq \frac{(z'_2 - z'_1)}{L} \exp(-kz'_1) = \frac{(\delta_- + \delta_+)}{L} \exp(-kz'_1) \qquad (11.25)$$

It can be shown that this expression is good to within a factor of 2 and is completely accurate in the limit as $k \to 0$, where $f_\delta \to (\delta_- + \delta_+)/L$. From this, it can be seen that f_δ is predominantly the dipole length $(\delta_- + \delta_+)$ normalized by the photoreceptor thickness. From eqn. (11.24) it can be seen that f_δ is always less than or equal to unity and approaches unity for all k in the limit as $z'_1 \to 0$ and $z'_2 \to L$ (i.e. full-length dipoles). For this reason, f_δ constitutes a proper generalization of the charge displacement process contained in the definition of the xerographic quantum efficiency. The above limits show that η_x, which is proportional to f_δ, reduces to the conventional photoconductivity definition of quantum efficiency in the limits of $k = 0$ or $\delta_- + \delta_+ \to L$.

The approximate form of f_δ given by eqn. (11.25) shows that the principal effect of an incomplete dipole, and $z'_1 > 0$ in particular, is to increase the overall coupling distance (between the dipoles and toner) from z to $z + z'_1$. The result is a shift in the onset of the high-frequency fall-off to slightly longer wavelengths. Such is the significance of $\exp(-kz'_1)$ appearing in eqn. (11.25). The other factor in eqn. (11.25), namely $(\delta_- + \delta_+)/L$, is a natural part of the quantum efficiency.

So far, we have discussed the electric field outside the photoreceptor only. To later quantify the latent-image formation process for large-amplitude images, it is essential to know the electric field inside the photoreceptor as well. Since the counter-electrode is removed under normal exposure conditions, it is enough to examine the case of $d \to \infty$. It will also be enough for illustrative purposes to consider now full-length image dipoles. Under these conditions, the z-component of the electric field just inside the photo-receptor surface can be written

References pp. 476—478

$$E_{p,z} = \frac{\Sigma_0}{\kappa_p \kappa_0} + \left[1 - \frac{1}{\kappa_p} f_k(z=0)\right] \frac{\Sigma_k}{\kappa_p \kappa_0} \cos kx \qquad (11.26)$$

This follows from Gauss' law, and since $d = \infty$, we have put $f_0 = 0$. The tangential component of E is continuous, so, near the surface ($z = 0$), it is still given by eqn. (11.19). As seen from eqn. (11.22), the value of b_k corresponding to $d = \infty$ and $z = 0$, is $b_k = f_k(z = 0)$. We shall make use of these results in later sections.

11.5 MACROSCOPIC IMAGE TRANSFORMATIONS IN CPX

In the previous sections, we have quantified the latent image forming process using a small-signal input image and the development process using a single toner. In this section, we describe the additional analytical details required to extend the formalism to handle large-signal input images and a multiplicity of toner for development. The objective is to transform input reflection density through a CPX system to output reflection density, and thereby clarify the dependence of the system input—output characteristic on the underlying toner and photoreceptor properties.

The input image and its subsequent representations are sketched in Fig. 11.7. Quantification of the transformation from one representation to the next is described below.

11.5.1 The input image

Typically, input images are manifest as reflection density variations, and a periodic input image can be expressed in the form

$$D_i(x, y) = D_{i,\min} + \frac{\delta D_{i,k}}{2}(1 + \cos kx) \qquad (11.27)$$

The modulation transfer function (MTF) of the optical system used to project the input image onto a photoreceptor always produces some additional modulation on the input image. But if we regard the terms in eqn. (11.27) as already corrected for this, then the exposure variation in the focal plane can be expressed in the form:

$$n_p(x, y) = n_{pm} 10^{-\delta D_{ik}(1 + \cos kx)/2} \qquad (11.28)$$

where n_{pm} corresponds to the peak exposure (in photons per unit area) in $D_{\min}$ areas.

It is apparent from this that the exposure variation is not a simple periodic function, unless δD_{ik} is small. But if we ignore the shape distortions, we can reconstruct the simple periodic form and still quantify the peak excursions

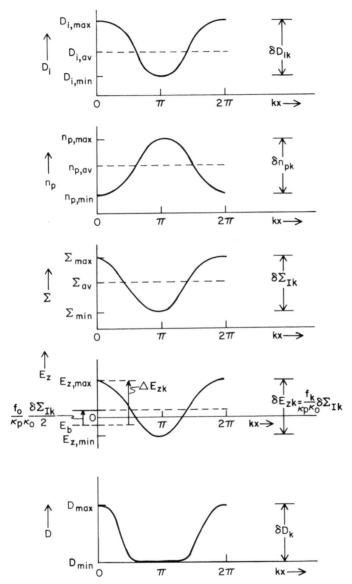

Fig. 11.7. The macroscopic representations of an image at various stages in a CPX system: D_i is the input reflection density; n_p is the exposure at the photoreceptor in photons per unit area; Σ is the equivalent surface charge density (or dipole strength) of the xerographic latent image; E_z is the z-component of the electric field acting on a toner; D is the output reflection density.

References pp. 476—478

correctly by replacing the variable part of eqn. (11.28) by

$$\delta n_p \simeq -n_{pm} \frac{(1 - 10^{-\delta D_{i,k}})}{2}(1 + \cos kx) \equiv \frac{\delta n_{pk}}{2}(1 + \cos kx) \tag{11.29}$$

We shall follow a similar scheme in the subsequent image transformations as well.

11.5.2 Latent image formation

In general, the problem of computing the macroscopic latent image for large exposures is tedious because it entails following the individual photo-generated carriers from the point where they are generated to the point where they are trapped. A good estimate of the net effect for $kL < 1$ however can be obtained by simply taking into account the field dependence of η_x locally. At the extrema ($kx = 0, \pi$) for example, we can obtain a good approximation to the local surface charge (or dipole strength) by integrating:

$$d\Sigma_I = -e\eta_x(E_p) \, dn_p \tag{11.30}$$

together with eqn. (11.26). E_p is the electric field inside the photoreceptor. To carry through the calculations, we would have to know $\eta_x(E_p)$, but to point up the salient features of the result it is enough to simply indicate the peak-to-peak charge variation by:

$$\delta\Sigma_{Ik} = \Sigma_{max}(n_{p,\,min}) - \Sigma_{min}(n_{pm}) \tag{11.31}$$

where Σ_{min}, as well as Σ_{max}, is a function of k. Note that the maximum surface charge corresponds to the minimum exposure and vice versa. To indicate the nature of the error that would result if the discharge were computed assuming the usual broad-area conditions ($f_k = f_0 = 0$), suppose the exposure is such that the integral of eqn. (11.30), denoted $\Delta\Sigma_I$, is of the order of $-\Sigma_c$, the corona charge. Then $E_p \simeq (\kappa_p)^{-1} f_k(z=0)\Delta\Sigma_I$, which is much larger than zero. The difference can be very large if $\eta_x(E_p)$ is very strongly field-dependent; as it is for many materials of commercial interest.

Now another frequently neglected but important aspect of latent image formation is the lateral motion of carriers along the free surface before they are trapped. The controlling mechanism is discussed in Section 11.6, where it is shown that the net macroscopic effect of any lateral motion is to attenuate the periodic amplitude by a factor, designated $A_k(t)$. It is less than unity and depends, in general, on the time lapse from exposure to development [cf. eqn. (11.60)]. After the initial (or unattenuated) charge amplitude expressed by eqn. (11.31) is multiplied by $A_k(t)$, the final peak-to-peak field variation which acts on the toner becomes

$$\delta E_{z,k} = f_k A_k(t) \frac{\delta\Sigma_{Ik}}{\kappa_p \kappa_0} \tag{11.32}$$

The corresponding absolute value of E_z which is correct at the extrema is given by

$$E_z = f_0 \left(\frac{\Sigma_{\min}(k)}{\kappa_p \kappa_0} - \frac{V_b}{L} + \frac{\delta \Sigma_{lk}}{2\kappa_p \kappa_0} \right) + f_k \frac{A_k}{2} \frac{\delta \Sigma_{lk}}{\kappa_p \kappa_0} \cos kx \qquad (11.33)$$

Note that the average surface charge is unaffected by lateral transport.

Equation (11.33) shows that E_z is a function of both the image charge (characterized by $\Sigma_{\min}$ and $\delta \Sigma_{lk}$) and the bias voltage V_b. Therefore, the adjustment of V_b strongly affects the resultant field pattern and the final output image. The question of optimally adjusting V_b for different applications is discussed in detail in ref. 6. Here, we simply consider the special case of reproducing or enhancing low-density input images in the presence of a broad white background. This case is of great practical interest because of its relevance to the reproduction of typewritten documents or lines.

To properly simulate this problem we must consider the transformation of two images at once — a local periodic pattern and a broad white area. It is sufficient, however, to assume $D_{\min}$ in the periodic region is identical to the density in the white background.

Now the net bias field corresponding to the broad "white" areas becomes:

$$E_b = f_0 \left(\frac{\Sigma_{\min}(k=0)}{\kappa_p \kappa_0} - \frac{V_b}{L} \right), \qquad (11.34)$$

while the field in the local $D_{\max}$ areas is larger than this by

$$\Delta E_{z,k} = \frac{(f_0 + f_k A_k)}{2} \frac{\delta \Sigma_{lk}}{\kappa_p \kappa_0} + f_0 \left(\frac{\Sigma_{\min}(k) - \Sigma_{\min}(k=0)}{\kappa_p \kappa_0} \right) \qquad (11.35)$$

To distinguish this from the total field variation given by eqn. (11.32) we use the Δ symbol.

It is of interest to examine this in the limits of small and large amplitude input densities. For small-amplitude input images, $\Sigma_{\min}(k) \to \Sigma_{\min}(k=0)$, and eqn. (11.35) reduces to

$$\Delta E_{z,k} = \frac{(f_0 + f_k A_k)}{2} 2.3 \frac{e}{\kappa_p \kappa_0} \eta_x(E_{p,\min}) n_{pm} \delta D_{i,k} \qquad (11.36)$$

To obtain this result, it is necessary to make use of eqn. (11.30) and the small-amplitude approximation of eqn. (11.29).

For a large-amplitude input density contrast, it is sufficient to approximate $\delta \Sigma_{lk}$ by

$$\frac{\delta \Sigma_{lk}}{\kappa_p \kappa_0} \simeq E_{p,\max}(n_{p,\min}) - E_{p,\min}(n_{pm}) \qquad (11.37)$$

References pp. 476—478

For $E_{p,\min} \ll E_{p,\max}$, eqn. (11.35) then becomes well approximated by

$$\Delta E_{z,k} \simeq \frac{(f_0 + f_k A_k)}{2} E_{p,\max}(n_{p,\min}) \qquad (11.38)$$

Equations (11.36) and (11.38) are the main results of this subsection. They exhibit the dependence of the field contrast available to drive development on the coupling coefficients (f_0 and f_k), maximum exposure (n_{pm}), xerographic quantum efficiency (η_x) and the maximum field inside the photoreceptor ($E_{p,\max}$). The latter is controlled only by the initial charging voltage or corona charge. Therefore, it follows that $\Delta E_{z,k}$ is independent of both n_{pm} and η_x for large-amplitude images, whereas it is dependent on their product for low-amplitude images. Thus η_x (which fully characterizes the photoreceptor from an imaging point of view) actually impacts the image reproduction characteristics of a system only for low input densities; and even then it is pertinent only in the most highly discharged areas of the photoreceptor. Consequently, the customary characterization of a photoreceptor in terms of its initial discharge characteristics actually provides very little information concerning the ability of a system to reproduce images.

In view of the above results, it can be said that a fundamental criterion for the selection of a photoreceptor is to find one for which $\eta_x(E_{p,\min})n_{pm}$ can be made large enough to reproduce low-contrast densities, while keeping $E_{p,\min}$ as far removed from $E_{p,\max}$ as possible. It may be recognized that this is another way of stating the optimization procedure discussed in ref. 6. To determine the actual size of $\eta_x n_{pm}$ required to reproduce low-density images, it is necessary to complete the final step of transforming the latent image to output density.

11.5.3 Development

The reflection density produced by opaque particles covering a "white" background is remarkably well approximated by:

$$D = \beta \bar{n} \bar{a}, \qquad (11.39)$$

where D is the local optical reflection density; $\bar{n}$ is the local particle concentration per unit area; $\bar{a}$ is their average projected area; and β is $\log_{10} e$ multiplied by a correction factor due to light scattered into a particle from the local region around it [28]. For typical toner on paper, $\beta \simeq 0.8$. This simple relation between density and number of accumulated toner particles is restricted to densities well below the saturation value corresponding to the front surface reflection density of the toner alone. It is typically valid, however, up to densities in excess of unity, which covers most practical circumstances.

Equation (11.39) provides a direct link between output density and single toner development. If carriers (such as beads in a magnetic brush) deliver a total of n_d toner particles per unit area to a given point close to the photoreceptor, where the average local electric field is $\bar{E}_z$, then the number of these which develop is simply $n_d P(\bar{E}_z)$. In practice, of course, the carrier may also physically contact previously developed toner particles and carry them away (called "scavenging"). In addition, the charge transported to the photoreceptor (and its counter charge left in the developer) may significantly alter the local electric field as development progresses (called image "neutralization"). But if both of these effects are neglected, then the local output density is simply given by

$$D = \beta \bar{a} n_d P(\bar{E}_z). \tag{11.40}$$

It might be argued that this equation is not relevant to practical systems because scavenging and neutralization are important in practice. On the other hand, it is apparent that both scavenging and neutralization tend to reduce the developed density below that given by eqn. (11.40). It therefore follows that eqn. (11.40) represents an upper bound to the developed density, which is an important limiting case. In practice, the effect of neutralization may be made as small as one likes by reducing the average toner charge, $\bar{Q}$. But, as we saw earlier (in Section 11.3) as $\bar{Q}$ falls below its optimal value, adhesion results in an ever-decreasing single-toner development gamma. The two effects together show that as $\bar{Q}$ decreases adhesion becomes increasingly important while the effect of neutralization vanishes. For this reason, it is appropriate to think of eqn. (11.40) as an adhesion-controlled limit. A practical example of this limit apparently arises when an imaging system fails as the toner charge falls too low. Presumably the trouble is a manifestation of some factor controlling single toner development rather than anything else. In fact, the only conceivable phenomena not quantified by eqn. (11.40) in the low $\bar{Q}$ limit is scavenging, but this is usually a second-order effect of minor importance.

To avoid unnecessary complexities which would extend the formalism beyond the scope of this book, we shall complete the transformation of a latent image to output density using eqn. (11.40). It turns out this is adequate to explain the inherent limitations on the image transformation characteristics of a system due to the underlying toner and photoreceptor properties.

It is evident from eqn. (11.40) that the transformation of a latent image to output density for this adhesion-controlled limit depends only on the single-toner development characteristic. It should be understood, however, that for a practical system $P(\bar{E}_z)$ must be an appropriate average of P_c and P_p (cf. Section 11.3), as determined by the actual distributions of toner size and delivery distances. Under the best circumstances, $P(\bar{E}_z)$ should be heavily weighted by proximity-type toner (P_p), with a typical result appearing such as the curve shown in Fig. 11.8. We assume the distribution of delivery distances are all small. Hence $P(\bar{E}_z)$ transforms the local average

References pp. 476—478

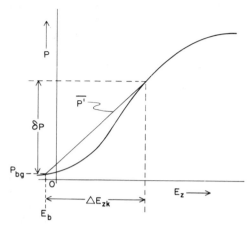

Fig. 11.8. Development characteristic of a single toner, illustrating changes in development probability (P) induced by changes in local $\bar{E}_z$.

field $\bar{E}_z$ into density (or P) on a local basis. An example of this transformation at a local maximum in the input density (i.e., $kx = 0$) is discussed below.

At $kx = 0$, we recall that $\bar{E}_z = E_b + \Delta E_{z,k}$, where E_b and $\Delta E_{z,k}$ are given by eqns. (11.34) and (11.36) or (11.38) respectively. E_b is the net development field in remote background areas and is a machine variable set by the bias voltage V_b. We shall discuss the problem of selecting this variable in a moment; but it obviously must be selected in the region where $P(E_b)$ is small. This is because the output background density is given by

$$D_{bg} = \beta n_d \bar{a} P(E_b). \tag{11.41}$$

At points where the input density is maximum, the corresponding output density rises above D_{bg} by the amount:

$$\delta D_k = \beta n_d \bar{a}[P(E_b + \Delta E_{z,k}) - P(E_b)] \equiv \beta n_d \bar{a} \bar{P}' \Delta E_{z,k} \tag{11.42}$$

where $\bar{P}'$ is the average slope of P between E_b and $E_b + \Delta E_{z,k}$. This is illustrated for a representative operating bias E_b in Fig. 11.8. Note that a change in E_b for a given input image would move $\Delta E_{z,k}$ along the $\bar{E}_z$ axis, the size of $\Delta E_{z,k}$ remaining fixed. The corresponding changes in $P(E_b)$, designated P_{bg}, and $\delta P = \bar{P}' \Delta E_{z,k}$ which result from a change in E_b may thus be visualized.

For the case of a small-amplitude input image, eqn. (11.36) may be substituted into eqn. (11.42), giving

$$\delta D_k = 1.15 \beta n_d \bar{a} \bar{P}'(f_0 + A_k f_k) \frac{e}{\kappa_p \kappa_0} \eta_x (E_{p,\min}) n_{pm} \delta D_{i,k}, \tag{11.43}$$

The coefficient of $\delta D_{i,k}$ defines a total-system transfer function; and the various factors in it bring into focus its dependence on the properties of the photoreceptor and developer. Of all the factors involved only $\bar{P}'$ is dependent on the bias field E_b.

The usual machine adjustment problem is to find a bias field (E_b) such that D_{bg} [given by eqn. (11.41)] is acceptably small, while assuring δD_k [given by eqn. (11.43)] is greater than $\delta D_{i,k}$, for all $\delta D_{i,k}$ greater than some specified amount. It is not possible to do this for all $\delta D_{i,k}$, because $\bar{P}' \to P'$ (E_b) → 0 for very small $P(E_b)$. Thus, some compromise is normally necessary and this is usually a sacrifice in the reproduction of low-contrast input densities in favor of low background. To minimize the range of $\delta D_{i,k}$ sacrificed, it is customary to select the exposure to maximize $\eta_x n_{pm}$, and hence maximize $\Delta E_{z,k}$. This makes it possible to also adjust E_b to maximize $\bar{P}'$, while $P'(E_b)$ and $P(E_b) \simeq 0$. Note also that there is no advantage in increasing $n_d \bar{a}$ beyond the minimum required to achieve D_{max}. It has the effect of increasing D_{bg} as well.

For the largest output contrast density possible, we substitute eqn. (11.38) for $\Delta E_{z,k}$ and eqn. (11.16) for $\bar{P}'$, giving

$$\delta D_{k,max} \simeq \beta n_d \bar{a} \frac{(f_0 + f_k A_k)}{2} \frac{E_{p,max}}{2\sigma(E^*)} \tag{11.44}$$

With $\beta n_d \bar{a}$ limited to 2 or 3 to assure low background, we obtain:

$$(f_0 + A_k f_k) E_{p,max} > 2\sigma(E^*) \tag{11.45}$$

as a necessary condition to achieve $\delta D_k > 1.0$. This establishes a useful relation between the photoreceptor properties, as contained in f_k, A_k, and $E_{p,max}$, and the developer properties, as contained in $\sigma(E^*)$ (cf. Section 11.3). For any given developer, eqn. (11.45) shows that the photoreceptor must be capable of supporting a minimum internal electric field, $E_{p,max}$; and at the same time, the photoreceptor must be thick enough to produce adequate coupling. (Recall from eqn. (11.23) that $f_k \propto L$.)

Equations (11.43)–(11.45) succinctly summarize the main results of this section. They stress the importance of independently maximizing $\eta_x(E_{p,min}) n_{pm}$ and $A_k f_k$ for the photoreceptor and P' for the developer. But to accomplish the latter, the bias voltage must be optimized and the average toner charge should be adjusted to minimize $\sigma(E^*)$.

11.5.4 Connection between physical and conventional image transformation descriptors

The above results describe the overall performance of an imaging system in physical terms. But to render the results more broadly useful as an interpretive tool, it remains to show how they relate to conventional image and electrical measurements. The system transfer function contained in eqn. (11.43), for a small-amplitude image, provides the desired connection. By examining its zero-frequency limit, the transfer function may be separated into the product of a system gamma, defined by:

$$\gamma_s \equiv \lim_{k \to 0} \frac{\delta D_k}{\delta D_{i,k}} = \frac{dD_0}{d\log_{10} n_p} = 2.3\beta n_d \bar{a}\bar{P}' f_0 \frac{e}{K_p K_0} \eta_x(k=0) n_p, \quad (11.46)$$

and an

$$\text{"MTF"} = \frac{f_0 + A_k f_k}{2f_0} \frac{\eta_x(k)}{\eta_x(k=0)}. \quad (11.47)$$

We have placed this "MTF" in quotation marks because it represents the relative response to a single input frequency rather than a Fourier spectrum. Thus, it is meaningful whether the system is linear or not. In fact, it has been derived here under highly non-linear circumstances. It should also be stressed that this "MTF" and the severity of the non-linearity it encompasses is a strong function of the bias voltage. In particular, it can be shown that the peculiar mixture of f_0 and f_k appearing in eqn. (11.47) is the result of a bias voltage selected to provide a low-density background. It can also be shown that with a sufficient forward bias (so $\bar{E}_z$ does not swing negative), $(f_0 + A_k f_k)/2 \to f_k A_k$ and the development part of the "MTF" becomes linear.

In xerography, the surface potential above a photoreceptor is often measured as a function of a broad-area input density or uniform log (exposure). The slope of this characteristic defines a latent image gamma. By making use of the connection between surface voltage and exposure provided in Section 11.6, cf. eqn. (11.66), it can be shown that the latent image gamma is given by

$$\frac{dV}{dD_{i,0}} = \frac{dV}{d\log_{10} n_p} = 2.3 L \frac{e}{K_p K_0} \eta_x(k=0) n_p. \quad (11.48)$$

Separating this from the total system gamma, given by eqn. (11.46), the remaining factor may be defined as a development gamma, designated γ_d,

$$\gamma_d \equiv \gamma_s \bigg/ \frac{dV}{dD_{i,0}} = \beta n_d \bar{a}\bar{P}' \frac{f_0}{L}. \quad (11.49)$$

It should be stressed that the specific separations of the gammas and "MTF" presented above are a natural consequence of the final image density being controlled by an electric field variation. The resulting formulae are therefore restricted to systems in which development is adhesion limited (subneutralized). Unfortunately, it is not possible in this short chapter to compare the present formulae with those which apply in the more general partially neutralized case. But as one might easily guess, in the extreme limit of complete neutralization, it is the charge contrast on the photoreceptor per unit input density ($d\Sigma_I/dD_{i,0}$) that more appropriately characterizes the latent image gamma rather than $dV/dD_{i,0}$. But since the two are related by the capacitance per unit area ($\kappa_p\kappa_0/L$), it is evident that $dV/dD_{i,0}$ is an important practical descriptor for the performance of a xerographic photoreceptor, regardless of the degree of image neutralization. In other words, it can be said that the charge contrast of a latent image is always important for development, but to convert it into a driving force ($Q\delta E$), the photoreceptor must have thickness. For this reason, $dV/dD_{i,0}$ is of direct importance for adhesion-limited development.

To assess the relative ability of a number of practical photoreceptors to function with an adhesion-limited developer, their maximum values of $dV/dD_{i,0}$ are compared in Section 11.8.

11.6 THE FUNDAMENTAL PROCESSES OF LATENT IMAGE FORMATION

In Section 11.2.2 the conversion of image photons into charges in PAPE and dipoles in CPX are characterized by quantum efficiencies η_p and η_x respectively. In general, both of these quantum efficiencies are comprised of three fundamental processes (generation, displacement and transfer *versus* trapping) which are basic "building blocks" from which η_p and η_x are constructed. In this section, we shall review the physical nature of these building blocks and discuss how they fit together in different ways to form the characteristic quantum efficiencies of PAPE and CPX. This will establish the basis for formulating the material requirements summarized in Section 11.7.

11.6.1 Energetics of the latent image forming process

An energy diagram which is useful for visualizing the fundamental processes comprising the formation of an electrophotographic latent image is illustrated in Fig. 11.9. This is an energy diagram for an electron in a layer of photoconductive material contiguous with a conductive substrate. On the opposite side of the photoconductor is a dielectric medium which may be a fluid or a solid. In the latter case, the photoconductor and dielectric together become a two-layer xerographic photoreceptor. The diagram is slanted (out of scale) to schematize the presence of internal electric fields, E_p in the photoconductor and E in the dielectric. In general, these fields are different; and the field in the photoconductor may be produced in part (or entirely) by a set of charges (shown as holes in the diagram) which are trapped at the photoconductor—dielectric interface.

A description of the allowed energy states in the photoconductor and dielectric is dependent on the mathematical formalism used to describe the movement of a carrier (electron or hole) through the system. In a

References pp. 476—478

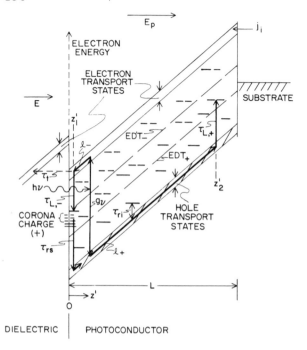

Fig. 11.9. A schematic energy diagram indicating allowed electronic energy states in a photoconductor and contiguous dielectric. g_ν, l_+ and $\tau_{L_\pm}$ represent the fundamental photogeneration, displacement and trapping processes of a hole and electron respectively. τ_t represents a possible transfer event competing with $\tau_{L,-}$.

"conventional" trap-controlled description [30, 31], the allowed energy states are broadly separated into "transport states" and "traps," with the motion of a carrier in the transport states being characterized via a well-defined mobility (μ). The traps, on the other hand, are characterized by a capture time (τ_i) from the transport states and a release time (τ_{ri}) for the return of a carrier back to the transport states. The notation "i" refers to the fact that the capture and release take place for the ith trap. Any spatial movement of a carrier during the capture and release events is neglected. The physical nature of the carrier motion within the transport states may be "free translational motion" or "hopping"; but, in the case of hopping, the transition (or tunneling) times between states must be short compared to measurable time intervals. It turns out that this is necessary to obtain a well-defined mobility, as discussed more fully later. Thus, any localized state, that is sufficiently isolated from neighboring states so that the transition time of a carrier to or from such a state is observably long, must be extracted from the transport states and treated as an isolated trap. In contrast with the above description, Scher and Montroll [32] consider localized states which are characterized by a release (or hopping) time and a "simultaneous" spatial displacement. The displacements are treated as unit "steps" on a regular lattice, and a complete release—displacement event is characterized by a distribution of hopping (or release) times. Thus, there is no separate identification of "transport states" and "traps" in this description. At the present time, it appears that the mathematical formalisms which follow from these different descriptions predict different transport behavior. The differences, however, appear as subtle as the distinction in the underlying physical bases, and further elucidation of the real consequences of "distributed traps" *versus* "stochastic hopping" remains as an interesting and challenging problem.

In the formulation of the electrophotographic quantum efficiencies to follow, the distributed-trap description will be utilized. It presents a simpler framework in which to describe the fundamental processes comprising the quantum efficiencies. It is also conveniently applicable to such materials as a-Se and a-As_2S_3 [33]. On the other hand, the "stochastic hopping" solution has been successfully used to interpret the transient photoresponse of a-As_2Se_3 [34], and certain organic materials [32] (see Chapters 3 and 7). A necessary modification in the analysis required to compute the electrophotographic quantum efficiencies in this case is indicated later.

The three fundamental processes comprising the formation of a latent image dipole are indicated in terms of transitions among the above states. Each of these processes are discussed in detail shortly, but their physical nature may be visualized in the following manner. The in-going photon creates a pair of mobile carriers which then drift through the transport states, in and out of "shallow" traps, until the carriers finally become "permanently" trapped (at z'_1 and z'_2) or they successfully arrive at the respective surfaces of the photoconductor. If a carrier (say the electron) does arrive at a boundary surface it may transfer into the adjoining medium or it may move laterally along the interface until it becomes "permanently trapped" at the interface. Quantification of the possible transfer and "permanent trapping" events (characterized by τ_t and τ_L respectively) are important in electrophotography, and examples are given later. But first it is essential to define what "permanent" means.

Once a carrier is captured by a trap whose mean release time is longer than the time allotted for development (or processing) of a latent image, the associated dipole becomes temporarily "fixed" for the image in question. Thus, the image processing time (T_p) is a useful discriminant in any given imaging application. Accordingly, we separate all traps into "fast" and "slow" types depending on whether the mean release time (τ_{ri}) is short or long compared to T_p. To assist visualization of this separation we have indicated "electrophotographic demarcation times" (EDT$_+$ and EDT$_-$) for the holes and electrons in Fig. 11.9. It should be stressed however that these demarcation "lines" are defined with respect to time not energy. This is an important distinction because τ_{ri} depends upon the matrix element

References pp. 476—478

connecting the traps to the transport states (or capture cross-section) as well as an activation energy. Thus the ordering of traps with respect to τ_{ri} can be drastically different from their ordering with respect to energy. But as long as one thinks of the energy diagram as applying to traps of a given capture cross-section (or matrix element), then it can be a useful visual aid.

11.6.2 The three fundamental processes

We now discuss the individual photogeneration, displacement and trapping *versus* transfer events on a more quantitative basis.

(i) Photogeneration

The photogeneration event alone is characterized by a probability (γ) that an absorbed photon is successfully converted into a separable electron and hole. A dependence of γ on the local electric field [35] is now well established for many materials and is usually accounted for in terms of a geminate recombination process. For a review of the underlying mechanisms responsible for the field dependence of the photogeneration process, and the general success of Onsager's theory [36] in this regard, the reader is referred to Chapters 3 and 7.

In terms of γ, the net local photogeneration rate at z' is given by:

$$g_\nu(z') = (1 - R_\nu)\gamma\alpha \exp(-\alpha z') F_0(t), \qquad (11.50)$$

where R_ν and α are the reflection and absorption coefficients for the incident light. F_0 is the incident light flux in photons per unit area.

(ii) Displacement

Given sufficient time subsequent to the generation event, the carriers will ultimately reach the surfaces of the photoconductor or be captured by a slow trap. With diffusion and surface recombination neglected, the displacements from a given generation point are given by the following Green's function solutions to the continuity equations [31]:

$$K_+(z_2', z') = \exp\left(-\int_{z'}^{z_2'} \beta_+(\mu)\, d\mu\right), \text{ for the hole;}$$

$$K_-(z_1', z') = \exp\left(+\int_{z'}^{z_1'} \beta_-(\mu)\, d\mu\right), \text{ for the electron;} \qquad (11.51)$$

where $\beta = [\mu\tau_L E_p(\mu)]^{-1} \equiv l^{-1}$ is the probability per unit displacement that a carrier will be captured by a slow trap; μ is the carrier mobility, and $\tau_L^{-1} = \Sigma_{i,\,\text{slow}}\, \tau_i^{-1}$ is the lifetime for capture by any slow trap. To avoid cumbersome notation the $+$, $-$ subscripts indicating specific reference to the hole and electrons respectively are left understood unless they affect the functional form of the equations in which they appear. Physically, K_+ and K_- represent the probabilities that the hole and electron will arrive at z_2' and z_1' respectively without being captured by a slow trap at all.

Since $\beta_+ dz_2'$ is the probability that a carrier will be trapped between z_2' and $z_2' + dz_2'$, given that it arrived at z_2' without being trapped, the expectation values for the fractional displacements defined by eqn. (11.24) are

$$\bar{f}_{\delta_1}(k, z') = \int_0^{z'} f_{\delta_1}(k, z_1')\beta_- K_- dz_1' + f_{\delta_1}(k, 0) K_-(0, z') \qquad (11.52)$$

$$\bar{f}_{\delta_2}(k, z') = \int_{z'}^{z} f_{\delta_2}(k, z_2')\beta_+ K_+ dz_2' + f_{\delta_2}(k, L) K_+(L, z')$$

These expressions are used later in the final formulation of the xerographic quantum efficiency. It should be pointed out here that probability distributions for z_1' and z_2' at time t (subsequent to generation) also appear in the Scher—Montroll formalism [32]. Thus, their solution of the displacement problem also can be used to compute the expectation of f_{δ_1} and f_{δ_2} at the moment of development ($t = T_p$).

Although the solution given above neglects diffusion it does accomodate a variable (non-vanishing) electric field. It can be shown however that diffusion generally produces very little effect. In fact, when the electric field is uniform, the continuity equation can be solved in closed form with diffusion and surface recombination included. The result has the effect of modifying the generation efficiency from γ to:

$$\gamma_{\text{eff}} \simeq \gamma \left[1 - \left(\frac{v_R}{v_R + \mu E_p} \right) \frac{\alpha}{\alpha + \frac{eE_p}{k_B T}} \right] \qquad (11.53)$$

where v_R is a "velocity" characterizing the recombination of carriers diffusing back into the upstream boundary (i.e., the dielectric boundary in Fig. 11.9 for the case of holes). This result is valid for $E_p > (k_B T/e\mu\tau_L)^{1/2}$. $k_B T$ in these expressions is Boltzmann's constant times temperature. This result is of some interest because the quantity in square brackets further reduces to $\mu E_p/(\mu E_p + v_R)$ for strongly absorbed light ($\alpha \gg eE_p/k_B T$). This shows that surface recombination eventually forces a fall-off in the net generation efficiency at sufficiently low electric fields. This may account for the apparent differences in the generation efficiencies for the holes and electrons that is sometimes observed [34, 37] when the electric field is reversed. (μ and v_R apply to different carriers in this case.)

A further restriction on the validity of eqn. (11.52) is that a sufficient time must lapse for the carriers to completely transit across the photoreceptor if they are not captured in a slow trap at all. Unfortunately it is not possible to quantify this condition in a simple way unless the carriers effectively sample each kind of fast trap several times. But when this does

References pp. 476—478

occur the mean transit time, designated t_T, is given by:

$$t_T = t_{T_0} + \sum_i N_i \tau_{ri}; \qquad (11.54)$$

where $t_{T_0} = L/\mu E_p$ corresponds to the transit time of a carrier which is not trapped at all, and $N_i = L/\mu \tau_i E_p$ is the expected number of times a carrier is captured in the ith kind of trap. When the latter is substituted into eqn. (11.54), it becomes:

$$t_T = \frac{L}{\mu E}\left[1 + \sum_i \frac{\tau_{ri}}{\tau_i}\right] = \frac{L}{\theta \mu E}; \qquad (11.55)$$

where $\theta \equiv (1 + \sum_i \tau_{ri}/\tau_i)^{-1}$ is the familiar θ-factor appearing in the definition of the drift mobility. It turns out, however, that eqn. (11.54), and the concept of a drift mobility as well, breaks down if $N_i \sim 1$ for any trap for which τ_{ri} is significant (meaning those near the EDT especially). It can be shown that the effect of infrequent trapping in different traps ($N_i \sim 1$), also leads to a broad dispersion in transit times which is much like the transit-time dispersion (and mobility failure) pointed out by Scher and Montroll [32] for stochastic hopping. A detailed comparison of dispersion under the different conditions, however, must await a more complete evaluation of the effect of distributed traps. As a working condition for the validity of eqn. (11.52), we may require $t_T < T_p$, where t_T is either a measured value (the time for the transient photocurrent to fall to half its peak value say) or eqn. (11.54) when it is applicable. We shall also use this condition in Section (11.7) as a materials criterion.

(iii) Interfacial transfer versus trapping

To quantify the transfer *versus* trapping options for a carrier that does arrive at the photoreceptor surface, we require a solution to the continuity equation that is appropriate to this region [31]. For the electron in Fig. 11.9, the probability of transfer into the dielectric after time t (measured from the moment of arrival at the surface) is given by:

$$T_-(t) = \frac{\tau_{LS-}}{\tau_{LS-} + \tau_{t-}}(1 - \exp(-t/\tau_{S-})); \qquad (11.56)$$

where τ_t^{-1} and τ_{LS}^{-1} are the probabilities per unit time for transfer and trapping respectively, and $\tau_S^{-1} \equiv \tau_{LS}^{-1} + \tau_t^{-1}$. The surface trapping time, τ_{LS}, is given by

$$\frac{1}{\tau_{LS-}} = \frac{\theta_{S-}}{\tau_{L-}} + \theta_{S-}v_{R-}\left(\frac{eE_p}{k_B T}\right); \qquad (11.57)$$

where θ_S is the fraction of total transport and fast traps in the surface region which are transport states; τ_{L-} is the bulk lifetime for capture of an electron by a slow trap, and v_{R-} is a recombination velocity characterizing the capture of electrons by any additional slow interface states. If the electrons have access to the deposited corona charge, for example, it can be shown that $v_{R-}(eE_p/k_B T) \gg \tau_{L-}^{-1}$.

The transfer probability τ_t depends on the nature of the transfer process, and there are a variety of possibilities. For example, if the dielectric layer next to the photoreceptor is a liquid, then transfer may occur via capture of the electron in some preferred interface state, followed by the desorption of an ion. A complete description of this particular transfer process is provided in ref. 31. Another interesting example which could occur in a two-layer photoreceptor is the thermionic transfer of an electronic charge into a solid dielectric. In this case, it can be shown that τ_t is given by:

$$\frac{1}{\tau_t} = \frac{\theta_S E_p}{k_B T N_c} A \exp\left(\frac{-\phi}{k_B T}\right) \frac{\mu_t E}{\mu_t E + v'_R} \tag{11.58}$$

where N_c is the total number of transport states per unit volume in the photoconductor; $A \simeq [4\pi m_\perp^* e(k_B T)^2/h^3]$ is the Richardson–Dushmann coefficient for thermionic emission ($m_\perp^*$ being the appropriate electron effective mass and h being Planck's constant); ϕ is the barrier height between the bottom of the transport states in the photoconductor and dielectric; μ_t is the carrier mobility in the dielectric and v'_R is a recombination velocity for a carrier backing into the interface from the dielectric side. This expression obviously applies to the special case of translational transport states in the photoconductor which overlap in energy with transport states in the dielectric. Alternative formulae for tunneling or thermally-assisted tunneling across the interface can also be written, but the above example of thermionic transfer is adequate to illustrate one way in which the distribution of states and inherent transport characteristics of the adjoining materials combine to control $T(t)$.

An important property of T_- (or T_+ for the hole) is that the destiny of a carrier remains undecided for times $t < \tau_S$. During this time, the carriers remain in the transport states (or fast traps) where they can be acted upon by the tangential electric field E_x [cf. eqn. (11.19)]. The effect of E_x is to move mobile carriers laterally and attenuate the image while it forms [38].

By solving for the tangential motion of the carriers in response to E_x, it can be shown that the amplitude of the latent image charge, $\delta\Sigma_{Ik}$ attenuates with time in accordance with:

$$\delta\Sigma_{Ik} = \int_0^t \delta\dot{\Sigma}_{Ik}(t') \exp\left[-\int_{t'}^t \theta\Sigma_S(t'')\mu_S \frac{kb_k}{\kappa_p\kappa_0} dt''\right] dt' \tag{11.59}$$

where $\delta\dot{\Sigma}_{Ik}$ is the time rate at which the periodic part of the dipole image forms; $\theta_S\Sigma_S(t)/e$ is the instantaneous concentration of accumulated electrons which actually reside in transport states; μ_S is the mobility of the

References pp. 476–478

carriers along the surface, and b_k is the coefficient given by eqn. (11.22) which characterizes the strength of the tangential electric field. Evaluation of this expression requires both the distribution of arrival times for the electrons at the photoconductor—dielectric interface and the instantaneous displacement of the holes moving toward the substrate. But to indicate an important aspect of the photoconductor—dielectric interface, it is enough to consider a δ-function arrival of the electrons and complete transit of the holes instantaneously; i.e. $\Sigma_S(t) = \Sigma_{I0}\delta(t-0)$, and $\delta\Sigma_{Ik} = \delta\Sigma_{Ik}(0)\delta(t-0)$; where Σ_{I0} and $\delta\Sigma_{Ik}(0)$ are the broad area (average) and periodic parts of the photogenerated charge, computed as if lateral carrier motion did not occur. In this case, eqn. (11.59) can be simply written $\delta\Sigma_{Ik} = \delta\Sigma_{Ik}(0)A_k(t)$, where

$$A_k(t) = \exp\left[-\frac{kb_k}{\kappa_p\kappa_0}\Sigma_{I0}\theta_S\mu_S\tau_S(1-\exp(-t/\tau_S))\right]. \tag{11.60}$$

Since the magnitude of the argument of this exponential is at least proportional to k, b_k being also proportional to k for $kL < 1$ and $\simeq 1$ otherwise, it can be seen that the effect of lateral motion is to limit resolution. By setting the argument equal to a constant (π say), one finds that wavelengths shorter than:

$$\lambda_c \simeq 2(\theta_S\mu_S\tau_S)E_{p0}\left[\frac{\Sigma_{I0}}{\Sigma_c}\right](1-\exp(-t/\tau_S)) \tag{11.61}$$

cannot be resolved. Here we have introduced $\kappa_p\kappa_0 E_{p0} = \Sigma_c$, where E_{p0} is the initial field inside the photoreceptor and Σ_c is the corona charge per unit area. For small $\delta D_{i,k}$, $\Sigma_{I0} \simeq \Sigma_c - \Sigma_{min}$; so $\Sigma_{I0} \simeq \Sigma_c$ for normal exposures. Thus, eqn. (11.61) implies λ_c less than a carrier schubweg along the surface $(\theta_S\mu_S\tau_S E_{p0})$ cannot be resolved. This has important material implications as discussed in Section 11.7.

11.6.3 The quantum efficiencies of CPX and PAPE

Using the above charge generation, displacement, and transfer events as "building blocks," we now synthesize the electrophotographic quantum efficiencies η_x and η_p. We consider first "specific" values which are conditional on the point of generation, at z' say, and the direction of the electric field. For the field direction shown in Fig. 11.9

$$\eta_x(k,z') \equiv \gamma_{eff}\bar{f}_\delta(k,z')(1-T_-). \tag{11.62}$$

This expression appears different from the original definitions [7] in that they are defined here for a specific absorption point (z'). In addition, both carriers are now generated with identical efficiencies, but the relevant γ_{eff} will depend on the directions of illumination and electric field. For the case shown in Fig. 11.9, γ_{eff} pertains to the hole; but if the field were reversed, or the direction of illumination were reversed for the same field directions,

γ_{eff} would pertain to the electron. The previous restriction to slowly varying images is also now removed by including the dependence of η_x on k explicitly. In the limit, as $k \to 0$, $\bar{f}_\delta \to \bar{f}_\delta(0, z') = [\bar{\delta}_+(z') + \bar{\delta}_-(z')]/L$, and the present generalized definition of η_x reduces to the original one.

The corresponding "specific quantum efficiency" for PAPE is given by

$$\eta_p(z') = \gamma_{\text{eff}}[K_-(0, z')T_- - K_+(L, z')T_+]. \tag{11.63}$$

To render the diagram in Fig. 11.9 applicable to this case, the substrate must be replaced by a dielectric fluid. T_+ can then be constructed by analogy with T_-. Note that the same processes build η_p as η_x. The main difference simply lies in whether or not transfer or trapping should be exploited.

The above quantum efficiencies are conditional on an absorption event. The net "average" quantum efficiencies are finally obtained by averaging eqns. (11.62) and (11.63) over the absorption point z'. The result depends upon the direction of illumination relative to the electric field. If their directions coincide as in Fig. 11.9, then

$$\eta_x(k) = (1 - R_\nu)\alpha \int_0^L \eta_x(k, z') \exp(-\alpha z') \, dz' \tag{11.64}$$

$$\eta_p = (1 - R_\nu)\alpha \int_0^L \eta_p(z') \exp(-\alpha z') \, dz'. \tag{11.65}$$

But if the direction of illumination is reversed with respect to the field then $\exp(-\alpha z')$ must be simply replaced by $\exp[-\alpha(L - z')]$.

11.6.4 Connection between η_x and the photoinduced discharge characteristic

We conclude this section on the fundamental processes of latent image formation by indicating certain conditions under which the xerographic quantum efficiency is simply related to a photo—induced—discharge characteristic (PIDC). In general, the connection between the two is simple only for an incremental uniform exposure. For then Δn_p photons per unit area will generate an effective surface charge density of $\Delta \Sigma_{I0} = e\eta_{x0}(E_p)\Delta n_p$; and the voltage drop across the photoreceptor is simply

$$\Delta V = \frac{e}{\kappa_p \kappa_0} \eta_{x0}(E_p) L \Delta n_p. \tag{11.66}$$

Here $\eta_{x0}(E_p)$ means $\eta_x(k = 0, E_p)$. Thus, a measurement of ΔV for a known Δn_p provides a direct measure of the overall broad area quantum efficiency. But it does not provide the generation efficiency (γ) and displacement efficiency ($\bar{f}_\delta$) separately. These can be separated only via a transit-time resolution measurement [34, 35]. ΔV also does not provide any information about transfer or lateral motion. The latter is typically observable only in a developed image.

References pp. 476—478

In the special case in which it is known that there is no range limitation (say by the observation of a transit time), then $\bar{f}_\delta = 1$, and the slope of the PIDC will give the generation efficiency (γ) directly. It is only necessary to make certain that the voltage at each exposure is measured after a transit time for the slowest carrier. The impact of slowly-moving carriers on the instantaneous shape of a PIDC has been discussed by Mort and Chen [39].

Unfortunately, when there is a range limitation (or even when there is multiple trapping in relatively-slow fast states) it can be a difficult task to separate the generation and displacement parameters (γ, β_+, β_-) from a PIDC measurement. An example which illustrates the procedure is a-As_2S_3 [33]. This is an extreme case, however, and is of little interest commercially — the range limitation is so severe that it precludes cyclic use. In more practical materials, where only a small amount of deep trapping is present, separation of the generation and displacement parameters is much easier — though extreme caution must be taken not to confuse "permanent" trapping from slow transit-time effects.

11.7 PHOTORECEPTOR MATERIAL REQUIREMENTS

During the early development of xerography, the only criterion used in the search for new photoconductive materials was high dark resistivity [40]. It was assumed that the resistivity had to be large enough so that the dielectric relaxation time (resistivity, ρ, times permittivity, $\kappa_p \kappa_0$) exceeded the development time. Subsequent work [41], however, shows that the dielectric relaxation time requirement is neither necessary nor sufficient. What is necessary instead is a high carrier range ($\mu \tau_L$) and low thermal generation rate (g_{th}). In addition, a relatively high carrier mobility ($\theta \mu$) and low equilibrium carrier concentration ($p/\theta = g_{th} \tau_L/\theta$) are also required. Here p is the free carrier density. Since the electrical conductivity is a measure of the product, $p\mu = g_{th} \mu \tau_L$, it provides no assurance by itself that the individual factors will fall in the correct range. An even more striking result is that the photoconductor must have blocking contacts and that all thermal carriers must be extracted over the useful thickness of the material. In other words, xerographic photoreceptors must be charge-depletion devices with a depletion layer thickness which is greater than the minimum required image dipole length (or device thickness). It also turns out that material conditions in the surface region of a photoreceptor must be fundamentally incompatible with the material conditions required in the bulk.

In this section, we reconstruct the logical developments which verify the above results and lead to a complete set of criteria for the selection of materials for efficient imaging. The essential inputs are the extreme fields inside the photoreceptor ($E_{p,max}$ and $E_{p,min}$), the minimum photoreceptor thickness or image dipole length (L) and the available image process or development time (T_p). As shown in Sections 11.3 and 11.5, $E_{p,max}$, $E_{p,min}$ and L are all determined by the developer properties, and T_p, of course, is determined by the system application. The outputs are conditions on g_{th}, $\theta\mu$ and τ_L, expressed in terms of the inputs ($E_{p,max}$, $E_{p,min}$, L and τ_L). The connection between the outputs and inputs are constructed by requiring that the latent image must be constructed before development begins, and that development must be completed before the latent image decays. To obtain the special surface conditions it is also necessary to include a resolution requirement.

We first examine the times required to complete each of the fundamental processes involved in the construction of the latent image. We then examine the additional dark conditions required to preserve the latent image long enough for it to be developed.

11.7.1 Photo-requirements

Each of the fundamental processes comprising the formation of a latent image must be accomplished in less than T_p.

(i) Generation

The time to generate enough carriers to decrease the internal field from $E_{p,max}$ to $E_{p,min}$ is given by

$$t_g = \frac{\kappa_p \kappa_0}{e(1-\exp(-\alpha L))(1-R_\nu)F_0} \int_{E_{p,min}}^{E_{p,max}} \frac{dE}{\gamma_{eff}(E)} \simeq$$

$$\simeq \frac{\kappa_p \kappa_0 (E_{p,max} - E_{p,min})}{e(1-\exp(-\alpha L))(1-R_\nu)F_0 \gamma_{eff}(E_{p,min})}. \tag{11.67}$$

It is evident that $t_g < T_p$ is effectively a condition on the exposure intensity F_0. It has been suggested that E_p could remain close to $E_{p,max}$ with a flash exposure, thus providing a higher effective conversion efficiency and less light [42]. To avoid field collapse in the generation region, however, it would require an exposure duration shorter than the transit time across the absorption depth (*i.e.* $t_g < (\alpha\mu E_{p,max})^{-1}$).

(ii) Displacement

To establish a temporal condition on the displacement event, it is assumed that each generation event shall contribute one full-length dipole. This assumption is made to exclude the possibility of synthesizing long dipoles from several short ones placed end to end. The latter is a physical possibility but it requires weakly absorbed light and many photons to make one dipole — a highly inefficient process.

For maximum efficiency then, it is evident that at least one member of any charge pair that is to expand into a full-length dipole must transit the photoconductor in less than the process time. In fact, this must be true for all dipoles contributing to a latent image. Hence, it must also be true for the slowest moving charge, which moves in a field of $E_{p,min}$. For the case in which the drift mobility (or θ-factor) concept applies, this condition can be written

$$t_T = \frac{L}{\mu\theta E_{p,min}} < T_p. \tag{11.68}$$

An obvious additional requirement is that the carrier must live long enough to accomplish the feat; i.e.

$$\frac{\tau_L}{\theta} > t_T. \tag{11.69}$$

It is of interest that this latter condition is identical to $\mu\tau_L E_{p,min} > L$, which is independent of the θ-factor. It should also be noted that this lifetime (or schubweg) condition implies $\bar{f}_\delta \to 1$, which can be verified from eqns. (11.51) and 11.52).

(iii) Transfer versus trapping

The desirability of the transfer event in favor of trapping or vice versa, depends upon the specific interface and application. At the surface of a photoactive pigment particle, for example, it is essential to favor transfer over trapping, meaning $T(t = T_p) \simeq 1$, for at least one of the charge carriers. It is also essential at internal boundaries in multicomponent xerographic photoreceptors such as photoactive particles dispersed in a solid matrix (or binder). On the other hand, there is always one boundary in a xerographic photoreceptor where the length of the image dipoles must terminate. At this boundary, it is essential to favor trapping over transfer, meaning $T(t = T_p) \ll 1$. Thus, there are two different situations to consider, which we shall refer to as "intentional transfer" $[T(T_p) \simeq 1]$ and "intentional trapping" $[T(T_p) \ll 1]$.

In each of the above cases, we must also consider the consequences of a time delay before trapping or transfer does occur. This is especially important at planar boundaries in a xerographic photoreceptor where lateral motion of the carriers may limit resolution. According to eqn. (11.61) $\lambda < \lambda_c$ cannot be resolved unless:

$$2\theta_S \mu_S \tau_S E_{p0} < \lambda_c, \text{ if } \tau_S < T_p \tag{11.70}$$

or

$$2\theta_S \mu_S T_p E_{p0} < \lambda_c, \text{ if } T_p < \tau_S. \tag{11.71}$$

This assumes general imaging applications which may include $\Sigma_{I0} \sim \Sigma_c$. Physically, the quantity $(\lambda_c/2\theta_S \mu_S E_{p0}) \equiv t_{\lambda_c}$ represents the time it takes for a carrier to drift along the interface a distance $\lambda_c/2$. Thus, eqns. (11.70) and (11.71) simply say that an image must be developed in less time than t_{λ_c}, unless the drift of a carrier is first terminated by transfer or trapping. (Recall that $\tau_S^{-1} \equiv \tau_t^{-1} + \tau_{LS}^{-1}$, so that τ_S is simply the shorter of τ_t and τ_{LS}.) We shall make use of the resolution constraint expressed by eqns. (11.70) and (11.71) in the discussion below on the material implications of transfer *versus* trapping.

For intentional transfer, it follows from eqn. (11.56) that both $\tau_t \ll \tau_{LS}$ and $\tau_t < T_p$ are required to assure $T(T_p) \simeq 1$. This means slow interface states must be avoided, and the transport states in the photoreceptor and dielectric must be reasonably well matched. The matching condition is not severe, however, though it does preclude barriers in excess of 0.5 eV say for $T_p \sim 1$ second. On the other hand, a much more severe matching condition arises for planar xerographic interfaces where lateral motion may limit resolution. Using $\tau_S^{-1} \equiv \tau_{LS}^{-1} + \tau_t^{-1}$, we know $\tau_{LS} \gg \tau_t$ implies $\tau_S \simeq \tau_t$, and $\tau_t < T_p$ means $\tau_S < T_p$. Whence, eqn. (11.70) implies $2\theta_S \mu_S E_{p0} \tau_t < \lambda_c$ (or, $\tau_t < t_{\lambda_c}$). To appreciate the significance of this requirement, it is useful to consider the special case in which the surface mobility is equal to the bulk mobility ($\theta_S \mu_S = \theta \mu$). In this case it can be seen that τ_t must be less than t_T at the highest field to resolve λ_c less than $2L$.

For intentional trapping [meaning $T(T_p) \ll 1$], it can be shown from eqn. (11.56) that we must satisfy either $\tau_{LS} < \tau_t$ or $T_p < \tau_t$ and τ_S (if $\tau_t \ll \tau_{LS}$). In general, these conditions mean transfer must be intentionally inhibited. Of course, this naturally occurs under the usual xerographic conditions where the field outside the photoreceptor vanishes. But more severe restrictions again arise as we next examine the above results in conjunction with the resolution criteria given by eqns. (11.70) and (11.71). We note first that if $T_p < \tau_S$, eqn. (11.71) applies directly. But when this is compared to eqn. (11.68), it can be seen that it would be impossible to resolve $\lambda < 2L$ unless $\theta_S \mu_S E_{p,\max} < \theta \mu E_{p,\min}$; i.e., the drift velocity along the surface at the highest field must be less than the drift velocity through the bulk at the lowest field. Turning to the other alternative possibility of $T_p > \tau_S$ we must now have $\tau_{LS} < \tau_t$ also to assure $T(T_p) \ll 1$. In this case, $\tau_S \simeq \tau_{LS}$ and eqn. (11.70) implies $2\theta_S \mu_S \tau_{LS} E_{p0} < \tau_c$. Comparing this with eqn. (11.69) we now find that $\lambda < 2L$ cannot be resolved unless $\theta_S \mu_S \tau_{LS} E_{p,\max} < \mu \tau E_{p,\min}$; i.e. the schubweg along the surface at the highest field must be less than the schubweg through the bulk at the lowest field.

Now the above are remarkable results because they show that intentional trapping at the free surface of a xerographic photoreceptor requires either (a) that the free surface must contain several orders of magnitude more traps than the bulk, or (b) that the carrier which is trapped at the free surface is necessarily incapable of also traversing the bulk. Stated differently, the latter case means the photoconductor must be unipolar and illuminated through the free surface.

References pp. 476—478

At first sight, it may seem that most photoreceptors may meet the conditions in case (b). But as we later examine the conditions required of a photoreceptor to accept a corona charge, we shall find that the free surface of the photoreceptor must contain sufficient slow surface traps to accommodate all the corona charge. It can further be shown that if the local concentration of these slow surface traps were extended into the bulk, the photoreceptor could not be discharged. Thus, it can be said that all corona chargeable photoreceptors effectively belong in the same class as case (a). It is of further significance in this regard that a corona charge also normally serves as effective slow traps (actually recombination centers) for the photogenerated carriers that approach the free surface from the interior. In fact, this is why resolution limitations due to lateral conduction are not normally present in xerography except when the photoreceptor is overcoated by an insulator which keeps the corona and image charge separated. Conversely, whenever the corona and image charge are separated by an insulator, as in Frost [9], Katsuragawa [57] and Canon NP [58] processes, it may be necessary to treat both surfaces of the insulator so that sufficient slow traps are made available to "pin down" both the corona and the image charge.

Finally, after corona chargeable photoreceptors are included with case (a) above, the only xerographic processes left that do not require an excess of slow traps at the surface are induction imaging and a persistent internal polarization [59] process whereby the sensitizing field used to separate the photogenerated carriers is not supplied by a corona. But, surprisingly enough, there is evidently no commercial system that exploits this interesting case.

11.7.2 Dark requirements

In general, the stability or lifetime of a latent image is controlled by thermally-generated discharge processes which can produce the same effect as the photogenerated dipoles. It is customary, however, to use the initial voltage decay rate in the dark, or the initial discharge current density $[j_D = (\kappa_p \kappa_0/L) dV/dt]$, as a quantitative measure of the thermally-generated discharge processes. A general criterion, for the stability of the latent image can thus be written:

$$j_D(E_{p,max}) T_p < 0.1 \, \kappa_p \kappa_0 E_{p,max}. \tag{11.72}$$

which simply asserts that no more than 10% dark discharge from the initial condition shall be allowed. To translate eqn. (11.72) into a useful condition for materials selection however, it is necessary to be specific about the nature of the physical processes which contribute to j_D.

Let us first examine what is wrong with the traditional assumption that $j_D = \sigma E_{p,max}$, where σ is the dark conductivity. Note that substitution of this j_D into eqn. (11.72) yields the traditional dielectric relaxation time criterion: $\tau_\kappa = (\kappa_p \kappa_0/\sigma) > 10 T_p$. Indeed, this is just the procedure normally followed to prove the dielectric relaxation time criterion. But to invoke this criterion, it must also be assumed that the contacts are "ohmic". Otherwise there would be no assurance that the current is really bulk limited. But Rose [23] has shown that a bulk-controlled current must transform from an ohmic current to a space-charge-limited current (SCLC) whenever $t_T < \tau_\kappa$; i.e.

$$j_D \doteq \begin{cases} \sigma E_{p,\max} = \kappa_p \kappa_0 E_{p,\max}/\tau_\kappa, \text{ for } \tau_\kappa < t_T \\ \dfrac{\kappa_p \kappa_0 \mu E_{p,\max}^2}{L} = \kappa_p \kappa_0 E_{p,\max}/t_T, \text{ for } t_T < \tau_\kappa \end{cases} \quad (11.73)$$

Substituting these into eqn. (11.72) gives $t_T > 10 T_p$ (for $t_T < \tau_\kappa$) and $t_T > \tau_\kappa 10 T_p$ (for $t_T > \tau_\kappa$). Clearly, neither of these is compatible with our earlier photo requirement of $t_T < T_p$ (cf. eqn. (11.68)).

Of course, it should be recalled at this point that actual occurrence of SCLC requires a sustained voltage across the sample, meaning the electrical circuit must be closed. And since the voltage above a photoreceptor is floating (or the circuit is open) under the conditions for which eqn. (11.72) applies (the usual xerographic operating condition), it can be argued that the substitution of the SCLC expression into eqn. (11.72) is not meaningful. On the other hand, it can be seen, on physical grounds, that the conditions which would produce SCLC if the voltage were sustained cannot be allowed either.

The significance of SCLC and ohmic conduction for the xerographic case can be envisioned with reference to Fig. 11.9. For simplicity, we may suppose the electrons are immobile, in which case the nature of the substrate contact is irrelevant. (The electrons it would supply could not move anyway.) An "ohmic contact" at the free surface may be simulated by considering the corona charge (manifest as holes) as occupying transport states and fast traps near the dielectric interface but inside the photoreceptor. They cannot occupy slow traps and still qualify as being available for conduction, as required by definition of an "ohmic" contact. In this case, the injected corona charge would become indistinguishable from photogenerated charge, and there would be enough of them present to completely discharge the photoreceptor in one transit time. Indeed, this is the meaning of SCLC; and the physical situation just described does correspond to $t_T < \tau_\kappa$. In the alternative case of $\tau_\kappa < t_T$, there would already be an equilibrium hole concentration greater than $\kappa_p \kappa_0 E_{p,\max}/L$ distributed through the bulk of the photoreceptor (before corona injection) and these carriers would not even have to move the full photoreceptor thickness to completely discharge the photoreceptor. Thus, ohmic contacts and bulk-limited dark discharge of a photoreceptor necessarily means complete discharge in one transit time or less. The same arguments would apply if the dark conduction were due to electrons and the substrate contact were ohmic.

As a converse of the above, it can now be seen that in order to prevent

References pp. 476—478

more than 10% discharge of the photoreceptor in one transit time, no more than 10% of the charge producing $\kappa_p \kappa_0 E_{p,\max}$ can be mobile (or occupy fast traps). Thus, the corona charge, and all the counter charge on which the field lines produced by the corona charge terminate, must occupy slow traps. In effect, this means the contacts must be injection limited (or "blocking"). Note also that this argument would still apply if the corona were replaced by a charged metal film. The free electrons in the metal would then have to be blocked outside the photoconductor to avoid complete collapse of the field in one transit time. Thus, we conclude that a blocking contact is a *necessary* condition for *efficient* latent image formation in xerography. The blocking contact is necessary to avoid complete collapse of the field in the dark while making it possible for each photogenerated carrier to completely traverse the photoconductor (or field region) in the light. In photoconductivity, a blocking contact is sometimes defined as a boundary across which charge is not allowed to move, thus precluding the possibility of a secondary photocurrent. However, it is also frequently used as one which is simply non-ohmic; *i.e.* it does not supply whatever charge is required to maintain a zero-field condition just inside the physical boundary of a material (as in SCLC) or space-charge neutrality throughout the bulk of a material. It is this non-ohmic sense that is the intended meaning of a blocking contact in this chapter.

To formulate a meaningful expression for the dark current when a photoconductor is fitted with blocking contacts, it is enough to examine the case of a unipolar material (hole transporting say). It is also enough to examine the transient current at times of the order of T_p.

With a blocking contact, a depletion layer always forms near the contact. But sufficiently far from the contact, equilibrium conduction eventually prevails [43]. In this equilibrium region (or before a field is applied at all), we can make use of the principle of detailed balance and write

$$g_{\text{th}} \equiv \sum_{i,\text{slow}} \frac{p_i}{\tau_{ri}} = \sum_{i,\text{slow}} \frac{p}{\tau_i} = \frac{p}{\tau_L} = \frac{p}{\theta} \times \frac{\theta}{\tau_L} \quad (11.74)$$

Here p_i is the concentration of holes in the ith kind of localized state; τ_{ri} is the mean release time from the same state; and p is the equilibrium concentration of carriers in the transport states. The summations extend only over slow traps. This definition of g_{th} is appropriate because it corresponds to the definition of τ_L, and the population of the slow traps will not change appreciably in times of the order of T_p. (Recall that $\tau_{ri} > T_p$ for slow traps by definition.) It should be noted that the definitions of both g_{th} and τ_L are unique to electrophotography and should not be confused with the more conventional definitions in semiconductor theory.

If we now envision a corona suddenly applied to the structure indicated in Fig. 11.9, all the mobile holes (p/θ) will be depleted from the transport and fast traps over a distance:

$$z_d' = \frac{\kappa_p \kappa_0 (E_{p,\max} - E_{p,i})}{e(p/\theta)} \tag{11.75}$$

where $E_{p,\max}$ is the maximum internal field next to the dielectric interface and $E_{p,i}$ is the field beyond the depletion layer. The current generated in the depletion layer and the field required to sustain it in the non-depleted region are determined by the following relations

$$j_D = eg_{th} z_d' = ep\mu E_{p,i} \tag{11.76}$$

It is assumed for the moment that there is negligible injection from the surface. When the above expressions are made consistent with the total voltage across the photoconductor, z_d', $E_{p,i}$ and $E_{p,\max}$ become uniquely determined. For our purposes, however, we need only examine the limiting solution for which the depletion layer extends completely through the photoreceptor. In this case, the current reaching the substrate saturates at:

$$j_D = eg_{th} L \tag{11.77}$$

and it can be shown that the necessary conditions for this current saturation are approximately given by:

$$\mu \tau_L E_p > L, \text{ if } \tau_\kappa > \tau_L, \tag{11.78}$$

or

$$\kappa_p \kappa_0 E_p > g\frac{p}{\theta} L, \text{ if } \tau_\kappa < \tau_L. \tag{11.79}$$

Note also, in the first case, that $\mu \tau_L E_p > L$ implies $t_T < \tau_L < \tau_\kappa$, and, in the second case, $\kappa_p \kappa_0 E_p > g(p/\theta)L$ implies $t_T < \tau_\kappa < \tau_L$. This shows that the appearance of a saturated generation-limited current (GLC) with blocking contacts is compatible with the photorequirements established earlier. Hence, it is an important dark characteristic for assuring maximum utilization of the primary photogenerated carriers (i.e. $\mu \tau_L E_p > L$).

Before formulating the maximum GLC allowed for xerography, it is of general interest to note that $\mu \tau_L E_p > L$ also implies a secondary photocurrent (or gain >1) under sustained voltage conditions. Since it has further been shown [55] that photocurrent gain depends only on the total number of recombination centers and their capture cross-section, independently of the contacts, it is concluded that ohmic contacts are not required in general to achieve a secondary photocurrent. These results appear inconsistent with the general argument of Rose [23] that ohmic contacts and SCLC in the dark are important characteristics for maximum photoresponse. On the other hand, eqns. (11.78) and (11.79) do imply SCLC would occur in the dark if the voltage is sustained and the contacts are ohmic. Thus the results described here are in agreement with Rose [23], except that the assumption of an ohmic contact is no longer required.

References pp. 476—478

Having established a dark current expression, eqn. (11.77), which is valid when the contacts are blocking, we next substitute it into eqn. (11.72), and obtain

$$eg_{th} L T_p < 0.1 \kappa_p \kappa_0 E_{p,\max} \tag{11.80}$$

For a given photoreceptor thickness, this is a condition on the maximum thermal generation rate. Note that it is independent of μ and τ_L and therefore no longer in conflict with the photorequirements. In fact, the basic photorequirement that caused trouble earlier ($t_T < T_p$, or, basically, $t_T < \tau_L$), now appears as one of the conditions for GLC. Thus, eqn. (11.80) arises as a useful replacement for the old dielectric relaxation criterion. This is why blocking contacts and GLC are important in xerography.

Now an important consequence of eqn. (11.80) and the photorequirements together is that they all become increasingly easy to satisfy as the photoreceptor becomes thinner. In fact, the realization of this provided the first clue that the required thickness of the photoreceptor must be determined by the electric field required to drive development. We now know, from the results in previous sections, how the constraint on the minimum photoreceptor thickness arises. We have seen that there is a minimum δE required to drive development, and for a fixed charge variation, this δE is proportional to the photoreceptor thickness. For the purpose of establishing the magnitude of the minimum thickness, we may then assume that the photoreceptor is charged to the maximum internal field that the photoconductor can withstand without producing excessive spurious discharge (or noise). Then the minimum required δE [namely $2\sigma(E^*)$], times a factor of 2, say, (to leave room for image neutralization) establishes a minimum required thickness. This is the thickness we must assume is known in the equations in this section in order to make them unique and meaningful.

In the case of a material for which the length of the latent image dipoles is limited by the depletion layer thickness (rather than $\mu \tau_L E_p$), it may be realized that any physical thickness of the photoreceptor greater than the depletion layer thickness is ineffectual (or irrelevant). Therefore, in such cases we may as well choose the maximum physical thickness to match the maximum depletion layer thickness. For this case, it can also be shown that $\tau_\kappa < \tau_L$, whence eqn. (11.79) applies as the condition required to reach the saturated-current limit given by eqn. (11.77).

Now, in converse to the above, if we regard L in eqn. (11.79) as the minimum photoreceptor thickness required for development and take E_p as the maximum surface field ($E_{p,\max}$), then eqn. (11.79) becomes a restriction on the equilibrium mobile carrier concentration (p/θ). Thus, eqns. (11.78) and (11.79) are both meaningful as xerographic material requirements, and it is significant that the two together cover all the possibilities for realizing a saturated GLC.

So far we have discussed only the bulk requirements that must be satisfied, and to establish these it was enough to consider perfectly blocking contacts. But in practice, contacts are never perfectly blocking. In fact they are often so "leaky" that it is difficult to actually observe the bulk GLC limit. To establish an upper bound on the allowed injection rate we may again invoke eqn. (11.72) with j_D replaced by an injection current density per unit area, designated J_i. Whence we obtain

$$J_i(E_p) < 0.1 \kappa_p \kappa_0 E_{p,\max}/T_p \qquad (11.81)$$

At the free surface, the injection current density can be more explicitly written $J_i = \kappa_p \kappa_0 E_{p,\max}/e\tau_{rs}$; where $\kappa_p \kappa_0 E_{p,\max}/e$ is the number of corona charges per unit area present and τ_{rs} is their mean release time from the surface states. Substituting this J_i into the limiting-current expression yields $\tau_{rs} > 10 T_p$. This means the corona charge must occupy slow traps, as expected.

Finally, it can be shown that if the above surface trap concentration were extended into the bulk of the photoconductor it would result in a range limitation ($\mu \tau E < L$) on the photodischarge process. Thus, corona charged photoreceptors require surface conditions which are fundamentally incompatible with the material conditions in the bulk. In view of this, it is remarkable that in the early development of xerography, a material (Se) was found which naturally provided the necessary surface traps, though it sometimes occurred only after "break-in".

TABLE 11.1

Photoreceptor material requirements

Photorequirements	Dark requirements
(1) generation time: $t_g < T_p$	(1) Blocking contacts (to mobile carriers)
(2) transit time : $t_T = \dfrac{L}{\mu E_{p,\min}} < T_p$	(2) Depletion: $\dfrac{p}{\theta} L < \kappa_p \kappa_0 E_{p,\max}$
(3) lifetime: $\tau_L > t_T$	(3) Bulk generation: $eg_{th} L T_p < 0.1 \kappa_p \kappa_0 E_{p,\max}$
(4) transfer *versus* trapping; (a) for intentional transfer $\tau_t < T_p$ and τ_{LS}, and $2\theta_S \mu_S \tau_t E_{p,\max} < \lambda_c$ (b) for intentional trapping $\tau_t \gg T_p$; $2\theta_S \mu_S T_p E_{p,\max} < \lambda_c$ or $\tau_t \gg \tau_{LS}$; $2\theta_S \mu_S \tau_{LS} E_{p,\max} < \lambda_c$	(4) Contact injection: $J_i T_p < 0.1 \kappa_p \kappa_0 E_{p,\max}$ (5) Corona traps: $\tau_{rs} > 10 T_p$

The results of this section are summarized in Table 11.1. Note that all the parameters which characterize the bulk and surface or contact conditions of the photoconductor (μ, τ_L, g_{th}, τ_{rs}, etc.) are entirely expressed in terms of quantities ($E_{p,\max}$, $E_{p,\min}$, L, T_p and λ_c) which, in turn, are basically determined by the developer, system process time and the required resolution.

References pp. 476—478

TABLE 11.2

Photodischarge characteristics of four practical photoreceptors. See text for discussion.

	Units	a-As$_2$Se$_3$	a-Se	1:1 TNF:PVK	ZnO-resin
$\max \dfrac{dV}{dD_{i,0}}$	V	1170[a]	930[a]	500[b]	800[c]
V_0	V	920	780	−800	−350
$V_{\min}$	V	190	100	−250	−70
$\Delta V_{\max} = V_0 - V_{\min}$	V	730	680	550	280
Thickness (L)	μm	60	60	17[g]	
Exposure	$\dfrac{\text{ergs}}{\text{m}^2} \times 10^{-4}$	5	16	20	~10^3 (300 lux s)
Light Source		3700 K WI	3700 K WI	550 nm	2900 K tungsten
$E_{p,\min} = \dfrac{V_{\min}}{L}$	$\dfrac{V}{m}$	3.2 × 10^6	1.7 × 10^6	15 × 10^6	
$t_T(E_{\min})$	s	0.05[d]	3 × 10^{-6}	0.4[e]	
τ_L	s	>1	2–50 × 10$^{-6\,f}$		

[a] Scharfe [44]
[b] Schaffert [45]
[c] Ohyama et al. [46]
[d] 5 times fastest transit. Fastest transit computed from data taken from Pai and Scharfe [47].
[e] 5 times fastest transit. Fastest transit computed from data taken from Gill [48].
[f] Tabak and Hillegas [49].
[g] Photoreceptor thickness guessed in this case.

The one material requirement not adequately emphasized in this summary is $\eta_x(E_{p,min})$. Although it is implicit in the generation time (t_g), its limitation on the latent image gamma ($dV/dD_{i,0}$) for small input contrast densities [cf. eqn. (11.48)] is normally the more demanding requirement.

To the authors knowledge, all photoreceptors of commercial interest satisfy the above conditions. The specific characteristics of several examples are discussed in the next section.

11.8 DISCUSSION OF PRACTICAL XEROGRAPHIC PHOTORECEPTORS

The properties of four commercial photoreceptors are now examined in terms of the theoretical formalism described herein. They consist of 2 chalcogenides (a-As_2Se_3, a-Se) one organic material (1:1 TNF:PVK) and one photoactive particle-binder structure (ZnO-resin). Pertinent data are most available for these photoreceptors and they reflect the wide variety of material and structural differences that are useful.

To construct a meaningful comparison between these photoreceptors, we may suppose that they are to be used to reproduce (or enhance) the same low density images — the usual copy machine requirement. We may also suppose that they must interact with the same developer. This is normally not the case; of course, and in cases where the corona charge is different, the same developer cannot even be used because the toner polarity would be wrong. In addition, the photoreceptor surface materials differ and this would alter the toner—photoreceptor adhesive interaction. But barring these differences, for which compensations may be made, in principle, it is of interest to examine how they would perform in interaction with an identical (or equivalent) developer.

Assuming also that development is done under adhesion-limited conditions, then the pertinent quantity for making a comparison is the maximum latent image gamma. These are listed in Table 11.2 and were computed from slopes of the voltage *versus* exposure curves provided by the indicated references. An example is illustrated in Fig. 11.10. The fact that the maximum gammas are comparable suggests that the developers with which the photoreceptors interact may also have similar properties (especially P').

The initial surface potentials (V_0) corresponding to the discharge characteristics from which the gammas were determined are believed typical of values actually used in commercial machines. Since it can be shown that the maximum $dV/dD_{i,0}$ increases monotonically with the initial surface potential, it is plausible to suppose that the operating potentials used are the highest possible before the appearance of "defect noise." This is consistent at least with the fact that charge acceptance of many photoreceptors is known to be limited by (or associated with) spurious discharge. To avoid

References pp. 476—478

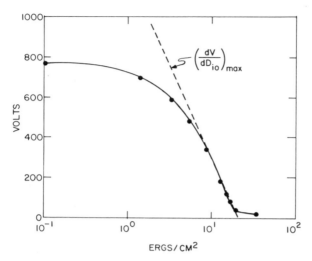

Fig. 11.10. Broad-area photodischarge characteristic of a-Se with a 3700K tungsten iodide light source. (From Scharfe [44].)

seeing the consequences of this in the output images, the photoreceptors must be operated somewhat below this level. Just how close to the noise limit practical systems are operated, however, has not been explicitly stated in the literature.

As pointed out by Schaffert [45], the latent image gamma of 1:1 TNF:PVK remains close to its maximum value over a broad exposure range; and from inspection of the discharge curves provided in ref. 45, it can be seen that the true maximum (for $V_0 = -800$ V) occurs close to a surface potential of -100 V. The larger (absolute) value of -250 V, indicated in Table 11.2, was selected to illustrate another advantage of operating a photoreceptor at the maximum possible surface potential. For example, from the transit time data on 1:1 TNF:PVK, provided by Gill [48], it can be computed that the "effective transit time" (explained later) is reduced from 2 seconds at $V_{min} = -100$ V to 0.4 seconds at $V_{min} = -250$ V. Thus, a clear advantage of operating at the higher V_{min} is a faster cycle rate, though it requires some sacrifice in the maximum voltage contrast ($\Delta V_{max} = V_0 - V_{min}$).

The "effective transit times" listed in Table 11.2 are 5 times the fastest transit in the cases of a-As_2Se_3 [47] and 1:1 TNF:PVK [48]. These are both broad dispersion materials and 5 times the fastest transit is long enough for about 90% of the carriers to complete their journey across the photoreceptor. To compute the transit time for 1:1 TNF:PVK it was necessary to guess the thickness of the specific photoreceptor for which the discharge characteristic is available. From other information provided in the same reference, 17 μm is a reasonable guess.

The lifetime for capture by slow traps (τ_L) has been measured only for

a-Se [49]. This is understandable because the effect of capture by slow release traps is difficult to distinguish from the effect of slow transits in materials with dispersive transport. Thus for a-As$_2$Se$_3$ and 1:1 TNF:PVK, it can only be said that τ_L is longer than the effective transit time. This much can be concluded from the low residual potentials that are obtained with strongly-absorbed light.

The exposure energies required to achieve the maximum latent image gammas vary greatly for the four photoreceptors. But this in itself does not have any particular significance with regard to the ability of a system to reproduce black and white images. In fact, for any photoreceptor for which reciprocity between time and intensity holds, it can be shown that the maximum latent image gamma is only a function of V_0 and V_{min} (the lowest voltage to which the photoreceptor is discharged), independent of the spectral sensitivity of η_x. However, the maximum latent image gamma is dependent on any field dependence of η_x, and is always smaller than:

$$\frac{dV}{dD_{i,0}}\bigg|_{V = V_{min}} = 2.3 |V_0 - V_{min}|$$

at the same V_0 and V_{min}. This upper bound obtains when η_x is independent of the electric field; and in no case is $dV/dD_{i,0}$, at the same V_0 and V_{min}, dependent on the spectral sensitivity of η_x. It is easy to verify that the latent image gammas listed in Table 11.2 are consistent with the above expression except for the ZnO-resin. In this case, the experimental value is slightly larger than the field-independent limit of 645 V, presumably because of a slight dependence of the discharge characteristic on exposure intensity.

The large exposure energy in the case of ZnO-resin is primarily due to the use of a "white" light source in conjunction with a nearly intrinsic (white) ZnO powder whose photoresponse starts at 380 nm. The use of a white powder is necessary if the developed image is fixed directly to the photoreceptor and a white background is desired. With dye sensitization, the "white light" exposure energy may be reduced by an order of magnitude; but the developed image in this case must be transferred to paper to obtain a white background.

The fundamentals of latent image formation in ZnO-resin remains obscure. Its discharge characteristics were first interpreted by Amick [50] in terms of a homogeneous space-charge (or depletion) layer formation — with the depletion-layer thickness shrinking below the physical-layer thickness at voltages below the neighborhood of 100 V. Although a wide variety of different ZnO-resin combinations appear to exhibit the depletion-layer characteristic, a subsequent measurement by Kiess [51], of the dark charging and photodischarge characteristics on the same sample provides at least one counter example. The charging characteristic at low voltages (<100 V) proved to be the normal (voltage independent) capacitor type —

instead of the $Q \propto V^{1/2}$ behavior expected for a depletion layer. At the same time, the photodischarge characteristic revealed a field-dependent "quantum efficiency" at low voltages; and above some threshold voltage (unspecified) the "quantum efficiency" became field independent. Now, it can be shown that a field-dependent quantum efficiency and a shrinking depletion layer produce very similar discharge characteristics, distinguishable only by careful scrutiny of the data at very low voltages. As a consequence, it appears that each particular ZnO-resin should be carefully re-examined for subtle differences in the low-voltage region. The depletion-layer cases should also be verified by measuring its charging characteristic — a point emphasized by Kiess [51].

By comparing the characteristics of ZnO-resin with ZnO single crystals, Kiess [51] also showed that there is no field dependence in the primary generation event in ZnO. Thus, the field dependence in the quantum efficiency (corresponding to our $\eta_{x,0}$) of the powder-resin must be due to limited carrier displacement — most likely due to trapping at interparticle barriers. In fact, Kiess [51] showed that the discharge rate at low surface potentials is plausible in terms of trapping. It is pertinent in this regard that the interfacial transfer probability $T(t)$ — necessary for particle-to-particle transfer in photoactive pigment-binder photoreceptors — is proportional to at least the first power of the field whenever the transfer time constant (τ_t) falls below the interfacial slow-trapping lifetime (τ_{LS}). This suggests that the characteristics of ZnO-resin photoreceptors may also be understandable in terms of an interparticle transfer probability. In addition, $T(t)$ is also dependent on time as well as the field and temperature, and this could conceivably explain many of the peculiar properties of particle-resin photoreceptors in general. For example, it could account for an intensity-dependent induction period (or reciprocity failure) and the apparent lack of a measurable transit time.

The first experiments which separate the dark discharge processes in a manner consistent with the material requirements formulated in the previous section was carried through for As_2Se_3 by Ing and Neyhart [52]. The bulk generation and surface injection discharge components were separated by measuring the total discharge rate as a function of the photoconductor thickness. For well-aged photoreceptors (presumably surface oxidized), surface injection was found negligible. Thus, the total discharge could be interpreted in terms of a field-assisted thermal generation of holes. Electron transport was ruled out because negatively-charged photoreceptors did not photodischarge with strongly-absorbed light.

By fitting the time dependence of the discharge rate of As_2Se_3 over a range of temperatures and fields, the following distribution of trapped holes, at $t = 0$, was determined

$$p_1(t = 0) = 0.35 \times 10^{13} \text{ cm}^{-3}, \text{ at } \mathcal{E}_{t,1} = 0.6 \text{ eV}$$

$$p_i(t=0) = 0.35 \times 10^{13}\, \frac{i}{2}\, \text{cm}^{-3}, \text{ at } \mathcal{E}_{t,i} = \mathcal{E}_1 + 0.02\,(i-1)$$

for $i = 2,3,4,\ldots 10$,

$$p_i(t=0) = 7 \times 10^{14}\, \text{cm}^{-3}, \text{ at } \mathcal{E}_{t,i} = 0.8, 0.82 \text{ and } 0.84 \text{ eV}.$$

The corresponding release times are given by $\tau_{ri} = \nu_i^{-1}\exp(\mathcal{E}_{t,i}/k_B T)$, where $\nu_i (=10^{11}\,\text{s}^{-1}$ for all traps) is the attempt-to-escape frequency and $\mathcal{E}_{t,i}$ is the activation energy. A common field dependence for all release times proved to be consistent with the data. The field dependence could be fit with a Poole–Frenkel-like expression ($\tau_{ri} \propto \exp \beta E^{1/2}/k_B T$) with $\beta = 2 \times 10^{-24}$ $\text{Jm}^{1/2}/\text{V}^{1/2}$ (about half the expected Poole–Frenkel value). The activation energies specified above correspond to a field of 10^7 V/m.

It can be shown that the solution of the dark discharge problem represented by the above distribution of generation centers is essentially unique. The only ambiguity concerns the density of very slow traps (defined by $\tau_{ri} \gg$ the longest measurement time) which cannot be separated from the attempt to escape frequency.

Following Ing and Neyhart, Schein [53] identified two discrete hole generation sites in a particular a-Se photoreceptor (containing 1.3 p.p.m. Cl and a 1 p.p.m. Fe) with zero-field activation energies of 0.6 and 0.7 eV. Their initial populations were 1.9×10^{12} and 3.5×10^{13} per cm^3 respectively; and their corresponding release times were 32 and 800 seconds at 19.7°C and 10^7 V/m. The field dependence of the release times from both sites are again Poole–Frenkel-like, though this time, in contrast with a-As_2Se_3, β is very close to the normal Poole–Frenkel value.

Hole injection from the corona charged surface of the above a-Se material is comparable to the bulk generation rate for a 60-μm-thick sample. Without prior surface treatment with nigrosine, or aging (presumably oxidation) the hole injection rate from the corona charged surface of a-Se is evidently much larger. In fact, without any surface treatment, or "break-in," the total dark decay may render an a-Se photoreceptor inoperative.

It was possible, for the above a-Se sample, to rule out electron injection from the substrate because negatively-charged samples failed to discharge with strongly-absorbed light. In general, however, the discharge properties of a-Se are strongly sensitive to impurities and ambipolar materials are not uncommon. In such cases, electron injection from the substrate is often significant as evidenced by the fact that charge acceptance is frequently sensitive to the substrate material.

Little detailed information is available concerning the specific effect of substrates on the total discharge rate of a-Se. It is known, however, that oxide layers thicker than 15 Å on aluminum prove to be adequate substrates for both charging polarities [54]. However, the fact that they work for the positively-charged mode is remarkable. One would expect the oxide to

References pp. 476–478

present a very large barrier for the escape of holes and this could lead (on cycling) to enhanced electron injection. In fact, a slightly enhanced dark decay on cycling (called "charge fatigue") is observed for oxide-coated aluminum substrates, and charge acceptance is limited to values corresponding to an internal electric field of 2×10^7 V/m. This is understandable if the probability of capture of an electron by a hole (accumulated inside the oxide) falls below unity at the maximum charging field. The remarkable thing is that the measured capture probability of a hole by an electron [55] (accumulated inside a phenoxy layer) does fall below unity at a field of 2×10^7 V/m. This could be a coincidence, or it could simply be that the capture cross-section of a coulombic center is about the same regardless of its polarity.

From what is known about the dark discharge properties of a-As_2Se_3 and a-Se so far, it is evident that much can be learned about the distribution of deep traps and the surface injection properties from a careful study of the time, temperature, field and thickness dependence of the dark discharge rate. A similar analysis of the dark discharge characteristics of practical organic and powder-binder photoreceptors is apparently unavailable. However, the total dark discharge rate of 1:1 TNF:PVK is known to be highly sensitive to the substrate material and corona charging conditions [45]. Presumably, a bulk generation component also exists, but it has not been separately identified. An increasing charge acceptance with decreasing layer thickness is indicative of a bulk process, but the effect is small and could conceivably be due to light from the corona. There is also a composition dependence (TNF to PVK ratio) on the charge acceptance [45], but this could also be a surface injection phenomenon. High concentrations of bulk states are expected to play a key role in the surface injection rate as well.

The dark discharge characteristics of ZnO single crystals have been analyzed by Kiess [56]. He also formulated a procedure for separating surface- and bulk-controlled processes under strong bulk-space-charge conditions. The procedure was demonstrated by comparing the dark discharge rates of an insulating crystal (which charged as a voltage-independent capacitor) with a conducting crystal (which charged as a depletion layer). The surface potential in both cases decayed in proportion to $\log(1 + t/\tau)$, indicating a uniform distribution of generation sites with respect to activation energy (assuming a constant attempt-to-escape frequency). The shortest release time (τ) was about 1 second for the insulating crystal and 60 seconds for the conducting crystal. In the latter case, the released species was necessarily the minority carrier (holes) from the bulk, whereas the carrier type and origin (surface *versus* bulk) could be either electrons or holes for the insulating crystal. The absence of any evidence of accumulated space charge at low voltages, however, strongly favors surface injection. The logarithmic decay law also strongly favors electron injection, because one would expect injection of holes from the substrate to follow a different discharge law.

Although Kiess's studies on ZnO single crystals have clearly demonstrated a strong dependence of the dark discharge processes on impurity content and surface conditions, it is impossible to apply the results to a powder-resin structure without knowing the specific nature of the individual crystallites. The interpretive difficulties are compounded by the fact that conducting particles may sometimes be converted into an insulating state by surface oxidation, making surface release of electrons and bulk generation of holes both significant contributors to the total dark discharge rate. Substrates are also known to sometimes affect the total discharge rate of ZnO resin photoreceptors, but separation of this effect from the bulk and free surface processes is clearly complex.

Much remains for a complete understanding of any specific photoreceptor, though significant advances have been made for the simpler homogeneous systems. Of course, it is not necessary to know the specific dark discharge mechanisms to the depth indicated above to predict the imaging performance of a system alone; but to understand the temporal limitations of a system and possible remedies, the more detailed information discussed here is clearly valuable.

11.9 CONCLUSIONS

The theory of charged pigment electrophotography, as reformulated herein, can be used effectively as a prototype for the analysis and evaluation of any physical imaging system in which development is quantized in units of a single particle. It is only necessary to identify (or postulate) the physical interaction (or forces) that control the appearance of a particle in the final image. The image-wise variation of this interaction and the subsequent formalism described in Sections 11.2, 11.3 and 11.5 will carry through in a similar manner. A general prediction is that the fluctuations (or noise) in the development interaction will always result in a fundamental limitation on the system performance.

In the theory of CPE in particular, we have seen that definition of the development interaction (QE) and an interfering force (adhesion) immediately led to the identification of the fluctuations in adhesion as the dominant factor limiting the performance of CPX. We have seen that adhesion noise determines a minimum size and charge of toner, and establishes a minimum strength of the electric field required to drive development. The latter, in turn, also determines a minimum thickness for the photoreceptor. A remarkable corollary to the above is that the size of adhesion in typical dry triboelectrified toner—carrier developers is evidently so large that the minimum required electric field is only marginally attainable.

References pp. 476—478

We have characterized the latent image forming interactions in CPE in general terms, by defining quantum efficiencies (η_p and η_x) for the change in charge on a particle or the creation of an electric dipole. We then showed how η_x and η_p can be synthesized from the same underlying generation, displacement and transfer *versus* trapping processes. Consideration of the physical conditions required to form a xerographic latent image and preserve it long enough for development then led to general criteria for the selection of materials to build a functional system. The three most notable implications of these criteria are: (1) The photoresponsive material must always be depleted of all mobile carriers in the sensitized state (*i.e.* when subjected to an electric field); and this rules out applicability of the traditional dielectric relaxation criterion and ohmic contacts. (2) The trap concentrations required at the surface of a photoconductor in typical xerographic systems greatly exceeds the concentrations that can be tolerated in the bulk. (3) A quantitative set of material criteria cannot be constructed without first specifying the thickness of the photoconductor and extrema in its internal electric field, both of which are determined by the developer.

We finally showed that several practical photoreceptors do produce about the same magnitude electric-field variation. Hence, they can all function, in principle, in interaction with the same (or equivalent) developer. Their electrical properties (in the dark) are also remarkably similar in a broad view; though they differ substantially in terms of the physical nature of the controlling charge-generation processes.

REFERENCES

1 C.F. Carlson, Electrophotography, U.S. Patent 2,297,691, (1942).
2 R.M. Schaffert, Electrophotography, revised edn., Focal Press, London, 1965; Halstead Press, New York, 1975.
3 J. Dessauer and H. Clark (Eds.), Xerography and Related Processes, Focal Press, London, 1965.
4 R.B. Comizzoli, G.S. Lozier and D.A. Ross, Electrophotography — A review, Proc. IEEE, 60 (1972) 348.
5 Special Issue on Electrographic Processes, IEEE Trans. Electron Devices, ED—19 (1972).
6 M.E. Scharfe and F.W. Schmidlin, Charged pigment xerography, Adv. Electron, Electron Phys., 38 (1975) 83.
7 F.W. Schmidlin, IEEE Trans. Electron Devices, ED—19 (1972) 448.
8 V. Tulagin and L.M. Carreira, U.S. Patent 3,384,565 (1968).
   V. Tulagin, J. Opt. Soc. Am., 59 (1969) 328.
9 R.W. Gundlach and C.J. Claus, Photogr. Sci. Eng., 7 (1963) 14.
10 R.W. Gundlach, Liquid development of electrostatic latent images, U.S. Patent 3,084,043, (1963).
11 J.T. Bickmore, M. Levy and J. Hall, Photogr. Sci. Eng., 4 (1960) 37.
12 E. Inoue, H. Kohado, K. Kosei and H. Sato, Denshi Shashin, 7 (1967) 71.
13 K.A. Metcalfe, J. Sci. Instrum., 32 (1955) 74.

14 C.J. Claus and E.F. Mayer, in J. Dessauer and H. Clark (Eds.), Xerography and Related Processes, Focal Press, London, 1965, Chap. XII.
15 H.G. Greig, U.S. Patent 2,811,465, (1957).
16 C.R. Mayo, U.S. Patent 3,245,823 (1966); 2,895,847 (1959).
17 R.W. Willmott, U.S. Patent 3,232,190, (1966).
18 C.J. Young and H.G. Greig, RCA Rev., 15 (1957) 471.
19 E. Giamo, U.S. Patent 2,786,440, (1957).
20 C.J. Young, U.S. Patent 2,786,441, (1957).
21 E. Wise, U.S. Patent 2,618,552, (1952).
22 L.E. Walkup, U.S. Patent 2,618,551, (1952).
23 A. Rose, Concepts in Photoconductivity and Allied Problems, Wiley, New York, 1936; IEEE Trans. Electron Devices, ED—19 (1972) 430.
24 M.D. Tabak, S.W. Ing and M.E. Scharfe, IEEE Trans. Electron Devices, ED—20 (1973) 132.
25 F.W. Schmidlin, 14th Ann. Fall Symp. SPSE, Wash. D.C., Oct. 23—26, 1974.
26 D.K. Donald, J. Appl. Phys., 40 (1969) 3013.
27 R.M. Schaffert, Electrophotography, Part II, Focal Press, London, 1965, Chap. 3.
28 J.A.C. Yule and W.J. Nielsen, Tech. Assoc. Graphic Arts Proc., 3 (1951) 65.
29 R.W. Gundlach, personal communication. See ref. 1 for discussion of physical basis of Gundlach's rule.
30 Results of this formalism pertinent to the interpretation of the transient photoresponse of As_2S_3 are discussed in ref. 33.
31 G.G. Hartmann and F.W. Schmidlin, J. Appl. Phys., 46 (1975) 266.
32 H. Scher and E. Montroll, Phys. Rev. B, 12 (1975) 2455.
33 S.W. Ing, Jr., J.H. Neyhart and F.W. Schmidlin, J. Appl. Phys., 42 (1971) 696.
34 M.E. Scharfe, Phys. Rev. B, 2 (1970) 5025.
35 M.D. Tabak and P.J. Warter, Jr., Phys. Rev., 173 (1968) 899.
36 L. Onsager, J. Chem. Phys., 2 (1934) 599.
37 D.M. Pai and R.C. Enck, Phys. Rev. B., 11 (1975) 5163.
38 F.W. Schmidlin and H.M. Stark, 22nd Ann. Conf. SPSE, Los Angeles, Calif., May 12—16, 1969.
39 I. Chen and J. Mort, J. Appl. Phys., 43 (1972) 1164.
40 C. Wood, in J. Dessauer and H. Clark (Eds.), Xerography and Related Processes, Focal Press, London, 1965, Chap. V.
41 F.W. Schmidlin, 26th Ann. Conf. SPSE, May 6—11, 1973.
42 I. Chen, R.L. Emerald and J. Mort, J. Appl. Phys., 44 (1973) 3490.
43 F.W. Schmidlin, G.G. Roberts and A.I. Lakatos, Appl. Phys. Lett., 13 (1968) 353.
44 M.E. Sharfe, personal communication.
45 R.M. Schaffert, IBM J. Res. Dev., 15 (1971) 75.
46 Y. Ohyama, T. Kurita and Y. Takahashi, J. Soc. Sci. Photogr. Jpn., 24 (1961) 22, from ref. 2, p. 249.
47 D.M. Pai and M.E. Scharfe, J. Non-Cryst. Solids, 8—10 (1972) 752.
48 W.D. Gill, J. Appl. Phys., 43 (1972) 5033.
49 M.D. Tabak and W.J. Hillegas, J. Vac. Sci. Technol., 9 (1971) 387.
50 J.A. Amick, RCA Rev., 20 (1959) 770.
51 H. Kiess, J. Appl. Phys., 40 (1969) 4054.
52 S.W. Ing, Jr. and J.H. Neyhart, J. Appl. Phys., 43 (1972) 2670.
53 L.B. Schein, Phys. Rev. B, 10 (1974) 3451.
54 W.D. Hope and M. Levy, in J. Dessauer and H. Clark (Eds.), Xerography and Related Processes, Focal Press, London, 1965, Chap. IV, pp. 109—112.
55 F.W. Schmidlin and S. Tutihasi, in J. Stuke and W. Brenig (Eds.), Amorphous and Liquid Semiconductors, Vol. I, Taylor and Francis, London, 1974, p. 383.

56 H. Kiess, On the dark decay of negatively charged ZnO single crystals. In W.F. Berg and K. Hauffe (Eds.), Current Problems in Electrophotography, deGruyter, Berlin, New York, 1972, pp. 89—101.
57 K. Yoshida, K. Kinoshita and T. Kawamura, Appl. Opt. Suppl. 3, (1969) 170.
58 M.J. Mitsui, IEEE Trans. Electron Devices, ED—19 (1972) 396.
59 H.P. Kallmann, J. Rennert and M. Sidran, Photogr. Sci. Eng., 4 (1960) 345.

AUTHOR INDEX

Numbers in *italics* refer to the pages where references are listed in full; numbers in parentheses are the actual reference numbers.

Abell, G.C., 359(89), *386*
Abkowitz, M., 86, 89, 90(21), 89(25), *114*, 291(177), 293(184), 294(187), *301, 302, 323(57), 334*
Abraham, A., 288, 290(152), *301*
Abrahams, E., 75, 88, 110(13), *114*
Adamczewski, I., 336, 341, 342, 349, 374, 375, 377(10), 336(11), 377(141, 142), *384, 387*
Adams, A.R., 46, 57(113), 50(133), *66*
Adams, E.N., 133(162), *151*
Adirovich, E.K., 22, 24(41), *25*
Aducci, F., 61(212), *69*
Ahmad, S.R., 27(3), *63*
Ahrenkiel, R.K., 61(205), *68*
Aihara, S., 403(28), *419*
Ainslie, N.G., 130(102), 130(103), *149*
Akamatu, H., 176(72), *184*
Akutagawa, W.M., 46, 47(107), *66*
Albrecht, A.C., 364(97), *386*
Aleksandrova, G.A., 130(111), 130(128), *149, 150*
Alexander, J.H., 186(7), *213*
Allemand, R., 380(160), *388*
Allen, A.O., 54(159), *67*, 318(47), *333*, 339, 341, 357(15), 341, 374(24), 344(43), 344, 370(44), 347, 357, 359(53, 54, 55), 349, 359(56), 370(129), 377(138, 139), *384, 385, 387*
Allen, C.W., 141(253), *152*
Allen, G.A., 130(120), *150*
Allen, M., 367(108), *386*
Allgaier, R.S., 59(185), *68*
Al-Sharbaty, A., 186, 191, 192, 194(9), 186, 195, 204(10), *213*, 241(73), *299*
Alvarez, L.W., 377, 380(148, 149), *387*
Ambegaokar, V., 88(23), *114*
Amick, J.A., 471(50), *477*
Amith, A., 62(216), *69*
Amodé, J.J., 413–417(48), *419*
Anderson, G., 56(170), *67*
Anderson, L.K., 414(50), *419*
Anderson, P.W., 295(192), *302*
Andreev, A.A., 60(187), *68*
Appling, A.J., 265(111), *300*

Arndt, R.A., 60(193), *68*, 172(61), *184*
Arnold, E., 408(39), *419*
Arnoldussen, T.C., 73(8), *114*, 197, 200(22), *213*, 239, 241, 245, 248, 250, 252(44), 241, 242, 245, 246, 248(64, 65), *298, 299*
Arthur, J.B., 34(54), *64*
Asaf, U., 363(165), *388*
Augustyniak, W.M., 83, 85(18), *114*
Austin, I.G., 253(88), 291(169), 293(181), *299, 301, 302*
Auth, J., 27(4), 62(217), *63, 69*
Avery, D.G., 35(52), *64*

Babaev, A.A., 253(85, 86), *299*
Baessler, H., 159(12), 159(13), 159(14), 163(34), 163(36), *182, 183*
Bagley, B.G., 267(114), *300*, 314(39), *333*, 373(132), *387*
Bakale, G., 339(17), 341(27), 346, 370(46, 47), 370, 372, 375(128), 371(171), 376(172, 173), *384, 385, 387, 388*
Balakin, A.A., 377(137), *387*
Baldini, G., 358(164), *388*
Balkanski, M., 135, 136(216), *152*
Banerji, J., 286, 288(143), *301*
Barbarie, A., 62(214), *69*
Bardeen, J., 3(6), *25*, 135(183), 135(184), *151*
Bardsley, W., 35(54), *64*
Barton, L.A., 122(54), 125, 126, 138(86), *148, 149*
Baru, V.G., 20(39), *25*
Batra, I.P., 40, 42, 43(86), 40, 42(87), 42(91), 42(94), 42(95), 42(96), *65*
Batt, R.H., 49(127), *66*, 328(64), *334*
Baxendale, J.H., 375(135), 376(136), *387*
Beck, G., 366, 377(106), *386*
Becker, M., 47(121), *66*
Bednarczyk, D., 60(194), *68*
Bednarczyk, J., 60(194), *68*
Beecham, D., 393(11), *418*
Beer, A.C., 59(181), *68*
Bekiarian, P.A., 12(28), *25*
Bell, M., 375(135), *387*

Beniski, G., 31(25), *64*
Bennett, A.J., 1(1, 2), *24*
Benoit a' la Guillaume, C., 27(2), *63*, 253(84, 87), *299*
Benson, M.J., 56(169), *67*
Bergmann, A., 27(3), *63*, 160(21), *183*
Berlman, I.B., 56(169), *67*
Berrihar, J., 161, 164(22), *183*
Berry, R.E., 61, 62(213), *69*
Berry, R.S., 174, 175(66), *184*
Berry, W.B., 141(253), *152*
Bers, A., 28(14), *63*
Berth, M., 405(32), *419*
Bertolini, G., 39(80), *65*
Bertram, W.J., 411(45), *419*
Bickmore, J.T., 423(11), *476*
Biktimirova, V.Kh., 271(116), *300*
Bishop, S.G., 253(82, 83), *299*
Blacknall, D.M., 130(117), *150*
Blakemore, J.S., 118(15), *147*
Blakney, R.M., 46(109), *66*, 257, 275, 277, 280, 281(93), *299*
Blanc, D., 350, 353(66), 377(143, 144, 145, 146, 147), 378(144), 378(153), *385, 387*
Blanc, J., 120(49), 120, 127, 128(50), 128(93), 129(98), 129(99), *148, 149*
Bleha, W.P., 141(260), *153*
Blossey, D.F., 49(132), *66*, 217, 234(3), 291(177), *297, 301*
Blount, G.H., 122(57), 122(58), 143, 144 (268), *148, 153*
Blumenfeld, S.M., 404(31), *419*
Boag, J.W., 50(140), *66*
Bode, D.E., 135(191), 141(247), *151, 152*, 394(12), *418*
Böer, K.W., 29, 31, 32, 33(18), *63*, 133(160), 133(180), 133(181), 138(227), 138(228), 138(229), *151, 152*, 393(9), *418*
Bogus, C., 51(144), 52(147), *67*
Bohun, A., 133(164), *151*
Bois, D., 130(137), *150*
Boltkas, B.I., 271(116), *300*
Booth, A.H., 133(157), *150*
Borchardt, E., 138(228), *152*
Borchardt, W., 138(228), 138(229), 138 (230), 138(231), *152*
Booz, J., 350, 353(65), 354(74), *385, 386*
Borders, J.A., 30(19), 30(21), *63*
Bordogna, J., 413—417(48), *419*
Boriev, I.A., 377(137), *387*
Borisova, Z.U., 271(116), *300*

Borkan, H., 141, 142(257), *153*
Boschi, R., 182(c), *182*
Botila', T., 211, 212(44), *214*
Bottoms, W.R., 135(219), *152*
Boucher, J., 33, 35(40), 35(52), *64*
Bouchriha, H., 161(23), *183*
Bowman, D.L., 128(97), *149*
Boyer, J., 377(143), *387*
Boyle, W.S., 408(40), *419*
Bradbury, M.E., 12(29), *25*
Braner, A.A., 133(161), *151*
Brattain, W.H., 135(183), *151*
Braun, C.L., 49(127), 49(128), *66*,167(43), *183*, 217, 234(2), *297*, 328(64), *334*
Braun, R., 50, 56(137), *66*
Bray, R., 29, 31(16), 35(53), *64*
Breckenridge, R.G., 117(3), *147*
Bree, A., 157(10), *182*
Briegleb, G., 364(100), *386*
Brinkerhoff, J., 371(168), *388*
Brissot, J.J., 405(32), *419*
Brodie, D.E., 32(34), 32(36), *64*, 210(39), *214*
Brodsky, M.H., 186(2), 186(3), 203(30), *213, 214*, 241, 244, 248, 253, 288(61), 286(142), *299, 301*
Broida, H.P., 53, 56(157), *67*
Broom, R.F., 130(118), *150*
Broser, I., 123(72), *149*
Brown, F.C., 30(19), 30(21), 35(60), 61(205) *63, 64, 68*
Brown, M.A.C.S., 35(54), *64*
Brownell, G.L., 383(162), *388*
Brückmann, G., 140(243), *152*
Bruschi, L., 344(42), 370(125), *385, 387*
Bryant, F.J., 123(66), 123(67), *148*
Bube, R.H., 27, 31, 45, 46, 61(1), 30, 32(22), 60(192), *62, 63, 68*, 72(1), 73(8), *114*, 117(1), 117(8), 117(9), 117(10), 117, 123(11), 117(12), 117(13), 118(24), 118, 125(26), 118, 125(26), 118, 125(27), 118(47), 120(49), 120, 127, 128(50), 121, 123(51), 122(52), 122(53), 122(54), 122(55), 122, 123(56), 122(57), 122(58), 122(59), 122, 123(60), 122, 123(61), 122(62), 122(63), 122(64), 122(65), 123(68), 123(71), 123(73), 123(74), 123(75), 124(77), 125(79), 125, 127(80), 125(82), 125, 126, 138(86), 128(87), 128(91), 128(92), 128(93), 128, 129(95), 129(98), 129(99), 130(112), 130(113), 130, 132(123), 132(143), 132(144), 132(145), 133(152), 133, 138, 139(153), 133(158), 133, 135(169), 133(170),

133(172), 133(178), 135(189), 135,
137, 141(190), 135(204), 135, 136(208),
135, 136(209), 135, 136(210), 135(218),
135, 136, 142(220), 135(221), 136(223),
138(226), 138, 139(237), 140, 145(241),
140, 145(242), 141(252), 143(267),
143, 144(270), 144(272), 144(274),
145(276), 145(277), 145(278), 145(279),
146(280), *147, 148, 149, 150, 151, 152,
153*, 197, 200(22), *213*, 238(37), 239,
241, 245, 248, 250, 252(44), 239(49),
241, 242, 245, 246, 248(64, 65), *298,
299*, 392, 394(5), 392, 393, 394(6), *418*
Buch, F., 146(280), *153*
Buch, T.M., 403(30), *419*
Buck, T.M., 61, 62(209), *69*
Buckley, R.R., 411(45), *419*
Burgos, J., 112, 113(42), *115*, 182(83), *184*
Burke, B.E., 28(14), *63*
Burke, T.J., 209(38), *213*
Burland, D.M., 172(62), *184*
Burnshtein, Z., 24(44), *25*, 164(39), *183*
Burnstein, E., 124(76), *149*
Burrell, G.I., 390(3), *418*
Burton, M., 56(166), *67*
Butcher, P.N., 291(172), *301*
Bykova, E.M., 27, 28(5), *63*

Cafarella, J.H., 28(14), *63*
Caminade, P., 350, 353(60), *385*
Camphausen, D.L., 210(41), *214*
Campos, M., 165(41), *183*
Canali, C., 43(100), 46, 47, 57(110), 57(178), *66, 68*
Cardon, F., 118, 125(26), 121, 123(51), *148*
Cardona, M., 28(12), *63*
Careri, G., 341(28), 370(124), *385, 387*
Carlson, C.F., 421(1), *476*
Carlson, R.O., 128(90), *149*
Carnes, J.E., 408, 412(43, 44), 411(46), *419*
Carreira, L.M., 422(8), *476*
Carter, G., 135(214), *152*
Carver, G.P., 59(185), *68*
Casey, H.C., Jr., 8(18), *25*
Cassanhiol, B., 241, 248, 249, 250, 253(68), *299*
Castro, G., 161(27), *183*
Caywood, J.M., 28, 55(10), 50, 52, 53(135), 52, 53(151), *63, 66, 67*, 164(38), *183*, 275, 280, 282(123), *300*
Cernogora, J., 253(84, 87), *299*

Chamberlin, R.R., 136(224), 142(265), 142(266), *152, 153*
Chance, R.C., 49(128), 50, 56(138), *66*, 167(43), *183*, 217, 234(2), *297*
Chartrand, M.G., 56(173), *67*
Chen, I., 40, 42(88), 43(102), *65, 66*, 287(147, 148), *301*, 458(39), 459(42), *477*
Cheroff, G., 32(35), *64*
Chiabrishvili, B.G., 130(131), *150*
Chiang, Y.S., 55(163), *67*
Chittick, R.C., 186(7), 186, 194(8), *213*
Choi, S., 159(16), 162(33), 171(57), 174, 175(67), *183, 184*
Chong, P., 343(32), *385*
Christensen, O., 27(2), *63*
Chybicki, M., 355(76, 77, 78), *386*
Cingolani, C., 61(212), *69*
Clar, E., 189(c), *182*
Clark, A.H., 207(33), 209(38), *214*
Clark, H., 421(3), *476*
Clarke, E.N., 135(185), *151*
Claus, C.J.,422, 469(9), 423(14), *476, 477*
Clauss, G., 60(196), *68*
Coelho, R., 54(160), *67*
Cohen, M.H., 243(74), 289(160), *299, 301*
Colbow, K., 83, 85(17), *147*
Collet, M.G., 408, 412(42), *419*
Collins, S.A., 294(186), *302*
Comizzolli, R.B., 421(4), *476*
Connel, G.A.N., 210(41), *214*
Conrad, E.E., 341, 346, 370(25), *285*
Canradi, J., 396(15), *418*
Consolo, S., 370(124), *386*
Coppage, F.N., 50, 51, 56(141), *66*, 112, 113(42), *115*, 167(46), *298*
Cooley, J.W., 95(29), *115*
Cooper, W.C., 237, 240, 285(33), *298*
Cornet, J., 240, 241(54), 241, 248, 249, 250, 252, 253(68), *298, 299*
Cosentino, L.S., 414(52), *420*
Cox, A.F.J., 123(66), 123(67), *148*
Creveceour, C., 293(183), *302*
Cronin, G.R., 130(114), *149*
Croitoru, N., 207(34), 209, 211, 212(37), 211, 212(44), *214*, 288(150), *301*
Crowder, B.L., 203(30), *214*
Crowell, C.R., 8(20), *25*
Crowell, M.H., 403(30), 408(39), *419*
Cunsolo, S., 344(41), *385*
Curie, P., 336(2), *384*
Cusano, D.A., 142(261), *153*
Cuter, M., 289(158), *301*

Cutler, M., 47(121), *66*
Czekulla, J., 364(100), *386*

Dalton, J.V., 403(30), *419*
D'Allessio, J.T., 56(166), *67*
Damask, A.C., 60(193), *68*, 172(61), *184*
Danishevskii, A.M., 27(4), *63*
Daurie, M., 35(52), *64*
Davern, W.E., 405(33), *419*
Davey, J.E., 211(43), *214*
Davies, D.K., 306(13), *333*
Davies, G.B., 293(185), *302*
Davis, E.A., 28, 45(15), 35(62), *63*, *65*, 190(16), 191(18), 203, 204(26), *213*, 222, 226, 227, 231(19), 238, 260(40), 288(149), *297*, *298*, *301*
Davis, H.J., 318(46), *333*
Davis, H.T., 35(67), 53, 54(155), *65*, *67*, 344(40), 368, 381(117), 370(121, 122, 123), 370(127), 372(131), 371(167), *385*, *386*, *387*, *388*
Davis, J.L., 141, 143(249), *152*
Davydov, A.S., 157(9), *182*
Day, G., 130(104), *149*
Day, P., 161(25), *183*
Debye, P., 375(134), *387*
deHaan, E.F., 403(29), *419*
deHaas, M.P., 375(133), *387*
Delacote, G., 51(145), 60(198), *67*, *68*, 161(23), 169(51), 172(62), *183*
Delcroix, J.L., 12(28), *25*
Dellonte, S., 159(15), *182*
Denisova, A.D., 118(38), *148*
Derenzo, S.E., 354(75), 377, 380(148, 149), *386*, *387*
Dessauer, J., 421(3), *476*
Devenyi, A., 207(34), *214*
Devins, J.C., 366(105), *386*
DeVore, H.B., 30(23), *64*, 133(148), *150*
Dewitt, H.J., 293(183), *302*
Diesel, T.J., 118(46), *148*
Dietrich, B.K., 367, 368(109), *386*
Dimond, N.A., 176(74), *184*
Dirksen, H.J., 118(21), *147*
Dittfield, H.J., 133(167), *151*
Dobrovolskii, V.N., 60(195), *68*
Dodelet, J.P., 217, 234(6), 234(30), *297*, *298*, 346, 370(49), 359(49), 359(88), 359(90), *385*, *386*
Doke, T., 357(84), *386*
Dolezalek, F.K., 36(70), 49, 51, 52(130), *65*, *66*, 112(43), *115*, 217, 224(1), 275, 280(118), *297*, *300*

Dolgoshein, B.A., 380(159), *388*
Domanevskii, D.S., 130(132), *150*
Donald, D.K., 436(26), *477*
Donelly, R.J., 371(167), *388*
Donovan, T.M., 186, 206, 207(4), 186, 206, 207, 209(6), 193(19), 207(35), 207, 208(36), *213*, *214*
Dow, J.D., 240(55), *298*
Dreeben, A.B., 125(79), 125, 127(80), *149*
Dresner, J., 60(192), *68*, 163(37), 172(61), *183*, *184*, 277, 280(125), *300*
Dretova, E.A., 130(110), *149*
Drew, R.E., 107(36), *115*
Drews, R.E., 237(36), *298*
Drift, A.v.d., 403(29), *419*
Duboc, C.A., 118(23), *148*
Duke, C.B., 1(1, 2), *24*
Dussel, G.A., 118, 125(27), 133, 138, 139(153), 133, 135(169), 133(170), *148*, *150*, *151*
Dy, K.S., 18, 20(36), *25*, 259, 271(105), *300*
Dyck, R.H., 407(38), *419*
Dzhafarov, T.D., 271(116), *300*

Eastman, D.E., 3(14), *25*
Eastman, P.C., 210(39), *214*
Ebert, H.G., 350, 353(64, 65), 354(74), *386*
Eckenbach, W., 211(42), *214*
Egawa, A., 364(101), *386*
Egiazaryan, G.A., 130(122), 130(129), *150*
Eisele, I., 60(188), *68*
Eisenberg, A., 237, 240(34), *298*
Ellis, B., 390(3), *418*
Ellis, G.W., 404(41), *419*
Emberson, D.L., 415(57), *420*
Emerald, R.L., 43(102), *66*, 107(36), *115*, 237(36), *298*, 308, 312, 319, 331(34, 35), *333*, 459(42), *477*
Emin, D., 61(205), *68*, 203, 204(29), *214*, 289(155, 156), *301*
Emtage, P.R., 18(33), *25*
Enck, R.C., 49, 56(129), *66*, 83, 85, 86(19), 108, 113(38), *114*, *115*, 217, 219, 222, 229—233, 275(8), 222, 239(24), 223, 239(26), *297*, *298*, 453(37), *477*
Engemann, D., 204(31), 204, 205(32), *214*
Engler, J., 380(156), *387*
Englman, R., 203(27), *213*

Ericson, G., 56(170), *67*
Erlich, P., 304(6), *332*
Ern, V., 161(23), *183*
Esaki, L., 239(51), *298*
Esbitt, A.S., 393(9), *418*
Espevik, S., 135, 137, 141(190), *151*
Esser, L.J.M., 408, 412(42), *419*
Essex, V., 344(38, 39), *385*
Ettenberg, M., 31(28), *64*
Evans, A.G.R., 32, 33, 57(37), *64*

Fagen, E.A., 6(16), *25*, 73(8), *114*, 197, 200(22), *213*, 239, 241(45), 239(50), 241, 242, 245, 246, 248(64, 65), *298*, *299*
Fahrenbruch, A.L., 140, 145(241), 140, 145(242), 146(280), *153*
Fang, P.H., 176(74), *184*
Fassbender, J., 31(30), *64*
Fayadh, J.M., 357(86), *386*
Fedorus, G.A., 138, 139(235), *152*
Felty, E., 286(141), *301*
Feng, D.F., 372(130), *387*
Ferrara, M., 61(212), *69*
Ferry, J.D., 281(133), *300*
Feucht, D.L., 3, 4, 8(13), *25*, 145(275), *153*, 400(23), *418*
Firle, T.E., 47(121), *66*
Fisher, F.D., 265, 266(113), 283, 285(132), *300*
Fisher, H., 56(164), *67*, 416(58), *420*
Fisher, J.E., 186, 206, 207(4), 186, 207(5), 186, 206, 207, 209(6), 193(19), 207(35), *213*, *214*, 204(31), 204, 205(32), *214*, 241, 244, 248, 253, 288(61), 253(89), *299*
Fistal, V.I., 130(127), *150*
Flagg, R., 377, 380(149), *387*
Flannery, J.B., Jr., 413—417(49), *420*
Flasck, R., 288(151), *301*
Fleming, R.J., 259, 271(106), *300*, 306, 309(12), *333*
Foner, S., 30(21), *63*
Forbes, L., 27, 28(5), *63*
Forgue, S.V., 401(26), *419*
Forster, R., 364(102), *386*
Fortin, E., 62(214), *69*
Fouchi, P.G., 217, 234(6), 234(30), *297*, *298*, 346, 370(48), 359, 370(87), *385*, *386*
Fourny, J., 161(23), 169(51), *183*
Fowler, I.L., 56(173), *67*
Fowler, R.H., 10, 11(23), *25*, 367(107), *386*

Fox, D.C., 135(211), *152*
Fox, S.J., 315(42), *333*
Franke, H.G., 56(165), *67*
Fraser, D.B., 142(263), *153*, 393(10), 414(53), *418*, *420*
Freeman, G.R., 55(162), *67*, 217, 234(6), 234(30), *297*, *298*, 339(20), 341(26), 346, 370(48, 49), 346, 354, 370(50), 357(86), 359, 370(87, 88), 359(90), *384*, *385*, *386*
Freeouf, J.L., 3(14), *25*
Frenkel, J., 157(8), *182*, 222, 225, 265(20), *298*, 314(37), *333*
Friedman, L., 60(201), 60(202), *68*, 156(6), 171(60), *182*, *183*, 203(28), *213*, 289(157), *301*
Friend, B., 380(156), *387*
Fritzsche, H., 238(43), 239, 241(45), 239(50), 243(74), 291(176), *298*, *299*, *301*
Fueki, K., 372(130), *387*
Fuhs, W., 211(42), *214*
Fuller, C.S., 118(35), *148*
Funabashi, K., 55(163), *67*, 318(48), *333*, 359(89), *386*

Gaeta, F.S., 341(28), *385*
Galassini, S., 133(179), *151*
Gangwer, T.E., 377(138, 139), *387*
Garbett, E.S., 293(181), *302*
Garlick, G.F.J., 133, 134(155), *150*
Garreta, O., 61(207), *68*
Gaylord, Th.K., 412, 414(47), *419*
Gazda, E., 341(23), 346, 376(51, 52), *384*, *385*
Geacintov, N.E., 162(30), 167(44), 176 (73), *183*, *184*
Geballe, T.H., 86, 88(20), *114*, 132(147), *150*, 285, 290(162), *301*
Geist, D., 28(13), *63*
Genzow, D., 27(4), *63*
Gerritsen, A.N., 357(81, 82), *386*
Gerritsen, H.J., 135(202), *151*
Geyer, R.D., 408(39), *419*
Ghosh, P.K., 53(154), *67*
Gianchino, J., 182(82), 182(83), *184*
Giamo, E., 423(19), *477*
Gibaud, R., 349, 350, 353(61), *385*
Gibbons, D.J., 43(101), 48(124), *66*
Gibbons, J.F., 39(81), 53(152), 57(177), *65*, *67*
Gibson, A.F., 27(3), 31(24), 35(49), 35 (54), *63*, *64*, 133, 134(155), *150*
Gilbert, S.L., 31(28), *64*

Gill, W.D., 42(92), 56(164), 67, 104(33), *115*, 145(276), 145(277), *153*, 261, 268(108), *300*, 308, 310, 312, 314, 321, 323, 331(24), 308, 310, 323(28), 311(31), 312, 319, 321, 331(33), 316(43), *333*, 468, 470(48), *477*
Glaberson, W.I., 370(126), *387*
Glaeser, R.M., 174, 175(66), *184*
Glarum, S.H., 174(65), *184*
Glinshuk, K.D., 118(38), 118(41), *148*
Gobrecht, H., 60(196), *68*
Goldstein, Y., 3(4), *24*
Golubovic, A., 176(74), *184*
Gomer, R., 344(36), 361, 363(93), 368 (114), 371(169), *385*, *386*, *388*
Gontar', V.M., 130(129), *150*
Gooch, C.H., 130(100), *149*
Goodman, A.M., 28, 55(11), 60(197), *63*, *68*
Goodrich, R., 401(26), *419*
Gorelenok, A.T., 130(141), *150*
Gorskii, S.E., 118(37), *148*
Gosar, P., 174, 175(67), *184*
Grammatica, S., 104, 106(34), *115*, 321(54), *334*
Grant, A.J., 241, 246, 250, 286, 288(60), 290(145), *299*, *301*
Gregg, E.C., 339(17), 341(27), *384*, *385*
Green, R.M., 56(173), *67*
Greene, D.B., 56(164), *67*
Greene, R.F., 141, 143(249), *152*
Grieg, H.G., 423(15), 423(18), *477*
Greiling, P.T., 31(29), *64*
Grigorovici, R., 194(20), 207(34), *213*, *214*
Gritsenko, Yu.I., 60(195), *68*
Grossweiner, L.I., 133(156), *150*, 304(4), *332*
Grosvalet, J., 61(207), *68*
Grover, N.B., 3(4), *24*, 33(39), *64*
Grubin, H.L., 8(22), *25*
Grunwald, H.P., 46(109), *66*, 257, 275, 277, 280, 281(93), *299*
Gudden, B., 35(55), *64*
Guenzer, C.S., 253(82), *299*
Gundlach, R.W., 422, 462(9), 422(10), 438(29), *476*, *477*
Gunn, J.B., 35(52), *64*
Gurney, R.W., 40(83), *65*
Gutkin, A.A., 130(130), 130(136), 130 (138), *150*
Gutmann, F., 304(2), *332*
Gzowski, O., 341, 376(29), *385*

Haake, C., 133(165), *151*

Haas, K.J., 135(211), *152*
Haberkorn, R., 166(42), *183*
Haden, C.R., 294(186), *302*
Haering, R.R., 133(162), *151*
Hahn, E.E., 117(3), *147*
Haisty, R.W., 130(101), 130(114), 134 (182), *149*, *151*
Hall, J., 423(11), *477*
Hall, J.A., 405(34), *419*
Halperin, A., 133(161), *151*
Halperin, B.I., 88(23), *114*
Halpern, B., 344(36), 361, 363(93), 368(114), 371(169), *385*, *386*, *388*
Ham, J.S., 60(187), *68*
Hamada, T., 357(84), *386*
Hamakawa, Y., 28(12), *63*
Hanson, D.M., 161(29), *183*
Harada, Y., 176(72), 181(80), 181(81), *184*
Harengel, W., 43(99), *65*, 166(42), *183*
Harmann, T.C., 398(21), *418*
Harnik, E., 33(39), *64*
Harper, J.G., 130, 132(123), 132(143), *150*
Harrison, S.E., 60(192), *68*
Hartke, J.L., 221(15), 227(27), 257, 275, 277, 282(92), *297*, *298*, *299*
Hartman, W.H., 141(260), *153*
Hartmann, G.G., 450, 452, 454(31), *477*
Hasegawa, K., 160(18), *183*
Hayashi, K., 306, 309, 324(19), *333*, 339(18, 19), *384*
Haynes, J.R., 33(41), 34, 35(45), 35(50), *64*
Heath, D.R., 130(119), *150*
Hecht, K., 35, 48(56), *64*
Heer, J., 32(35), *64*
Heijne, L., 72(5), *114*
Heiland, G., 135(193), 135(200), *151*
Heilmeier, G.H., 60(192), *68*
Heim, V., 253(89), *299*
Heine, V., 3(7), *25*
Heinrich, V.E., 141, 142(257), *153*
Heleskivi, J., 144(269), *153*
Helfrich, W., 38(76), 40, 41(85), 42, 51(98), *65*
Hemenger, P.M., 59(182), *68*
Hemila, S.O., 123(71), *149*
Henisch, H.K., 8(19), *25*
Henson, B.L., 344(37), *385*
Herrmann, K.H., 39(80), 60(187), *65*, *68*, 175(68), *184*, 308, 310(25), 308(26), *333*
Herring, C., 132(146), *150*

Hermann, K.H., 27(4), *63*
Heukeroth, U., 33(41), *64*
Hill, R.M., 314(38), *333*
Hillegas, W.J., 277, 280(126), *300*, 468, 471(49), *477*
Hillman, E.E., 167(945), *183*
Hilsum, C., 130(100), *149*
Hilton, A.R., 130(116), *149*
Hilton, E.A., 135, 136(208), *152*
Hine, G.J., 383(162), *388*
Hintenberger, H., 135(186), *151*
Hirooka, T., 181(81), *184*
Hirsch, J., 36(73), 38(79), *65*, 286, 288 (143), *301*, 306, 308(14), *333*
Hirth, H., 46(115), 46(118), *66*
Hlavka, J., 62(219), *69*
Ho, C.T., 123(74), 133, 138, 139(153), *149, 150*
Hodby, J.W., 30(19), 30(21), *63*
Hoegl, H., 104(32), *115*, 309, 324, 326(9), *333*
Hoesterey, D.C., 46(114), *66*, 171(59), *183*
Hoffman, B., **135(201)**, *151*
Hofmann, W., 380(156), *387*
Hofstein, S.R., 42(97), *65*
Holeman, B.R., 130(100), *149*
Holmberg, S., 73(8), *114*, 197, 200(22), *213*, 241, 242, 245, 246, 248(64, 65), *299*
Holmsharo, R.T., 415(57), *420*
Holroyd, R.A., 365(103, 104), 367(108), 367, 368(109), 367(110), 370(129), 377(138, 139), *386, 387*
Holstein, T., 60(202), *68*, 171(60), 174(64), *183, 184*, 289(155), *301*, 315(40), *333*
Honig, A., 83, 85, 86(19), *114*
Hoogenstraaten, W., 133(159), *151*
Hope, W.D., 473(54), *477*
Hopfield, J.J., 83, 85(18), *114*
Hornbeck, A., 35, 46(57), *64*
Hornbeck, J.A., 35(50), *64*
Hornig, J.F., 41, 42, 43(89), 49(27), *65, 66*, 161(27), *183*, 328(64), *334*
Houser, N.E., 364, 366(98), 364(99), *386*
Howard, W.E., 239(51), 241(63), *298, 299*
Howe, S., 35, 39, 53, 54, 57(65), *65*
Howell, J., 24(42), *25*
Hower, S., 369(166), *388*
Hruby, A., 288, 290(152), *301*
Huang, C.I., 62(216), *69*, 130(135), *150*
Huang, I.L., 259, 271(105), *300*
Hudson, D.E., 54(158), *67*
Hudson, D.D., 344(45), *385*

Hughes, D.M., 135(214), *152*
Hughes, R.C., 51(146), 56(171), *67*, 306, 309, 313(17), 313, 329(36), 325, 328(59), 329(65), *333*
Hull, G.W., 132(147), *150*
Hummel, A., 339, 341, 357(15), 341, 374(24), 343, 359(33), 349, 359(56), 375(133), 358(163), *384, 385, 387, 388*
Hurst, G.S., 378(150), *387*
Huth, F., 130(106), *149*
Hutson, A.R., 28(14), *63*
Hwang, C.J., 31(27), *64*

Ibrugimov, V.Yu., 130(126), *150*
Ikeda, M., 324, 327(58), *334*
Im, H.B., 138, 139(237), *152*
Infelta, P., 375(133), *387*
Ing, S.W., 50(134), 55(163), *66, 67*, 217, 237, 238, 286(10), 222, 225, 238(18), 297, 426(24), 451—458(33), 472(52), *477*
Inkson, J.C., 3(9), *25*
Inokuchi, H., 36(70), *65*, 155(2), 168(48), 169, 170(52), 170, 174(53), 170(54), 172(61), 176(72), 181(80), 181(81), *182, 183, 184*
Inoue, E., 423(12), *476*
Inoue, T., 130(121), 130(125), *150*
Inuishi, Y., 27(3), *63*, 306, 309, 324(19), *333*, 343(32), *385*
Ioanid, G., 211, 212(44), *214*
Isawa, S., 288(151), *301*
Iseler, G.W., 125(82), *149*
Ishihara, Y., 169(50), *183*
Ishii, S., 135(192), *151*
Iwasaki, N., 179(79), *184*
Iwashima, S., 170, 174(53), *183*

Jachym, A., 349(60), *385*
Jachym, B., 336(11), 349(60), *384, 385*
Jacobi, W., 349' 359(57), *385*
Jaffe, G., 336(3, 4), 336, 341(5), 336, 349(6), 336(7), *384*
Jahns, A., 349, 359(57), *385*
Jarnagin, R.C., 111(40), *115*, 159(17), *183*, 230(28), *298*, 364, 366(98), 364(99), *386*
Jayaraman, S., 27(3), *63*
Jimenez, R.L., 141(260), *153*
Johnson, V.A., 27(6), *63*
Jonath, A.P., 130(112), *149*
Jones, C.E., 130(116), *149*
Jonscher, A.K., 72(3), 89(24), *114*,

291(166), 291(173, 174), *301*
Jortner, J., 27(3), *63*, 160(21), 162(33), 171(57), 171(58), *183*, 203(27), *213*, 336, 369(13, 14), *384*
Joshida, A., 288(154), *301*
Juska, G.B., 235, 240(31), 237, 240, 257, 258, 275, 277, 280(32), 275, 280(119), 275(121), 275, 280(122), *298*, *300*

Kafalsas, J.A., 125(82), *149*
Kagan, M.B., 130(130), *150*
Kajiwara, T., 36(70), *65*, 169, 170(52), *183*
Kallmann, H.P., 123(70), *148*, 163(35), 176(71), 176(73), 177(78), 182(82), *183*, *184*, 462(59), *478*
Kamura, Y., 170(54), *183*
Kanajawa, K.K., 40, 42, 43(86), 42(92), 42(94), 42(96), *65*, 311(31), 322, 323(55), *333*, *334*
Kaneda, T., 60(190), *68*
Kanev, S., 140(238), 140(239), 140(240), 140, 145(241), *152*
Kang, C.S., 122, 123(56), *148*
Karkhanin, Yu.I., 130(107), 130(134), *149*, *150*
Karl, N., 50, 51(142), 53(153), *66*, *67*
Karpovich, I.A., 141(256), *153*, 210(40), *214*
Kastalskii, A.A., 27(3), *63*
Katon, J.E., 304(3), *332*
Katz, J.L., 171(56), *183*
Katz, M.J., 135(211), *152*
Kaufmann, W.M., 393(9), *418*
Kaul, R., 8(22), *25*
Kawaji, J., 135(203), *151*
Kawamura, T., 462(57), *478*
Kazmerski, L.L., 141(253), *152*
Kearns, D.R., 161(28), *183*
Keating, P.N., 133(163), *151*
Keezer, R.C., 285(138), *301*
Keim, H., 380(156), *387*
Keller, H., 250(78), *299*
Kelley, R.H., 123(69), *148*
Keneman, S.A., 412, 414(48), *419*
Kepler, R.G., 35(68), 36(70), 50, 51, 56(141), *65*, *66*, 112, 113(42), *115*, 156(3), 160 (19), 160(20), 167(46), 169(49), *182*, *183*, 304(8), *332*
Kestner, N.R., 336, 369(14), *384*
Kevan, L., 60(188), *68*, 367(112), 372(130), *386*, *387*
Kholev, B.A., 130(130), *150*
Kiess, H., 471, 472(51), 474(56), *477*, *478*

Kikoin, I.K., 61(206), *68*
Kikuchi, R., 46, 47(107), *66*
Killesreiter, H., 50, 56(137), *66*, 159(12), 163(34), *182*, *183*
Kim, C.W., 405(33), *419*
Kimmit, M.F., 27(4), *63*
Kinder, J., 166(42), *183*
Kiner, J., 43(99), *65*
Kino, G.S., 34, 35(46), 36, 53(71), 57(176), *64*, *65*, *67*
Kinoshita, K., 462(57), *478*
Kipperman, A.H.M., 59(182), *68*
Kiseleva, N.K., 239(46), *298*
Kiuchi, Y., 403(27, 28), *419*
Klasens, H.A., 118(25), *148*
Klassen, N.V., 354(72), *386*
Kleinheins, G., 54(160), *67*
Knies, G., 380(157), *387*
Knight, J., 306, 309(15), *333*
Knights, J.E., 222, 226, 227, 231(19), *297*
Knotek, M.L., 186, 206, 207, 209(6), 207, 208(36), *213*, *214*
Kobayashi, A., 135(203), *151*
Kobayashi, S., 60(190), *68*, 135(192), *151*, 357(85), *386*
Kobayashi, T., 170, 174(53), *183*
Kochi, M., 181(81), *184*
Koepp, R., 350, 353(65), *385*
Kohado, H., 423(12), *476*
Kolchanova, N.M., 130(110), 130(126), *149*, *150*
Kolomiets, B.T., 238, 239(41), 239(46), 239, 241, 242, 250, 253(47, 48), 253(81, 85, 86), 258(99), 282, 285(136), *298*, *299*, *300*
Kondrasiuk, J., 170(55), *183*
Konno, S., 357(84, 85), *386*
Korn, A.I., 60(193), *68*, 172(61), *184*
Korn, D.M., 27, 28(5), *63*
Kornilov, B.V., 118(36), 118(37), 118(39), 118(40), 118(42), 118(43), 118(44), 118(45), 130(110), 130(128), *148*, *149*, *150*
Korsunskaya, N.E., 138(236), *152*
Kortsch, U., 359(88), *386*
Kosei, K., 423(12), *476*
Kosek, F., 237(36), *298*
Kosonocky, W.F., 408, 412(44), 411(46), *419*
Kovac, M.G., 411(46), *419*
Kramer, B., 123(70), *148*
Kramer, B.M., 57(179), *68*
Kramers, H.A., 350(68), *385*

Kressel, H., 31(28), *64*
Kruse, P.W., 390(4), *418*
Kryszewski, M., 306, 308(11), *333*
Kubota, S., 357(84), *386*
Kulp, B.A., 123(69), *148*
Kümmel, U., 133(180), 133(181), *151*
Kurita, T., 468(46), *477*
Kurosawa, T., 174(63), *184*
Kwok, H., 132(144), 132(145), *150*

Labes, M.M., 304(6), 308, 319(23), *332, 333*
Labuda, E.F., 403(30), *418*
Ladu, M., 378(154, 155), *387*
Lagnado, I., 142(264), *153*
Lakatos, A.I., 3, 8, 9(15), *25*, 36(72), *65*, 86, 89, 90(21), *114*, 261(107), 291(177), 293(184), 294(187), 296(197), *300, 301, 302*, 307, 310, 312, 330, 331(21), 323(57), 330(67), *333*, 464(43), *477*
Lakova, M., 140(240), *152*
Lambe, J., 133(171), *151*
Lampert, M.A., 38(78), *65*, 125, 126(85), 133(175), *149, 151*
Landel, R.F., 288(133), *300*
Langer, J.S., 88(23), *114*
Langevin, P., 50(139), *66*, 112(41), *115*, 329(66), *334*
Langmann, U., 31(26), *64*
Lanyon, H.P.D., 257, 275, 277(91), *299*
Lappe, F., 32(38), *64*
Lardon, M., 310, 326(27), *333*
Lark-Horovitz, K., 27(6), *63*
Laval, M., 380(160), *388*
Lavine, J.M., 59(184), *68*
Lawrence, R., 35(49), *64*
Lax, M., 50(139), *66*, 74, 75(10), 74, 81(11), 75(12), 78, 80, 87(15), 105(39), *114, 115*, 291(170, 171), *301*
Lazarescu, M., 288(150), *301*
Lea, D.E., 336, 350(9), *384*
Leach, S., 363(95), *386*
Leadbetter, A.J., 265(111), *300*
Lebedenko, V.N., 380(159),.*388*
Lebedev, A.A., 130(136), 130(138), *150*
Lebedev, E.A., 258(99), 282, 285(136), *300*
LeBlanc, O.H., Jr., 35, 36, 53, 55(64), 35 (68), 35(69), 60(200), *65, 68*, 156, 158(3), 156(5), 171(56), *182, 183*, 343(30), *385*
LeComber, P.G., 53, 54(156), *67*, 89(25), *114*, 185(1), 186, 188, 201(11), 186, 188, 194, 201(12), 186, 187, 194, 202 (13), 188(15), *213*, 244(76), 279(128,

129), *299, 300*, 369(118), 371(170), *386, 388*
Lee, C.H., 27(3), *147*
Lee, E.H., 141(252), *152*
Lehmann, H., 31(30), *148*
Lekner, J., 344(36), 368(114), 370(119), *385, 386, 387*
Lell-Doller, E., 310, 326(27), *333*
Lember, L., 20(38), *25*
Lemke, H., 38, 42(77), 47(122), 47(123), *65, 66*
Lenkeit, S., 350, 353(64), *385*
Lerge, J.P., 135(215), *152*
Leston, G.M., 46(114), *66*, 171(59), *183*
Levine, J.D., 3(21), *25*
Levine, P., 160(21), *183*, 411(46), *419*
Levinson, J., 161, 164(26), 164(39), *183*
Levinson, L., 24(44), *25*
Levinstein, H., 117(4), 135(191), *147, 151*, 390, 395—397(2), 397(18), *418*
Levy, J.L., 60(189), *68*
Levy, M., 423(11), 473(54), *476, 477*
Lewis, A., 194(21), *213*
Lewis, G.N., 363(96), *386*
Li, H.T., 218, 275(11), *297*
Li, S.S., 62(216), *69*, 130(35), *150*
Liang, W.Y., 239(53), *298*
Lichjensteiger, M., 142(264), *153*
Lidholt, R., 56(170), *67*
Liebson, S.H., 135(205), *151*
Lifschits, T.M., 27, 28(5), *63*
Lile, D.L., 62(218), *69*
Lin, A.L., 128, 130, 133, 134, 135(96), *149*
Lind, E.L., 122(59), 122(62), 122(64), 122(65), 125, 127(80), *148, 149*
Lindquist, P.F., 145(278), 145(279), *153*
Lipinski, A., 172(j), *172*
Lipkin, D., 363(96), *386*
Lipskis, K., 144(271), *153*
Litovchenko, N.M., 118(38), 118(41), *148*
Lock, M.W.B., 306, 309(15), *333*
Loeb, L.B., 12(27), *25*
Loebner, E.E., 118(46), *148*, 415(56), *420*
Loferski, J.J., 417(59), *420*
Lohman, R.D., 414(52), *420*
Lomnes, R., 132(142), *150*
Long, D., 398(20), *18*
Lorenz, M.R., 125(81), *149*
Louie, S.G., 370, 380(149), *387*
Louis, E., 3(11), *25*
Loveland, R.J., 53, 54(156), *67*, 185(1),

186, 191, 192, 194(9), 186, 195, 204
(10), *213*, 241(73), *299*, 369(118), 371
(170), *386, 388*
Lozier, G.S., 421(4), *476*
Lucovsky, G., 125(84), *149*, 221(16), 282–
286(135), 285(138, 139, 140), *297,
300, 301*
Ludwig, G.W., 35(50, 51), *64*
Ludwig, P.K., 56(166), *67*
Lummis, F.L., 141(251), *152*
Lupu, N.Z., 60(191), *68*
Luschik, Ch.B., 133(166), *151*
Lyons, L.E., 157(10), *182*, 304(2), *332*
Lyubin, V.M., 238, 239(41), *298*

MacDonald, H.E., 60(192), *68*, 120, 127,
128(50), 128(91), 128(92), 128(93),
128, 129(95), *148, 149*
MacDonald, J.R., 59(183), *68*
MacDonald, R.E., 56(164), *67*
Mackey, R.C., 56(169), *67*
MacMillan, H.F., 125(82), 125(83), *149*
Madan, A., 186, 188, 194, 201(12), 186,
187, 194, 202(13), *213*, 244(76), 279
(129), *299, 300*
Magee, J.L., 231(29), *298*
Main, C., 197(23), *213*, 241, 250, 252, 288
(56), 241, 245, 248, 250, 252, 288(57,
58), *298, 299*
Malm, H.L., 56(173), *67*
Mamontova, T.N., 253(81, 85, 86), *299*
Manadaliev, M., 60(187), *68*
Mandics, J., 367(111), *386*
Manfredotti, C., 133(179), *151*
Mankarious, R.G., 141(255), *153*
Many, A., 3(4), *24*, 22(40), 24(44), *25*, 29,
31(17), 33(39), 35(47), 38, 42, 48(57),
40, 41, 48(84), *63, 64, 65*, 161, 164(26),
164(39), *183*
Mariam, F.G., 377, 380(149), *387*
Mark, P., 38(78), 40, 41(85), *65*, 235(207),
135, 136(212), 135(217), 135(219),
151, 152
Markevich, I.V.,138(236), *152*
Marshall, J.M., 241, 250, 252, 288(56), 257,
258, 275, 277(95), 258, 259(100), 265,
266, 267, 281(112), 265, 266(113),
283, 285(132), 288(153), *298, 300,
301*, 318(51), *333*
Martin, E.H., 36(73), *65*, 306, 308(14),
333
Martini, M., 35, 39, 45, 47, 55, 56, 57(59),
46, 47, 57(110), 47(119), 47(120), 56
(175), *64, 66, 67*

Martinuzzi, S., 135(215), *152*
Maruyama, Y., 55(163), *67*, 168(48), 170,
174(53), 170(54), 172(61), 176(72),
179(79), *183, 184*, 318(48), *333*
Mast, T.S., 354(75), 377, 388(149),
386, 387
Masuda, K., 18(34), *25*
Mather, D.P., 408(39), *419*
Matheson, M., 363(95), *386*
Mathieu, J., 350, 353(66), 353(70), 377
(143, 144, 145, 146, 147), 378(144),
378(153), *385, 387*
Matsumato, S., 306, 309(18), *333*
Matthews, H.E., 130, 132(123), 132(143),
138, 139(237), *150, 152*
Matulionis, A., 235, 240(31), 237, 240,
257, 258, 275, 277, 280(32), *298*
Mayer, E.F., 423(14), *477*
Mayer, J.W., 35, 39, 45, 47, 55, 56, 57(59),
46, 47(108), 52, 53(151), *64, 66, 67*
Mayo, C.R., 423(16), *477*[
Mazzi, G., 370(125), *387*
Mazzoldi, P., 370(124), *387*
McCreary, R.D., 339(17), *384*
McGarrity, J.M., 56(172), *67*
McGlauchlin, L.D., 390(4), *418*
McGlynn, S.P., 172(i), *172*
McGroddy, J.C., 27(2), *63*
McIntyre, R.J., 396(15), *418*
McKim, F.S., 61, 62(209), *69*
McMath, T.A., 47(119), 47(120), *66*
McQuistan, R.B., 390(4), *418*
Mead, C.A., 2(3), *24*, 3(12, 14), *25*, 50, 52,
53(135), 52, 53(151), *66, 67*, 275,
280, 282(123), *300*
Mechior, H., 142(263), *153*
Mehal, E.W., 130(101), *149*
Mehl, W., 318, 322(49), *333*
Mehta, R.R., 414(51), *419*
Meier, H., 55(163), *67*
Melchior, H., 393(10), 395(14), *418*
Melngailis, I., 398(19, 21), *418*
Melnick, D.A., 135(198), *151*
Melved, D.B., 135(199), *151*
Melz, P.J., 49(131), *66*, 217, 234(5), *297*,
327, 328(62), *344*
Memelink, O.W., 118(21), *147*
Menendez, J., 159(15), *182*
Menezes, C.A., 239, 241, 245, 248, 250,
252(44), *298*
Meray-Horvath, L., 406(35), *419*
Metcalfe, K.A., 423(13), *476*
Metcalfe, K.E., 306(16), *333*
Mey, W., 175(68), *184*

Meyer, J., 28(13), *63*
Meyer, J.E., Jr., 406(35), *419*
Meyer, L., 54(161), *67*, 343(34, 35), 344 (40), 370(121, 122, 123), 37(167), *385, 386, 388*
Mezrich, R.S., 414(52), *420*
Michel-Beyerle, M.E., 43(99), *65*, 166(42), *183*
Micheletti, F.B., 135, 136(212), 135(217), *152*
Mikhailov, I.D., 306(16), *333*
Miller, A., 75, 88, 110(13), *114*
Miller, G.R., 265, 266, 267, 281(112), *300*
Miller, L.D., 133, 138, 139(153), *150*
Miller, L.S., 35, 39, 53, 54, 57(65), *65*, 369(166), *388*
Miller, P.H., 135(197), *151*
Milnes, A.G., 3, 4, 8(13), *25*, 145(175), *153*, 400(23), *418*
Minafra, A., 61(212), *69*
Minanni, T., 288(154), *301*
Minday, R.M., 35(67), 53, 54(155), *65, 67*, 318(46), *333*, 368, 381(117), 370(127), 372(131), *386, 387*
Minden, H.T., 135(187), *151*
Minomura, S., 36(70), *65*, 169, 170(52), *182*
Mirdzhalilova, M.A., 130(126), *150*
Mitchell, D.L., 253(83), *299*
Mitchell, K., 146(280), *153*
Mitsui, M.J., 462(58), *478*
Miyajima, M., 357(84), *386*
Möbius, D., 159(14), *182*
Modi, J.K., 400, 403(24), *418*
Mohler, F.L., 350, 352(62), *385*
Moll, J.L., 34, 35(46), *64*
Mollot, F., 253(84, 87), *299*
Mollwo, F., 135(196), 135(200), 135(201), *151*
Montroll, E.W., 76(14), 90, 95, 98, 99(26), 90, 91, 99(28), *114, 115*, 258, 260—263, 267, 268, 277(104), *300*, 450, 451, 453, 454(32), *477*
Mooradian, A, 285(138), *301*
Moore, A.R., 28(14), 46(112), *63, 66*
Morehead, F.F., 124(78), *149*
Morel, D.L., 175(68), *184*
Morgan, K., 60(198), *68*, 173(c), *173*
Morimoto, K., 324, 327(58), *334*
Morin, F.J., 118(35), *148*
Morris, G.C., 182(d), *182*
Morris, R., 52(148), *150*
Morrison, S.R., 135(184), *151*
Mort, J., 28, 55(9), 28(14), 36(72), 38, 42, 43(74) 43(102), 46, 48(105), *63, 65, 66*, 104, 106(34), *115*, 238(42), 261 (107), 261, 265, 268(110), 280(130), 287(148), 296(197), *298, 300, 301*, 307, 310, 312, 330, 331(21), 308, 312, 319, 331(34, 35), 321(54), 330(67), 331(69), *333, 334*, 458(39), 459(42), *477*
Moss, T.S., 62(215), *69*, 390(3), *418*
Mott, N.F., 7(17), *25*, 28, 45(15), 35(62), 40(83), *63, 65*, 190(16), 191(18), 203, 204(26), *213*, 238, 260(40), 267(115), 288(149), 289(158, 159), 291(168), 291(169), 295 (188, 189, 190, 191), 295 (193, 194), *298, 300, 301, 302*
Möstl, K., 46(116), *66*
Moustakas, T.D., 241, 245, 246, 248, 250, 252, 286, 288(59), 241, 246, 250, 286, 288(60), 290(145), *299, 301*
Mozumder, A., 217, 234(7), 231(29), *297, 298*, 360(91, 92), *386*
Mukerjee, S.N., 400, 403(24), *418*
Mulder, B.J., 150(11), *182*
Muller, F., 363(95), *386*
Müller, G.O., 38, 42(77), 47(123), *65, 66*
Muller, R.A., 354(75), 377, 380(148), *386, 387*
Muller, R.S., 141(254), *153*
Mullins, F.D., 167(45), *183*
Munn, R.W., 156, 174(7), 176(69), 176(70), *182, 184*
Murakami, Y., 324, 327(58), *334*
Murrell, J.N., 156(4), *182*
Murygin, V.I., 130(122), 130(129), *150*
Myers, M., 286(41), *298*

Nadya, F., 27, 28(5), *63*
Nagle, E.M., 414(52), *420*
Nakada, I., 169(50), *183*
Nakagawa, Y., 239, 241, 245, 248, 250, 252(44), *298*
Nakato, Y., 364(101), *386*
Nasledov, D.N., 130(126), 130(130), *150*
Nathan, M.I., 203(30), *213*
Negreskul, V.V., 253(81), *299*
Neuffer, D., 380(157), *387*
Neugebauer, C.A., 141(259), *153*
Neukermans, A., 34, 35(46), *64*
Newman, R., 118(28), 118(30), 118(31), 118(32), 118(33), 128(89), *148, 149*
Neyhart, J.H., 217, 237, 238, 286(10), *297*, 451, 458(33), 472(52), *477*
Nickson, R., 380(156), *387*
Nicholas, K.H., 133(168), 138(233),

138(234), *151, 152*
Nickel, B., 177(77), *184*
Nicoll, F.H., 415(54), *420*
Nicoll, F.M., 144(273), *153*
Niekisch, E.A., 29, 31, 32, 33(18), *63*, 133(149), 133(150), 133(151), *150*
Nielsen, P., 291(178), *301*
Nielsen, W.J., 444(28), *477*
Nikuradse, A., 380(161), *388*
Ning, J.H., 27, 28(5), *63*
Nishi, Y., 34, 35(46), *64*
Nishino, T., 28(12), *63*, 130(133), *150*
Noble, P.J.W., 407(37), *419*
Noda, S., 367(112), *386*
Norris, C.B., Jr., 39(81), 57(177), 57(180), *65, 67, 68*
Noskov, M.M., 61(206), *68*

Oberländer, S., 133(160), 138(229), *151, 152*
Obraztsov, A.A., 271(116), *300*
O'Dwyer, J.J., 18(33), *25*
Ohki, K., 172(f), *172*
Ohno, K., 181(80), *184*
Ohyama, M., 130(121), 130(125), *150*
Ohyama, Y., 468(46), *477*
Okamura, S., 339(18, 19, 21), *384*
Olness, D., 159(17), *183*
Omelianovski, E., 130(112), *149*
Omelyanovskii, E.M., 130(127), *150*
Onn, D.G., 15(30), 18(35), 18, 20(36), *25*
Onsager, L., 17(32), *25*, 49(126), *66*, 108 (37), *115*, 222, 228(21), *298*, 325, 327(60), *334*, 350, 359(69), *385*, 452 (36), *477*
Orshinsky, S.R., 243(74), *299*
Ottaviani, G., 43(100), 46, 47, 57(110), 56(175), 57(178), *66, 67, 68*
Owen, A.E., 197(23), *213*, 241, 250, 252, 288(56), 241, 245, 248, 250, 252, 288 (57, 58), 257, 258, 275, 277(95), 258, 259(100), 265, 266(113), 283, 285(132), 288(153), 291(175), *298, 299, 300, 301*, 318(51), *333*
Ozaki, M., 364(101), *386*

Pai, D.M., 49, 56(129), 50(134), *66*, 108, 113(38), *115*, 217, 234(4), 217, 219, 222, 229—233, 275(8), 220, 275(12), 222, 225, 238, 275(18), 258, 259, 261 (101), 277, 281, 282(127), 296(196), *297, 300, 302*, 307, 308, 310, 312, 316, 325(22), 315(41), *333*, 423(37), 468, 470(47), *477*

Pao, C.S., 349, 350, 353(58), *385*
Papadakis, A.C., 42(90), 43(101), *65, 66*
Patan, J.P., 350, 353(66), 377(147), *385, 387*
Paul, W., 210(41), *214*
Pauling, L., 3(5), *25*
Peacock, R.N., 141(260), *153*
Pehl, R.H., 57(180), *68*
Peka, G.P., 130(134), *150*
Pell, E.M., 59(184), *68*, 117(5), *147*
Pellegrini, B., 3(10), *25*
Pelliccioni, M., 374(154, 155), *387*
Pence, I.W., 31(29), *64*
Penchina, C.M., 211(43), *214*
Penney, T., 241, 246, 250, 286, 288(60), *299*
Perlmutter, A., 123(70), *148*
Pervova, L.Ya., 130(127), *149*
Pethig, R., 60(198), *68*, 173(c), *173*
Petritz, R.L., 141(251), *152*
Pettit, G.D., 130(103), 130(115), *149*, 186(3), *213*
Pfister, G., 104, 106(34), 106(35), *115*, 217, 234(4), 258, 262, 271(97), 258, 262, 267, 268, 271(98), 258, 265(102), 258, 261, 264, 277, 278, 280(103), 271(117), *297, 300*, 321(54), *334*
Phillips, J.C., 3(8), *25*
Phipps, P.B.P., 122, 123(56), 122(58), *148*
Picus, G., 124(76), *149*
Pike, G.E., 291(167), *301*
Pike, W.S., 406(35), *419*
Pilloff, H.S., 364(97), *386*
Pinard, P., 130(137), *150*
Pisarev, A.F., 380(158), *387*
Pisarev, V.F., 380(158), *387*
Pochettino, A., 155(1), *182*
Pohl, R.O., 281(134), *300*
Pohl, R.W., 35(55), *64*
Pokrovskii, Y., 27(2), *62*
Polder, D., 132(141), *150*
Pollak, M., 86, 88(20), 88(23), *114*, 285, 290(162), 291(163, 164, 165), 291 (167), *301*, 323(56), *334*
Pollack, S.A., 56(169), *67*
Pope, M., 18(34), *25*, 112, 113(42), *115*, 162(30), 162(32), 163(35), 167(44), 176(71), 176(73), 177(78), 182(82), 182(83), *183, 184*
Popeska, C., 288(150), *301*
Portis, A.M., 141, 143(250), *152*
Powell, C.F., 35, 46(57), *64*
Pratt, R.G., 135(188), *151*

Price, M.G., 161(25), *183*
Prock, A., 50, 56(138), *66*
Pruett, H.D., 53, 56(157), *67*
Prunier, J., 380(160), *388*
Purohit, R.K., 400(22), *418*
Putley, E.H., 59(181), *68*, 397(16, 17), *418*
Putseiko, E.K., 135(195), *151*

Quaranta, A.A., 46, 47, 57(110), 56(175), 57(178), *66, 67, 68*
Quinn, R.K., 289(156), *301*

Rabie, S., 118(48), *148*
Radu, R.K., 130(138), *150*
Rakavy, G., 22(40), *25*, 40, 41, 48(84), *65*
Raman, R., 172(i), *172*
Ramy, J.P., 355(79), *386*
Randall, J.T., 133(154), *150*
Ranicar, J.H., 306, 309(12), *333*
Rasburn, E.J., 376(136), *387*
Rashevskaya, E.P., 130(27), *150*
Ray, B., 142(262), *153*
Read, W.T., 118(14), *147*, 243(75), *299*
Redaelli, G., 56(175), *67*
Redfield, A.G., 60(203), 61(205), *68*
Redfield, D., 240(55), *298*
Redington, R.W., 404(31), *419*
Regensburger, P.J., 218, 275(11), 221(15), 287(146), *297, 301*, 307, 308, 310, 312, 314, 316, 324, 330(20), *333*
Reif, F., 54(161), *67*, 343(34, 35), *385*
Reiser, A., 306, 309(15), *333*
Rembaum, A., 304(5), *332*, 308, 310(25), *385*
Rennert, J., 462(59), *478*
Reucroft, P.J., 167(45), *183*, 304(7), *332*
Revenko, G.S., 380(158), *387*
Ricateau, P., 12(29), *25*
Rice, S.A., 159(16), 171(56), 171(58), *183*, 336, 369(12, 14), 344(36, 40), 368(114), 370(121, 122, 123), 371(167), *384, 385, 386, 387, 388*
Rice, S.O., 79(16), *114*
Richards, E.W.T., 355(80), *386*
Riedl, H.R., 241, 248, 250(66), *299*
Riehl, N., 159(13), *182*
Rizakhanov, M.A., 210(40), *214*
Rizzo, A., 133(179), *151*
Roberts, G.G., 3, 8, 9(15), *25*, 464(43), *477*
Roberts, I., 361(94), *386*
Robertson, N., 203(28), *213*
Robinson, A.L., 122(55), 135, 136(209), 143, 144(268), *148, 152, 153*
Robinson, J.E., 59(183), *68*
Robinson, M., 346, 354, 370(50), *385*
Robinson, M.G., 359, 370(87), *386*
Robson, P.N., 32, 33, 57(37), *64*
Roccella, M., 378(154), *387*
Rockstad, H.K., 239, 288(52), 288(151), *298, 301*
Rodgers, R.L., 411(46), *419*
Rodinov, B.U., 380(159), *388*
Rojas, L.F., 159(15), *182*
Rose, A., 43, 45, 46, 49(103), 46(104), *66*, 72(2), *114*, 117(2), 118, 133(16), 118(17), 118(18), 118(19), 133(173), 133(174), 133(175), 135(202), *147, 151*, 200(25), *213*, 238(38), *298*, 390(1), *418*, 425, 465(23), *477*
Rosental, A., 20(38), 24(43), *25*
Rosi, F.D., 120(49), *148*
Rosier, L.L., 27, 28(5), *63*
Ross, D.A., 421(4), *476*
Rossier, D., 240, 241(54), 241, 248, 249, 250, 252, 253(68), *298, 299*
Rossiter, E.L., 275(122), *300*
Roush, M.L., 56(172), *67*
Roy, S.B., 135(219), *152*
Rubin, V.S., 130(122), 130(129), *150*
Ruch, J.G., 36, 53(71), 57(176), *65, 67*
Ruggiero, L., 133(179), *151*
Rukhlyadev, Yu.V., 239, 241, 242, 250, 253(47, 48), *298*
Rumin, N., 118(48), *148*
Ruppel, W., 135(202), *151*
Russell, B.R., 59, 60(186), *68*, 117(3), *147*
Russell, R.L., 365(104), *386*
Rutherford, E., 340, 342(22), *384*
Ryan, F.M., 60(187), *68*
Ryerson, R.J., 135(189), *151*
Ryvkin, S.M., 27(4), 27, 31, 62(6), *63*, 72(4), *114*, 117(6), 135(194), *147, 151*, 238(39), 252(79, 80), *298, 299*

Sadasiv, G., 406(35), *419*
Sah, C.T., 27, 28(5), *63*
Sakai, E., 56(173), *67*
Sakamoto, M., 135(192), *151*
Sakalas, A.P., 144(271), *153*, 275(121), *300*
Salaneck, W., 16, 18, 19(31), *25*
Salo, T., 144(269), *153*
Salvan, F., 27(2), *63*
Sanderson, A.C., 122(57), *148*
Santini, M., 344(42), 370(125), *385, 387*

Sato, H., 324, 327(58), *334*, 423(12), *476*
Saura, J., 123(75), *149*
Scales, J.L., 33(44), *64*
Schachter, H., 52(150), *67*
Schadt, M., 60(198), *68*, 172(61), *184*
Schaffert, R.M., 310, 326(28), *333*, 421(2), 436, 439(27), 468, 470, 474(45), *476, 477*
Schampers, P.D.M., 403(29), *419*
Scharfe, M.E., 39, 40(82), 49(125), *65, 66*, 217, 237, 258, 259, 261, 265, 268(9), 220, 257, 275(13), 237(35), 258, 261, 265(96), 258, 259, 261(101), *297, 298, 300*, 311(30), 315(41), 318(44), *333*, 421, 423(6), 426(24), 451, 453, 457(34), 468(44), 468(47), *476, 477*
Scharn-Horst, K.P., 241, 248, 250(66), *299*
Schechtman, B.H., 40, 42(87), 42(94), *65*, 322, 323(55), 328(63), *334*
Schein, L.B., 473(53), *477*
Scher, H., 38, 42, 43(74), *65*, 74, 75(10), 78, 80, 87(15), 86, 89, 90(21), 88(22), 90, 95, 98, 99(26), 90(27), 90, 91, 99(28), 105(39), *114, 115*, 258, 262, 267, 268, 271(98), 258, 260—263, 267, 268, 277 (104), 291(170, 171), 294(187), *300, 301, 302*, 318(45), 323(57), 331(69), *333, 334*, 450, 451, 453, 454(32), *477*
Schiller, R., 367(111), *386*
Schilling, R.B., 52(150), *67*
Schlesinger, M., 98, 99(31), *115*
Schmeing, H.S., 56(165), *67*
Schmid, E., 53(153), *67*
Schmidlin, F.W., 3, 8, 9(15), *25*, 217, 237, 286(10), 296(197), *297, 302*, 421, 423(6), 421, 423, 456(7), 435(25), 450, 452, 454(31), 455(38), 458(41), 474(55), *476, 477*
Schmidt, L.D., 35(67), 53, 54(155), *65, 67*, 318(46), *333*, 368, 381(117), 370 (127), 372(131), *386, 387*
Schmidt, W.F., 52(149), 54(159), *67*, 217, 234(7), *297*, 318(47), *333*, 339, 341(16), 343, 359(33), 344(43), 344, 370(44), 346, 370(46, 47), 347, 357, 359(53, 54, 55), 350, 353(67), 354(72), 367(113), 370, 372, 375(128), 378(151), 371(171), 376(172, 173), *384, 385, 386, 387, 388*
Schmidt-Parce Fall, W., 380(156), *387*
Schmit, J.L., 398(20), *418*
Schneider, W.G., 42, 51(98), *65*
Schnürer, E., 56(174), *67*
Schön, M., 118(22), *147*

Schott, M., 60(198), *68*, 161, 164(22), 168(47), 172(61), *183, 184*
Schottmiller, J., 282—286(135), *300*
Schubert, G., 56(174), *67*
Schultz, B.H., 31(31), *64*
Schütze, W., 378(152), *387*
Schwartz, L.M., 41, 42, 43(89), *65*
Schwarz, H.A., 367, 368(109), *386*
Schwarz, K.W., 35, 36, 55, 56(66), *65*
Schwemin, A.J., 377, 380(149), *387*
Sclar, N., 124(76), *149*
Seager, C.H., 61(205), *68*, 289(156), *301*
Sealer, D.A., 411(45), *419*
Seanor, D.A., 304(1), *322*
Searle, T.M., 253(88), *299*
Sebenne, C., 135, 136(216), *152*
Seccombe, S.D., 27, 28(5), *63*
Secker, P.E., 344(38, 39), *385*
Seeger, M., 53(153), *67*
Segall, B., 125(81), *149*
Segar, A., 380(156), *387*
Sehr, R., 400, 402(25), *418*
Seibt, W., 42(93), *65*
Seker Dzijski, V., 140(238), 140(239), *152*
Seki, H., 35(63), 40, 42, 43(86), 40, 42(87), 42(91), 42(94), 42(95), 51(43), *65, 67*, 222(23), 222(25), 261, 268 (109), *298, 300*, 308, 310, 323(29), 311, 315, 317(32), *333*
Seki, K., 181(80), *184*
Selway, P.R., 130(119), *150*
Sequin, C.H., 411(45), *419*
Shallcross, F.V., 141, 142(257), 141(258), *153*, 393(7), 411(46), *418, 419*
Shaposhnikova, T.A., 130(130), 130(136), *150*
Sharma, B.L., 400(22), 400, 403(24), *418*
Sharma, B.S., 414(51), *420*
Sharma, R., 50, 52(136), *66*, 162(31), *183*
Sharp, J.H., 43(102), *66*, 330(68), *334*
Shattuck, M.D., 326(61), *334*
Shaw, M.P., 8(22), *25*
Shaw, R.F., 239(53), *298*
Shear, H., 135, 136(208), *152*
Sheinkman, M.K., 138, 139(235), 138(236), *152*
Shelykh, A.I., 60(187), *68*
Shibamura, H., 357(84), *386*
Shilo, V.P., 239, 241, 242, 250, 253(47, 48), *298*
Shimizu, K., 403(27, 28), *419*

Shimoda, K., 60(190), *68*
Shinshaka, K., 359(88), *386*
Shockley, W., 33(41), 33, 34(42), 34, 35 (45), *64*, 118(14), *147*, 243(75), *299*, 370(120), *387*
Shulman, R.G., 47(121), *66*
Sidorov, V.I., 27, 28(5), *63*
Sidran, M., 462(59), *478*
Siebrand, W., 156, 174(7), 176(69), 176(70), *182, 183, 184*
Sigmon, T.W., 39(81), 53(152), 57(180), *65, 67, 68*
Silbey, R., 171(58), *183*
Silinsh, E.A., 165, 167(40), *183*
Silver, M., 15(30), 18(34), 18(35), 18, 20 (36), 20(37), 24(42), *25*, 50, 52(136), 52(148), 52(150), *66, 67*, 111(40), *115*, 159(17), 162(31), *183*, 230(28), 259, 271(105), *298, 300*
Silverman, J., 56(172), *67*, 341, 346, 370 (25), *385*
Simhony, M., 38, 42, 48(75), *65*
Simmons, J.G., 197(24), *213*, 241, 246, 247, 248, 250(67), 241, 242(69), 241 (70), 241, 242(71), 241, 245, 247(72), *299*
Sirucek, I., 62(219), *69*
Skarman, H.S., 136(224), 142(265), 142 (266), *152, 153*
Sklensky, A.F., 118(47), *148*
Skvortsov, I.M., 130(111), *149*
Slade, M.L., 237(26), *298*
Smejtek, P., 18, 20(36), *25*
Smith, G.C., 30, 61(20), 60(199), 61(204), *63, 68*, 172(61), *184*
Smith, G.E., 408(30), *419*
Smith, J.E., Jr., 203(30), *214*
Smith, R.A., 73(9), *114*, 140(244), 140 (245), *152*
Smith, R.W., 28(14), 46(112), *63, 66*, 133(173), 133(175), 133(177), *151*
Smits, R.G., 377, 380(148, 149), *387*
Smollett, M., 135(188), *151*
Smorgonskaya, E.A., 253(86), *299*
Snowden, D.P., 141, 143(250), *152*
Sochard, I., 61, 62(213), *69*
Solomon, P.R., 8(22), *25*
Sommer, G., 50, 51(142), *66*
Sommer, W.T., 368(115), *386*
Sommers, H.S., Jr., 61(211), 61, 62(213), *69*
Somorjai, G.A., 135(222), *152*
Sonnonsteine, T.J., 175(68), *184*
Sorrows, H.E., 141(251), *152*

Soudain, J., 377(147), *387*
Sowada, U., 370, 372, 375(128), 371(171), *387, 388*
Spear, W.E., 35, 38, 39, 53, 55, 56, 57(58), 35, 57(61), 35, 39, 53, 54, 57(65), 36 (70), 46, 48(105), 46(111), 46, 57(113), 48(124), 49, 51, 52(130), 50(133), 53 (154), 53, 54(156), *65, 66, 67*, 89(25), 112(43), *114, 115*, 185(1), 186, 191, 192, 194(9), 186, 195, 204(10), 186, 188, 201(11), 186, 188, 201(11), 186, 188, 194, 201(12), 186, 187, 194, 202 (13), 186, 187, 194, 202(14), 188(15), *213*, 217, 224(1), 241(73), 244(77), 244(76), 257, 275, 277(90, 91), 275, 280(118), 279(128, 129), 281(131), 291(179), *297, 299, 300, 301*, 369 (118), 369(166), 371(170), *386, 388*
Spicer, W.E., 193(19), *213*
Spitzer, W.G., 3(12), *25*, 47(121), *66*
Sproull, R.L., 59(184), *68*
Staffev, V.I., 130(122), 130(129), *150*
Stahel, E., 377(140), *387*
Stark, H.M., 455(38), *447*
Starr, L.H., 35, 46(57), *64*
Steinberger, I.T., 363(165), *388*
Steingart, M., 370(126), *387*
Steingrabber, O.J., 56(169), *67*
Sterling, H.F., 186(7), *213*
Stern, F., 253(89), *299*
Stewart, W.C., 414(52), *420*
Stille, R., 57(179), *68*
Stössel, W., 32(33), *64*
Stratton, R., 130(101), *149*
Strauss, A.J., 125(82), *149*
Street, R.A., 203, 204(26), 253(88), 293 (185), *298, 299, 302*
Stringfellow, G.B., 122, 123(60), 122, 123(61), *148*
Strom, K., 293(182), *302*
Stöckman, F., 29, 55(16), 46(118), *63, 66*
Stoica, T., 211, 212(44), *214*
Stojanov, V., 140(238), 140(239), 140 (240), *152*
Stone, J.L., 294(186), *302*
Stowac, L., 288, 290(152), *301*
Stuke, J., 211(42), *214*, 250(78), *299*
Su, J.L., 34, 35(46), *64*
Süptitz, P., 33(41), *64*
Suzuki, A., 172(g), *172*
Swan, D.W., 357(83), *386*
Swiatek, J., 306, 308(11), *333*
Swicord, M., 159(17), *183*

Sworakowski, J., 165, 167(40), *183*
Sze, S.M., 8(20), *25*
Szepes, Z., 415(55), *420*
Szymanski, A., 170(55), *183*, 306, 308(11), 308, 319(23), *333*

Tabak, M.D., 39, 40(82), 49(125), *65*, *66*, 220, 257, 275(13), 221(14), 221(16), 221, 222, 225, 236, 257, 275(17), 257, 258, 275, 277(94), 277, 280(126), 282—286(135), 282, 284(137), *297*, *299*, *300*, *301*, 315(41), 318(50), *333*, 426(24), 452, 457(35), 468, 471(49), *477*
Tachiya, M., 360(92), *386*
Takada, K., 339(21), *384*
Takagaki, T., 339(18, 19, 21), *384*
Takeda, S.S., 364, 366(88), *386*
Takahashi, T., 357(84), *386*
Takahashi, Y., 468(46), *477*
Talalakin, Ga.N., 130(110), 130(136), 130 (138), *149*, *150*
Tallan, N.M., 60(191), *68*
Tanagida, T., 130(133), *150*
Tanaka, M., 288(154), *301*
Tannhauser, D.S., 60(191), *68*
Tantalo, P., 61(212), *69*
Taroni, A., 42(93), 43(100), *65*, *66*
Tauc, J., 61(210), *69*, 132(139), 132(140), *150*, 191(17), *213*, 237(36), *298*
Tauchert, W., 367(110, 113), 368(116), 376(172), *386*, *388*
Tausend, A., 60(196), *68*
Taylor, G.W., 197(24), *213*, 241, 246, 247, 248, 250(67), 241, 242(69), 241(70), 241, 242(71), 241, 245, 247(72), *299*, 350, 352(62), 350(63), *385*
Taylor, P.C., 293(182), *302*
Taylor, W., 285(138), *301*
Tefft, W.E., 46(106), *66*
Temnitskii, Y.N., 20(39), *25*
Terenin, A.N., 135(195), *151*
Tereshko, G.N., 130(128), *150*
Terelecki, J., 341, 376(29), 343(31), *385*
Teucher, I., 161, 164(26), *183*
Tewari, P.H., 55(162), *67*, 341(26), *385*
Theobald, J.K., 12(26), *25*
Thomas, A., 371(168), *388*
Thomas, D.G., 83, 85(18), *114*
Thomas, G., 380(160), *388*
Thomas, J.K., 366, 367(106), *386*
Thomas, P.A., 135, 136(216), *152*
Thomsen, S.M., 122(52), 136(223), *148*, *152*
Thomson, G.P., 12(25), *25*

Thomson, J.J., 12(25), *25*, 336(1), *384*
Tien, H.T., 177(76), *184*
Title, R.S., 186(2), 186(3), *213*
Tkachev, V.D., 130(132), *150*
Tobolsky, A.V., 237, 240(34), *298*
Tödheide-haupt, U., 46(115), *66*
Tokarsky, R.W., 32(34), 32(36), *64*
Tokumam, Y., 130(108), *149*
Tombs, T., 173(d), *173*
Tompsett, M,F., 408—412(41), 411(45), *419*
Tooke, C.C., 130(119), *150*
Torres, L., 377(145), *387*
Tove, P.A., 56(170), *67*
Tretyak, O.V., 130(107), 130(124), *149*, *150*
Triebwasser, S., 32(35), *64*
Trofimenko, A.P., 138, 139(235), *152*
Tsarenkov, B.V., 130(131), *150*
Tscholl, E., 137(225), *152*
Tsu, R., 239(51), 241(63), *298*, *299*
Tsubomura, H., 176(75), *184*, 364(101), *386*
Tukey, J.W., 95(29), *115*
Tulagin, V., 422(8), *476*
Turner, W.J., 130(103), 130(115), *149*
Tutihasi, S., 271(117), 293(180), *300*, *301*, 474(55), *477*
Twose, W.D., 295(193), *302*
Tyler, W.W., 118(28), 118(29), 118(30), 118(31), 118(32), 118, 128(34), 128 (88), 128(89), *148*, *149*
Tyndall, A.M., 35, 46(57), *64*
Tyrrell, M., 380(156), *387*

Ullmaier, H., 349, 350, 353(59), *385*
Ullmaier, H.A., 354(71), *385*

Vahtra, V., 326(61), *334*
Vala, M.T., 171(58), *183*
Vancu, A., 194(20), *213*
Van de Graff, R.J., 35, 46(57), *64*
Van der Bizl, H.J., 336, 341(8), *384*
Van der Leeden, G.A., 59(182), *68*
Van der Pauw, L.J., 31, 33, 57(32), 59(182), *64*, *68*, 132(141), *150*
Van Heck, H.F., 60(194), *68*
Van Heyningen, R., 35(60), 46(117), *64*, *66*
Vannikov, A.V., 306, 308(10), 306(16), *333*
Van Roosbroeck, W., 8(18), *25*, 33, 34(43), 61, 62(208), 62(216), *64*, *68*, *69*
Vasanelli, L., 133(179), *151*
Vasilchenko, W., 146(280), *153*
Vass, Sz., 367(111), *386*

Vaubel, G., 159(13), 159(14), 163(36), *182, 183*
Vaughn, R.A., 185(1), *213*
Vavilov, V.S., 117(7), *147*
Vecht, A., 393(8), *418*
Vengris, S., 275, 280(119), 275(121), 275, 280(122), *300*
Vermande, P., 377, 378(144), 377(146), 378(153), *387*
Vermeil, C., 363(95), *386*
Vescan, L., 209, 211, 212(37), 211, 212(44), *214*
Vescar, L., 288(150), *301*
Vilkotskii, V.A., 130(128), *150*
Viscakas, J., 144(271), *153*, 235, 240(31), 237, 240, 257, 258, 275, 277, 280(32), 275, 280(119), 275(121), *298, 300*
Voigt, J., 133(160), 133(167), *151*
Volger, J., 141, 143(238), *152*
Von der Linde, D., 56(170), *67*
Voos, M., 27(2), *63*
Vorobev, Yu.V., 130(107), *149*

Wahlig, C., 59, 60(186), *68*
Wakayama, N., 161, 164(24), *183*
Walker, A.C., 27(4), *63*
Walkup, L.E., 423(22), *477*
Walsh, D., 27(3), *63*
Walsh, E.J., 403(30), *419*
Ward, A., 282—286(135), *300*
Ward, A.L., 33(44), *64*
Wardman, P., 375(135), *387*
Warfield, G., 275(120), *300*
Warman, J.M., 375(133), *387*
Warter, P.J., 221, 222, 225, 236, 257, 275 (17), 222(22), 275(124), *297, 298, 300*, 452, 457(35), *477*
Watanabe, V., 27(3), *63*
Watkins, B.G., 141(254), *153*
Watson, F.H., 341, 374(24), *385*
Watters, R.L., 35(50, 51), *64*
Waxman, A., 135(213), 141, 142(257), *152, 153*
Webb, P.P., 56(173), *67*, 396(15), *418*
Weckler, G.P., 407(38), *419*
Weckler, O.P., 407(36), *419*
Wegener, D., 380(156), *387*
Wei, J.C., 366(105), *386*
Weichman, F.L., 132(142), *150*
Weigl, J.W., 310, 326(27), *333*
Weimer, P.K., 141, 142(257), *153*, 401(26), 406(35), *419*
Weisberg, L.R., 128(94), 129(98), 129(99), *149*

Weiser, K., 186(3), *213*, 241, 245, 246, 248, 250, 252, 286, 288(59), 241, 246, 250, 286, 288(60), 241, 244, 248, 253, 288(61), 241, 244, 250, 253(62), 253 (89), 286(142), 290(145), 295(195), *299, 301, 302*, 319, 320, 326(52), 320, 326(53), *334*
Weiss, G.H., 76(14), *114*
Weisz, S.Z., 38, 42, 48(75), *65*, 159(15), *182*
Wendland, P.H., 395(13), *418*
Wendt, F.S., 414(52), *420*
Wendt, M., 27(4), *63*
Westbury, R.A., 237, 240, 285(33), *298*
Weston, W., 18(34), *25*
Westphal, W.C., 34, 35(45), *64*
Wilkins, M.H.F., 133(154), *150*
Willard, T., 380(156), *387*
Williams, D.F., 60(198), *68*, 161, 164(24), 172(61), *183, 184*
Williams, D.J., 217, 234(4), *297*
Williams, E.W., 130(117), *150*
Williams, F., 339(18), *384*
Williams, M.L., 281(133), *300*
Williams, R., 10(24), *25*, 28, 55(8), *63*, 135(206), *151*, 163(37), *183*
Willmott, R.W., 423(17), *477*
Wilson, E.G., 361(94), *386*
Wilson, R.H., 404(31), *419*
Winokur, P.S., 56(171), *67*
Winter, K., 380(156), *387*
Wintle, H.J., 42(96), *65*
Wise, E., 423(21), *477*
Witte, R.S., 56(169), *67*
Wlerick, G., 118(20), *147*
Wolff, N.E., 318, 322(49), *333*
Wood, C., 458(40), *477*
Woodall, J.M., 130(105), *149*
Woodbury, H.H., 118(28), 118(29), 118(31), 118(32), 118, 128(34), 125(81), 128(88), *148, 149*
Woods, J., 133(168), 138(232), 138(233), 138(234), *151, 152*
Woods, J.F., 130(102), 130(105), 141(246), 141(251), *149, 152*
Wright, D.A., 138(232), *152*
Wright, G.B., 285(138), *301*
Wu, C., 135, 137, 141(190), 135, 136, 142(220), *151, 152*

Yakovlev, B.S., 377(137), *387*
Yahagi, K., 306, 309(18), *333*
Yamashita, J., 174(63), *184*
Yamazawa, Y., 339(18), *384*

Yaroshetskii, I.D., 27(4), *63*
Yates, D.A., 211(43), *214*
Yguerabide, J., 56(165), *67*
Yndurain, F., 3(11), *25*
Yoffe, A.D., 239(50), 293(185), *298, 302*
Yoshida, K., 462(57), *478*
Yoshida, O., 403(28), *419*
Yoshino, K., 27(3), *63*, 306, 309, 324(19), *333*
Yoshimura, S., 160(18), *183*
Young, C.J., 423(18), 423(20), *477*
Young, L.A., 12(29), *25*
Yukhnevich, A.V., 28(12), *63*
Yule, J.A.C., 444(28), *477*

Zaklad, H., 377, 380(148, 149), *387*
Zaklad, Z., 354(75), *386*
Zallen, R., 49(132), *66*, 107(36), *115*, 217, 234(3), 237(36), *297, 298*
Zanarini, G., 42(93), 43(100), 56(175), *65, 66, 67*
Zanio, K.R., 35, 39, 45, 47, 55, 56, 57(59), 46, 47(107), 46, 47, 57(110), *64, 66*
Zavadskii, Yu.I., 113(44), 118(45), 130(111), *148, 149*
Zavetova, N., 288, 290(152), *301*
Zeller, R.C., 281(134), *300*
Zessoules, N., 371(168), *388*
Zimmerman, W., 32(33), *64*
Ziman, J., 73(7), *114*
Zitter, R.N., 62(216), *69*
Zohta, Y., 130(109), *149*
Zuleeg, R., 400, 402(25), *418*
Zullinger, H.R., 57(180), *68*
Zvonkov, B.N., 141(256), *153*, 210(40), *214*

SUBJECT INDEX

absorbing boundary, 98 ff.
absorption spectrum,
    anthracene, 157, 162
    a-Se, 220
    a-Si, 192
    napthacene, 179
    orthorhombic S, 107
    TMPD, 365
a.c. conductivity, 81ff., 86ff., 290ff., 322
    a-As_2Se_3, 292, 294
    a-Si, 86, 89
    CdS, Al-doped, 294
    contact resistance, 294
    extended states, 290ff.
    hopping conduction, 86ff., 292ff.
    poly(N-vinyl carbazole): trinitrofluor-
        enone charge transfer complex, 90, 323
    temperature dependence, 292
activation energies,
    amorphous chalcogenides, 249ff.
    a-As_2Se_3, 273ff.
    a-Se, 276ff.
    a-Si, 195ff.
    poly(N-vinyl carbazole): trinitrofluor-
        enone charge transfer complex, 314ff.
accumulation layer, 4ff.
anthracene,
    absorption spectrum, 157, 162
    band gap energy, 163
    cyclotron resonance, 172
    drift mobility, 169ff.
    electron affinity, 161
    energy level diagram, 178
    exciton—exciton interaction, 159ff.
    exciton-free charge interaction, 161
    exciton-states, 156ff.
    exciton—surface interaction, 158ff.
    exciton-trapped charge interaction, 161
    hole mobility, 60, 171ff.
    hopping transport, 169ff.
    intrinsic photogeneration, 161ff.
    ionization energy, 161
    Langevin recombination, 167ff.
    mobility, pressure dependence of, 170

anthracene (continued),
    mobility, temperature dependence of, 169ff.
    mobility, theoretical, 173ff.
    Onsager mechanism, 167, 234
    photoemission, 163ff.
    photoconductivity, intensity dependence of, 159ff.
    photoconductivity, spectral dependence of, 158, 160
    photogeneration process, 167ff.
    recombination coefficient, 167
    recombination process, 165ff.
    singlet—singlet annihilation, 160
    singlet—singlet interaction, 160
    trapping process, 165ff.
    two-photon absorption, 160
As_2S_3, amorphous,
    carrier range, 238, 286, 458
    CuI_2-doped, 286
    optical gap, 286
    xerographic discharge, 287
As_2Se_3, amorphous,
    conductivity, 267
    drift mobility, 255ff., 265
    mobility, activation energy of, 273ff.
    mobility, thickness and field dependence of, 271ff.
    optical quenching, 253
    photogeneration, 237ff.
    radiative recombination, 253
    Scher—Montroll model, 268ff.
    stochastic hopping transport, 268ff.
    universality of current, 103, 268ff.
    xerographic properties, 468ff.
As—Se alloys, drift mobility, 283ff.
As—Te alloys, 239ff.
As_2SeTe_2, amorphous, dark and photoconductivity, 241ff.
As_2Te_3, amorphous, 252ff.
    photoconductivity, 288ff.
$As_{30}Te_{48}Ge_{12}Si_{10}$, amorphous, dark conductivity, 266
Auger recombination coefficient, 123ff.

biphotonic excitation,
   anthracene, 160
   a-Se, 222ff.
   TMPD, 364ff.
blocking contact, 8, 39
Bose—Einstein distribution, 110

capture cross-section,
   CdS, 123
   GaAs, 121
   InP, 121
carrier range, 48
CdS, crystalline,
   capture cross-section, 123
   carrier concentration, intensity dependence of, 132ff.
   Cu and Ga impurities, 125ff.
   heterojunctions, 145ff.
   optical quenching, 122ff.
   photochemical changes, 137ff.
   photoconductivity, spectral response of, 126ff.
   photothermoelectric effects, 142ff.
   sintered and polycrystalline, 136ff.
   solution-sprayed films, 136, 142ff.
   thermally-stimulated conductivity, 138
CdSSe, crystalline,
   ionization energy of sensitizing centers, 124
   thermally-stimulated conductivity, 133ff.
CdTe, crystalline, Cl and Ga impurities, 125
chalcogenides, a.c. conductivity, 290ff.
   density of states, 243ff.
   decay times, 252ff.
   mobility edge, 295ff.
   photoconductivity activating energies, 241ff.
   radiative recombination, 253
   steady-state photoconductivity, 238ff.
charge-coupled devices, 408ff.
charged pigment xerography, 428ff.
chemisorption, 135ff.
clearing-field method, 340ff.
columnar ionization, 349
conductivity,
   a-As_2Se_3, 267
   a-As_2SeTe_2, 241ff.
   a-$As_{30}Te_{48}Ge_{12}Si_{10}$, 266
   a-$Se_{40}Te_{40}As_{20}$, 248ff.
   a-Si, 188ff.
   radiation-induced, 337ff.
contacts, 2ff.
coupling coefficient, 436ff.
covalently bonded material, 3, 72, 117

cyclotron resonance, anthracene, 172

dangling bond, 204
Davydov splitting, 157
Debye length, 21, 24
Dember effect, anthracene, 176
density of states, chalcogenides, 243ff.
   a-Si, 187ff.
depletion layer, 5ff., 15, 465
detectivity, 391
development, 444ff.
   adhesion-controlled, 431ff.
   density function, 431ff.
   viscosity-controlled, 431ff.
dielectric relaxation time, 7, 33
drift mobility, 35ff., 168ff., 189, 260ff., 306ff.
   anthracene, 47, 168
   a-As_2Se_3, 255ff.
   a-Se, 275ff.
   a-Si, 188
   definition, 36, 168, 261
   experimental technique, 35ff., 168ff.
   high electric field effects, 372ff.
   hydrocarbons, 370ff.
   liquified gases, 370
   liquified rare gases, 369ff.
   monoclinic Se, 278
   N-isopropyl carbazole, 104ff.
   organic crystals, 172
   polymers, 308ff.
   poly(N-vinyl carbazole), 310ff.
   poly(N-vinyl carbazole):trinitrofluorenone charge transfer complex, 310ff.
   shutter method of measurement, 343
   trinitrofluorenone, 312ff.
   tetramethylsilane, 371
dye-sensitization,
   anthracene, 177ff.
   poly(N-vinylcarbazole), 324

effective mass,
   anthracene, 172
   liquid He, 23
Einstein relation, 112
electroabsorption, poly(N-vinylcarbazole): trinitrofluorenone charge transfer complex, 326ff.
electrolytic electrodes, 18
electron affinities, molecular crystals, 182
electron attachment, 376
electron back-scattering, 12ff.
electron coupling, 175
electronegativity, 2

electrophoresis, 427ff.
energy level diagram,
   anthracene, 178
   a-Si, 198
   photoreceptors for xerography, 450
   poly(N-vinyl carbazole):trinitro-
      fluorenone charge transfer com-
      plex, 332
escape cone, 14
exciton states, 156ff.

ferroelectric photoconductor device, 413ff.
field effect,
   a-Se, 291
   a-Si, 187
fluorescence, 51, 84
Fokker—Planck equation, 367
free ion yield, 347ff., 357ff.
   cyclohexane, 348
   diethylether, 348
   liquids, 358
   neopentane, 348

GaAs, amorphous, photoconductivity, 212
GaAs, crystalline,
   capture, cross-section, 121
   photoconductivity spectral response,
      131
   thermally-stimulated conductivity, 130ff.
GaP, amorphous,
   photoconductivity, 211
   light decay curve, 84
GaSb crystalline, donor behavior, 125ff.
Gaussian packet, 90, 95, 101
Ge, amorphous,
   electrolytic, 210ff.
   evaporated, 210ff.
   photoconductivity, 207ff.
geminate recombination, 49, 349ff.
   anthracene, 167
   a-Se, 222ff.
   poly(N-vinyl carbazole):trinitrofluor-
      enone charge transfer complex, 324ff.
glow discharge decomposition, 186

Hall angle, 59
Hall effect, measurement technique, 58ff.
Hall mobility,
   anthracene, 171ff.
   CdS, polycrystalline, 141ff.
   PbS, 137
   poly(N-vinyl carbazole):iodine charge
      transfer complex, 310
Haynes—Shockley experiment, 34

heterojunction photodiode, 400
   Cu_2S-CdS, 145ff.
   photovoltaic cell, 146
n-hexane,
   conductivity, dosage dependence, 340
   drift mobility, 340
   ionization current, 339ff., 352ff.
   linear energy transfer, 379ff.
   purification, 380ff.
high electric field effects, 372ff.
hopping activation energy, 243ff.
hopping mobility, definition, 100ff.
hopping times, 76ff., 91ff.
   distribution function, 78ff.
hopping transport, 74ff.
   anthracene, 173ff.
   a-As_2Se_3, 256ff.
   a-Se, 275ff.
hot electrons, 17

image converters, 415
image intensifiers, 415
image neutralization, 445
image potential barrier, 14
image storage, 412ff.
imaging systems, 400ff.
impurity effects, a-Se, 282ff.
infrared emission, 123
infrared detectors, 394ff.
   PbS, 140ff.
injecting contact, 6ff., 42
input image, 440
InP, crystalline, Cu-doped,
   capture cross-section, 121
   sensitized photoconductivity, 121ff.
InSb, amorphous, photoconductivity,
   211ff.
interfacial transfer, 454ff.
interfacial trapping, 454ff.
ionic materials, 3, 117
ionization chambers, 373ff.
ionization current,
   CS_2, 352ff.
   liquid argon, 354
   liquid nitrogen, 355ff.
   n-hexane, 353
   n-heptane, 355
   n-pentane, 353
ionization potential,
   argon, 363
   molecular crystals, 182
   xenon, 363

jumping distance, 373

jumping frequency, 373

Langevin recombination, 40, 49ff., 111ff.
   anthracene, 52, 112, 167
   hydrocarbons, 375ff.
   poly(N-vinyl carbazole):trinitrofluor-
      enone charge transfer complex, 329
   orthorhombic S, 52
   theoretical, 111ff.
latent image formation, 442ff.
leucobase of malachite green, 322
lifetime,
   imperfection control, 118
   minority carrier, 31, 47
   Si, Zn-doped, 420ff.
light sensors, 390ff.
light sources, 55ff.
localized hopping transition, 243ff.
localization radius, 104
   N-isopropyl carbazole, 104
   trinitrofluorenone, 104, 320
localized states, 74ff., 187ff., 243ff.
linear energy transfer, 337ff.
luminescence centers, 124

magnetic field effects, 58ff.
   Hall effect, 58ff.
   photoHall effect, 58ff.
   photoelectromagnetic effect, 61ff.
memory devices, optical, 412ff.
mobility,
   ambipolar, 34
   drift, see drift mobility
   edge, 242, 295ff.
   Hall, see Hall mobility
modulated excitation, 31ff.
momentum exchange scattering, 16
MOS bucket brigade, 412
multivacancy complexes, 204

napthacene,
   absorption spectrum, 179
   bandgap energy, 163
   photocurrent spectral response, 180
Nernst—Einstein relation, 346, 375
N-isopropyl carbazole, 104, 261
noise equivalent power, 390

ohmic contact, 4, 42
Onsager mechanism, 17, 49ff., 107ff.,
      167, 216ff., 350ff.
   anthracene, 167, 234
   a-Se, 108, 350ff.
   As_2S_3, crystalline, 234

Onsager mechanism (continued),
   hopping transport, 107ff.
   polar liquids, 351
   poly(N-vinyl carbazole), 325
   poly(N-vinyl carbazole):trinitrofluor-
      enone charge transfer complex,
      327ff.
optical gap, a-Si, 192
optical memory devices, 412ff.
optical quenching, 120

pair luminescence, 81ff., 86
photoactive pigment electrophotography,
   428ff.
PbS,
   drift mobility, 141ff.
   Hall mobility, 137
   hole density, intensity dependence of,
      141
   infrared detector, 140
   polycrystalline, 140ff.
photoadsorption effects, 135ff.
photoconductivity,
   anthracene, 156ff., 161ff.
   $a-As_2Se_3$, 237ff.
   a-GaAs, 212
   a-GaP, 211
   a-Ge, 206ff.
   a-InSb, 211
   a-Se, 216ff.
   a-Si, 186ff.
   CdS, 126ff.
   GaS, 130ff.
   napthacene, 180
   Si, Zn-doped, 119
   violanthrene, 179ff.
   vitreous Se, 235ff.
photochemical changes, CdS, 137ff.
photodiode arrays, 407ff.
photo-Hall effects, GaAs, 129
photoinduced discharge, 457ff.
   a-Se, 469ff.
photoelectromagnetic effect, 61ff.
photoemission, external, 180
   anthracene, 180
   napthacene, 181
photoemission, internal, 10ff., 55, 366
   anthracene, 24
   hydrogen, 19
   image potential barrier, 14ff.
   liquid xenon, 368
   non-polar liquids, 367
   poly(N-vinyl carbazole), 18, 330ff.
   theoretical, 10ff., 15ff., 18ff.

photoemission, internal (continued)
   vitreous Se, 235ff.
photoluminescence, GaP, 85
photoreceptors, practical, 469ff.
photothermoelectric effects, 128ff.
   Si, crystalline, 132
plumbicons, 403
polyethylene terepthalate, mobility, 313
poly(N-vinyl carbazole),
   drift mobility, 307ff.
   mobility, concentration dependence, 311ff.
   molecular structure, 307
   photoemission, 18ff.
   universality, 330
poly(N-vinyl carbazole):trinitrofluorenone charge transfer complex,
   drift mobility, 312ff.
   hopping transport, 104, 315ff.
   mobility activation energy, 315
   Onsager mechanism, 234, 323ff.
   optical absorption, 320
   thermally-stimulated current, 316
   universality, 261, 317
Poole—Frenkel mechanism, 167, 222ff., 245, 325
   a-As_2Se_3, 473
   a-Se, 227, 473
pulsed excitation, 30ff.

quantum efficiency,
   a-As_2Se_3, 236
   anthracene, 167
   a-Se, 216ff.
   determination of, 48ff.
   poly(N-vinyl carbazole), 325
   poly(N-vinyl carbazole):trinitrofluorenone charge transfer complex, 327ff.
   vitreous Se, 235

radiation detectors, 377ff.
radiative recombination, 84, 253
random walk, 74ff.
   continuous time, 76ff., 91
rare gas liquids, 369ff.
recombination centers, 118
recombination coefficient,
   anthracene, 167
   a-Si, 52
   poly(N-vinylcarbazole), 329
recombination kinetics, a-Si, 199ff.
recombination lifetime, a-Si, 193
recombination process, a-Si, 202ff.

recombination theory, 197ff., 241ff.
reflectivity, orthorhombic S, 107
relative biological efficiency, 379ff.
response time, 392
Richardson constant, 8

scattering probability,
   back, 16
   isotropic, 12
   inelastic, 16
   momentum exchanging, 16
scavenging, 445
Schottky barrier, 9, 142
Se, amorphous,
   absorption coefficient, 220
   biphotonic generation, 229ff.
   drift mobility, electrons, 280ff.
   drift mobility, holes, 255, 276ff.
   drift mobility, activation energies, 276ff.
   impurity effects, 282ff.
   initial separation, 229ff.
   Onsager mechanism, 228ff.
   photogeneration of electrons, 233ff.
   photogeneration of holes, 216ff.
   preparation, 217
   surface recombination, 222ff.
Se, monoclinic, 278
Seebeck effect, 128
sensitizing centers, 118, 124
Se, vitreous, 235
$Se_{40}Te_{40}As_{20}$, amorphous,
   photoconductivity, 243
   recombination theory, 249
Shockley—Read recombination, 118
Si, amorphous,
   conductivity, 189
   density of states, 187ff.
   dependence of electrical properties on deposition conditions, 189
   drift mobility, 188
   energy level diagram, 197ff.
   hopping transport, 188, 201ff.
   localized state, 198ff.
   optical absorption, 191
   photoconductivity, 186ff., 190, 206
   photoconductivity, intensity dependence of, 197
   photoconductivity, spectral dependence of, 190
   photoconductivity, temperature dependence of, 195ff.
   photoluminescence, 204ff.
   preparation, 186
   recombination kinetics, 199ff.

Si, amorphous, (continued)
  recombination lifetime, 193
  recombination process, 202ff.
Si, crystalline, Zn-doped,
  electron lifetime, 121
  excitation spectrum, 119
singlet exciton, 157ff.
Smoluchowski equation, 108, 110
solar cells, 415ff.
  GaAs, 417
  Si, 416
space-charge-limited current, 7, 20ff., 40ff., 164
  anthracene, 167
  a-Se, 43ff.
  liquid He, 23
  napthacene, 165ff.
  S, 43
  Si, surface barrier structure, 43
stochastic hopping, 76ff.
Stoke's shift, 203
surface recombination, 223ff.
surface recombination velocity, 33

Te glasses, $Te_{80}X_{20}$(X= As, Ge, and Si), 252ff.
thermally-stimulated conductivity, 134
  GaAs, 130
  poly(N-vinyl carbazole):trinitrofluorenone charge transfer complex, 316
thermally-stimulated Hall effect, GaAs, 130
thermionic emission, 18
time-of-flight technique, see transit-time technique
TMPD,
  absorption spectrum, 365
  photocurrent spectral response, 365
TNF, 2,4,7-trinitro-9-fluorenone,
  drift mobility, 312ff.
  drift mobility, concentration dependence of, 318ff.
  hopping transport, 104, 312ff.
  molecular structure, 307
toner,
  contact, 433ff.
  proximity, 433
  development characteristics, 445ff.
transit-time techniques, 33ff., 256ff.
  double shutter technique, 54
  interrupted transit, 48
transition rate, intersite, 83
trap density, energy distribution of,
  technique for measuring, 32
trap depth, Ge, Si, 47

trapping time,
  definition, 45
  orthorhombic S, 48
trap release time, definition, 45

universality, 96, 261, 268

vidicons,
  infrared-sensitive, 405ff.
  Sb_2Se_3, 401
  Si diode, 402
violanthrene A,
  absorption spectrum, 180
  photocurrent spectral response, 180

work function, 3

xenon liquid,
  photoionization, 362
  photoyield spectral response, 362
xerographic discharge, 218, 287
xerographic process, 426ff.
x—y-addressed system, 406ff.

ZnO resin, 471
ZnSe, crystalline, luminescence photoconductivity, 123
ZnSSe, crystalline, hole ionization energy, 124

RETURN TO ➡ **PHYSICS LIBRARY**
351 LeConte Hall   642-3122

LOAN PERIOD 1 **1-MONTH**	2	3
4	5	6

ALL BOOKS MAY BE RECALLED AFTER 7 DAYS
Overdue books are subject to replacement bills

DUE AS STAMPED BELOW

AUG 2 6 1997	JUL 1 1 2011	
APR 2 7 1998		
Rec'd UCB PHYS		
OCT 0 1 1998		
SEP 2 5 1998		
Rec'd UCB PHYS		
JUL 1 2 2000		
MAY 1 6 2001		
APR 2 8 2002		
AUG 2 7 2002		
AUG 1 3 2004		

FORM NO. DD 25

UNIVERSITY OF CALIFORNIA, BERKELEY
BERKELEY, CA 94720